Prestressed Concrete Design to Eurocodes

Ordinary concrete is strong in compression but weak in tension. Even reinforced concrete, where steel bars are used to take up the tension that the concrete cannot resist, is prone to cracking and corrosion under low loads. Prestressed concrete is highly resistant to stress, and is used as a building material for bridges, tanks, shell roofs, floors, buildings, containment vessels for nuclear power plants and offshore oil platforms. With a wide range of benefits such as crack control, low rates of corrosion, thinner slabs, fewer joints and increased span length; prestressed concrete is a stronger, safer, more economical and more sustainable building material.

The introduction of the Eurocodes has necessitated a new approach to the design of prestressed concrete structures and this book provides a comprehensive practical guide for professionals through each stage of the design process. Each chapter focuses on a specific aspect of design

- fully consistent with Eurocode 2, and the associated parts of Eurocodes 1 and 8
- examples of challenges often encountered in professional practice worked through in full
- detailed coverage of post-tensioned structures
- extensive coverage of design of flat slabs using the finite element method
- examples of pre-tensioned and post-tensioned bridge design
- an introduction to earthquake resistant design using Eurocode 8.

Examining the design of whole structures as well as the design of sections through many fully worked numerical examples which allow the reader to follow each step of the design calculations, this book will be of great interest to practising engineers who need to become more familiar with the use of the Eurocodes for the design of prestressed concrete structures. It will also be of value to university students with an interest in the practical design of whole structures.

Prabhakara Bhatt is an honorary Senior Research Fellow at the School of Engineering, University of Glasgow, U.K. and author or editor of eight other books, including *Programming the Dynamic Analysis of Structures*, and *Reinforced Concrete, 3rd Edition*, both published by Taylor & Francis

PRESTRESSED CONCRETE DESIGN TO EUROCODES

Prabhakara Bhatt

Department of Civil Engineering

University of Glasgow

CRC Press
Taylor & Francis Group
Boca Raton London New York

CRC Press is an imprint of the
Taylor & Francis Group, an **informa** business

A SPON PRESS BOOK

CRC Press
Taylor & Francis Group
6000 Broken Sound Parkway NW, Suite 300
Boca Raton, FL 33487-2742

First issued in paperback 2019

ISBN-13: 978-0-415-43911-4 (hbk)
ISBN-13: 978-0-367-86547-4 (pbk)

British Library Cataloguing in Publication Data
A catalogue record for this book is available from the British Library

Library of Congress Cataloging in Publication Data
Bhatt, P.
Prestressed concrete design to eurocodes / P. Bhatt.
p. cm.
Includes bibliographical references and index.
1. Prestressed concrete construction--Standards--Europe. I. Title.
TA683.9.B492 2011
624.1'8341202184--dc22
2010050299

**Visit the Taylor & Francis Web site at
http://www.taylorandfrancis.com**

**and the CRC Press Web site at
http://www.crcpress.com**

DEDICATED WITH LOVE AND AFFECTION TO

OUR GRANDSONS

KIERON ARJUN BHATT

DEVAN TARAN BHATT

CONTENTS

12. Design of slabs   261

PREFACE

The object of this book is to assist students and professional engineers to design prestressed concrete structures using Eurocodes 1, 2 and 8. It is assumed that the reader is familiar with the design of reinforced concrete structures. The book is suitable not only for professional engineers familiar with prestressed concrete design but also for use in university courses as sufficient introductory material has been included.

Eurocodes related to design in structural concrete are very substantial documents and the present book has been kept to a manageable length by confining the attention to rules related mainly to prestressed concrete design. All important clauses are discussed in detail and their use illustrated by copious examples. Complete calculations are given for all numerical examples.

Grateful acknowledgements are extended to the following individuals and organizations for help in writing the book

- Mr. Paul Stainrod and his staff from CCL Ltd for many of the photographs reproduced in the book.
- Mr. Hans-Rudolf Ganz, Ms. Renata Kunz-Tamburrino and Ms. Christine Mueller-Sinz from VSL Ltd systems for many of the photographs reproduced in the book.
- Ms. Birgit Etzold-Carl of Dywidag Prestressing systems for many of the photographs reproduced in the book.
- Mr. Ken McColl, Computer Manager, Department of Civil Engineering, Glasgow University, for assistance with computational matters.
- Professor Simon Wheeler, Head, Department of Civil Engineering, Glasgow University, for support and encouragement
- Dr. Lee Cunningham, Senior Structural Engineer, Blackpool Council, for technical help.
- My family: Sheila, Arun, Ranjana, Sujaatha and Amit, for constant encouragement and moral support.
- British Standards Institution for permission to quote from Eurocodes.

Permission to reproduce from British Standards is granted by BSI. British Standards can be obtained in PDF or hard copy formats from the BSI online shop: www.bsigroup.com/Shop or by contacting BSI customer services for the hard copies only: Tel: +44(0) 20 8996 9001, Email:cservices@bsigroup.com

Prabhakara Bhatt
2nd October 2010 (Mahatma Gandhi's birthday)

CHAPTER 1

BASIC CONCEPTS

1.1 INTRODUCTION

The idea of applying forces to neutralize stresses caused by external loads is an old one. For example, as shown in Fig. 1.1, spokes of a bicycle wheel are pre-tensioned to prevent them from buckling under compressive loads applied by the rider. Similarly as shown in Fig. 1.1, stay cables, also called guy ropes, used to hold masts are pre-tensioned so that they are effectively under compressive forces induced by wind loads.

The above examples refer to cases where it is necessary to prevent compressive forces from developing, which causes the member to be ineffective.

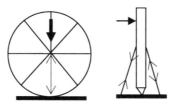

Fig. 1.1 Examples of pre-tensioning

A common example of where it is necessary to apply pre-compression rather than pre-tension is the case of wooden casks shown in Fig. 1.2. In the case of wooden casks made from individual wooden staves, the latter are held together by metal bands. The tension in the metal bands applies compression between the stays so that the cask retains its liquid contents without leaking.

Another example of where pre-compression is applied is shrink-fitting a sleeve on to a gun barrel as shown in Fig. 1.3. When the gun is fired, the explosion causes high tensile stresses in the barrel (Fig. 1.3a). The close-fitting sleeve is heated and is shrunk on to the barrel. This causes tensile stresses in the sleeve (Fig. 1.3b) but compressive stresses in the barrel (Fig. 1.3c), which reduces the tensile stresses due to explosion.

1.2 PRESTRESSED CONCRETE

As is well known, although concrete is very strong in compression, with high strength concrete having cylinder strength exceeding 100 N/mm^2 (about 120 N/mm^2 in cube strength), it is very weak in tension, with a tensile strength approximately only 10% of compressive strength. In reinforced concrete, structures, steel is used to carry the tension developed due to applied loads.

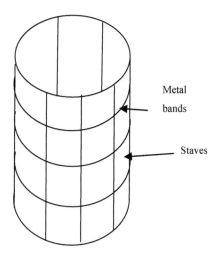

Metal
bands

Staves

Fig. 1.2 Examples of pre-compression

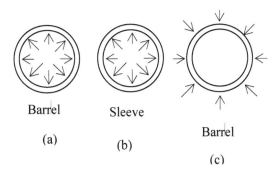

Barrel   Sleeve

(a)   Barrel

(b)

(c)

Fig. 1.3 Stresses in gun barrel and sleeve

However, this does not solve all the problems associated with low concrete tensile strength. For example consider a simply supported beam under vertical loads. Vertical cracks develop due to tensile stresses caused by bending and inclined cracks due to shear stresses. The tensile cracks due to bending occur even under working loads. This has the following major disadvantages.

- Cracked concrete simply adds to the dead weight without participating in resisting the loads.
- A structure with cracked concrete is less stiff than one with uncracked concrete and will therefore deflect more under load. In order to prevent problems with excessive deflection at serviceability loads, it will therefore be necessary to use a correspondingly larger section, leading to larger

dead loads. This is especially the case say in bridge structures, where a major part of the total load is due to dead load of the structure.
- Cracked concrete is more prone to corrosion of steel reinforcement.
- Cracking could lead to failure under serviceability limit state design
- Cracking reduces the ability of concrete to resist shear stresses.
- Cracking is irreversible.

The above problems can be overcome, if concrete can be subjected to external compressive forces so as to neutralize tensile stresses caused by applied loads. Concrete which has been subjected to pre-compression is known as prestressed concrete.

A prestressed concrete structure can be defined as a concrete structure where external compressive forces are intentionally used to overcome tensile stresses caused by unavoidable loads due to gravity, wind, etc. In other words, it is pre-compressed concrete, meaning that compressive stresses are introduced into areas where tensile stresses might develop under working load and this pre-compression is introduced even before the structure begins its working life.

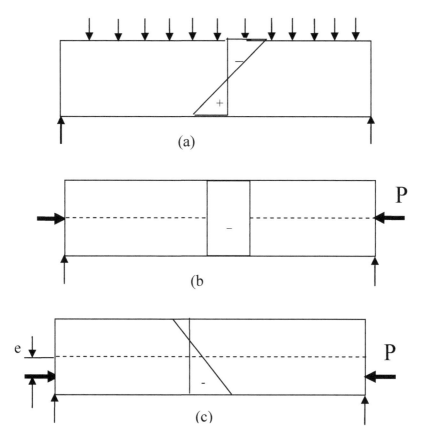

Fig. 1.4 A simply supported beam with prestress

For example consider a simply supported beam supporting loads as shown in Fig. 1.4a. At mid-span, the stress due to bending causes compression at the top fibre and tension at the bottom fibre. If a compressive force is applied at the centroidal axis as in Fig. 1.4b, it sets up uniform compression throughout the beam cross section. The compressive stress does neutralize the tensile stresses at the bottom portion of the beam caused by bending but it has the disadvantage of increasing the total compressive stresses at the top face. If, however, the compressive force is applied towards the bottom face at an eccentricity of **e** from the centroidal axis, then in addition to an axial force of P, a bending moment equal to Pe of a nature **opposite** to that caused by external loads is created. Thus with proper manipulation of values of P and e, one can create compressive stresses at the bottom face and tensile stresses at the top face which are exactly opposite to that caused by external loads. This idea is known as load balancing and will be discussed in detail in Chapter 5.

1.3 ECONOMICS OF PRESTRESSED CONCRETE

As prestressed concrete is generally free from cracks at the serviceability limit stage, it is stiffer than a corresponding reinforced concrete beam, resulting in a possible reduction in depth of up to 20%. The reduction in dead weight is particularly important in the case of large span structures. In fact for spans exceeding about 20 *m*, for economic design, use of prestressed concrete becomes obligatory. Unlike reinforced concrete members, in the case of prestressed concrete members, cracking is reversible in the sense that when the load is removed, cracks close. In addition, absence of cracks at working loads and reduction in dead weight lead to less maintenance and reduction in foundation costs.

CHAPTER 2

TECHNOLOGY OF PRESTRESSING

2.1 METHODS OF PRESTRESSING

There are basically two methods of applying prestress to concrete. They are:
- *Pre-tensioning*: This method is used to manufacture a large number of similar precast prestressed concrete products such as simply supported beams, one-way spanning slabs and railway sleepers. This is a factory-based operation.
- *Post-tensioning*: This method is an onsite operation and is generally used to prestress simply supported and continuous beams, two-way spanning slabs, statically indeterminate structures such as frames and bridge structures.

2.2 PRE-TENSIONING

Pre-tensioning is a factory-based operation which is particularly suitable for producing a large number of similar precast prestressed units such as bridge beams, double T beams for floors, floor slabs, railway sleepers, etc. Fig. 2.1 shows a line diagram of a pre-tensioning bed. Fig. 2.2 shows a photograph of a prestressing bed. Steel cables which are normally 7-wire strands as shown in Fig. 2.3, are stressed between heavy abutments. The number of cables, their location in the cross section and the tension in the cables depend on the design requirements.

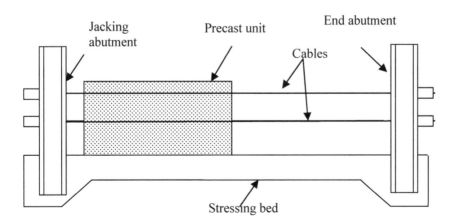

Fig. 2.1 Prestressing bed

Fig. 2.2a Photograph of a pre-tensioning bed

Photo courtesy of CCL

As shown in Fig. 2.4, the process consists of the following steps.

- Any reinforcing steel such as links are threaded through the high-tensile steel 'cables' and the cables are tensioned or 'jacked' to the desired level of tension between abutments. The cable is anchored using a simple barrel and wedge device as shown in Fig. 2.5. The wedge normally consists of two or three segments with a collar held by an 'O' ring to keep them in the same relative position. In order to improve the grip, the wedges have groves on the inside surface which is in contact with the cable. Because the cables are tensioned before concrete is cast, the name pre-tensioning is used for this process.

- The formwork is built round the steel cables

- Concrete is placed in the moulds around the steel

- Concrete is allowed to cure to gain the desired level of strength. This is often speeded up using steam curing. This also enables the prestressing bed to be reused quickly for another job. The side shutters are usually stripped when the concrete has attained a cylinder strength of about 10 MPa.

Fig. 2.2b Placing concrete after pre-tensioning the cables

Photo courtesy of CCL

Fig. 2.2c Close-up of anchors and stressing of a cable

Photo courtesy of CCL

Fig. 2.3 Seven-wire cables.   Photo courtesy of CCL

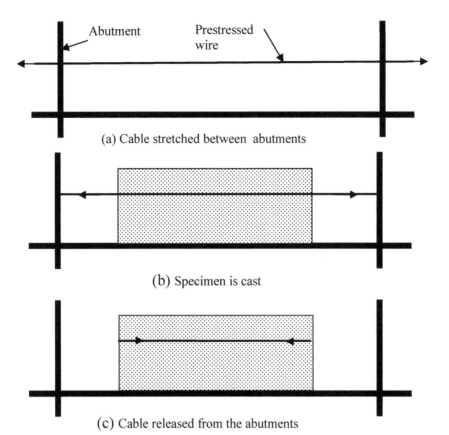

(a) Cable stretched between abutments

(b) Specimen is cast

(c) Cable released from the abutments

Fig. 2.4 Three stages of pretensioning

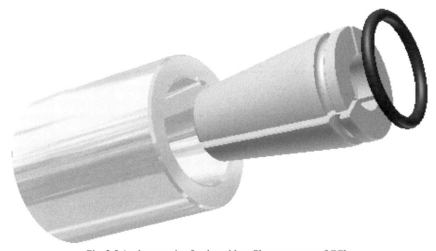

Fig. 2.5 Anchor or grips for the cables. Photo courtesy of CCL

- When the concrete strength reaches about 30 MPa, the cables are released from their abutments. The cables want to return to their unstressed state but are prevented by concrete. In this process, the cables grip the concrete and apply precompression.

In practice a large number of identical units are cast at the same time using what is known as the long-line production method because the length of the casting beds can be as long as 200 m. Cables can be stressed either individually or as a group so that all cables are stressed to the same level. If the cables are stressed individually using jacks, then they are released by either flame cutting or sawing.

However if all cables are stressed to the same level then the stressing is done by using long-stroke jacks and header plates as shown in Fig.2.6. In this case the cables are released slowly. It is important to release the cables as gradually as possible as otherwise the dynamic forces associated with sudden release can damage the bond between the cables and the surrounding concrete.

2.2.1 Debonding/Blanketing of Strands

If all the cables are stressed to the same value, then at the ends of simply supported beams, there is the danger of serious cracking at the top face. In order to prevent this problem, as shown in Fig. 2.7, depending on design requirements, lengths of some of the cables towards the support are wrapped in plastic tubing to prevent bond between steel and concrete and thus to make those cables inactive over the covered length. This is known as debonding or blanketing of cables. This is a very simple technique for varying the total prestressing force at a cross section.

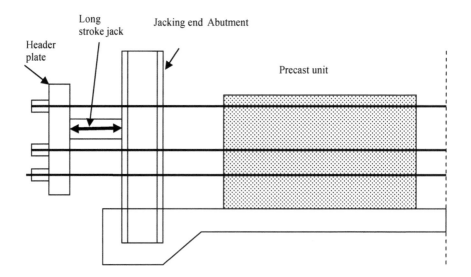

Fig 2.6 Detail of header plate system

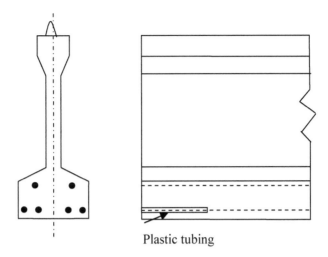

Plastic tubing

Fig. 2.7 Debonding or blanketing of bottom layer strands

2.2.2 Deflecting/Draping/Harping of Strands

In some cases, instead of varying the number of cables at a cross section, the eccentricity of individual cables can be varied by pulling down the cable at specific points as shown in Fig.2.8. This process is known as deflecting, draping or harping of strands. This is the less preferred method of altering the net eccentricity of the prestressing force, although it is commonly used in North American practice.

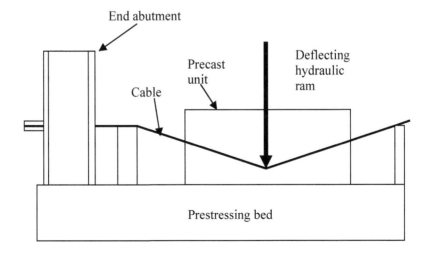

Fig. 2.8 Schematic arrangement for deforming tendons

2.2.3 Loss of Prestress at Transfer

It is worth noting that if the total force in the prestressing cables when they are stressed between the abutments is P_{jack}, when they are released from the abutments and allowed to grip the concrete, thus transferring the force from the abutments to the concrete specimen, the specimen contracts. Because of the full bond between concrete and steel, steel also suffers the same contraction, leading to a certain loss of stress from the stress at the time of jacking. This is known as *loss of prestress at transfer* and is generally of the order of 10%. Thus

$$P_{transfer} \approx 0.9 \, P_{jack}$$

Where $P_{transfer}$ = Total force in the cable after initial loss of stress due to compression of concrete,
P_{jack} = total force used at the time of jacking.

2.2.4 Transmission Length

The transfer of force between concrete and steel takes place gradually. The force transfer takes place due to

- *Bond:* It is very important therefore to ensure that the 'cable' is clean and free from loose rust and the concrete is well compacted.

- *Friction and wedging action:* The cable is stretched and therefore has a reduced diameter due to the Poisson effect. However, when the force is released, the wire regains its original diameter. As shown in Fig. 2.9, this creates a certain amount of wedging action and frictional forces also come in to play. This is known as the Hoyer effect.

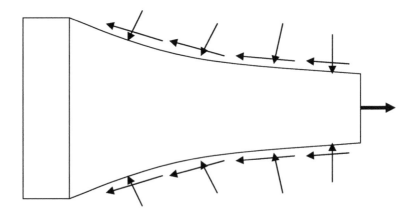

Fig. 2.9 Hoyer effect

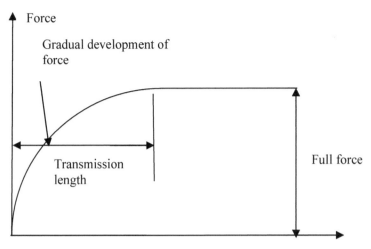

Fig. 2.10 Transmission length

As shown in Fig. 2.10, because the end of the cable is free from force, the force is zero at the ends of the cable and gradually builds up to its full value over a certain length when the bond and friction effects are sufficient to prevent the cable from slipping from concrete. The length over which the force finally builds up from zero to full value is known as the *transmission length*, ℓ_{pt}. If the effective depth is d, generally after a distance of $\ell_{disp} = \sqrt{(\ell_{pt}^2 + d^2)}$, the stress distribution can be assumed to be linear. The transmission length ℓ_p varies depending on the surface characteristics of the cables, the diameter of the cable and the strength of concrete. It is generally of the order of about 50 diameters for 7-wire strand. The upper bound of the transmission length is given by $\ell_{pt2} = 1.2\ell_{pt}$, where

$$\ell_{pt} = \alpha_1 \, \alpha_2 \, \phi \, \frac{\sigma_{pm0}}{f_{bpt}} \tag{2.1}$$

where

$\alpha_1 = 1.0$ for gradual release and 1.25 for sudden release

$\alpha_2 = 0.25$ for tendons with circular cross section and 0.19 for 3- and 7-wire tendons

ϕ = nominal diameter of tendon

σ_{pm0} = tendon stress just after release

f_{bpt} = bond stress = $\eta_{p1} \, \eta_1 \, f_{ctd}$ (t)

$\eta_{p1} = 2.7$ for indented wires and 3.2 for 3 and 7-wire tendons

$\eta_1 = 0.7$ normally, but if a good bond conditions can be ensured then $\eta_1 = 1.0$

From equation (3.3) in chapter 3, $f_{ctd} = 0.467 \, f_{ctm}$ (2.2)

$$f_{ctm} = 0.3 f_{ck}^{0.667} \leq C50/60$$
$$= 2.12 \ln(1.8 + 0.1 f_{ck}) > C50/60 \tag{2.3}$$

$$f_{ctd}(t) = (\exp\{s[1 - (\frac{28}{t})^{0.5}]\})^{\alpha} f_{ctd} \tag{2.4}$$

where s = depends on the type of cement and varies from 0.20 to 0.38.

$\alpha = 1$ for $t < 28$ days and 0.67 for $t \geq 28$ days.

2.2.4.1 Example of Calculation of Transmission Length

Calculate the transmission length for the given data:

f_{ck} = C40/60, 15.2 mm diameter 7-wire tendon,

time of release of tendons, t = 7 days; s = 0.2 for class R cement. From equations (2.3) and (2.4),

$$f_{ctm} = 0.3 \times 40^{0.667} = 3.5 \text{MPa}$$
$$f_{ctd} = 0.467 \times 3.5 = 1.64 \text{MPa}$$

Using s = 0.2, t = 7 days, $\alpha = 1$

$$f_{ctd}(t) = (\exp\{0.2[1 - (\frac{28}{7})^{0.5}]\})^{1} \times 1.64 = e^{-0.2} \times 1.64 = 1.34 \text{MPa}$$

f_{bpt} = bond stress = $\eta_{p1} \eta_1 f_{ctd}(t)$

η_{p1} = 3.2 for 7-wire tendons, η_1 = 0.7 normally, but if a good bond conditions can be ensured then $\eta_1 = 1.0$. From equation (2.5),

$$f_{bpt} = 3.2 \times 0.7 \times 1.34 = 3.0 \text{ MPa}$$

From equation (2.1),

$$\ell_{pt} = \alpha_1 \alpha_2 \phi \frac{\sigma_{pm0}}{f_{bpt}}$$

$\alpha_1 = 1.0$ for gradual release, $\alpha_2 = 0.19$ for 7-wire tendons

Φ = nominal diameter of tendon = 15.2 mm

Taking f_{pk} for prestressing steel = 1860 MPa, $f_{p0.1k} = 0.88 \, f_{pk}$, $k_2 = 0.9$,

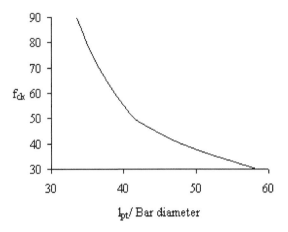

Fig. 2.11 Transmission length vs. f_{ck}

stress in tendons at jacking = 0.9 $f_{p0.1k}$ = 0.9 ×0.88 × 1860 = 1473 *MPa*

Assuming 10% loss at release, σ_{pm0} = 0.9 × 1473 = 1326 *MPa*

$$\ell_{pt} = \alpha_1 \, \alpha_2 \, \phi \, \frac{\sigma_{pm0}}{f_{bpt}} = 1.0 \times 0.19 \times \phi \times \frac{1326}{3.0} = 84\phi = 1276 \text{ mm}$$

Fig. 2.11 shows the variation of transmission length for 7-wire tendons as a multiple of bar diameter with compressive strength. The assumptions made are:

stress at transfer = 1326 MPa, time at transfer t = 28 days,

good bond conditions η_1 = 1.0, gradual release α_1 = 1.0

If η_1 = 0.7, increase the transmission length by a factor of 1.4.

If time of release is other than 28 days, multiply the given transmission length by

the reciprocal of $(\exp\{s[1-(\frac{28}{t})^{0.5}]\})^{\alpha}$.

2.3 POST-TENSIONING

One of the limitations of the pre-tensioning system is that the cables need to remain straight because the cable is pre-tensioned. This limitation can be overcome if ducts fixed to reinforcement are laid to any desired profile and the cable is placed inside the ducts. After this, concrete is cast and once it reaches the desired strength, the cable is tensioned and anchored using external anchors rather than relying on the bond between steel and concrete as in the case of pre-tensioning. This is the basic idea of post-tensioning. It consists of three stages, as shown in Fig. 2.12.

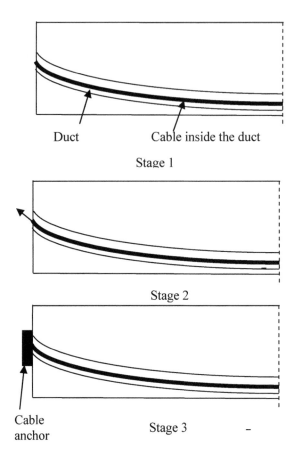

Fig. 2.12 Three stages of post-tensioning

- Metal ducts with a wall thickness of about 0.25 mm or corrugated high density polyethylene (HDPE) ducts with a wall thickness of about 3 to 6 mm are fixed to the required profile by tying to the reinforcement cage at about 0.5 m to 1.0 m intervals with the permanent anchorages also positioned at the ends of the duct. It is important that the duct is strong enough to prevent distortion during concreting and also to withstand pressure during grouting. Fig. 2.13 shows a photograph of ducts fixed to the reinforcement. The cross-sectional area of the ducts is approximately twice the area of the tendons. The cable can be threaded through the ducts after concreting or the cables can be pre-placed in the ducts while fixing them. The cables can be placed inside the ducts either by pulling the cables using a winch as shown in Fig. 2.14 or by pushing the cables as shown in Fig. 2.15.

- Concrete is cast.

Fig. 2.13 Ducts fixed to reinforcement
Photo courtesy DYWIDAG Systems International

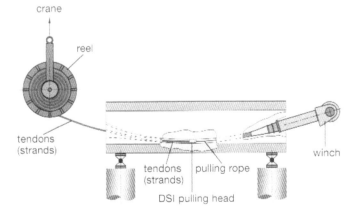

Fig. 2.14 Cable from the reel is pulled inside the ducts by a winch

Diagram courtesy DYWIDAG Systems

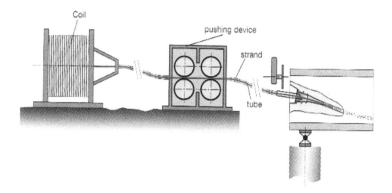

Fig. 2.15 Cable from the coil is pushed inside the ducts by a pushing device.

Diagram courtesy DYWIDAG Systems

- When the concrete has hardened, the cable is tensioned and anchored to the concrete using permanent anchors. In most cases stressing is done from 2 to 7 days after casting.
- Finally, the duct is filled with a colloidal grout under pressure in order to establish bond a between concrete and steel and also as protection against corrosion.

Because the cables are tensioned after the concrete has hardened, this system is known as post-tensioning.

2.3.1 Post-Tensioning Anchors

There are various types of anchors used in practice but they are generally of two types, viz.
- A threaded nut system as shown in Fig. 2.16. The main advantage of this system is that load can be applied in stages to suit design considerations or losses can be taken up at any time prior to grouting. The anchorage is completely positive and *there is no loss of prestress at transfer stage.*
- A wedge system which is similar to that in the case of pre-tensioning. In general, cables are stressed from one end only. In DYWIDAG system as shown in Fig. 2.17, anchors at the ends from where stressing is done consist of a cast tube unit to guide the strands into the duct. A taper hole bearing plate in which the strands are anchored by hardened steel wedges sits against this tube unit as shown in Fig. 2.3. The number of strands that can be anchored varies from 4 to 37.
- At the non-stressing end, the cables are anchored using 'basket-type dead-end' anchors shown in Fig. 2.18. A reinforcement cage is used to provide proper spacing to individual strands. In this case the stressing force is transmitted to concrete by bond and these are much cheaper to use than standard anchors. Buried anchors require that the tendons should be prefabricated and in place

before the concrete is cast. On the other hand, with the wedge system of anchoring, cables can be threaded thorough the ducts after the concrete has hardened. Generally for cables less than about 50 m in length or if the profile of the cable is flat, single-end stressing is used. In other cases, double-end stressing is normally used, with independent jacks at each end acting simultaneously. Fig 2.19 shows a prestressing jack used for stressing the cables. The jacks can weigh about 250 kg for single strands and about 2500 kg for larger tendons.

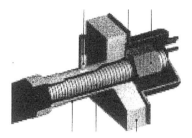

Fig. 2.16 Threaded anchor

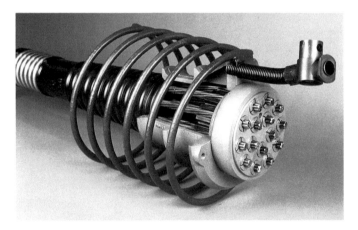

Fig. 2.17 Stressing end anchor. Photo Courtesy DYWIDAG

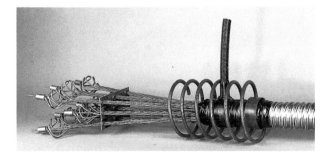

Fig. 2.18 Dead end anchor. Photo Courtesy DYWIDAG

- At construction joints, it is necessary to have a means of connecting the stressed cables in the cast section to the new set of cables in the still to be cast

Fig. 2.19 Stressing the cable   Photo Courtesy CCL

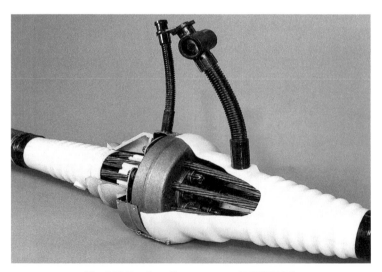

Fig. 2.20 Coupler.   Photo courtesy of DYWIDAG

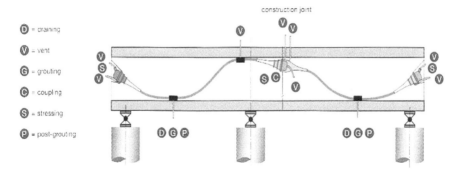

Fig. 2.21 Construction joint

Diagram courtesy DYWIDAG International

section. This is done using couplers. Fig. 2.20 shows a typical coupler. Fig. 2.21 shows the cable arrangement at a construction joint.

Once the cables are stressed, cement grout is pumped through the ducts to establish bond between the cables and the concrete in the structure. Fig. 2.21 shows the position of draining at low points in the duct and venting of air at high points of the duct to ensure that the whole duct is filled with grout as a protection for cables against rust.

2.3.2 Loss of Prestress at Transfer

It is useful to note that in post-tensioning, as the strands are being tensioned, concrete compresses at the same time so that there is no loss of prestress due to compression of concrete as in the case of pre-tensioning. However, in this case, *there is loss of prestress at the transfer stage because of the slip between the cable and the wedge before the wedges bite in.*

2.3.3 External Prestressing

One of the disadvantages of traditional post-tensioning is that there is no guarantee that the ducts are properly filled with grout to prevent corrosion and if the steel corrodes, it cannot be replaced. In order to overcome these problems, external prestressing is used. The cables are 'external' to the concrete of the beam and the eccentricity is varied using saddles at appropriate places to obtain the required profile. There is no bond between steel and concrete and it also avoids serious congestion of steel inside the concrete. This is similar to the use of deflected tendons in pre-tensioning. This system allows replacement of cables as required. This system can be used for applying prestress to an existing structure in order to improve its performance. Especially in the case of box girders used in bridge construction, the cables are inside the box and deviation of the cables is achieved using intermediate diaphragms or saddles as shown in Fig. 2.22. The cables are protected from weather by some form of grease.

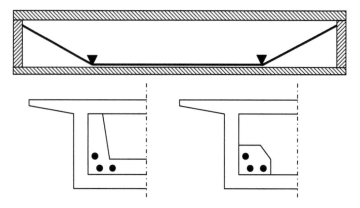

Fig. 2.22a 'External' tendons in a box girder

Fig. 2.22b 'External' tendons in a box girder

Photo courtesy of CCL Ltd.

2.3.4 Unbonded Systems

In order to avoid placing any reliance on grouting to prevent corrosion of the cables, in some systems cables which are encased in a greased plastic tube as shown in Fig. 2.23 are used. In this system there are no ducts as in normal post-tensioning work. The cables are cast in the normal way and prestressed once the concrete has hardened. The advantage of this '*unbonded*' system is speed of construction, as no grouting is done. This system is extensively used in the construction of prestressing slabs.

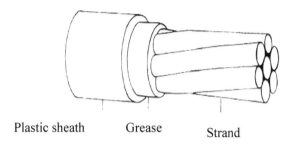

Plastic sheath   Grease   Strand

Fig. 2.23 Unbonded cable

CHAPTER 3

MATERIAL PROPERTIES

3.1 PROPERTIES OF CONCRETE

High-strength concrete is a major constituent of all prestressed concrete structures. The two major aspects of importance are

- Compressive and tensile strengths
- Deformational properties such as elastic modulus, variation of creep and shrinkage deformations with time.

The deformational properties are particularly important as they have an important effect on how much prestress is retained in the long term. Properties of concrete are given in Table 3.1 of the Eurocode 2.

3.2 COMPRESSIVE STRENGTH OF CONCRETE

Compressive strength of concrete is specified in terms of the characteristic cylinder strength f_{ck} at 28 days. Characteristic strength is defined as the strength below which not more than 5% of the test results fall. Characteristic strength f_{ck} is related to mean strength f_{cm} and standard deviation by the relationship

characteristic strength f_{ck} = mean strength, f_{cm} − 1.64 standard deviation

In the Eurocode 2, the standard deviation is taken as a constant value of approximately 5 *MPa* so that

characteristic strength f_{ck} = mean strength f_{cm} − 8 *MPa*

The characteristic cylinder strength f_{ck} is approximately related to the corresponding cube strength $f_{ck, cube}$ by

$$f_{ck} \approx 0.8 \, f_{ck \, cube}$$

Strength classes for concrete are quoted in terms of $f_{ck}/f_{ck, cube}$. Thus concrete grade C40/50 refers to a concrete whose characteristic cylinder strength is 40 *MPa* and the corresponding cylinder strength is 50 *MPa*.

The strength of concrete at an age of less than 28 days is required for demoulding, release of prestress, etc. The strength of concrete at time t in days cured at 20°C is given by

$$f_{ck} (t) = \beta_{cc} (t) \, f_{cm} - 8 \text{ MPa,} \qquad 3 < t < 28 \text{ days}$$

$$\beta_{cc} (t) = \exp\{s \times [1 - (\frac{28}{t})^{0.5}]\} \qquad\qquad (3.1)$$

where s = coefficient depending on the type of cement. s = 0.38, 0.25, 0.20 respectively for cements of classes S, N and R.

$$f_{ck}(t) = f_{ck}, \qquad t \geq 28 \text{ days}$$

If t ≤ 3 days, $f_{ck}(t)$ must be obtained by testing specimens.

Fig. 3.1 shows the variation of β_{cc} with time for the three classes of cement.

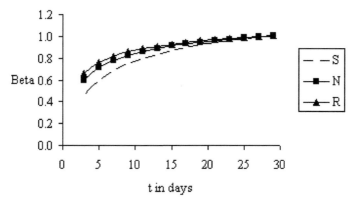

Fig. 3.1 Variation of β with time t

3.3 TENSILE STRENGTH OF CONCRETE

The mean value tensile strength of concrete f_{ctm} is related to cylinder compressive strength f_{ck} by the equation

$$\begin{aligned} f_{ctm} &= 0.30\, f_{ck}^{(2/3)}, & C \leq 50/60 \\ f_{ctm} &= 2.12\ln(1.8 + 0.1 \times f_{ck}), & C > 50/60 \end{aligned} \tag{3.2}$$

The characteristic tensile strength $f_{ctk,\,0.05}$ (or 5% fractile) is related to mean strength f_{ctm} by the equation

$$f_{ctk,\,0.05} = 0.7\, f_{ctm}, \qquad f_{ctd} = f_{ctk,\,0.05}/(\gamma_m = 1.5) = 0.467\, f_{ctm} \tag{3.3}$$

Fig. 3.2 shows the variation of f_{ctm} and $f_{ct0.05}$ with f_{ck}.

Up to 28 days it may be assumed that the variation of tensile strength with time follows the same pattern as the compressive strength.

If the tensile strength of concrete determined from the split cylinder test is $f_{ct,\,sp}$, the axial tensile strength f_{ct} is equal to 0.9 $f_{ct,\,sp}$.

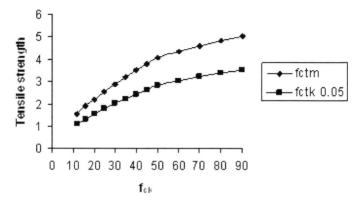

Fig. 3.2 Variation of tensile strength with compressive strength

3.4 DEFORMATIONAL PROPERTIES

The main deformational properties of interest are the tangent and secant elastic moduli, creep and shrinkage deformation with time.

3.4.1 Elastic Moduli

The mean *secant modulus* of elasticity E_{cm} in *GPa* is related to the mean cylinder strength f_{cm} in *MPa* by

$$E_{cm} = 22 (0.1 \times f_{cm})^{0.3} \tag{3.4}$$

The mean tangent modulus of elasticity E_c in GPa is related to the mean secant modulus E_{cm} by

$$E_c = 1.05 \, E_{cm} \tag{3.5}$$

Since in prestressed concrete stressing is done before 28-day strength has been attained, in certain calculations, the initial tangent modulus is of interest.

The variation of E_{cm} with time t can be estimated by

$$E_{cm}(t) = [f_{cm}(t)/f_{cm}]^{0.3} \, E_{cm} \tag{3.6}$$

3.4.2 Creep Coefficient

When a load is applied to a material like steel, it deforms and if there is no variation in load, the deformation remains constant with time. Concrete is different. Like steel, concrete deforms as soon as the load is applied. This is known as immediate elastic deformation. But unlike steel, as shown in Fig. 3.3, if the load is left in place, the displacement gradually increases with time, reaching a value as large as three to four times that of immediate elastic deformation. This inelastic deformation under sustained load is known as creep deformation. ***Creep is defined as the increase of strain with time when the stress is held constant.*** As a rule, an increase in water/cement ratio or an increase in cement content increases creep. On the other hand an increase in aggregate content decreases creep deformation.

If at any time, a part of the load is removed, there is an immediate decrease in strain due to elastic recovery and a gradual ***incomplete*** recovery due to creep. This is shown by the dashed line in Fig. 3.3.

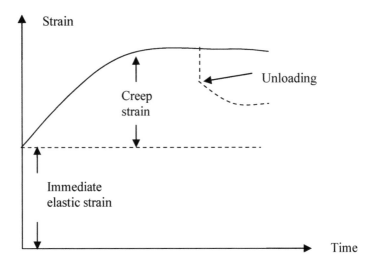

Fig. 3.3 Long-term stress-strain behaviour of concrete

When creep is taken into account, its design effects are always evaluated under quasi-permanent combination of actins irrespective of the design situation considered, i.e. persistent, transient or accidental.

The *total creep* strain ε_{cc} (∞, t_0) of concrete due to the constant compressive stress of σ_c applied at the concrete age of t_0 is given by

$$\varepsilon_{cc}(\infty, t_0) = \varphi(\infty, t_0) \times \frac{\sigma_c}{E_c} \tag{3.7}$$

where $\varphi(\infty, t_0)$ is the *final creep coefficient*, which can be determined from Figure 3.1 in Eurocode 2, provided the value of σ_c does not exceed 0.45 f_{ck} (t_0). E_c is the tangent modulus.

$$\phi(\infty, t_0) = \phi_{RH} \times \frac{16.8}{\sqrt{f_{cm}}} \times \frac{1}{(0.1 + t_0^{0.20})}$$

$$\phi_{RH} = 1 + \frac{1-0.01 \times RH}{0.1 \times h_0^{0.333}}, \qquad f_{cm} \leq 35 MPa$$

$$\phi_{RH} = (1 + \frac{1-0.01 \times RH}{0.1 \times h_0^{0.333}} \alpha_1)\alpha_2, \qquad f_{cm} > 35 MPa$$

$$\alpha_1 = (\frac{35}{f_{cm}})^{0.7}, \qquad \alpha_2 = [\frac{35}{f_{cm}}]^{0.2}$$

$$f_{cm} = f_{ck} + 8 \; MPa$$

$$t_0 = t_{0,T}(\frac{9}{2+t_{0,T}^{1.2}}+1)^{\alpha} \geq 0.5, \qquad \alpha = \{-1(S), 0(N), 1(R)\} \tag{3.8}$$

where RH = relative humidity in %,

t_0 = age at the time of loading,

$h_0 = 2A_c/u$ mm,

A_c = cross sectional area,

u = perimeter of the member in contact with the atmosphere

S, R, and N refer to different classes of cement

The final creep coefficient $\varphi(\infty, t_0)$ is dependent on several factors such as ambient humidity, composition of concrete and dimensions of the member. Table 3.1 gives the value of $\varphi(\infty, t_0)$ for a specific case.

Example: Determine the final creep coefficient for a rectangular specimen of dimensions 400 *mm* × 800 *mm* and made from concrete grade C40/50 using Class S cement and with two longer sides and one short side exposed to outside conditions with a relative humidity of 80% and loaded at 3 days with no allowance for temperature adjustment.

RH = 80, $A_c = 400 \times 800 = 32 \times 10^4$ mm^2, u = 400 + 2 × 800 = 2000 mm,

$h_0 = 2A_c/u = 160$ mm, $f_{cm} = 40 + 8 = 48$ MPa.

$$\alpha_1 = (\frac{35}{48})^{0.7} = 0.80, \; \alpha_2 = (\frac{35}{48})^{0.2} = 0.94, \; \alpha = 1 \text{ for Class S cement.}$$

$$\phi_{RH} = (1 + \frac{1-0.01 \times 80}{0.1 \times 160^{0.333}} \times 0.80) \times 0.94 = 1.22$$

$$t_0 = t_{0,T}(\frac{9}{(2+t_{0,T}^{1.2})} + 1)^{\alpha} = 3 \times (\frac{9}{(2+3^{1.2})} + 1)^{-1} = 1.2$$

$$\phi(\infty, t_0) = 1.22 \times \frac{16.8}{\sqrt{48}} \times \frac{1}{(0.1 + 1.2^{0.20})} = 2.60$$

Table 3.1 Values of φ (∞, t_0) for f_{ck} = 40 MPa, Class N cement

	RH = 50%			RH = 80 %		
	Inside, dry conditions			Outside, humid conditions		
	h_0 mm			h_0 mm		
Age at loading	50	150	600	50	150	600
1	4.3	3.6	3.1	3.0	2.7	2.5
7	3.0	2.5	2.1	2.1	1.9	1.7
28	2.3	2.0	1.6	1.6	1.5	1.3
90	1.9	1.6	1.3	1.3	1.2	1.1

Table 3.2 Value of k_σ in terms of f_{ck}

$\dfrac{\sigma_c}{f_{ck}(t_0)}$	k_σ
0.5	1.078
0.6	1.252
0.7	1.455
0.8	1.691
0.9	1.964
1.0	2.282

If the compressive stress applied at the age of t_0 exceeds 0.45 $f_{ck}(t_0)$ as can happen during prestress transfer stage, then the final creep will be larger and the final creep coefficient $\varphi(\infty,t_0)$ is multiplied by a factor given by

$$k_\sigma = \exp[1.5 \times (\frac{\sigma_c}{f_{ck}(t_0)} - 0.45)] \qquad (3.9)$$

Table 3.2 gives the value of k_σ.

The creep coefficient at any age t can be calculated from the formula

$$\varphi(t, t_0) = \varphi(\infty, t_0) \times \beta_c(t, t_0)$$

$$\beta_c(t, t_0) = (\frac{(t-t_0)}{(\beta_H + t - t_0)})^{0.3}, \qquad \alpha_3 = \sqrt{\frac{35}{f_{cm}}}$$

$$\beta_H \;=\; 1.5[1+(0.012RH)^{18}]h_0 + 250 \;\le 1500, \qquad\qquad f_{cm} \le 35 \text{ MPa}$$

$$\beta_H \;=\; 1.5[1+(0.012RH)^{18}]h_0 + 250\alpha_3 \;\le 1500\,\alpha_3, \qquad f_{cm} > 35 \text{ MPa} \qquad (3.10)$$

3.4.3 Shrinkage

Creep and shrinkage are both affected by the same parameters. Like creep, shrinkage strain is also a function of many variables such as relative humidity, surface exposed to atmosphere, compressive strength of concrete and type of cement. The total shrinkage is made up of two parts:

- Plastic shrinkage: This takes place in the first few hours after placing concrete.

- Drying shrinkage: Mainly due to loss of water by evaporation.

Like creep, shrinkage is not an entirely reversible process. Although a higher aggregate content reduces shrinkage, the type of aggregate, whether shrinkable or not, has a major effect on the total shrinkage. Similarly, the presence of reinforcement has a major effect on the total amount of shrinkage that can take place.

Table 3.3 Cement type and α coefficients

Cement type	α_{ds1}	α_{ds2}
S	3	0.13
N	4	0.12
R	6	0.11

Table 3.4 Variation of k_h with h_0

h_0	k_h
100	1.0
200	0.85
300	0.75
≥ 500	0.70

The total shrinkage strain ε_{cs} can be calculated from the formulae

$$\varepsilon_{cs} \;=\; \varepsilon_{cd0} \times k_h + 2.5(f_{ck}-10)\times10^{-6}$$

$$\varepsilon_{cd0} = 0.85[(220 + 110\alpha_{ds1}) \times \exp(-\alpha_{sd2} \times 0.1 \times f_{cm})] \times 1.55[1 - (0.01RH)^3] \times 10^{-6}$$

$$(3.11)$$

where RH = relative humidity in %, $f_{cm} = f_{ck} + 8$, $h_0 = 2A_c/u$,

A_c = cross-sectional area

u = perimeter of the member in contact with the atmosphere

Table 3.3 gives the values of parameters α_{ds1} and α_{ds2} as a function of the type of cement. Table 3.4 gives the k_h as a function of h_0. Table 3.5 gives the values of ε_{cd0} for three classes of cement as a function of the relative humidity.

Table 3.5 Values of drying shrinkage $\varepsilon_{cd0} \times 10^6$

$f_{ck}/f_{ck, cube}$		RH in %					
MPa		20	40	60	80	90	100
20/25	S	500	471	395	246	136	ALL
	N	616	582	487	303	168	
	R	845	798	668	416	231	
40/50	S	385	363	304	189	105	ZEROS
	N	485	458	383	239	132	
	R	678	640	536	334	185	
60/75	S	297	280	235	146	81	
	N	381	360	301	188	104	
	R	544	514	430	268	149	
80/95	S	229	216	181	113	63	
	N	300	283	237	148	82	
	R	437	412	345	215	119	
90/105	S	201	190	159	99	56	
	N	266	251	210	131	73	
	R	391	369	309	193	107	

Example: Calculate the shrinkage strain ε_{cs} for a 500×1000 member made from CEM class N cement at a relative humidity of 80, compressive strength of 40/50, with two longer sides and one shorter side exposed to atmosphere.

$\varepsilon_{cdo} = 239 \times 10^{-6}$, $A_c = 2 \times 500 \times 1000 = 10 \times 10^5$ mm^2

$u = 2 \times 1000 + 500 = 2500 \; mm$

$h_0 = 2A_c/u = 800mm, \quad k_h = 0.70,$

$\varepsilon_{cs} = 239 \times 10^{-6} \times 0.70 + 2.5 \times (40 - 10) \times 10^{-6} = 242 \times 10^{-6}$

3.5 STRESS-STRAIN RELATIONSHIP

A typical stress-strain curve for concrete under compression is shown in Fig. 3.4.

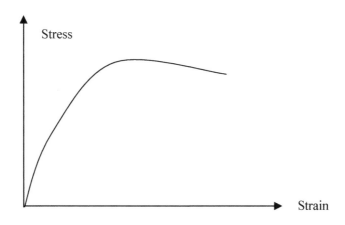

Fig. 3.4 Stress –strain curve in compression

Up to about 40% of the compressive strength, the relationship is approximately linear. After this level of stress, the stiffness decreases gradually up to about 70% of the compressive strength. Beyond that the stiffness decreases quite rapidly. Failure occurs when the maximum compressive strain is reached. Low-strength concretes are much more ductile compared with high-strength concretes. As a rule, the higher the compressive strength, the lower is the ductility.

For the ***design of cross sections***, either a parabola-rectangle combination or bi-linear stress-strain relationship as shown in Figs. 3.5 and 3.6 respectively can be used. The maximum stress permitted is f_{cd}, where

$$f_{cd} = \alpha_{cc} \frac{f_{ck}}{\gamma_m} \tag{3.12}$$

where α_{cc} = coefficient taking account of long-term effects of compressive strength and unfavourable effects resulting from the way load is applied.

α_{cc} = normally 1.0 but can vary from 0.8 to 1.0.

γ_c = material safety factor to be used in ultimate limit state calculation. It is equal to 1.5 for persistent and transient loading and 1.2 for accidental loading.

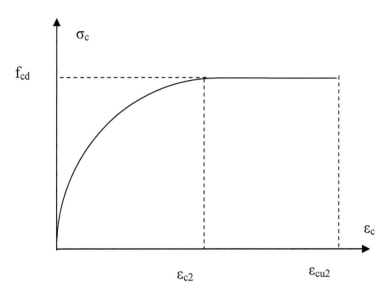

Fig. 3.5 Parabolic-Rectangular stress-strain relationship

3.5.1 Parabolic-Rectangular Relationship

Provided $f_{ck} \leq 90$ MPa, the parabolic relationship as shown in Fig. 3.5 is given, by

$$\sigma_c = f_{cd}\,[1-\{1-\frac{\varepsilon_c}{\varepsilon_{c2}}\}^n\,], \qquad 0 \leq \varepsilon_c \leq \varepsilon_{c2}$$

$$= f_{cd}, \qquad\qquad\qquad \varepsilon_{c2} \leq \varepsilon_c \leq \varepsilon_{cu2} \tag{3.13}$$

$$\varepsilon_{c2} = 2.0\times10^{-3}, \qquad\qquad\qquad\qquad \sigma_c \leq 50\,\text{MPa}$$

$$= [2.0\ +0.085\,(f_{ck}\ -\ 50)^{0.53}\,]\times10^{-3}, \qquad \sigma_c > 50\,\text{MPa} \tag{3.14}$$

$$\varepsilon_{cu2} = 3.5\times10^{-3}, \qquad\qquad\qquad\qquad \sigma_c \leq 50\,\text{MPa}$$

$$= [2.6\ +35\times\{\frac{90-f_{ck}}{100}\}^4\,]\times10^{-3}, \qquad \sigma_c > 50\,\text{MPa} \tag{3.15}$$

$$n = 2.0, \qquad\qquad\qquad\qquad \sigma_c \leq 50\,\text{MPa}$$

$$= 1.4+23.4\times\{\frac{90-f_{ck}}{100}\}^4, \qquad \sigma_c > 50\,\text{MPa} \tag{3.16}$$

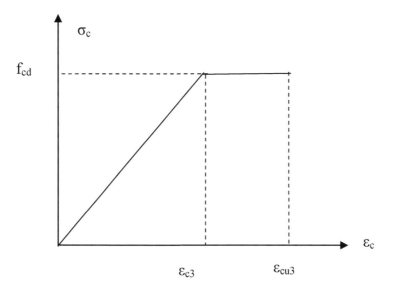

Fig. 3.6 Bi-linear stress - strain relationship

3.5.2 Bi-linear Relationship

The parabolic - rectangular relationship can be simplified for computational purposes and replaced by a bi-linear relationship as shown in Fig. 3.6.

Provided $f_{ck} \leq 90$ MPa, the bi-linear relationship is given by

$$\sigma_c = f_{cd} \frac{\varepsilon_c}{\varepsilon_{c3}}, \qquad\qquad 0 \leq \varepsilon_c \leq \varepsilon_{c3}$$

$$= f_{cd}, \qquad\qquad\qquad \varepsilon_{c3} \leq \varepsilon_c \leq \varepsilon_{cu3} \tag{3.17}$$

$$\varepsilon_{c3} = 1.75 \times 10^{-3}, \qquad\qquad \sigma_c \leq 50\,\text{MPa}$$

$$= [1.75 + 0.55 \times \{\frac{f_{ck} - 50}{40}\}] \times 10^{-3}, \qquad \sigma_c > 50\,\text{MPa} \tag{3.18}$$

$$\varepsilon_{cu3} = 3.5 \times 10^{-3}, \qquad\qquad \sigma_c \leq 50\,\text{MPa}$$

$$= [2.6 + 35 \times \{\frac{90 - f_{ck}}{100}\}^4] \times 10^{-3}, \qquad \sigma_c > 50\,\text{MPa} \tag{3.19}$$

3.5.3 Confined Concrete

In order to increase ductility in the case of earthquake resistant structures, zones where plastic hinges can form are enclosed by closely spaced closed links. This increases the compressive strength as well as the ultimate strain in concrete.

Depending on the confining stress $\sigma_2 = \sigma_3$, the ultimate strength $f_{ck,\,c}$ and strains $\varepsilon_{c2,\,c}$ and $\varepsilon_{cu2,\,c}$ are given by

$$f_{ck,\,c} = f_{ck}\,(1.0 + 5\,\alpha), \qquad \alpha = \sigma_2/\,f_{ck} \le 0.05$$

$$f_{ck,\,c} = f_{ck}\,(1.125 + 2.55\,\alpha), \qquad \alpha = \sigma_2/\,f_{ck} > 0.05$$

$$\varepsilon_{c2,\,c} = \varepsilon_{c2}\,\alpha^2, \qquad \varepsilon_{cu2,\,c} = \varepsilon_{cu2} + 0.2\alpha \tag{3.20}$$

3.6 PERMISSIBLE STRESSES IN CONCRETE

Compressive stress in concrete is limited in order to prevent the development of longitudinal cracks, micro cracks or high levels of creep. Permissible stresses in concrete at the transfer and serviceability limit state are as follows.

- At transfer, the stress is limited to $0.6f_{ck}$, where f_{ck} is the cylinder compressive strength at the time of stressing in post-tensioned members or at force transfer in the case of pre-tensioned members.

- At service, the stress should be limited to $0.6f_{ck}$ where f_{ck} is the 28-day cylinder compressive strength. In order to ensure that creep deformation is linear, compressive stress under ***quais-permanent loads*** should be limited to $0.45f_{ck}$.

The tensile stress should be limited to f_{ctm} using the appropriate value of f_{ck} depending on whether it is transfer conditions or serviceability limit state.

$$f_{ctm} = 0.30\,f_{ck}^{(2/3)}, \qquad C \le 50/60$$

$$f_{ctm} = 2.12\ln(1.8 + 0.1 \times f_{ck}), \qquad C > 50/60 \tag{3.21}$$

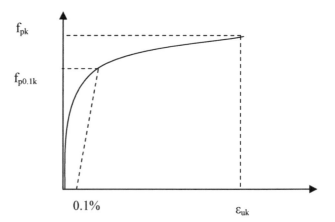

Fig. 3.7 Stress–strain curve for prestressing steel

3.7 PRESTRESSING STEEL

Prestressing steel is usually cold-drawn high tensile steel wires or alloy steel bars and is stress relieved by heating the strand to about 350°C and allowed to cool slowly to increase ductility. It has a characteristic ultimate strength of about 1800 MPa compared with 500 MPa for high yield reinforcing bars. Fig. 3.7 shows a typical stress–strain curve, where f_{pk} is the characteristic strength and $f_{p0.1k}$ is the stress at 0.1% strain. The common form of prestressing steel in practice is a 7-wire strand made from 6 wires spun round a straight central wire as shown in Fig. 2.3.

Overall nominal diameter varies from 12.5 mm to 18 mm. In order to reduce the overall diameter, the normal strand can be passed through a die to compress the cable and reduce the voids. The final shape of the individual wires is trapezoidal rather than circular as shown in Fig. 3.8. This type of cable is called '*Drawn*' and for the same nominal diameter, it has higher amount of steel. The design value of Young's modulus for a strand can be taken as an average value of 195 GPa.

For design purposes, a simplified relationship can be used as shown in Fig. 3.9. The material safety factor γ_m is taken as 1.15 and the value of $f_{p0.1k}$ can be taken as approximately equal to 0.85 f_{pk}. The design value f_{pd} of steel stress is given by

$$f_{pd} = f_{p0.1k}/\gamma_s \approx 0.85\ f_{pk}/1.15 \approx 0.74\ f_{pk} \qquad (3.22)$$

Various types of strands are available depending on the manufacturers. Table 3.6 shows the properties of 7– wire strands.

Table 3.6 Properties of 7– wire strands

Properties	STRAND TYPE			
	N	S	N	S
Nominal diameter, mm	13	13	15	15
Nominal area, mm^2	93	100	140	150
Tensile strength, MPa	1860	1860	1860	1860
Minimum breaking load, kN	173	186	260	279
Maximum relaxation %	2.5	2.5	2.5	2.5

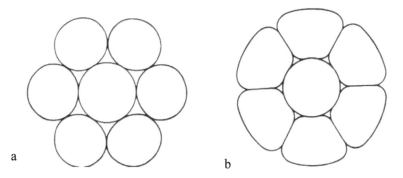

Fig. 3.8 Cross−section of a standard and a drawn strand.

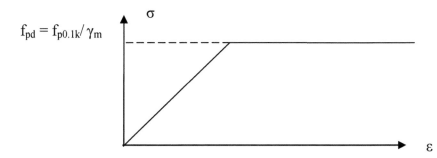

$$f_{pd} = f_{p0.1k}/\gamma_m$$

Fig. 3.9 Idealized design stress−strain curve

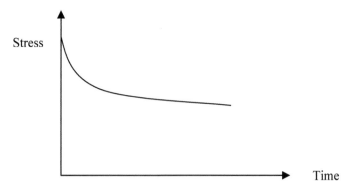

Fig. 3.10 Variation of stress with time under constant strain

3.8 RELAXATION

Just as concrete exhibits time dependent deformation due to creep, steel also exhibits a property called ***relaxation***. If the ***strain*** in steel is maintained constant, then the ***stress*** required to maintain that strain reduces with time as shown in Fig. 3.10. Relaxation is thus *loss of stress under constant strain*. Generally tests are conducted for 1000 hours (about 42 days) at an initial stress of 70% of the tensile strength at 20°C. The final long-term relaxation loss is expressed as a multiple of the 1000-hour loss at 20°C designated as ρ_{1000}.

Two types of heat treatment are used to improve the elastic and 'yield' properties. They are:

- Normal relaxation or stress-relieved strand. The strand is heated to about 350°C and allowed to cool slowly.

- Low relaxation or strain tempered strand. The strand is heated to about 3500 C *while the strand is under tension and allowed to cool slowly.* Such steel has a relaxation stress loss of about 25% of the stress− relieved strand.

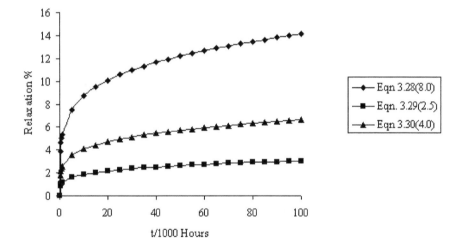

Fig 3.11 Relaxation curves for different types of steel

There are three classes of relaxation:
Class 1: wire or strand−ordinary relaxation, $\rho_{1000} = 8\%$
Class 2: wire or strand−low relaxation, $\rho_{1000} = 2.5\%$
Class 3: hot rolled and processed bars, $\rho_{1000} = 4\%$
Long−term loss can be calculated from the equations for t = half a million hours, which is equal to approximately 57 years.

$$\frac{\Delta\sigma_{pr}}{\sigma_{pi}} = 5.39 \times \rho_{1000} \times e^{6.7\,\mu} \times (\frac{t}{1000})^{0.75(1-\mu)} \times 10^{-5} \qquad \textbf{Class 1}$$

$$\frac{\Delta\sigma_{pr}}{\sigma_{pi}} = 0.66 \times \rho_{1000} \times e^{9.1\mu} \times (\frac{t}{1000})^{0.75(1-\mu)} \times 10^{-5} \qquad \textbf{Class 2}$$

$$\frac{\Delta\sigma_{pr}}{\sigma_{pi}} = 1.98 \times \rho_{1000} \times e^{8.0\mu} \times (\frac{t}{1000})^{0.75(1-\mu)} \times 10^{-5} \qquad \textbf{Class 3} \quad (3.23)$$

σ_{pi} = for post-tensioned structures, absolute value of initial prestress and for pre-tensioned structures jacking stress minus immediate elastic loss due to concrete compression.

$\Delta\,\sigma_{pr}$ = value of relaxation loss

$\mu = \sigma_{pi}/f_{pk}$

f_{pk} = characteristic value of the tensile strength of the prestressing steel.

Fig. 3.11 shows the relaxation curves for different types of strands.

3.9 MAXIMUM STRESS AT JACKING

The maximum stress $\sigma_{p,\,max}$ in the tendon is limited at the transfer stage to

$\sigma_{p,\,max}$ = min (0.8 f_{pk}, 0.9 $f_{p0.1k}$)

Taking $f_{p0.1k} \approx 0.85\ f_{pk}$, $\sigma_{p,\,max} = 0.9\ f_{p0.1k} \approx 0.77\ f_{pk}$ (3.24)

This value will govern the maximum permissible stress in the cable at the time of jacking.

3.10 LONG−TERM LOSS OF PRESTRESS

Long−term deformation of concrete due to creep and shrinkage and the loss of stress in steel due to relaxation contribute to the loss of prestress. The long-term loss can be as high as 25% of the initial stress. Thus

P $_{Service}$ $\approx$ 0.75 P$_{jack}$

P $_{Service}$ = total prestress remaining in the long term under working load conditions,

                    after all the losses have taken place

P$_{jack}$ = total load in the cables at the time of jacking.

It should be noted that these long−term losses of prestress are common to both pre− and post−tensioning systems.

3.11 REFERENCES TO EUROCODE 2 CLAUSES

In this chapter, various aspects of steel and concrete have been described. The following are the appropriate references to the clauses in the code EC2. The reader should refer to the code clauses for complete information.

Concrete: 3.1
Strength: 3.1.2
Elastic deformation: 3.1.3
Creep and shrinkage: 3.1.4
Design compressive and tensile strengths: 3.1.6
Stress–strain relations for the design of cross sections: 3.1.7
Confined concrete: 3.1.9
Prestressing steel: 3.3
Properties: 3.3.2
Strength: 3.3.3
Design assumptions: 3.3.6
Maximum prestressing force: 5.10.2

CHAPTER 4

SERVICEABILITY LIMIT STATE DESIGN OF PRE-TENSIONED BEAMS

4.1 DESIGN OF PRESTRESSED CONCRETE STRUCTURES

Structures have to be designed to satisfy two basic limit state criteria. They are

- Ultimate limit state (ULS): This limit state requires that structures should be able to safely resist the maximum loads that are likely to act. This is basically to do with safety. It is appreciated that at ultimate load large deflections and wide cracks are possible but the structure will not fall down. It is for this reason that, especially in earthquake zones, ductile structures are highly desirable.

- Serviceability limit state (SLS): This limit state requires that at working loads structures will resist loads without excessive deflection, unacceptable crack widths, unpleasant vibration, etc.

Before going into details of how to design a prestressed concrete structure, it is important to note the main difference in the approaches to the design of a reinforced concrete structure and a prestressed concrete structure.

A reinforced concrete structure is generally designed to satisfy the ultimate limit state and checked to make sure that the structure behaves satisfactorily at serviceability limit state. On the other hand a **prestressed concrete structure** is generally designed to satisfy the serviceability limit state requirements and then checked to ensure that the structure has adequate safety at ultimate limit state. The reason for this difference is that in prestressed concrete structures, serviceability limit state conditions are much more critical than the conditions at ultimate limit state, which is the opposite of what happens in the design of reinforced concrete structures. Thus generally but not always structures designed for serviceability limit state also satisfy the ultimate limit state criteria but not the other way round.

4.2 BEAM DESIGN BASED ON ENGINEERS' THEORY OF BENDING

Design of prestressed concrete is normally done for the serviceability limit state for bending. It is therefore useful to recapitulate some elementary aspects of elastic stress determination under the action of axial force and bending moment. This is best illustrated through an example.

4.2.1 Sign Convention

The following sign convention will be adopted in all the examples in this book.

Stresses: compressive is negative, tensile is positive.

Force: prestressing force is positive compressive.

Eccentricity: positive below the neutral axis.

Bending moment: Sagging moment is positive, hogging moment is negative; bending moment diagram is always drawn on the tension side.

Shear force: 'clockwise' shear stress is positive.

4.2.2 Example of Beam Design Based on Engineers' Theory of Bending

Example: Calculate the stresses in a simply supported prestressed beam of 12 m span. The cross section is a double T-section as shown in Fig. 4.1.

Top flange = 2400 × 100 mm thick, two webs each: 150 mm thick,

Overall depth: 750 mm

It is given that P_{jack} = 1111.1 kN and eccentricity e = 447 mm. Both P and e are constant over the entire span. It can be assumed that short-term loss is 10% and long-term loss is 25% of P_{jack}. The beam carries a live load of 8.0 kN/m². It is required to calculate the stresses at mid-span and support sections both at transfer and at SLS conditions.

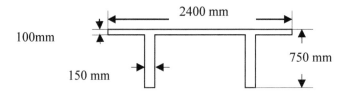

Fig. 4.1 Double T-beam

Solution:

Calculate the section properties:

Area of cross section A:

A = 2 × 750 × 150 + (2400 – 2 ×150) × 100 = 435.0 × 10³ mm²

Distance to centroid from the soffit y_{bar}:

Taking moments about the soffit,

A × y_{bar} = 2× 750× 150 × (750/2) + (2400 – 2 ×150) × 100× (750 – 100/2)

= 231.375 × 10⁶ mm³

$y_{bar} = 231.375 \times 10^6 / (435.0 \times 10^3) = 532$ mm

Distances to the bottom and top fibres from the centroidal axis:

$y_b = 532$ mm, $y_t = 750 - 532 = 218$ mm

Second moment of area I about the horizontal centroidal axis:

$I = 2 \times [150 \times 750^3/12 + 750 \times 150 \times (750/2 - 532)^2]$

$\quad + (2400 - 2 \times 150) [150^3/12 + 150 \times (750 - 100/2 - 532)^2]$

$\quad = 2.219 \times 10^{10}$ mm^4

Section moduli to top and bottom fibres:

$Z_t = I/y_t = 2.219 \times 10^{10} /218 = 101.8 \times 10^6$ mm^3

$Z_b = I/y_b = 2.219 \times 10^{10} /532 = 41.73 \times 10^6$ mm^3

Self weight of beam:

Self weight of beam /m = Area of cross section $\times$ Unit weight of concrete

where Area of cross section is in m^2 and Unit weight of concrete in in kN/m^3

Taking Unit weight of concrete = 25 kN/m^3,

Self weight/m = $(435.0 \times 10^3 \times 10^{-6}) \times 25 = 10.875$ kN/m

Live load:

Live load /m = 8.0 kN/m^2 $\times$ Width of top flange = $8.0 \times 2.4 = 19.2$ kN/m

Load at serviceability limit state:

SLS load = Live load + Self weight = $19.2 + 10.875 = 30.075$ kN/m

General formula for bending stress calculation:

Stress at top fibre = $(-P/A + Pe/Z_t) \times \gamma - M/Z_t$

Stress at bottom fibre = $(-P/A - Pe/Z_b) \times \gamma + M/Z_b$

P = Prestress, γ = Load factor

Note that stress due to M and that due to Pe are of opposite nature.

Stress calculation at the time of transfer:

P at transfer, $P_t \approx$ 90% $P_{jack} = 0.9 \times 1111.1$ kN = 1000.0 kN

$\gamma = \gamma_{superior} = 1.05$

Eccentricity e = 447 mm

External load acting is load due to self weight only = 10.875 kN/m

Bending stresses at mid–span at transfer:

Moment at mid-span, M = $10.875 \times 12^2/8 = 195.75$ kNm

$\sigma_{top} = (- P_t /A + P_t e/Z_t) \times \gamma_{superior} - M/Z_t$

$\quad = [-1000.0 \times 10^3/ (435.0 \times 10^3) + (1000.0 \times 10^3 \times 447)/ (101.8 \times 10^6)] \times 1.05$

$\quad\quad - (195.75 \times 10^6)/ (101.8 \times 10^6)$

$\quad = -2.41 + 4.61 - 1.92 = 0.28$ MPa

$\sigma_{bottom} = - P_t /A - P_t e/Z_b + M/Z_b$

$\quad\quad = [-1000.0 \times 10^3/ (435.0 \times 10^3) + (1000.0 \times 10^3 \times 447)/ (41.7 \times 10^6)] \times 1.05$

$\quad\quad\quad + (195.75 \times 10^6)/ (41.7 \times 10^6)$

$\quad\quad = -2.41 - 11.26 + 4.69 = - 8.98$ MPa

Bending stresses at support at transfer:

At support M = 0. Since P and e are the same throughout the span, stress due to P and Pe are the same as at mid-span. Therefore,

$\sigma_{top} = -2.41 + 4.61 = 2.20$ MPa

$\sigma_{bottom} = -2.41 - 11.26 = - 13.67$ MPa

Notice that under *transfer* conditions, the stress state at the *support* is more critical than at mid-span mainly because the stress due to selfweight reduces the net bending moment at a section. The beam *hogs up during transfer state*.

Stress calculation at service:

P at service stage, $P_s \approx 75\% P_{jack} = 0.75 \times 1111.1$ kN = 833.33 kN

$\gamma = \gamma_{inferior} = 0.95$

Eccentricity e = 447 mm

SLS load = Live load + Self weight = 19.2 + 10.875 = 30.075 kN/m

Bending stresses at mid-span at service:

Moment at mid-span = $30.075 \times 12^2/8 = 541.35$ kNm

$\sigma_{top} = (- P_s /A + P_s e/Z_t) \times \gamma_{Inferior} - M/Z_t$

$\quad = [-833.33 \times 10^3/ (435.0 \times 10^3) + (833.33 \times 10^3 \times 447)/ (101.8 \times 10^6)] \times 0.95$

$\quad\quad - (541.35 \times 10^6)/ (101.8 \times 10^6)$

$\quad\quad = -1.82 + 3.48 - 5.32 = - 3.66$ MPa

$\sigma_{bottom} = - P_s /A - P_s e/Z_b + M/Z_b$

$\quad\quad = [-833.33 \times 10^3/ (435.0 \times 10^3) + (833.33 \times 10^3 \times 447)/ (41.7 \times 10^6)] \times 0.95$

$\quad\quad\quad + (541.35 \times 10^6)/ (41.7 \times 10^6)$

$\quad\quad = -1.82 - 8.49 + 12.98 = 2.67$ MPa

Bending stresses at support at service:

M = 0 at support. Since P and e are the same throughout the span, stress due to P is the same as at mid-span. Therefore,

At top = −1.82 + 3.48 = 1.66 MPa

At bottom = −1.82 − 8.49 = − 10.31 MPa

Clearly, under **service** conditions, the stress state at **mid-span** is more critical than at support. The beam *sags down during service conditions.*

4.3 DEVELOPMENT OF SLS DESIGN EQUATIONS

It was shown in the example in 4.2.2, that at transfer, due to the high value of the prestress and as only the self weight of the beam is acting, the beam hogs up. Therefore the critical conditions are:

- tension is critical at the top of the beam

- compression is critical at the bottom of the beam.

Therefore, at any cross section,

At transfer:

$$[-\frac{P_t}{A} + \frac{P_t e}{Z_t}]\gamma_{Superior} - \frac{M_{self\,weight.}}{Z_t} \leq f_{tt} \qquad \text{at top face} \qquad (4.1)$$

$$[-\frac{P_t}{A} - \frac{P_t e}{Z_b}]\gamma_{Superior} + \frac{M_{self\,weight.}}{Z_b} \geq f_{tc} \qquad \text{at bottom face} \qquad (4.2)$$

where f_{tt} and f_{tc} are the permissible stresses at transfer in tension and compression respectively. The *first subscript t stands for Transfer* and the second subscript *t stands for tension* and c *for compression*.

Since

$P_t = \alpha\, P_{jack}$, where $\alpha \approx 0.90$ because of 10% loss at transfer

$P_s = \beta\, P_{jack}$ where $\beta \approx 0.75$ because of 25% long term loss in prestress

$P_s = \eta\, P_t$, where $\eta = \beta/\alpha \approx 0.83$

Putting $P_s = \eta\, P_t$, the above two equations can be converted to

$$[-\frac{P_s}{A} + \frac{P_s e}{Z_t}] \leq \frac{\eta}{\gamma_{Superior}}[f_{tt} + \frac{M_{self\,weight.}}{Z_t}] \qquad (4.3)$$

$$[-\frac{P_s}{A} - \frac{P_s e}{Z_b}] \geq \frac{\eta}{\gamma_{Superior}}[f_{tc} - \frac{M_{self\,weight.}}{Z_b}] \qquad (4.4)$$

Dividing throughout by P_s,

$$-\frac{1}{A}+\frac{e}{Z_t}\leq\frac{\eta}{\gamma_{Superior}}\langle\frac{M_{self\ weight.}}{Z_t}+f_{tt}\rangle\frac{1}{P_s} \qquad (4.5)$$

$$-\frac{1}{A}-\frac{e}{Z_b}\geq\frac{\eta}{\gamma_{Superior}}\langle-\frac{M_{self\ weight.}}{Z_b}+f_{tc}\rangle\frac{1}{P_s} \qquad (4.6)$$

At service:

At service, due to the reduced value of the prestress and the higher value of the SLS load acting on the beam, the beam sags down. Therefore the critical conditions are:

- compression is critical at the top of the beam

- tension is critical at the bottom of the beam.

$$[-\frac{P_s}{A}+\frac{P_s e}{Z_t}]\gamma_{Inferior}-\frac{M_{service}}{Z_t}\geq f_{sc} \qquad \text{at top face} \qquad (4.7)$$

$$[-\frac{P_s}{A}-\frac{P_s e}{Z_b}]\gamma_{Inferior}+\frac{M_{service}}{Z_b}\leq f_{st} \qquad \text{at botom face} \qquad (4.8)$$

$M_{service}$ = Moment at serviceability limit state, i.e. it includes selfweight, additional dead loads, superimposed dead load, live loads, etc.

f_{st} and f_{sc} are the permissible stresses at service in tension and compression respectively. The *first subscript s stands for service* and the second subscript *t stands for tension* and **c** *for compression*.

Dividing throughout by P_s,

$$-\frac{1}{A}+\frac{e}{Z_t}\geq\frac{1}{\gamma_{Inferior}}\langle\frac{M_{service}}{Z_t}+f_{sc}\rangle\frac{1}{P_s} \qquad (4.9)$$

$$-\frac{1}{A}-\frac{e}{Z_b}\leq\frac{1}{\gamma_{Inferior}}\langle-\frac{M_{service}}{Z_b}+f_{st}\rangle\frac{1}{P_s} \qquad (4.10)$$

The prestress and eccentricity must be so chosen such that for a given section and loads acting, the stress limitations given by all four equations must be satisfied.

4.3.1 Example of SLS design equations

Example: For the beam cross section and loads given in the example in section 4.2.2, determine the required prestress and eccentricity. Assume

- prestress and eccentricity are constant along the span.

- the permissible stresses at transfer: $f_{tt} = 2.6$ MPa, and $f_{tc} = -15.0$ MPa
- the permissible stresses at service: $f_{st} = 3.5$ MPa, and $f_{sc} = -24.0$ MPa

Note: f_{tc} and f_{sc} are both negative because they are limiting compressive stresses.

- Load factors for prestress: $\gamma_{Superior} = 1.05$, $\gamma_{Inferior} = 0.95$
- loss of 10% at transfer and 25% at service, $\eta = P_s/P_t = 0.75/0.9 = 0.83$

Solution:

From section 4.2.2,

Area of cross section $A = 435.0 \times 10^3$ mm^2

Section moduli: $Z_t = 101.8 \times 10^6$ mm^3, $Z_b = 41.73 \times 10^6$ mm^3

Self weight $= 10.875$ kN/m, SLS load $= 30.075$ kN/m

Span of beam $= 12$ m.

$1/A = 1/ (435 \times 10^3) = 229.885 \times 10^{-8}$

$1/Z_t = 1/ (101.8 \times 10^6) = 0.982 \times 10^{-8}$

$1/Z_b = 1/(41.73) \times 10^6 = 2.396 \times 10^{-8}$

Bending moments at mid-span:

$M_{self\ weight} = 10.875 \times 12^2/8 = 195.75$ kNm

$M_{service} = 30.075 \times 12^2/8 = 541.35$ kNm

Bending stresses at mid-span:

$M_{self\ weight} / Z_t = 195.75 \times 10^6/ (101.8 \times 10^6) = 1.92$ MPa

$M_{self\ weight} / Z_b = 195.75 \times 10^6/ (41.73 \times 10^6) = 4.69$ MPa

$M_{service} / Z_t = 541.35 \times 10^6/ (101.8 \times 10^6) = 5.32$ MPa

$M_{service} / Z_b = 541.35 \ x10^6/ (41.73 \times 10^6) = 12.97$ MPa

At support bending moment is zero both at transfer and at service.

Note that there are no signs associated with the above values as the proper signs are included in the equations.

Substituting the above values in the formulae, we have

At transfer: Because prestress and eccentricity remain constant along the span, at transfer the conditions are critical at support and not at mid-span.

At top: Substituting in equation (4.5),

$(-229.885 + 0.982\ e) \times 10^{-8} \leq (0.83/1.05)\{1.92 + 2.6\}(1/P_s)$ at mid-span

$(-229.885 + 0.982\ e) \times 10^{-8} \leq (0.83/1.05)\{0 + 2.6\}(1/P_s)$ at support

As the stress state at support is more critical than at mid-span, the equation corresponding to support is the governing equation.

$$(-229.885 + 0.982\ e) \leq 2.055\ (10^{-8}/P_s) \qquad\qquad (4.5a)$$

At bottom: Substituting in equation (4.6),

$$(-229.885 - 2.396\ e) \times 10^{-8} \geq (0.83/1.05)\{-5.32 - 15.0\}(1/P_s) \text{ at mid-span}$$

$$(-229.885 - 2.396\ e) \times 10^{-8} \geq (0.83/1.05)\{-15.0\}(1/P_s) \text{ at support}$$

As the stress state at support is more critical than at mid-span, the equation corresponding to support is the governing equation.

$$(-229.885 - 2.396\ e) \geq -11.857\ (10^8/P_s) \text{ at support} \qquad\qquad (4.6a)$$

Service: At service the conditions are always critical at mid-span due to increase in the applied load and reduction of prestress due to long term losses.

At top: Substituting in equation (4.9),

$$(-229.885 + 0.982\ e) \times 10^{-8} \geq (1/0.95)\{5.32 - 24.0\}(1/P_s) \text{ at mid-span}$$

$$(-229.885 + 0.982\ e) \geq -19.663\ (10^8/P_s) \qquad\qquad (4.9a)$$

At bottom: Substituting in equation (4.10),

$$(-229.885 - 2.396\ e) \times 10^{-8} \geq (1/0.95)\{-12.97 + 3.5\}(1/P_s) \text{ at mid-span}$$

$$(-229.885 - 2.396\ e) \geq -9.97\ (10^8/P_s) \qquad\qquad (4.10a)$$

Summarizing:

$$(-229.885 + 0.982\ e) \leq 2.055\ (10^{-8}/P_s) \text{ at support} \qquad\qquad (4.5a)$$

$$(-229.885 - 2.396\ e) \geq -11.857\ (10^8/P_s) \text{ at support} \qquad\qquad (4.6a)$$

$$(-229.885 + 0.982\ e) \geq -19.663\ (10^8/P_s) \text{ at mid-span} \qquad\qquad (4.9a)$$

$$(-229.885 - 2.396\ e) \geq -9.97\ (10^8/P_s) \text{ at mid-span} \qquad\qquad (4.10a)$$

Note that if $(10^8/P_s) = 0$, then

From (4.5a) and (4.9a), $e = Z_t/A = 234.2$ mm

From (4.6a) and (4.10a), $e = -Z_b/A = -95.95$ mm

4.3.2 Magnel Diagram

Equations (4.5a) to (4.10a) are four linear inequalities in e and 108/Ps. When the four straight lines corresponding to the four equations with equality sign are drawn, they enclose an area inside which all the four inequalities are satisfied. In other words, this is the feasible area for choice of Ps and e and any combination of Ps and e is acceptable from the stress calculation point of view, provided the system can accommodate the chosen value of eccentricity. This diagram is known as the Magnel diagram. Fig. 4.2 shows the Magnel diagram.

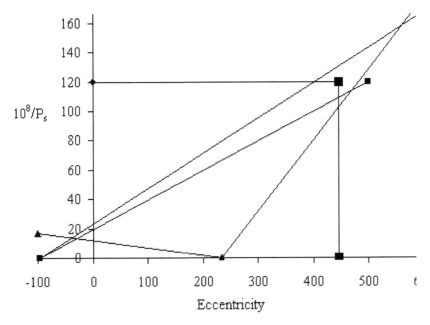

Fig. 4.2 Magnel diagram

Two points corresponding to each straight line can be obtained and the lines drawn.

From (4.5a), $10^8/P_s = 0$, e = 234.2

$$10^8/P_s = 174.9, e = 600.0$$

From (4.6a), $10^8/P_s = 0$, e = −95.945

$$10^8/P_s = 119.96, e = 500.0$$

From (4.9a), $10^8/P_s = 0$, e = 234.2

$$10^8/P_s = 16.69, e = −100.0$$

From (4.10a), $10^8/P_s = 0$, e = −95.945

$$10^8/P_s = 167.40, e = 600.0$$

It is worth pointing out that with the inequality signs, the valid areas are:

- For (4.5a), only the area to left of the line
- For (4.6a), only the area above the line
- For (4.9a), only the area to right of the line
- For (4.10a), only the area below the line

4.3.3 Choice of Prestress and Eccentricity

Once the Magnel diagram is drawn, suitable values of P and e can be chosen from the feasible area. In choosing a suitable value of prestress and eccentricity, it is useful to remember the following points.

- It is a *good* idea to work to as large a value of $10^8/P_s$ as possible as this will reduce the prestressing force and generally lead to an economical solution.

- It is *not* a good idea to work near a vertex, as this will not allow any flexibility in positioning the strands.

- Since during transfer operation, tensile stresses are set up at the top face of the beam, for safety reasons, it is necessary to keep some strands near the top face of the beam. This will also assist in tying links acting as shear reinforcement.

- At any section it is necessary to maintain as much prestress as possible. This will assist in improving the shear capacity of the section and decrease the need for shear links and thus aid in the ease and economy of construction.

- In all standard precast sections, the positions where a strand can be placed are predetermined by the position of the holes in the heavy steel end plates. It is therefore important that for a given value of P, a certain amount of flexibility in the choice of e is necessary.

In this example assume that the maximum breaking load per 12.7 mm 7-wire strand with f_{pk} equal to 1860 MPa and a cross sectional area of 112 mm^2 is 208 kN. The maximum permitted load P_{max} at jacking is about 80% of the breaking load. From equation (3.24), $P_{max} = 0.77 \times 208 = 160.4$ kN

From the section properties, the distance from the centroidal axis to the bottom fibre is 533 mm. Assuming a cover of say 50 mm, the maximum eccentricity is about 480 mm. From the Magnel diagram, the corresponding maximum value of $10^8/P_s$ is 135. Depending on the positioning of cables, it might not be possible to design to this value.

Therefore the minimum value of $P_s = (10^8/135) \times 10^{-3} = 740.7$ kN

Prestressing force at jacking: assuming loss of prestress at service stage as 25% of jacking force, then force at jacking

$P_{jack} = P_s/0.75 = 740.7/0.75 = 987.7$ kN

Maximum number of strands required = 987.7/160.4 = 6.2

In order to maintain symmetry, say a total of eight strands with four strands per web is required.

Assuming that strands can be provided in the web at two strands per layer and the layers are spaced at 50 mm intervals with the first layer at 60 mm from the soffit, then the value of eccentricity is

$e = y_b - 60 - 50/2 = 533 - 60 - 25 = 448$ mm

From the Magnel diagram, for e = 448 mm, $10^8/P_s = 120$,

$P_s = (10^8/120) \times 10^{-3} = 833.3$kN

$P_{jack} = P_s/0.75 = 833.3/0.75 = 1111$ kN

Using a total of eight strands, the force per strand is $1111/8 = 139$ kN < 160.4 kN

Therefore the final choice is $P_s = 833.3$ kN, e = 447 mm.

4.3.4 Stress Check

Although the Magnel diagram ensures that all stress constraints are satisfied, it is useful to run a separate check to ensure no errors have been committed.

At transfer: Check using equations (4.1) at the top face and (4.2) at the bottom face

$P_t = P_s/\eta = 833.33/0.83 = 1004.0$ kN, e = 447 mm, $\gamma_{Superior} = 1.05$, $\eta = 0.83$

Using the values calculated previously

$A = 435.0 \times 10^3$ mm^2, $Z_t = 101.8 \times 10^6$ mm^3, $Z_b = I/y_b = 41.73 \times 10^6$ mm^3

At mid-span, $M_{self\,weight}/Z_t = 1.92$ MPa, $M_{self\,weight}/Z_b = 4.69$ MPa

Substituting the above values in the equations, the stresses at transfer at top and bottom are:

At support: top = 2.20 MPa, bottom = -13.7 MPa

At mid-span: top = 0.28 MPa, bottom = -9.0 MPa

As can be checked, these stresses do not violate values of

$f_{tt} = 2.6$ MPa and $f_{tc} = -15.0$ MPa.

At service: Check using equations (4.7) at top face and (4.8) at bottom face:

$P_s = 833.33$ kN, e = 447 mm, $\gamma_{Inferior} = 0.95$, $\eta = 0.83$

Using the values calculated previously

$A = 435.0 \times 10^3$ mm^2, $Z_t = 101.8 \times 10^6$ mm^3, $Z_b = I/y_b = 41.73 \times 10^6$ mm^3

At mid-span, $M_{service}/Z_t = 5.32$ MPa, $M_{service}/Z_b = 12.97$ MPa

Substituting the above values in the equations, the stresses at transfer at top and bottom are:

At support: top = 1.8 MPa, bottom = -11.4 MPa

At mid-span: top = -3.7 MPa, bottom = 2.7 MPa

As can be checked these stresses do not violate values of

$f_{st} = 3.5$ MPa and $f_{sc} = -24.0$ MPa.

Thus all stress conditions are satisfied.

4.3.5. De-bonding

In the example in section 4.3.1, it was assumed that both P and e remain constant throughout the whole span. However this is not normally the case. If we assume de-bonding, then clearly we can accept that both P and e can vary from section to section. In this example purely for illustrative purposes, let us calculate the permissible values of prestressing forces and corresponding eccentricities for only two sections. The necessary equations are taken from calculations in section 4.3.1.

Support Section:

Transfer: These are same as equations (4.5a) and (4.6a) from section 4.3.1:

$$(-229.885 + 0.982 \text{ e}) \leq 2.055 \ (10^{-8}/P_s) \text{ at support} \qquad (4.5a)$$

$$(-229.885 - 2.396 \text{ e}) \geq -11.857 \ (10^8/P_s) \text{ at support} \qquad (4.6a)$$

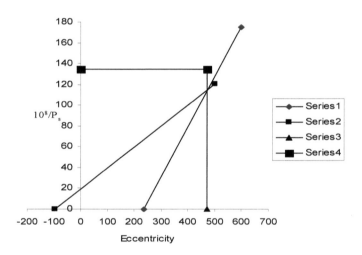

Fig. 4.3 Magnel diagram for support section

Service:

There is no need to consider this case as it is never critical because of loss of prestress from transfer to service situation. Thus there are only two valid equations for the support section and the feasible area is an open section as shown in Fig. 4.3.

Mid-span:

Transfer:

Substituting the values corresponding to mid-span in equations (4.5) and (4.6), the equations corresponding to top and bottom are respectively

Top: $(-229.885 + 0.982 \ e) \times 10^{-8} \leq (0.83/1.05)\{1.92 + 2.6\} \ (1/P_s)$

Simplifying

$$(-229.885 + 0.982 \ e) \leq 3.57 \ (10^8/P_s) \tag{4.5b}$$

Bottom: $(-229.885 - 2.396 \ e) \times 10^{-8} \geq (0.83/1.05)\{-5.32 - 15.0\} \ (1/P_s)$

Simplifying

$$(-229.885 - 2.396 \ e) \geq -16.06 \ (10^8/P_s) \tag{4.6b}$$

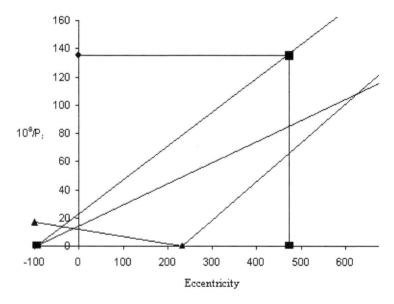

Fig. 4.4 Magnel diagram for mid-span section

Service: These are same as equations (4.9a) and (4.10a) from section 4.3.1:

$$(-229.885 + 0.982 \ e) \geq -19.663 \ (10^8/P_s) \text{ at mid-span} \tag{4.9a}$$

$$(-229.885 - 2.396 \ e) \geq -9.97 \ (10^8/P_s) \text{ at mid-span} \tag{4.10a}$$

Thus one Magnel diagram for each chosen section can be drawn. If Magnel diagrams for several sections are all drawn on the same figure, then they can be used to decide on a pattern of de-bonding. One effect of accepting the possibility of de-bonding is clear from Fig. 4.4. The feasible regions for mid-span and support sections are much larger than the feasible region common to both sections. This indicates two things, viz.

- that a much wider choice of P and e is possible than when P and e have to remain constant over the whole span.

- Clearly one can operate with much smaller values of P than when P and e are constant.

4.3.6 Choice of Prestress and Eccentricity at Different Sections

Once the Magnel diagram is drawn, a suitable value of P and e can be chosen from the feasible area. The maximum number of strands is required at the mid-span section. This section is therefore designed first. After that, at other sections, using the appropriate Magnel diagram, the number of strands that can be debonded can be determined.

Mid-span section: Using the data from section 4.3.3 and if all the strands can be provided in one row, then

$e = y_b - 60 = 532 - 60 = 472$ mm

The maximum valid value of $10^8/P_s$ from the Magnel diagram for the mid-span section in Fig. 4.4 is 135.

Therefore the maximum value of $P_s = (10^8/135) \times 10^{-3} = 741$ kN

Prestressing force at jacking: assuming loss of prestress at service stage as 25% of jacking force, then force at jacking

$P_{jack} = P_s/0.75 = 741 0.75 = 988$ kN

Maximum number of strands required = 988/160.4 = 6.2

Provide the strands in two rows. Therefore

$e = y_b - 60 - 50/2 = 532 - 60 - 25 = 447$ mm

The corresponding value of $10^8/P_s$ from the Magnel diagram is 125. This leads to

$P_s = (10^8/125) \times 10^{-3} = 800.0$ kN

$P_{jack} = P_s/0.75 = 800.0/0.75 = 1067$ kN

Number of strands required = 1067/160.4 = 6.7, i.e. four strands per web in order to maintain symmetry.

Therefore the final choice is $P_s = 800.0$ kN, e = 447 mm.

Support section: As can be seen from Fig. 4.3, the point corresponding to

$10^8/P_s = 135$ and e = 447 is inside the Magnel diagram for the support section. Therefore no debonding is necessary in this case.

4.4 INITIAL SIZING OF SECTION

Before the Magnel diagram can be drawn to determine the required prestressing force and its eccentricity, one needs to know the cross sectional properties of the chosen section. Manufacturers of precast pretensioned sections provide aids to choose a section to carry a specified load over a given span. However, Magnel equations can be used to arrive at the required section moduli as well.

From section 4.3, the stress limitation criteria at transfer and service conditions are given respectively by the equation pairs (4.3) and (4.4) and (4.7) and (4.8).

Eliminating P_s and e from (4.3) and (4.7),

$$\frac{M_{service}}{Z_t} - \frac{\eta M_{self\,weight.}}{Z_t} \leq \eta f_{tt} - f_{sc}$$

$$Z_t \geq \frac{M_{service} - \eta M_{self\,weight}}{\eta f_{tt} - f_{sc}} \tag{4.11}$$

Since $M_{service} = M_{self\,weight} + M_{dead\,loads} + M_{live}$, where $M_{dead\,loads}$ are dead loads acting in addition to the selfweight, the above equation can be written as

$$Z_t \geq \frac{M_{live} + M_{dead\,loads} + (1-\eta)M_{self\,weight}}{\eta f_{tt} - f_{sc}} \tag{4.12}$$

Eliminating P_s and e from (4.4) and (4.8),

$$\frac{M_{service}}{Z_b} - \frac{\eta M_{self\,weight.}}{Z_b} \leq f_{st} - \eta f_{tc}$$

$$Z_b \geq \frac{M_{service} - \eta M_{self\,weight}}{f_{st} - \eta f_{tc}} \tag{4.13}$$

$$Z_b \geq \frac{M_{live} + M_{dead\,loads} + (1-\eta)M_{self\,weight}}{f_{st} - \eta f_{tc}} \tag{4.14}$$

Note that because $\eta \approx 0.8$, the effect of ignoring the $M_{self\,weight}$ has a small effect on the required section moduli. This simplifies the calculations as at the start of calculations, $M_{self\,weight}$ is not known. Once a section is chosen, the self weight can be included to arrive at a better estimation of the required section moduli.

4.4.1 Example of Preliminary Sizing

Choose a suitable single T-section as shown in Fig. 4. 5. Assume $\eta = 0.83$,

Permissible stresses: $f_{tt} = 2.6$ MPa, $f_{tc} = -15.0$ MPa, $f_{st} = 3.5$ MPa, $f_{sc} = -24.0$ MPa

Moment excluding self weight at service on a simply supported span of 12 m is 400 kNm

Substituting in the equations (4.12) and (4.14) and ignoring self weight,

$$Z_t \geq [\frac{400 \times 10^6 + (1-0.83) \times 0}{0.83 \times 2.6 - (-24)} = 15.3 \times 10^6 \text{ mm}^3]$$

$$Z_b \geq [\frac{400 \times 10^6 + (1-0.83) \times 0}{3.5 - 0.83 \times (-15.0)} = 25.1 \times 10^6 \text{ mm}^3]$$

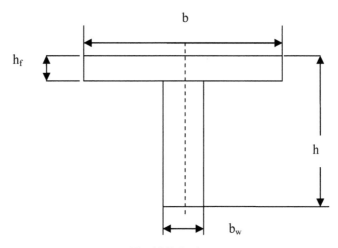

Fig. 4.5 T- Section

For a T-section the section properties can be calculated from the following equations:

$$\alpha = \frac{b_w}{b}, \qquad \beta = \frac{h_f}{h}$$

$$A = bh[\alpha + \beta(1-\alpha)]$$

$$y_{bar} = h \frac{[\frac{\alpha}{2} - +\beta(1-\alpha)(1-\frac{\beta}{2}]}{\alpha + \beta(1-\alpha)}$$

$$I = \frac{bh^3}{12}[\alpha + (1-\alpha)\beta^3 + \alpha(0.5 - \frac{y_{bar}}{h})^2 + \beta(1-\alpha)(1-0.5\beta - \frac{y_{bar}}{h})^2]$$

$$Z_b = \frac{I}{y_{bar}}, \qquad Z_t = \frac{I}{(h - y_{bar})} \tag{4.15}$$

Choose a section for $\alpha = 0.125$, $\beta = 0.125$

A = 0.2344 bh, y_{bar} = 0.296 h, z_b = 0.2055 bh^2, z_t = 0.0864bh^2

Choosing b = 2500 mm, from equation 4.11 and equation 4.12

z_b = 0.0309 $bh^2 \geq$ 25.1 × 10^6 mm,   h ≥ 570 mm

z_t = 0.0734 $bh^2 \geq$ 15.3 × 10^6 mm,   h ≥ 289 mm

Choosing b = 2500 mm, h = 600 mm,

b_w = 312 mm, h_f = 75 mm, A = 351.6 × 10^3 mm^2,

Taking unit weight of concrete per metre at 25 kN/m^3, the self weight per linear metre of the cross section is 8.79 kN/m.

Mid-span moment over a 12 m span due to self weight = 158.22 kNm. Using this value of self weight, section properties can be revised:

$$Z_t \geq [\frac{400 \times 10^6 + (1-0.83) \times 158.22 \times 10^6}{0.83 \times 2.6 - (-24)} = 16.4 \times 10^6 \ mm^3]$$

$$Z_b \geq [\frac{400 \times 10^6 + (1-0.83) \times 158.22 \times 10^6}{3.5 - 0.83 \times (-15.0)} = 26.8 \times 10^6 \ mm^3]$$

Choosing b = 2500 mm,

z_b = 0.0309 $bh^2 \geq$ 26.8 × 10^6 mm,   h ≥ 589 mm

z_t = 0.0734 $bh^2 \geq$ 16.4 × 10^6 mm,   h ≥ 299 mm

Therefore the choice of h = 600 mm appears to satisfy the section moduli requirements.

4.5 COMPOSITE BEAM SECTION

It is common in bridge construction to use precast pretensioned beams of various shapes to be placed on piers and using permanent formwork to cast an in-situ slab connecting the beams as shown in Fig. 4.6. If the beams are spaced close together, then glass-fibre-reinforced cement (GRC) panels are used as permanent formwork. On the other hand if the beams are widely spaced then ribbed glass reinforced plastic (GRP) panels which can span up to 4 m are used. The beams are normally spaced at 1 m centres. This is an efficient form of construction requiring minimum false work to complete the job. Before the slab has hardened, the precast section resists the self weight, the weight of wet concrete of slab, the weight of permanent formwork and associated construction loads. However once the concrete has hardened, the slab acts as the top flange of an I-beam and all live loads are resisted by the I-beam rather than by the precast section alone. This reduces the stresses in the precast beam due to live loads. It is naturally necessary that slip between the precast unit and cast in-situ slab is prevented in order to develop composite action. As shown in Fig. 4.7, if a slip occurs between the in-situ slab and the precast beam, then composite action is destroyed. This results in the precast beam and cast in-situ slab acting as two independent beams placed one on top of another rather than as a composite beam. This can be seen in the discontinuity of strain at a cross section. This results in considerable increase in stress in the precast beam. If composite action is retained then the strain distribution in the cross section will be continuous with comparatively low stresses in the precast beam

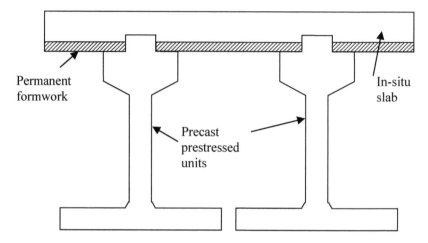

Fig. 4.6 Precast prestressed beams with an in-situ slab

4.5.1 Magnel Equations for Composite Beam

In developing the equations for drawing the Magnel diagram in the case of composite beams, it is necessary to remember that

- Prestress is applied to the precast section. Stresses due to prestress act only on the precast section. Thus in expressions like P/A and $(Pe)/Z$, the values of A and Z refer to the **precast section.** Similarly in the equation $M_{selfweight}/Z$, the section modulus Z refers to the precast section.

- The precast section resists both self weight as well as the weight of the *wet concrete* of the slab and the weight of permanent formwork.

- The composite section resists all the live load, super dead load and additional dead load from parapets, etc.

- When the cast in-situ slab shrinks, because of the bond between the precast unit and the slab, the top face of the precast unit has to contract as well. This naturally forces the precast beam to bend, resulting in stresses in the precast beam due to shrinkage of slab.

- When a cross section heats or cools, thermal gradients are set up in the cross section. This induces stresses in the cross section.

At transfer: Since only the precast section is present, the equations are identical to equations (4.5) and (4.6) of section 4.3.

At service: Equations have to reflect the fact that two different types of sections resist load:

$$
(-\frac{P_s}{A} + \frac{P_s e}{Z_t}) \gamma_{Inferior} - \frac{\{M_{self\,weight} + M_{slab+formwork}\}}{Z_t}
$$

$$
- \frac{M_{'live'}}{Comp.Z_{top\,of\,precast}} - Shrinkage\,stress + Thermal\,stress \geq f_{sc}
$$

(4.16)

$$
(-\frac{P_s}{A} - \frac{P_s e}{Z_b}) \gamma_{Inferior} + \frac{\{M_{self\,weight} + M_{slab+Formwork}\}}{Z_b}
$$

$$
+ \frac{M_{'live'}}{Comp.Z_b} + Shrinkage\,stress + Thermal\,stress \leq f_{st}
$$

(4.17)

Note: When calculating the stresses due to 'live' loads, we are interested in the stresses in the precast beam only. Thus the stress at the top is given by

$M_{live} \times$ (Depth of precast beam - Y_{comp})/I_{Comp},

where Y_{comp} = Position of centroidal axis of composite beam from the soffit.

It has to be remembered that 'live' loads include any additional dead loads, superimposed dead loads and live loads such as those due to vehicular and pedestrian loads on a bridge structure.

The Magnel equations for composite sections are

$$
-\frac{1}{A} + \frac{e}{Z_t} \geq \frac{1}{\gamma_{Inferior}} [\frac{\{M_{self\,weight} + M_{slab}\}}{Z_t} + \frac{M_{'live'}}{Comp.Z_{top\,of\,precast}}
$$

$$
+ Shrinkage\,stress + Thermal\,stress + f_{sc}]\frac{1}{P_s}
$$

(4.18)

$$
-\frac{1}{A} - \frac{e}{Z_b} \leq \frac{1}{\gamma_{Inferior}} [-\frac{\{M_{self\,weight} + M_{slab}\}}{Z_b} - \frac{M_{'live'}}{Comp.Z_b}
$$

$$
- Shrinkage\,stress + Thermal\,stress + f_{st}]\frac{1}{P_s}
$$

(4.19)

The calculation of shrinkage stress is considered in section 4.5.2 and the calculation of thermal stress is detailed in section 4.7.

4.5.2 Shrinkage Stress Calculation

When an in-situ slab is cast on top of a precast beam, the slab shrinks due to the concrete curing and drying. Because of the bond between the slab and the precast beam, the beam needs to shorten at the top fibre. In this process the slab forces the beam to shorten and bend and thus inducing shrinkage stresses in the precast beam.

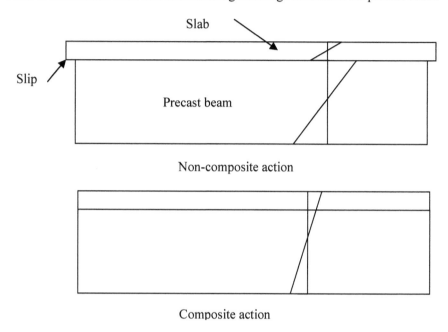

Fig. 4.7 Strain distribution at a cross section in a composite and non-composite beam

Because of the restraint provided by the beam, the final shrinkage of the slab is smaller than free shrinkage of the slab. The stresses due to shrinkage can be calculated as follows.

Stage 1: As shown in Fig. 4.8a, decouple the slab and beam and allow the slab to shrink freely. Let the strain due to shrinkage be $\varepsilon_{shrinkage}$.

Stage 2: As shown in Fig. 4.8b, apply external tensile force F to the slab to neutralize the free shrinkage strain of the slab. This sets up a tensile stress σ in the slab:

$$\sigma = \varepsilon_{shrinkage} \times E_c$$

where E_c = Young's modulus of slab concrete

$$F = A_{slab} \times \sigma = A_{slab} \times \varepsilon_{shrinkage} \times E_c$$

where A_{slab} = Area of cross section of the slab.

The slab is subjected to a uniform tensile stress of σ. There is no stress in the precast beam.

Stage 3: As shown in Fig. 4.9, with the force acting on the slab, bond the slab to the beam and release the force F. Releasing the force is equivalent to applying a compressive force to the composite beam. The force F acts at an eccentricity of a, where a = Distance between the centroidal axes of the composite beam and slab.

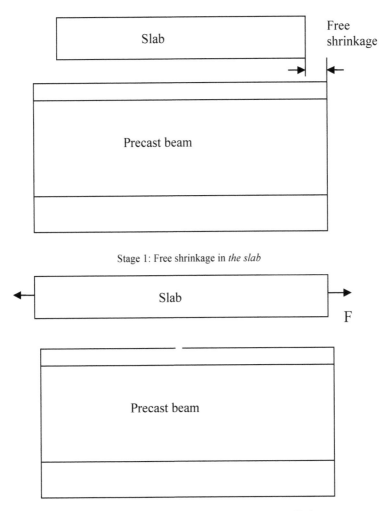

Stage 1: Free shrinkage in *the slab*

Stage 2: Free shrinkage in slab compensated by external tensile force

Fig. 4.8 Stages 1 and 2 in shrinkage stress calculation

Total ***stress in the slab*** is due to stages 2 and 3:

$$\sigma_{top} = \frac{F}{A_{slab}} - [\frac{F}{A_{comp.beam}} + \frac{Fa}{Z_{top\ of\ comp.beam}}] \times \frac{E_{cSlab}}{E_{cBeam}} \qquad (4.20)$$

$$\sigma_{bottom} = \frac{F}{A_{slab}} - [\frac{F}{A_{comp.beam}} + \frac{Fa}{Z_{Comp.beam\ to\ bottom\ of\ slab}}] \times \frac{E_{cSlab}}{E_{cBeam}}$$

$$(4.21)$$

Stresses in the precast beam due to shrinkage of slab at stage 3 only:

$$\sigma_{top} = -\frac{F}{A_{comp.beam}} - \frac{Fa}{Z_{Comp.beam\ to\ top\ of\ precast\ beam}} \qquad (4.22)$$

$$\sigma_{bottom} = -\frac{F}{A_{comp.beam}} + \frac{Fa}{Z_{bottom\ of\ comp.beam}} \qquad (4.23)$$

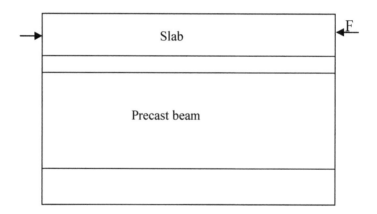

Fig. 4.9 Stage 3 in shrinkage stress calculation

4.5.3 Example of Shrinkage Stress Calculation

Example: Calculate the stresses due to shrinkage using the following data.

Precast pretensioned inverted T-beam:

Bottom flange: 750 mm wide × 300 mm thick

Web: 300 mm thick, Overall depth: 1100 mm, y_{bar} from soffit: 434 mm

Area of cross section, $A_{beam} = 465.0 \times 10^3$ mm^2

Second moment of area, $I_{beam} = 4.96 \times 10^{10}$ mm^4

$Z_b = I/y_{bar} = 114.29 \times 10^6$ mm^3, $Z_t = I/(1100 - y_{bar}) = 74.48 \times 10^6$ mm^3

Beam is made from grade C50/60. $f_{cm} = 50 + 8 = 58$ MPa

From (3.4), $E_{cm\ beam} = 22 \times (\frac{f_{cm}}{10})^{0.3} = 37.28$ GPa

In-situ concrete slab:

1000 mm wide × 160 mm thick

Area of cross section, $A_{slab} = 160.0 \times 10^3$ mm^2

Second moment of area, $I_{slab} = 0.034 \times 10^{10}$ mm^4

Slab is made from grade C25/30, $f_{cm} = 25 + 8 = 33$ MPa

From (3.4), $E_{cm\ slab} = 22 \times (\frac{f_{cm}}{10})^{0.3} = 31.48$ GPa

Modular ratio $= E_{cm\ slab}/E_{cm\ beam} = 31.48/37.28 = 0.844$

Shrinkage strain: Assume outside condition with a relative humidity of 80%. Concrete is made with cement CEM Class N.

Solution:

a. Section properties of composite beam:

For elastic analysis, for a common value of Young's modulus for beam and slab of $E_{cm} = 37.28$ GPa, the modified cross sectional properties of the slab are:

Modified width of slab = Actual width × Modular ratio = $1000 \times 0.844 = 844$ mm

Modified area of cross section, $A_{slab} = 135.04 \times 10^3$ mm^2

Modified second moment of area, $I_{slab} = 0.029 \times 10^{10}$ mm^4

Composite beam properties:

Area of cross section: $A_{comp} = A_{slab}$ modified $+ A_{beam}$

$$= 135.04 \times 10^3 + 465.0 \times 10^3 = 600.04 \times 10^3 \text{ mm}^2$$

$y_{bar\ comp}$ from soffit:

$$y_{bar\ comp} = \frac{A_{slab\ modified} \times (1100 + 160/2) + A_{beam} \times 434}{A_{comp}}$$

$$= \frac{[135.04 \times 1180 + 465 \times 434] \times 10^3}{600.04 \times 10^3} = 602 \text{ mm}$$

Second moment of area:

$$I_{comp} = I_{beam} + A_{beam} \times (602 - 434)^2$$

$$+ I_{slab\ modified} + A_{slab\ modified} \times (1100 + 160/2 - 602)^2$$

$$= 4.962 \times 10^{10} + 465 \times 10^3 \times 168^2 + 0.029 \times 10^{10}$$

$$+ 135.04 \times 10^3 \times 578^2 = 10.82 \times 10^{10} \text{ mm}^4$$

Section moduli:

$$Z_{t \text{ comp}} = I_{comp}/(1100 + 160 - 602) = 164.44 \times 10^6 \text{ mm}^3$$

$$Z_{Comp} \text{ to top of precast or bottom of slab} = I_{comp}/(1100 - 602)$$
$$= 217.27 \times 10^6 \text{ mm}^3$$

$$Z_{b \text{ comp}} = I_{comp}/(602) = 179.73 \times 10^6 \text{ mm}^3$$

b. Shrinkage strain:

The total shrinkage strain ε_{cs} can be calculated from (3.11) as follows.

$$\varepsilon_{cs} = \varepsilon_{cd0} \times k_h + 2.5 (f_{ck} - 10) \times 10^{-6}$$

$$\varepsilon_{cd0} = 0.85[(220 + 110\,\alpha_{ds1}) \times \exp\{-\alpha_{sd2} \times 0.1 \times f_{cm}\}] \times 1.55\{1 - (0.01RH)^3\} \times 10^{-6}$$

From Table 3.3, $\alpha_{ds1} = 4$, $\alpha_{ds2} = 0.12$

Assuming only top and bottom surfaces are exposed to atmosphere,

$A_c = 1000 \times 160 = 160 \times 10^3 \text{ mm}^2$, $u = 2 \times 1000 = 2 \times 10^3 \text{ mm}$

$h_0 = A_c/u = 80$ mm. From Table 3.4, $k_h = 1.0$

$f_{ck} = 25$ MPa, $f_{cm} = 33$ MPa, RH = 80

$\varepsilon_{cd0} = 176 \times 10^{-6}$, $\varepsilon_{cs} = 213 \times 10^{-6}$

c. Calculate F:

E_{cm} for slab concrete $= 31.48 \text{ GPa} = 31.48 \times 10^3 \text{ MPa}$, $\varepsilon_{shrinkage} = 213 \times 10^{-6}$

$\sigma = 213 \times 10^{-6} \times 31.48 \times 10^3 = 6.7$ MPa

$F = A_{slab} \times \sigma = 1000 \times 160 \times 6.7 = 1072$ kN

Note: A_{slab} is the actual area, not the equivalent area discussed before.

This force acts at the centroid of the slab. $a = 1100 + 160/2 - 602 = 578$ mm

$F \times a = 619.16$ kNm

d. Calculate the stresses:

Using equations (4.20) and (4.21), total **stresses in the slab** due to stages 2 and 3:

$$\sigma_{top} = \frac{1072 \times 10^3}{160 \times 10^3} - [\frac{1072 \times 10^3}{600.04 \times 10^3} + \frac{619.16 \times 10^6}{164.44 \times 10^6}] \times 0.844 = 2.0 \text{ MPa}$$

$$\sigma_{bottom} = \frac{1072 \times 10^3}{160 \times 10^3} - [\frac{1072 \times 10^3}{600.04 \times 10^3} + \frac{619.16 \times 10^6}{217.27 \times 10^6}] \times 0.844 = 2.8 \text{ MPa}$$

Using equations (4.22) and (4.23), **stresses in the precast beam** due to shrinkage of slab due to stage 3 only:

$$\sigma_{top} = -\frac{1072 \times 10^3}{600.04 \times 10^3} - \frac{619.16 \times 10^6}{217.27 \times 10^6} = -4.6 \, \text{MPa}$$

$$\sigma_{top} = -\frac{1072 \times 10^3}{600.04 \times 10^3} + \frac{619.16 \times 10^6}{179.73 \times 10^6} = 1.7 \, \text{MPa}$$

4.5.4 Magnel Diagrams for a Composite Beam

Example: Draw the Magnel diagram for the composite beam considered in section 4.5.3. It is used to carry an 18 kN/m live load over a simply supported beam span of 24 m. Use the following data. Ignore, for simplicity, stresses induced due to thermal gradients.

a. Precast beam properties:

Area of cross section, $A_{beam} = 465.0 \times 10^3 \, \text{mm}^2$

$Z_b = 114.29 \times 10^6 \, \text{mm}^3$, $Z_t = 74.48 \times 10^6 \, \text{mm}^3$

$1/A = 1/(465 \times 10^3) = 215.05 \times 10^{-8}$, $1/Z_b = 1/(114.29 \times 10^6) = 0.875 \times 10^{-8}$

$1/Z_t = 1/(74.48 \times 10^6) = 1.343 \times 10^{-8}$.

b. Composite beam properties:

Z_{Comp} to top of precast $= 217.27 \times 10^6 \, \text{mm}^3$, $Z_{b \, comp} = 179.73 \times 10^6 \, \text{mm}^3$

c. Permissible concrete strength (See section 3.6, Chapter 3):

The permissible stresses are at transfer: $f_{ck} = 25/30 \, \text{MPa}$

From (3.21), $f_{tt} = f_{ctm} = 0.30 \times 25^{2/3} = 2.6 \text{MPa}$

$f_{tc} = -0.6 \times 25 = -15.0 \, \text{MPa}$

The permissible stresses are at service: $f_{ck} = 40/50 \, \text{MPa}$

From (3.21), $f_{st} = f_{ctm} = 0.30 \times 40^{2/3} = 3.5 \text{MPa}$

$f_{tc} = -0.6 \times 40 = -24.0 \, \text{MPa}$

d. Load factors for prestress: $\gamma_{Superior} = 1.05$, $\gamma_{Inferior} = 0.95$

e. Loss of prestress: loss of 10% at transfer and 25% at service,

$$\eta = P_s/P_t = 0.75/0.9 = 0.83$$

f. Shrinkage stresses: $\sigma_{top} = -4.6 \, \text{MPa}$, $\sigma_{bottom} = 1.7 \, \text{MPa}$

Magnel diagram for mid-span:

Unit weight of concrete $= 25 \, \text{kN/m}^3$

$q_{self \, weight} =$ Area of beam ($= 4.65 \times 10^5$) $\times 10^{-6} \times 25 = 11.625 \, \text{kN/m}$

$M_{self\ weight} = 11.625 \times 24^2/8 = 837$ kN/m

$\sigma_{top} = 837 \times 10^6/Z_t = 11.24$ MPa, $\sigma_{bottom} = 837 \times 10^6/Z_b = 7.32$ MPa

q_{slab} = Area of slab (= 1.6×10^5) $\times 10^{-6} \times 25 = 4.0$ kN/m

Allow 10% additional load for formwork:

$q_{slab\ +\ Formwork} = 1.1 \times 4.0 = 4.4$ kN/m

$M_{slab\ +\ formwork} = 4.4 \times 24^2/8 = 316.8$ kN/m

$\sigma_{top} = 316.8 \times 10^6/Z_t = 4.25$ MPa, $\sigma_{bottom} = 316.8 \times 10^6/Z_b = 2.77$ MPa

$q_{live} = 18.0$ kN/m, $M_{live} = 18.0 \times 24^2/8 = 1296.0$ kN/m

$\sigma_{top} = 1296 \times 10^6/Z_{t\ comp}$ to top of precast $= 5.96$ MPa

$\sigma_{bottom} = 837 \times 10^6/Z_{b\ comp} = 7.21$ MPa

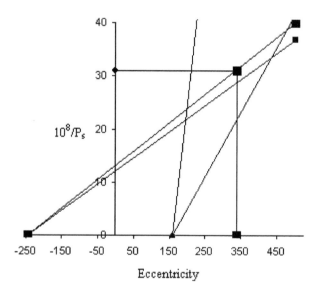

Fig. 4.10 Magnel diagram for mid-span section

At transfer: Substituting in equations (4.5) and (4.6) and using a load factor of $\gamma_{superior} = 1.05$,

$$-215.05 + 1.343\,e \leq [\frac{0.83}{1.05}\langle 11.24 + 2.6\rangle = 10.94]\frac{10^8}{P_s}$$

$$-215.05 - 0.875\,e \geq [\frac{0.83}{1.05}\langle -7.32 - 15.0\rangle = -17.64]\frac{10^8}{P_s}$$

At service: Substituting in equation (4.18) and (4.19) and using a load factor of $\gamma_{Inferior} = 0.95$,

$$-215.05 + 1.343e \geq [\frac{1}{0.95}(11.24 + 4.25 + 5.96 + 4.6 - 24.0) = 2.16]\frac{10^8}{P_s}$$

$$-215.05 - 0.875e \geq [\frac{1}{0.95}(-7.32 - 2.77 - 7.21 - 1.7 + 3.5) = -16.32]\frac{10^8}{P_s}$$

Fig. 4.10 shows the Magnel diagram for the mid-span section.

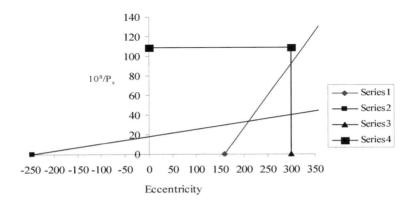

Fig. 4.11 Magnel diagram for support section

Magnel diagram for support:

At transfer: Substituting in equations (4.1) and (4.2) and using a load factor of $\gamma_{superior} = 1.05$,

$$-215.05 + 1.343e \leq [\frac{0.83}{1.05}\langle 2.6\rangle = 2.06]\frac{10^8}{P_s}$$

$$-215.05 - 0.875e \geq [\frac{0.83}{1.05}\langle -15.0\rangle = -11.86]\frac{10^8}{P_s}$$

Fig. 4.11 shows the Magnel diagram for the support section.

4.5.5 Choice of Prestress and Eccentricity at Different Sections

Assume that in the bottom flange, there are four rows of 12 strands per row spaced vertically at 50 mm intervals with the first row at 60 mm from the soffit.

Mid-span section: If all the strands can be provided in the bottom row, then

$e = y_b - 60 = 434 - 60 = 374$ mm

From the Magnel diagram for the mid-span section in Fig. 4.10, the corresponding maximum valid value of $10^8/P_s$ is 33.0.

Therefore the maximum value of $P_s = (10^8/33.0) \times 10^{-3} = 3030$ kN

Prestressing force at jacking: assuming loss of prestress at service stage as 25% of jacking force, then force at jacking

$P_{jack} = P_s/0.75 = 3030/0.75 = 4040$ kN

Maximum number of strands required $= 4040/160.4 = 25.2$

All these strands cannot be provided in the bottom row only. Therefore the effective eccentricity will be smaller than 374 mm. Therefore $10^8/P_s$ will be smaller than 33.0. Assuming 28 strands will be required, which will be provided in the bottom three rows in the order 12, 12 and 4, the net eccentricity is given by

$e = y_b - \{12 \times 60 + 12 \times 110 + 4 \times 160\}/28 = 338$ mm

From the Magnel diagram for the mid-span section in Fig. 4.10, the corresponding maximum valid value of $10^8/P_s$ is 31.0. Therefore the maximum value of

$P_s = (10^8/31.0) \times 10^{-3} = 3226$ kN

Prestressing force at jacking: assuming loss of prestress at service stage as 25% of jacking force, then force at jacking $P_{jack} = P_s/0.75 = 3226/0.75 = 4301$ kN

Maximum number of strands required $= 4301/160.4 = 26.8$

Therefore the final choice is $P_s = 3226$ kN, $e = 338$ mm and this choice is inside the mid-span Magnel diagram.

Support section: As can be seen from Fig. 4.11, the point corresponding to

$10^8/P_s = 31$ and $e = 338$ mm is well outside the Magnel diagram for the support section. Therefore debonding is necessary in this case. By trial and error it is found that if only four strands in the second and third rows are retained and the rest debonded, then

$e = y_b - \{0 \times 60 + 4 \times 110 + 4 \times 160\}/8 = 299$ mm

From mid-span for 28 strands, $10^8/P_s = 31.0$. For only eight strands,

$10^8/P_s = 31.0 \times \{28/8\} = 108.5$, $P_s = (10^8/108.5) \times 10^{-3} = 921.7$ kN

From the Magnel diagram for the support the final choice is $P_s = 921.7$ kN, $e = 299$ mm and this choice is inside the support Magnel diagram.

4.6 CRACKING

Eurocode 2 adopts two criteria for limiting crack width. They are:

- Decompression: This requires that all parts of the tendon or duct lie at least 25 mm within the concrete in compression.

- Limiting crack widths to a specified value

For members with bonded tendons cracking is checked only for frequent load combinations. Table 4.1 gives the criterion to be adopted.

At transfer, if the tensile stress in the top fibre does not exceed mean tensile strength of the concrete f_{ctm}, there is no need to check for crack widths.

Decompression: Decompression can be ensured by using the following equation in drawing the Magnel diagram. $Z_{b\,25}$ refers to the section modulus at 25 mm from the soffit. The stress at that level is taken as zero.

$$-\frac{1}{A}-\frac{e}{Z_b} \leq \frac{1}{\gamma_{Inferior}} \langle -\frac{M_{Frequent}}{Z_{b\,25}}+0\rangle \frac{1}{P_s} \tag{4.24}$$

Table 4.1 Permissible maximum crack widths

Exposure Class	Maximum width of crack	Remarks
X0, XC1	0.2 mm	Acceptable appearance
XC2 –XC4	0.2 mm	Also check decompression under quasi-permanent combination of loads
XD1, XD2, XS1-XS3	Decompression	

4.7 THERMAL STRESS CALCULATION

As was stated in section 4.5.1, when considering the state of stress at the serviceability limit state, not only the stresses due to external loading but also self equilibrating stresses induced due to the shrinkage of the cast in-situ slab and the stresses induced due to thermal gradients in the cross section resulting from differential heating and cooling of the different parts of the cross section have to be included.
Depending on the type of construction, Eurocode 1 gives temperature differentials to be taken into account when determining the stresses caused by these differentials. The three forms of construction considered as shown in Fig. 4.12 are:
- Solid slab
- 'I-beam'
- Box beam

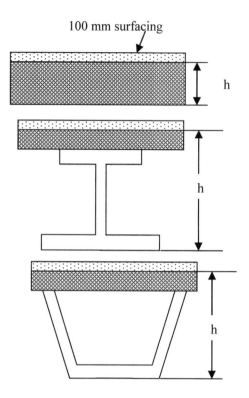

Fig. 4.12 Three different forms of concrete cross sections

4.7.1 Heating

Fig. 4.13 shows the variation of temperature differences in the cross section due to heating of the cross section. In all cases it is assumed that there is a 100 mm surfacing. Table 4.2 shows the temperature differentials to be considered.
$h_1 = 0.3h \leq 150$ mm, $h_2 = 0.3h$ but ≥ 100 mm and ≤ 250 mm,
$h_3 = 0.3h$ but $\leq (100$ mm + surfacing depth in mm)

Table 4.2 Temperature differentials: Heating

h, mm	ΔT_1	ΔT_2	ΔT_3
≤ 200	8.5	3.5	0.5
400	12.0	3.0	1.5
600	13.0	3.0	2.0
≥ 800	13.0	3.0	2.5

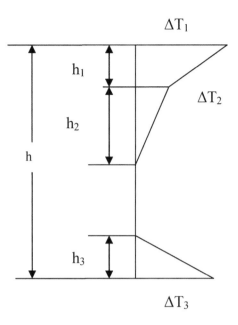

Fig. 4.13 Temperature difference due to heating

4.7.2 Cooling

Fig. 4.14 shows the variation of temperature differences in the cross section due to cooling of the cross section. In all cases it is assumed that there is a 100 mm surfacing. Table 4.3 shows the temperature differentials to be considered.
$h_1 = h_4 = 0.22\,h \le 250$ mm, $h_2 = h_3 = 0.25\,h$ but ≥ 200 mm

Table 4.3 Temperature differentials: Cooling

h, mm	ΔT_1	ΔT_2	ΔT_3	ΔT_4
≤ 200	−2.0	−0.5	−0.5	−1.5
400	−4.5	−1.4	−1.0	−3.5
600	−6.5	−1.8	−1.5	−5.0
800	−7.6	−1.7	−1.5	−6.0
1000	−8.0	−1.5	−1.5	−6.3
≥ 1500	−8.4	−0.5	−1.0	−6.5

4.7.3 Calculation of Stresses due to Thermal Gradients

The procedure for calculating stresses due to temperature differentials is similar to the calculation of shrinkage stresses detailed in section 4.5.3. Each element is allowed free expansion or contraction. External forces are applied to restrain the change of length. The total axial force and bending moment applied to the section are calculated. External axial force and moment are applied to the whole beam cross section to counteract the axial force and moment due to the external forces applied to restrain free change of length. Stress at any section is the sum of the stress due to external force applied to restrain free change of length plus the axial stress and bending stress caused by forces applied to the whole beam cross section.

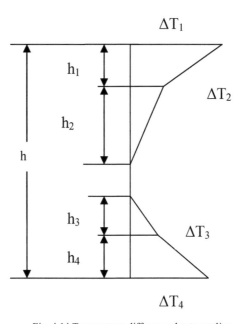

Fig. 4.14 Temperature difference due to cooling

4.7.4 Example of Thermal Stress Calculation

Calculate the stress distribution in the box girder shown in Fig. 4.15 due to
- Heating
- Cooling

Assume: Coefficient of thermal expansion $\alpha_T = 10 \times 10^{-6}/°C$,
         Young's modulus, $E_c = 34$ GPa.

$\alpha_T \times E_c = 0.34$ MPa

The compressive stress σ due to restraining the expansion due to a temperature change of ΔT is given by

$\sigma = -\alpha_T \times E_c \times \Delta T$

Area of cross section, A:

$A = (1500 - 2 \times 200) \times 250 + (900 - 2 \times 200) \times 300 + 1500 \times 2 \times 200$
$\quad = 1.025 \times 10^6$ mm^2

Taking moments about the soffit, the first moment of area

$A \times y_b = (1500 - 2 \times 200) \times 250 \times (1500 - 250/2)$
$\qquad + (900 - 2 \times 200) \times 300 \times (300/2)$
$\qquad + 1500 \times 2 \times 200 \times (1500/2) = 850.625 \times 10^6 \text{ mm}^3$

Centroidal distance from soffit $y_b = 850.625 \times 10^6/ (1.025 \times 10^6) = 830$ mm
Centroidal distance from top $y_t = 1500 - 830 = 670$ mm
Second moment of area, I:
$I = (1/12) \times \{(1500 - 2 \times 200) \times 2503 + (900 - 2 \times 200) \times 3003 + 15003 \times 2 \times 200\}$
$\qquad + [(1500 - 2 \times 200) \times 250 \times (1500 - 250/2 - y_b)^2$
$\qquad + (900 - 2 \times 200) \times 300 \times (300/2 - y_b)^2$
$\qquad + 1500 \times 2 \times 200 \times (1500/2 - y_b)^2] = 26.994 \times 10^{10} \text{ mm}^4$

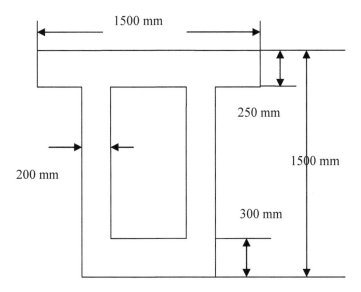

Fig. 4.15 Beam cross section

4.7.4.1 Thermal Stress Calculation: Heating

h = 1500 mm, 0.3h = 450 mm, h_1 = 0.3h ≤150 mm
h_2 =0.3h but ≥ 100 mm and ≤ 250 mm
h_3 = 0.3h but ≤ (100 mm + surfacing depth in mm)
Therefore take: h_1 = 150 mm, h_2 = 250 mm, h_3 = 200 mm
From Table 4.2, ΔT_1 = 13.0°C, ΔT_2 = 3.0°C, ΔT_3 = 2.5°C
Corresponding restraining stresses are: $\sigma = -\alpha_T \times E_c \times \Delta T$
$\sigma = -0.34 \times \Delta T = (-4.42, -1.02, -0.85)$ MPa
Table 4.4 shows the restraining stresses at various levels in the cross section.
    Having obtained the restraining stresses, the next step is to calculate the forces
F in different segments of the cross section using the formula
F = Average stress × Area
The values of the forces are shown in Table 4.5.

Table 4.4 Restraining stresses in the cross section: heating

Level from top, mm	σ, MPa
0	−4.42
150	−1.02
250	−0.61
400	0
1300	0
1500	−0.85

Table 4.5 Restraining forces in the cross section: Heating

σ at top MPa	σ at bottom MPa	area	F, kN
−4.42	−1.02	1500 × 150	−612.00
−1.02	−0.61	1500 × 100	−122.25
−0.61	0	2 × 200× 150	−18.36
0	−0.85	900 × 200	−76.50

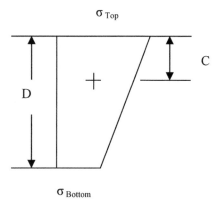

Fig. 4.16 Position of the resultant

Having calculated the restraining forces, the next step is to calculate the moment due to the restraining forces about the centroidal axis of the beam. Table 4.6, shows the details of the calculation of the moment due to force F. Note that if the stress distribution is as shown in Fig. 4.16, the resultant is at a distance C from the top:

$$C = \frac{D}{3} \times \frac{(2\sigma_{Bottom} + \sigma_{Top})}{(\sigma_{Bottom} + \sigma_{Top})} \tag{4.24}$$

Table 4.6 Calculation of moment: Heating

F, kN	D mm	C mm	Lever arm mm	Moment kNm
–612.00	150	59	670 – 59 = 611	–373.70
–122.25	100	46	670–150–46 =474	–57.95
–18.36	150	50	670–250–50=370	–6.79
–76.50	200	133	–(830–200 + 133) = –763	58.40
Σ=–829.11				Σ= –380.04

The last step is the calculation of the final stress in the cross section as the sum of the initial restraining stress plus the stress due to axial force and moment in order to produce a self-equilibrating stress system. Table 4.7 shows the details of stress calculation. Fig. 4.17 shows the final stresses.

Table 4.7 Thermal stress calculation: Heating

Position from top	σ MPa	$-\Sigma\,F/A$ MPa	y mm	$-(\Sigma\,M/I)y$	Final stress MPa
0	–4.42	0.81	670	0.94	–2.67
150	–1.02	0.81	520	0.73	0.52
250	–0.61	0.81	420	0.59	0.79
400	0	0.81	270	0.38	0.79
670	0	0.81	0	0	0.81
1300	0	0.81	–630	–0.88	–0.07
1500	–0.85	0.81	–830	–1.17	–1.21

Note: y_b = 830 mm, y_t = 670 mm

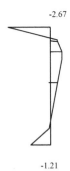

-2.67

-1.21

Fig.4.17 Final stresses: heating

4.7.4.2 Thermal Stress Calculation: Cooling

The procedure to be adopted is similar to the calculation of stresses due to heating.
$h = 1500$ mm, $0.2h = 300$ mm, $h_1 = h_4 = 0.2h \leq 250$ mm
$0.25h = 375$ mm, $h_2 = h_3 = 0.25h$ but ≥ 200 mm
Therefore $h_1 = h_4 = 250$ mm, $h_2 = h_3 = 375$ mm
From Table 4.3, $\Delta T_1 = -8.4°C$, $\Delta T_2 = -0.50°C$, $\Delta T_3 = -1.0°C$, $\Delta T_4 = -6.5°C$
The corresponding restraining stresses are:
$\sigma = -\alpha_T \times E_c \times \Delta T = -0.34 \times \Delta T = (2.86, 0.17, 0.34, 2.21)$ MPa
    Table 4.8 shows the restraining stresses in the cross section and the corresponding forces and moments are shown in Table 4.9 and Table 4.10 respectively. Table 4.11 shows the final stresses due to temperature differentials due to cooling.

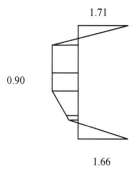

1.71

0.90

1.66

Fig.4.18 Final stresses: Cooling

Table 4.8 Restraining stresses in the cross section: Cooling

Level from top	σ
0	2.86
250	0.17
625	0
875	0
1200	0.30*
1250	0.34
1500	2.21

* By interpolation

Table 4.9 Restraining forces in the cross section: Cooling

σ at top, MPa	σ at bottom, MPa	Area, mm^2	F, kN
2.86	0.17	1500 × 250	568.13
0.17	0	2 × 200× 350	12.75
0	0.30	2 × 200× 325	19.50
0.30	0.34	900 × 50	14.40
0.34	2.21	900 × 250	286.88

Table 4.10 Calculation of moment: Cooling

F, kN	D mm	C mm	Lever arm mm	Moment kNm
568.13	250	88	670 − 88 = 582	330.65
12.75	375	117	670–250–125 = 295	3.76
19.50	325	217	−(830–300–325+ 217) = −422	−8.23
14.40	50	26	−(830–300+26) = −556	−8.01
286.88	250	156	−(830–250 + 156) = −736	−211.14
Σ= 901.66				Σ= 107.03

Note: y_b = 830 mm, y_t = 670 mm

Table 4.11 Thermal stress calculation: Cooling

Position from top	σ MPa	−Σ F/A MPa	y mm	−(Σ M/I)y	Final stress MPa
0	2.86	−0.88	670	−0.27	1.71
250	0.17	−0.88	420	−0.17	−0.88
625	0	−0.88	45	−0.02	−0.90
875	0	−0.88	−205	0.08	−0.80
1200	0.30*	−0.88	−530	0.21	−0.37
1250	0.34	−0.88	−580	0.23	−0.31
1500	2.21	−0.88	−830	0.33	1.66

4.8 DETAILING

Pretensioned tendons should be detailed as follows:
Horizontal spacing $\geq$ Maximum (2 φ, d_g + 5mm, 20 mm)
Vertical spacing $\geq$ Maximum (2 φ, d_g)
φ = tendon diameter, d_g = maximum aggregate size

4.9 REFERENCE TO EUROCODE 1 AND EUROCODE 2 CLAUSES

In this chapter, reference to following clauses in Eurocode 2 and Eurocode 1 has been made. The reader should refer to the code for full details.

Maximum prestressing force: 5.10.2.1

Limitation of concrete stresses: 5.10.2.2

Load factors on prestressing force: 5.10.9

Crack widths: 7.3.1

Detailing: 8.10.1.2

Thermal gradients: Eurocode 1: Actions on structures, parts 1 to 5: Thermal actions

Temperature difference components: 6.1.4

CHAPTER 5

BONDED POST-TENSIONED STRUCTURES

5.1 POST-TENSIONED BEAMS

Post-tensioned beams are a very common form used in both statically determinate and statically indeterminate structures. In post-tensioned simply supported beams, prestress is constant but the eccentricity is varied generally in a parabolic fashion so that the stresses due to prestress counteract the stresses due to self weight and other external loads. Unlike pretensioned beams where the cables are generally horizontal unless they are draped and the prestress and eccentricity along the span is varied using the technique of debonding, in a post-tensioned beam, the prestress is kept constant along the span but the eccentricity is varied by varying the profile of the cable. The value of the prestress and a range of eccentricities at mid-span are obtained in the usual way by drawing the Magnel diagram and choosing an appropriate value of Ps. Keeping this value of Ps constant over the entire span, the eccentricity is varied in a parabolic manner so that the stress constraints are satisfied.

5.2 CABLE PROFILE IN A POST-TENSIONED BEAM

The equations to be satisfied at transfer and service stages were derived in Chapter 4, section 4.3. They are repeated here for convenience.

At transfer:

$$[-\frac{P_s}{A} + \frac{P_s e}{Z_t}] \leq \frac{\eta}{\gamma_{Superior}}[f_{tt} + \frac{M_{self\,weight.}}{Z_t}] \qquad \text{(4.3 repeat)}$$

$$[-\frac{P_s}{A} - \frac{P_s e}{Z_b}] \geq \frac{\eta}{\gamma_{Superior}}[f_{tc} - \frac{M_{self\,weight.}}{Z_b}] \qquad \text{(4.4 repeat)}$$

At service:

$$[-\frac{P_s}{A} + \frac{P_s e}{Z_t}]\gamma_{Inferior} - \frac{M_{service}}{Z_t} \geq f_{sc} \qquad \text{(4.7 repeat)}$$

$$[-\frac{P_s}{A} - \frac{P_s e}{Z_b}]\gamma_{Inferior} + \frac{M_{service}}{Z_b} \leq f_{st} \qquad \text{(4.8 repeat)}$$

Since the prestress P_s is constant along the span and is predetermined, the object is to determine the value of the range of eccentricities at each section in order to satisfy stress limitations.

The above equations can be recast as follows:

$$e \le \frac{Z_t}{P_s} \{ \frac{P_s}{A} + \frac{\eta}{\gamma_{superior}} (\frac{M_{self weight.}}{Z_t} + f_{tt}) \}$$ (5.1)

$$e \le \frac{Z_b}{P_s} \{ -\frac{P_s}{A} + \frac{\eta}{\gamma_{Superior}} (\frac{M_{self weight.}}{Z_b} - f_{tc}) \}$$ (5.2)

$$e \ge \frac{Z_t}{P_s} \{ \frac{P_s}{A} + \frac{1}{\gamma_{Inferior}} (\frac{M_{service.}}{Z_t} + f_{sc}) \}$$ (5.3)

$$e \ge \frac{Z_b}{P_s} \{ -\frac{P_s}{A} + \frac{1}{\gamma_{Inferior}} (\frac{M_{service.}}{Z_b} - f_{st}) \}$$ (5.4)

The smaller of the values of eccentricity at a section given by equations (5.1) and (5.2) and the larger of the values of eccentricity given by equations (5.3) and (5.4) define the range of the eccentricity possible at a section. Section 5.2.1 gives a numerical example.

5.2.1 Example of Permitted Cable Zone

An unsymmetrical I-section shown in Fig. 5.1 is used as a simply supported beam to span 30 m. The beam supports, in addition to its self weight, an 8 kN/m superimposed dead load and an 80 kN/m live load. It is made from C40/50 concrete. The concrete grade at the time of stressing can be taken as C25/30. Determine the value of the prestressing force required and the range of eccentricities permitted at sections along the span.

a. Section properties:

Area $= 2500 \times 250 + (3200 - 250) \times 200 + (1000 - 250) \times 450$

$\qquad = 1.5525 \times 10^6 \ mm^2$

First moment of area about the soffit:

$A \times y_{bar} = 2500 \times 250 \times 2500/2 + (3200 - 250) \times 200 \times (2500 - 200/2)$

$\qquad + (1000 - 250) \times 450 \times 450/2 = 2273.19 \times 10^6 \ mm^3$

$y_{bar} = (2273.19 \times 10^6)/ (1.5525 \times 10^6) = 1464 \ mm$

$y_b = 1464 \ mm, \ y_t = 2500 - y_{bar} = 1036 \ mm$

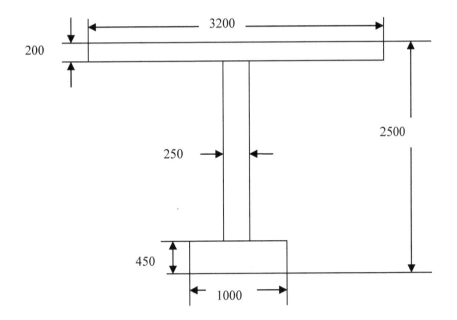

Fig. 5.1 An unsymmetrical I-beam (dimensions in mm).

$I = 250 \times 2500^3 /12 + 250 \times 2500 \times (2500/2 - y_{bar})^2$

$\quad + (3200 - 250) \times 200^3/12 + (3200 - 250) \times 200 \times \{2500 - 200/2 - y_{bar})^2$

$\quad + (1000 - 250) \times 450^3/12 + (1000 - 250) \times 450 \times (450/2 - ybar)^2$

$I = 13.9681 \times 10^{11}$ mm^4

$Z_t = I/y_t = 1348.5 \times 10^6$ mm^6, $Z_b = I/y_b = 953.98 \times 10^6$ mm^6.

b . Permissible concrete strength (see section 3.6, Chapter 3):

The permissible stresses at transfer are: Taking $f_{ck} = 25/30$ MPa

From (3.21), $f_{tt} = f_{ctm} = 0.30 \times 25^{2/3} = 2.6$MPa

$\qquad f_{tc} = -0.6 \times 25 = -15.0$ MPa

The permissible stresses at service are:Taking $f_{ck} = 40/50$ MPa

From (3.21), $f_{st} = f_{ctm} = 0.30 \times 40^{2/3} = 3.5 MPa$

$\qquad f_{tc} = -0.6 \times 40 = -24.0$ MPa

c. Load factors for prestress: $\gamma_{Superior} = 1.1$, $\gamma_{Inferior} = 0.90$

d. Loss of prestress: Loss of 10% at transfer and 25% at service,

$\qquad \eta = P_s/P_t = 0.75/0.9 = 0.83$

f. Loads:

Unit weight of concrete = 25 kN/m^3

Self weight = $(1.5525 \times 10^6 \times 10^{-6}) \times 25 = 38.81$ kNm

Super imposed dead load = 8 kN/m

Live load = 80 kN/m

Solution: Draw the Magnel diagram for the mid-span section.

Bending stresses at mid-span:

Self weight moment = $38.81 \times 30^2/8 = 4366.125$ kNm

Stresses due to self weight:

$$\sigma_{Top} = 4366.125 \times 10^6/ (1348.5 \times 10^6) = 3.2 \text{ MPa},$$
$$b_{ottom} = 4366.125 \times 10^6/ (953.98 \times 10^6) = 4.6 \text{ MPa}$$

Load at service = $(38.81 + 8.0 + 80) = 126.81$ kN/m

Moment at service = $126.81 \times 30^2 /8 = 14266.125$ kNm

Stresses at service:

$\sigma_{Top} = 14266.125 \times 10^6/ (1348.5 \times 10^6) = 10.6$ MPa,

bottom = $14266.125 \times 10^6/ (953.98 \times 10^6) = 15.0$ MPa

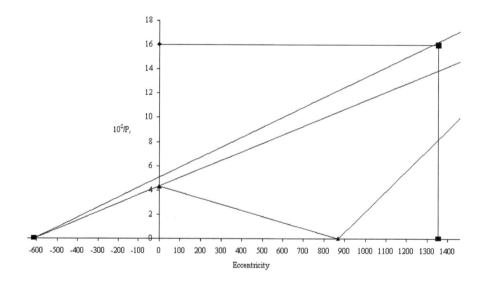

Fig. 5.2 Magnel diagram for mid-span

5.2.1.1 Magnel Equations

$1/A = 64.41 \times 10^{-8}$, $1/Z_t = 0.074 \times 10^{-8}$, $1/Zt = 0.105 \times 10^{-8}$

−Substituting the appropriate values in the relevant equations (4.5), (4.6), (4.9) and (4.10) from Chapter 4, section 4.3

$(-64.41 + 0.074\ e) \times 10^{-8} \leq [(0.83/1.1)\{3.2 + 2.6\} = 4.38]\ (1/P_s)$ (4.5b)

$(-64.41 - 0.105\ e) \times 10^{-8} \leq [(0.83/1.1)\{-4.6 - 15.0\} = -14.79]\ (1/P_s)$ (4.6b)

$(-64.41 + 0.074\ e) \times 10^{-8} \leq [(1/0.90)\{10.6 - 24.0\} = -14.89]\ (1/P_s)$ (4.9b)

$(-64.41 - 0.105\ e) \times 10^{-8} \leq [(1/0.90)\{-15.0 + 3.5\} = -12.78]\ (1/P_s)$ (4.10b)

Using the above equations, the Magnel diagram for the mid-span section can be drawn as shown in Fig. 5.2.

5.2.1.2 Determination of Maximum Eccentricity

In practice strands are grouped into cables with suitable anchorages. The variety of cables and the associated anchorages depend on the manufacturer. For example VSL anchorages are available for cables containing 1, 2, 3, 4, 7, 12, 15, 19, 22, 27, 31, 37, 43 and 55 strands. The CCL anchorages allow for cables of 4,7,12, 19 and 22 strands of 13 mm diameter or 4, 7, 12, 19, 27 and 37 strands of 15 mm diameter. The cables are placed inside ducts before concreting and the ducts are grouted afterwards to prevent corrosion of the strands. During tensioning the strands in a cable tend to bunch together onto the tension side of the duct. This alters the position of the centroid of the tensile force with respect to the centroid of the duct as shown in Fig. 5.3. The shift depends primarily on the diameter of the duct and the number of strands in the cable. Manufacturers provide data on this. Table 5.1 shows the data for VSL cables.

For example for a cable containing 15 strands in a duct whose external diameter is 82 mm (internal diameter 75 mm), the shift is 10 mm. Thus if the cover to the duct is say 60 mm, the centre of the tension will be at $(60 + 82/2 + 10) = 111$ mm from the soffit. The maximum available eccentricity is

$e_{max} \approx y_{bar} - 112 \approx 1464 - 111 = 1353$ mm.

From the Magnel diagram a value of $10^8/P_s = 16$ is suitable.

$P_s = (10^8/16) \times 10^{-3} = 6250$ kN

$P_{jack} = P_s/0.75 = 8333$ kN

Assume 15.2 mm strands with a cross sectional area of 140 mm^2 and a yield stress $f_{pk} = 1860$ MPa. The maximum allowable load P_{max} per strand at stressing from (3.24) is

$P_{max} = 140 \times 0.77 \times 1860 \times 10^{-3} = 200.5$ kN

Number of strands = $P_{jack}/200.5 = 8333/206.3 = 42$. Assuming 15 strands per cable, the total number of cables is 3. It is not necessary to tension all the strands in a cable but care has to be taken to ensure that the strands which are stressed remain in a symmetrical fashion in the anchor.

Table 5.1 Shift in eccentricity

No. of strands	Anchor unit	Corrugated steel strip sheath: Range of Internal Diam./External diam.	Range of Shift mm
1	6–1	25/30	5
2	6–2	40/45	9
3	6–3	40/45	6
4	6–4	45/50	7
5–7	6–7	50/57 – 55/62	9–7
8–12	6–12	65/72– 75–82	9–11
13–15	6–15	80/87	13–10
16–19	6–19	85/92–90/97	25–18
20–22	6–22	100/107	14–22
23–27	6–27	100/107–110/117	13–16
28–31	6–31	110/117–120/127	15–19
32–37	6–37	120/127–130/137	16–22
38–43	6–43	140/147	21–25
44–55	6–55	150/157–160/167	23–29

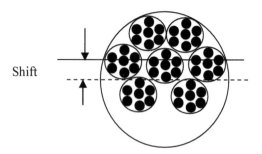

Shift

Fig. 5.3 Shift in the centroid of the tensile force

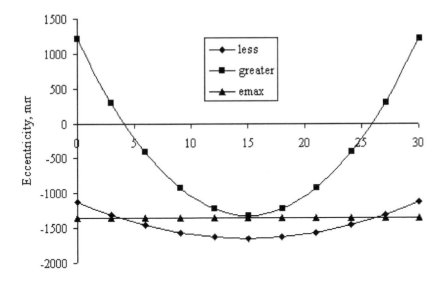

Fig. 5.4 Permissible cable zone.

Table 5.2 Permitted value of eccentricity

x	e≥	e ≤	Gap
0	-1210	1120	2330
3	-296	1311	1607
6	414	1352	938
9	921	1352	431
12	1225	1352	127
15	1327	1352	25
18	1225	1352	127
21	921	1352	431
24	414	1352	938
27	−296	1311	1607
30	−1210	1120	2330

5.2.1.3 Determination of Cable Zone

Once the value of P_s is determined, then keeping the value of P_s constant the eccentricity at different sections along the span can be determined.

Moment M at a distance x from left hand support = $0.5qx (L- x)$

where L = span, q = uniformly distributed live load

Taking e as positive below the neutral axis, the limits for e at sections along the span are shown in Table 5.2. The maximum permitted value of e is 1352 mm as calculated before. The maximum and minimum values of eccentricity permitted are plotted in Fig. 5.4. As can be seen, the width of the permitted zone in which the cables can be laid is very narrow at mid-span and widens towards the supports. This means that the cables can be taken at a steep angle at supports, which will benefit shear resistance. The stress limitation criteria will be satisfied as long as all the cables remain inside the cable zone. The cables can be 'peeled off' to suit anchoring at the ends. It is normal to thicken the end section of the web to accommodate the anchors.

5.2.1.4 Detailing of post-tensioned tendons

Pretensioned ducts should be detailed as follows:
Horizontal spacing $\geq$ Maximum (φ, d_g + 5mm, 50 mm)
Vertical spacing $\geq$ Maximum (φ, d_g, 40 mm)
φ = duct diameter, d_g = maximum aggregate size
Normally ducts should not be bundled except in the case of a pair of ducts placed vertically one above the other.

5.3 CONCEPT OF EQUIVALENT LOADS

A prestressed concrete element free from external forces is a self-equilibrated system consisting of two stressed elements, viz. concrete in compression and steel in tension. Therefore the forces acting on concrete is simply the reverse of the forces acting on steel.

(i) As an example consider a prestressed beam with a straight tendon as shown in Fig. 5.5. If the tensile force in steel is F, the force acting on concrete is a compressive force F.   The force F is known as the equivalent load.

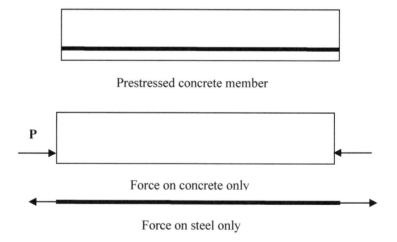

Prestressed concrete member

P

Force on concrete only

Force on steel only

Fig. 5.5 Equivalent loads of a prestressed beam with a straight tendon

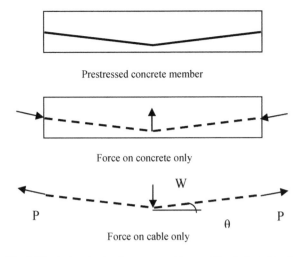

Prestressed concrete member

Force on concrete only

Force on cable only

Fig. 5.6 Equivalent loads of a prestressed beam with a deflected tendon

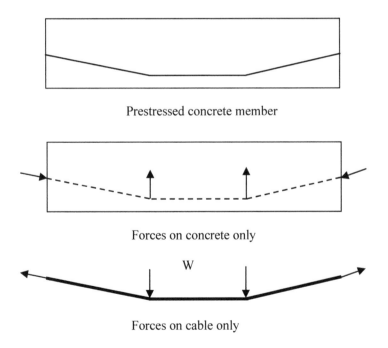

Prestressed concrete member

Forces on concrete only

Forces on cable only

Fig. 5.7 Equivalent loads of a prestressed beam with a deflected tendon

(ii) If the cable, instead of being straight, is deflected, as shown in Fig. 5.6, then in order to keep the tendon in the deflected shape, a concentrated load W need to be applied at the deflected point. If the tensile force in the tendon is P, then the force W to maintain the cable in the deflected shape is given by

W = 2 P sin θ,   θ = Inclination of the cable to the horizontal.

The forces acting on concrete are two inclined compressive forces at the end and a concentrated load W acting upwards at the middle.

(iii) If the cable has two points of deflection as shown in Fig. 5.7, then the forces acting on the tendon to maintain the deflected shape are two concentrated loads W at points of deflection and tensile forces at the ends. The forces acting on concrete are simply the opposite of the forces acting on steel.

(iv) If the cable is curved instead of consisting of linear segments, as shown in Fig. 5.9 (see p. 92), then the force acting along the length of the cable is a distributed load along with the tensile force at the ends.   The forces acting on concrete are equal and opposite of those acting on the cable.

5.3.1 General Equation for Equivalent Loads

Consider a cable with a constant tension of P as shown in Fig. 5.8.

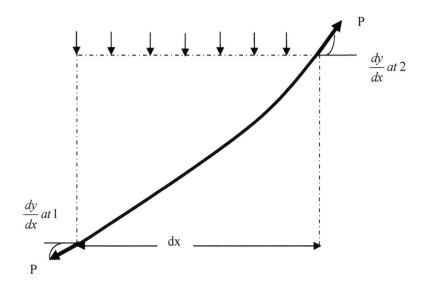

Fig. 5.8 Distributed loads on a cable

  Taking two sections Δx apart, the difference in the vertical component in the force betwe sections must be equal to q Δx, where q is a distributed load. Since the curvature of the ca fairly low, we can use the approximation sin θ ≈ tan θ = dy/dx. Therefore

$$q\,\Delta x = P[\frac{dy}{dx} \text{ at end 2} - \frac{dy}{dx} \text{ at end 1}]$$

$$q = P \frac{[\frac{dy}{dx} \text{ at end 2} - \frac{dy}{dx} \text{ at end 1}]}{\Delta x} = P \frac{\frac{d^2y}{dx^2} \Delta x}{\Delta x}$$

$$q = P \frac{d^2y}{dx^2}$$

$q = P \times$ Curvature of the cable. (5.5)

It has to be remembered that the distributed force q acts normal to the cable. However if the dip is less than Span/12, then for all practical purposes, the force q can be considered to be vertical.

If distributed loads act on the cable along its length, as can happen due to frictional forces, then the tension in the cable is not constant. In addition the distributed force along the cable length has a downward vertical component. Taking these factors into account, equation (5.5) is modified as follows.

$$q \Delta x = \{P + \frac{dP}{dx} \Delta x\} \{\frac{dy}{dx} + \frac{d^2y}{dx^2} \Delta x\} - P \frac{dy}{dx} - \frac{dP}{dx} \frac{dy}{dx} \Delta x$$

Ignoring terms of second order and simplifying

$$q = P(x) \frac{d^2y}{dx^2}$$ (5.6)

Note that q is not a constant but a function of x.

5.3.2 General Equation for Distributed Loads for Parabolic Cable Profile

Frequently the profile of the cable in a statically determinate post-tensioned beam is parabolic. If the eccentricity of the cable profile is parabolic, the general equation for the shape is given by

$e = Ax^2 + B x + C$

where A, B, C are constants and x is the distance from one end.

The curvature is given by

$$\frac{d^2e}{dx^2} = 2A$$

The curvature is constant and hence the distributed load q is constant along the length of the cable

(i) Symmetric cable: If the cable is symmetric as shown in Fig. 5.9, and defined by

$e = e_1$ at $x = 0$ and $x = L$

$e = e_2$ and $de/dx = 0$ at $x = 0.5L$

$A = -4 (e_2 - e_1)/L^2$, $B = 4 (e_2 - e_1)/L$, $C = e_1$

Curvature $= 2A = -8\,(e_2 - e_1)/L^2$

Then the distributed load q is given by

$$q = \frac{8}{L^2} P(e_2 - e_1) \tag{5.7}$$

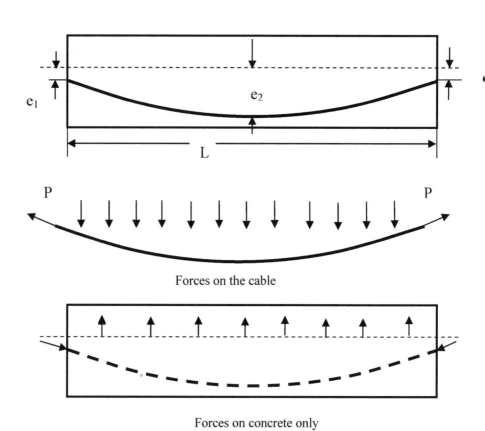

Forces on the cable

Forces on concrete only

Fig. 5.9 Equivalent loads of a prestressed beam with a curved tendon

(ii) Unsymmetrical cable: If the cable is unsymmetrical as shown in Fig. 5.10 and defined by

$$e = e_1 \text{ at } x = 0, \quad e = e_2 \text{ at } x = 0.5L, \quad e = e_3 \text{ x} = L$$

$$A = -4\{e_2 - (e_1 + e_3)/2\}/L^2, \; B = \{4e_2 - e_3 - 3e_1\}/L, \; C = e_1$$

$$\text{Curvature} = 2A = -8\,\{e_2 - (e_1 + e_3)/2\}/L^2$$

The distributed load q is given by

$$q = \frac{8}{L^2}\{e_2 - \frac{(e_1 + e_3)}{2}\} \qquad (5.8)$$

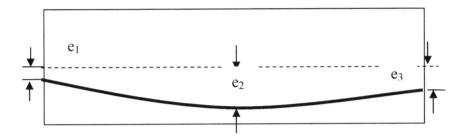

Fig. 5.10 Unsymmetrical cable

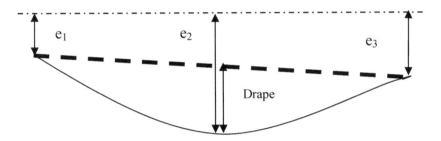

Fig. 5.11a Drape for e_1, e_2 and e_3 positive.

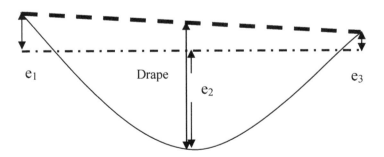

Fig. 5.11b Drape for e_1, e_3 negative but e_2 is positive.

5.3.3 Drape of the Cable

The term $[e_2 - (e_1 + e_3)/2]$ is known as the drape of the cable. It is the net eccentricity at the centre measured from a line joining the ends of the cable. Assuming that the eccentricity is **positive** when lying **below** the centroidal axis of the beam, the drape of the cable when all three values of eccentricity e_1, e_2 and e_3 are positive and when only e_2 is positive are shown respectively in Fig. 5.11a and Fig. 5.11b.

5.4 LOAD BALANCING

The idea of equivalent loads leads to a very simple concept known as Load Balancing. Consider the I-beam considered in Section 5.2.1. At service, the loads acting are

Self weight = 38.81 kNm

Super imposed dead load = 8 kN/m

Live load = 80 kN/m

Span L = 30 m

If it is assumed that say 70% of the total load at service is balanced by prestress, then

$$q = 0.70 \times (38.81 + 8 + 80) = 8 \, P_s e/L^2$$

If a load factor of $\gamma_{inferior} = 0.9$ is applied to prestress, $P_s \, e = 11095.875$ kNm.

If the maximum permissible eccentricity is assumed to be say 1350 mm, then

$$P_s = 8219 \text{ kN}$$

$$P_{jack} \approx P_s / 0.75 \approx 10960 \text{ kN}$$

If the allowable load per strand at jacking is say 200.5 kN, then the number of strands required is 11096/200.5 = 60

This approach does not check if any of the stress criteria are violated but it is a very useful way of being able to visualize the prestress as essentially counteracting the applied loads on the beam.

The concept of equivalent load is used extensively in the design of statically indeterminate prestressed concrete structures. This will be dealt in detail in Chapter 6.

5.5 REFERENCE TO EUROCODE 2 CLAUSES

Detailing of post-tensioned ducts: Clause 8.10.1.3

CHAPTER 6

STATICALLY INDETERMINATE POST-TENSIONED STRUCTURES

6.1 INTRODUCTION

In the case of statically determinate structures, when a beam is post-tensioned, no reactive forces are created at the supports. Therefore the moment at a section due to prestress is simply the product of the prestressing force and the eccentricity at the section. This is not true in the case of statically indeterminate structures. The reason for this is that the act of post-tensioning alters the reactions at the supports and hence the moment induced at a section due to prestress. The net moment due to prestress at a section is the sum of the product of prestressing force and the eccentricity at the section *plus the moment induced by the reactions*. At a section the moment given by the product of the prestressing force and the eccentricity is called as the primary moment and the moment induced by the reaction is called the secondary moment. Although the moment induced by the reactions is called the secondary moment, it is not necessarily a negligible quantity. The example in the next section illustrates this point.

6.1.1 Primary and Secondary Moments

Consider a statically determinate simply supported beam and a statically indeterminate two span continuous beam shown in Fig. 6.1. For simplicity assume that the beams are prestressed by a prestress of constant eccentricity.

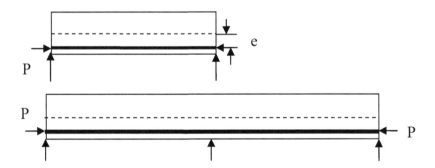

Fig. 6.1 Statically determinate and indeterminate prestressed beams

The prestress can be replaced by end moments as shown in Fig. 6.2. When the simply supported beam is prestressed it hogs up but the reactions are not affected. On the other hand, in the case of the continuous beam, the central reaction prevents the beam from deflecting at the centre. The resulting moment distribution due to the end moments of Pe can be determined from matrix analysis. Assuming uniform flexural rigidity EI and span length L, the stiffness matrix in terms of the rotations at the supports θ_1, θ_2 and θ_3 is given by

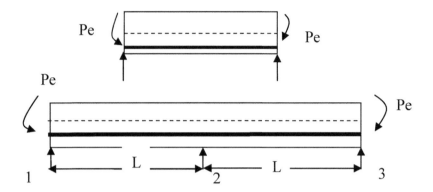

Fig. 6.2 Beams under end moments due to prestress

$$\frac{EI}{L}\begin{bmatrix} 4 & 2 & 0 \\ 2 & 8 & 2 \\ 0 & 2 & 4 \end{bmatrix}\begin{bmatrix} \theta_1 \\ \theta_2 \\ \theta_3 \end{bmatrix} = Pe\begin{bmatrix} -1 \\ 0 \\ 1 \end{bmatrix}$$

Using the Gaussian elimination method, reduce the stiffness matrix to the upper triangular form:

$$\frac{EI}{L}\begin{bmatrix} 4 & 2 & 0 \\ 0 & 7 & 2 \\ 0 & 0 & 3.4286 \end{bmatrix}\begin{bmatrix} \theta_1 \\ \theta_2 \\ \theta_3 \end{bmatrix} = Pe\begin{bmatrix} -1 \\ 0.5 \\ 0.8571 \end{bmatrix}$$

Solving, $\theta_1 = -0.25$ (L/EI), $\theta_2 = 0$, $\theta_3 = 0.25$ (L/EI)
The final bending moments are $M_{12} = -Pe$, $M_{21} = -0.5Pe$, $M_{23} = 0.5Pe$, $M_{32} = Pe$
The reactions are $V_1 = 1.5$ Pe/L, $V_2 = -3$ PE/L, $V_3 = 1.5Pe/L$.
Fig. 6.3 shows the three bending moment diagrams. They are
- Due to end moments on the continuous beam with zero deflection at support. This is the final bending moment due to prestress and is the one used in design.
- Bending moment at any section due to the product of P and e at the section. This is the primary moment. This is the only moment that exists in a statically determinate beam. In a statically indeterminate beam the primary moment alone cannot be used in design.
- Secondary moment due to the reactions developed in the continuous beam.

As can be seen the final bending moment due to prestress is the sum of the primary moment and the secondary moment.

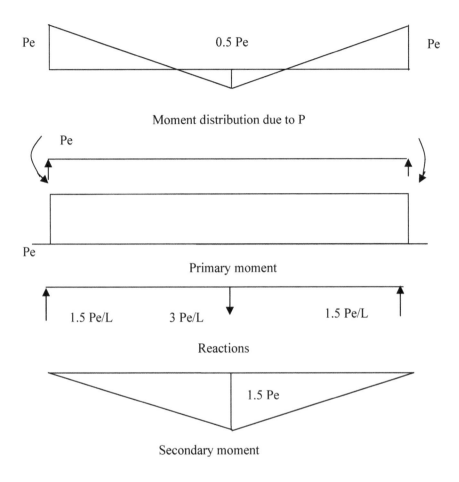

Fig. 6.3 Final, primary and secondary bending moments.

6.1.2 Prestressing a Propped Cantilever

As another example to demonstrate the concept of primary and secondary moments, consider a propped cantilever prestressed by a force P of constant eccentricity e as shown in Fig. 6.4.

Primary moment: If the prop is removed, then the primary moment is a constant value of Pe causing tension at the top face. The downward deflection at the prop is $\Delta_{Primary} = Pe \dfrac{L^2}{2EI}$

If R is the prop force, then the upward deflection due to the prop is

$$\Delta_R = R \dfrac{L^3}{3EI}$$

Equating the deflection at the tip due to Pe and due to R, for zero deflection at the prop,

$$R = 1.5 \, Pe/L$$

The secondary moment due to the prop R is linear and at the support is 1.5Pe causing tension at the bottom face.

The final moment due to prestress is the sum of primary and secondary moments, as shown in Fig. 6.4.

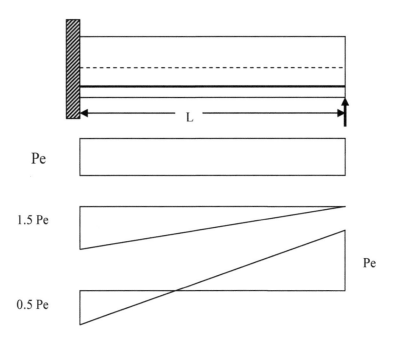

Fig. 6.4 Primary, secondary and final bending moments.

6.2 ANALYSIS TO DETERMINE THE MOMENT DISTRIBUTION DUE TO PRESTRESS

The determination of the bending moments due to prestress in a statically indeterminate structure is best done using the concept of equivalent loads discussed in section 5.3, Chapter 5. The profile of cables in post-tensioned structures consists of a series of parabolic segments with convex segments over supports and concave segments in the spans. It is useful to develop equations for equivalent loads for various cable profiles.

6.2.1 Equivalent Loads for a Cable Profile of a Single Parabola

Although a single parabolic profile without convex segments is not common, in order to simplify calculations, frequently the actual profile is approximated to a single parabola. A general equation to a parabola is given by

$$e = Ax^2 + Bx + C$$

where A, B, C are constants.

Note that all parabolic profiles have constant curvature and hence result in uniformly distributed equivalent load.

Assuming eccentricity is positive below the centroidal axis, for a cable of parabolic profile with eccentricities of:

$e = -e_1$ at x = 0, e = e_3 at x = $(1-\lambda)$ L and e = $-e_2$ at x = L
Then the profile of the cable is

$$e = -e_1 + B(\frac{x}{L}) + A(\frac{x}{L})^2$$

Using the 'boundary' conditions, the constants A and B can be determined:

$$A = \frac{(1-\lambda)(e_1 - e_2) - (e_1 + e_3)}{\lambda(1-\lambda)}$$

$$B = -\frac{(1-\lambda)^2 (e_1 - e_2) - (e_1 + e_3)}{\lambda(1-\lambda)}$$

$$\text{Curvature} = \frac{2}{L^2}[\frac{(1-\lambda)(e_1 - e_2) - (e_1 + e_3)}{\lambda(1-\lambda)}]$$

$$= -\frac{2}{L^2}[\frac{e_3 + \lambda e_1 + (1-\lambda)e_2}{\lambda(1-\lambda)}]$$

The slope at x = 0 and x = L are

$$\frac{de}{dx} = -\frac{(1-\lambda)^2 (e_1 - e_2) - (e_1 + e_3)}{\lambda(1-\lambda)L} \qquad \text{at x = 0}$$

$$\frac{de}{dx} = \frac{(1-\lambda^2)(e_1 - e_2) - (e_1 + e_3)}{\lambda(1-\lambda)L} \qquad \text{at x = L}$$

Total deviation in the direction of the cable from x = 0 to x = L is

$$\text{deviation} = (\frac{de}{dx} \text{ at x = 0}) - (\frac{de}{dx} \text{ at x = L}) = \frac{2}{L}\frac{\{e_3 + \lambda e_1 + (1-\lambda)e_2\}}{\lambda(1-\lambda)} \qquad (6.1)$$

Note that the term $\{e_3 + \lambda e_1 + (1-\lambda) e_2\}$ is the deflection of the cable at x = $(1-\lambda)$ L from a line joining the ends of the cable. This term is known as drape.

$$\text{Curvature} = -\frac{2}{L^2}[\frac{e_3 + \lambda e_1 + (1-\lambda)e_2}{\lambda(1-\lambda)}] = -\frac{2}{L^2}[\frac{\text{drape}}{\lambda(1-\lambda)}] \qquad (6.2)$$

Equivalent uniformly distributed load q = P × Curvature

In particular if $\lambda = 0.5$, then

$$e = -e_1 + (3e_1 + 4e_3 + e_2)\frac{x}{L} - (2e_1 + 4e_3 + 2e_3)(\frac{x}{L})^2$$

$$\frac{de}{dx} = \frac{(3e_1 + e_2 + 4e_3)}{L} \qquad \text{at } x = 0$$

$$\frac{de}{dx} = -\frac{(e_1 + 3e_2 + 4e_3)}{L} \qquad \text{at } x = L$$

Letting drape, $s = e_3 + 0.5(e_1 + e_3)$, the total deviation in direction of the cable from $x = 0$ to $x = L$ is

$$\text{deviation} = (\frac{de}{dx}\text{ at } x = 0) - (\frac{de}{dx}\text{ at } x = L) = \frac{8}{L}\{e_3 + \frac{(e_1 + e_2)}{2}\} = \frac{8}{L}\text{drape} \qquad (6.3)$$

$$\text{curvature} = \frac{d^2e}{dx^2} = \frac{8}{L^2}[e_3 + \frac{(e_1 + e_2)}{2}] = \frac{8}{L^2}\text{drape} \qquad (6.4)$$

This profile is suitable for simply supported beams and also as an approximation in the case of continuous structures.

6.2.2 General Equation for Equivalent Load for a Cable Profile Consisting of Three Parabolic Segments

Consider a cable profile made up of three parabolas in the span. Starting from the left hand support parabolas 1 and 2 are concave segments and towards the end of the support is parabola 3 with a convex segment. Continuity of both slope and eccentricity are maintained at the junction of the segments.

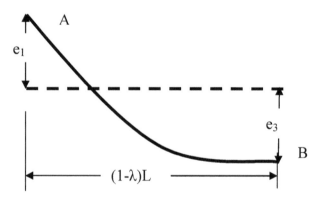

Fig. 6.4 Parabolic segment 1

Parabola 1: As shown in Fig. 6.4, the first segment parabola AB has a negative (i.e. above the centroidal axis) eccentricity of e_1 at A and a positive (i.e. below the centroidal axis) of e_3 at B. The length of the segment is $(1-\lambda)$ L where L is the span. The equation of the parabola is given by

$$e = A_1 x^2 + B_1 x + C_1$$

where A_1, B_1, C_1 are constants.

The boundary conditions are:

At A: $x = 0$, $e = -e_1$

At B: $x = (1-\lambda)$ L, $e = e_3$ and de/dx= 0

Solving for the constants:

$$C_1 = -e_1, \quad B_1 = 2\frac{(e_1 + e_3)}{(1-\lambda)L}, \quad A_1 = -\frac{(e_1 + e_3)}{(1-\lambda)^2 L^2}$$

Substituting for the constants,

$$e = -e_1 - \frac{(e_1 + e_3)}{(1-\lambda)^2} \frac{x}{L} [\frac{x}{L} - 2(1-\lambda)]$$

$$\frac{de}{dx} = -\frac{(e_1 + e_3)}{(1-\lambda)^2} \frac{2}{L} [\frac{x}{L} - (1-\lambda)]$$

At x = 0, $\quad \dfrac{de}{dx} = 2\dfrac{(e_1 + e_3)}{(1-\lambda)L}$

$$\frac{d^2 e}{dx^2} = -\frac{2}{L^2} \frac{(e_1 + e_3)}{(1-\lambda)^2}, \quad 0 \le x \le (1-\lambda)\,L \tag{6.5}$$

Parabola 2: As shown in Fig. 6.5, for the second segment parabola BC has a positive (i.e. below the centroidal axis) eccentricity of e_3 at B and a negative (i.e. above the centroidal axis) of $(e_2 - h_1)$ at B. The length of the segment is $(\lambda - \beta)$ L where L is the span. The equation of the parabola is given by

$$e = A_2 x^2 + B_2 x + C_2$$

where A_2, B_2, C_2 are constants.

The boundary conditions at B are:x = 0, e = e_3 and de/dx = 0. This condition ensures continuity of slope and deflection at the junction of parabolas 1 and 2.

At C: $x = (\lambda - \beta)$ L, $e = -(e_2 - h_2)$

Solving for the constants:

$$C_2 = e_3, \quad B_2 = 0, \quad A_2 = -\frac{(e_3 + e_2 - h_2)}{(\lambda - \beta)^2 L^2}$$

Substituting for the constants,

$$e = e_3 - \frac{(e_3 + e_2 - h_2)}{(\lambda - \beta)^2}(\frac{x}{L})^2 \ , \quad 0 \le x \le (\lambda - \beta) L$$

$$\frac{de}{dx} = -\frac{2x}{L^2}\frac{(e_3 + e_2 - h_2)}{(\lambda - \beta)^2}$$

$$\frac{d^2e}{dx^2} = -\frac{2}{L^2}\frac{(e_3 + e_2 - h_2)}{(\lambda - \beta)^2}$$

From parabola 3, h_2 can be shown to be $h_2 = (e_3 + e_2)\dfrac{\beta}{\lambda}$

Substituting for h_2,

$$\frac{de}{dx} = -\frac{2x}{L^2}\frac{(e_3 + e_2)}{\lambda(\lambda - \beta)}$$

Slope at C: $x = (\lambda - \beta) L$, $\dfrac{de}{dx} = -\dfrac{2}{L}\dfrac{(e_3 + e_2)}{\lambda}$

$$\frac{d^2e}{dx^2} = -\frac{2}{L^2}\frac{(e_3 + e_2)}{\lambda(\lambda - \beta)} \tag{6.6}$$

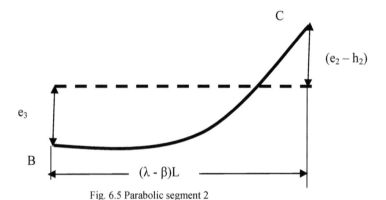

Fig. 6.5 Parabolic segment 2

Parabola 3: As shown in Fig. 6.6, the third segment parabola CD has a negative (i.e. above the centroidal axis) eccentricity of $(e_2 - h_2)$ at C and e_2 at D. The length of the segment is βL, where L is the span. The equation of the parabola is given by

$$e = A_3x^2 + B_3x + C_3$$

where A, B, C are constants.

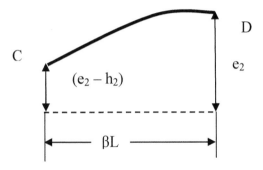

Fig. 6.6 Parabolic segment 3

The boundary conditions are

At C: x = 0, e = $- (e_2 - h_2)$

Slope of segment 3 must be equal to slope at C of parabolic segment 2. This ensures continuity of parabolas 2 and 3 at their junction. Therefore

$$\frac{de}{dx} = -\frac{2}{L}\frac{(e_3 + e_2 - h_2)}{(\lambda - \beta)}$$

At D: x = βL, e = $-e_2$ and de/dx = 0

There are thus four conditions to be satisfied, leading to

$$C_3 = -(e_2 - h_2), \quad B_3 = -2\frac{h_2}{\beta L}, \quad A_3 = \frac{h_2}{(\beta L)^2}$$

$$h_2 = (e_3 + e_2)\frac{\beta}{\lambda}$$

Substituting for the constants,

$$e = -(e_2 - h_2) + h_2\frac{x}{\beta L}[\frac{x}{\beta L} - 2], 0 \le x \le \beta L$$

$$\frac{de}{dx} = h_2\frac{2}{\beta L}[\frac{x}{\beta L} - 1] = (e_3 + e_2)\frac{2}{\lambda L}[\frac{x}{\beta L} - 1]$$

At x = 0, $\quad \frac{de}{dx} = -\frac{2}{\lambda L}(e_3 + e_2)$

$$\frac{d^2e}{dx^2} = \frac{2}{L^2}\frac{(e_3 + e_2)}{\lambda \beta} \tag{6.7}$$

The radius of curvature R is therefore given by

$$R = \frac{L^2}{(e_3 + e_2)} \frac{\lambda\beta}{2}$$

Depending on the diameter of the cable, R can be fixed and hence the value of β can be calculated.

This profile is suitable for the end span of a continuous beam. Summarising, the equations for slopes and curvature are:

Segment 1: $\dfrac{d^2e}{dx^2} = -\dfrac{2}{L^2}\dfrac{(e_1+e_3)}{(1-\lambda)^2}$, $\quad \dfrac{de}{dx} = -\dfrac{(e_1+e_3)}{(1-\lambda)^2}\dfrac{2}{L}[\dfrac{x}{L}-(1-\lambda)]$ $\qquad$ (6.8)

Segment 2: $\dfrac{d^2e}{dx^2} = -\dfrac{2}{L^2}\dfrac{(e_3+e_2)}{\lambda(\lambda-\beta)}$, $\quad \dfrac{de}{dx} = -\dfrac{2x}{L^2}\dfrac{(e_3+e_2)}{\lambda(\lambda-\beta)}$ $\qquad$ (6.9)

Segment 3: $\dfrac{d^2e}{dx^2} = \dfrac{2}{L^2}\dfrac{(e_3+e_2)}{\lambda\beta}$, $\quad \dfrac{de}{dx} = (e_3+e_2)\dfrac{2}{\lambda L}[\dfrac{x}{\beta L}-1]$ $\qquad$ (6.10)

Note that curvatures are constant but slopes vary linearly.

Similarly, ignoring the signs, the equations for slopes are

Segment 1: $\dfrac{de}{dx} = 2\dfrac{(e_1+e_3)}{(1-\lambda)L}$ at A and $\dfrac{de}{dx} = 0$ at B $\qquad$ (6.11)

Segment 2: $\dfrac{de}{dx} = 0$ at B and $\dfrac{de}{dx} = \dfrac{2}{\lambda L}(e_3+e_2)$ at C $\qquad$ (6.12)

Segment 3: $\dfrac{de}{dx} = \dfrac{2}{\lambda L}(e_3+e_2)$ at C and $\dfrac{de}{dx} = 0$ at D $\qquad$ (6.13)

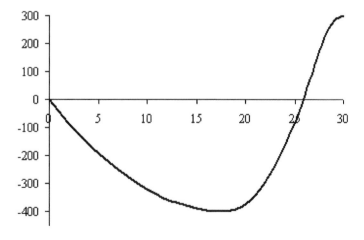

Fig. 6.7 Cable profile made up of three parabolic segments.

Example: span $L = 30$ m, $(1 - \lambda)L = 18$ m, $\beta L = 3$ m. $e_1 = 0$ mm, $e_2 = 300$ mm, $e_3 = 400$ mm.

$(1 - \lambda) \times 30 = 18$, $\lambda = 0.4$. $\beta \times 30 = 3$, $\beta = 0.1$

Fig. 6.7 shows a plot of the profile of the cable.
  The curvatures and the radius of curvature of the three segments are

Segment 1: 2.47×10^{-3} m^{-1}, 405.0 m (Concave)

Segment 2: 12.96×10^{-3} m^{-1}, 77.1 m (Concave)

Segment 3: 38.89×10^{-3} m^{-1}, 25.7 m (Convex)

Fig. 6.7 shows a plot of the profile of the cable.
The slopes are
Segment 1: At A, $de/dx = 44.44 \times 10^{-3}$, At B: $de/dx = 0$
Segment 2: At B: $de/dx = 0$, At C: $de/dx = 116.67 \times 10^{-3}$
Segment 3: At C: $de/dx = 116.67 \times 10^{-3}$, At D: $de/dx = 0$

6.2.3 General Equation for Equivalent Load for a Cable Profile Consisting of Four Parabolic Segments

The three-parabolic-segment cable profile is suitable for end spans of continuous beams. For interior spans it is necessary to have a convex segment at either end of the span. Therefore consider a cable profile made up of four parabolas with continuity of both slope and eccentricity at their junctions as follows.

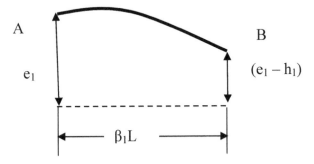

Fig. 6.8 Parabolic segment 1

Parabola 1: As shown in Fig. 6.8, the first segment parabola AB has a negative (i.e. above the centroidal axis) eccentricity of e_1 at A and $(e_1 - h_1)$ at B. The length of the segment is $\beta_1 L$, where L is the span. The equation of the parabola is given by

$$e = A_1 x^2 + B_1 x + C_1$$

where A_1, B_1, C_1 are constants.

The boundary conditions are:

At A: $x = \beta_1 L$, $e = -e_1$ and $de/dx = 0$

At B: $x = 0$, $e = -(e_1 - h_1)$

Solving for the constants:

$$C_1 = -(e_1 - h_1), \quad B_1 = -2\frac{h_1}{\beta_1 L}, \quad A_1 = \frac{h_1}{(\beta_1 L)^2}$$

$$e = -(e_1 - h_1) - 2h_1 \frac{x}{\beta_1 L} + h_1 (\frac{x}{\beta_1 L})^2$$

$$\frac{de}{dx} = -2\frac{h_1}{\beta_1 L} \text{ at B}, \quad \frac{d^2 e}{dx^2} = 2\frac{h_1}{(\beta_1 L)^2} \tag{6.14}$$

Parabola 2: As shown in Fig. 6.9, the second segment parabola BC has a negative (i.e. above the centroidal axis) eccentricity of $(e_1 - h_1)$ at B and a positive (i.e. below the centroidal axis) of e_3 at C. The length of the segment is $(1 - \lambda - \beta_1) L$ where L is the span. The equation of the parabola is given by

$$e = A_2 x^2 + B_2 x + C_2$$

where A_2, B_2 and C_2 are constants.

The boundary conditions are:

At C: $x = 0$, $e = e_3$ and $de/dx = 0$.

At B: $x = (1 - \lambda - \beta_1) L$, $e = -(e_1 - h_1)$ and $\dfrac{de}{dx} = -2\dfrac{h_1}{\beta_1 L}$. This will ensure

continuity at B between parabolic segments 1 and 2.

The constants are

$$C_2 = e_3, \quad B_2 = 0, \quad A_2 = -\frac{(e_1 + e_3 - h_1)}{(1 - \lambda - \beta_1)^2 L^2}$$

Substituting for the constants,

$$e = e_3 - \frac{(e_1 + e_3 - h_1)}{(1 - \lambda - \beta_1)^2} (\frac{x}{L})^2, \quad 0 \le x \le (1 - \lambda - \beta_1) L$$

For continuity of slope at B, the slope at $x = (1 - \lambda - \beta_1) L$ of parabolic segment 2 is the same as the slope at B of parabolic segment 1. Therefore

$$\frac{de}{dx} = -\frac{2}{L}\frac{(e_1 + e_3 - h_1)}{(1 - \lambda - \beta_1)} = -\frac{2}{L}\frac{h_1}{\beta_1}$$

$$h_1 = (e_3 + e_1)\frac{\beta_1}{(1 - \lambda)} \qquad (6.15)$$

Substituting for $h_1 = (e_3 + e_1)\dfrac{\beta_1}{(1 - \lambda)}$, the slope at B is given by

$$\frac{de}{dx} = -\frac{2}{(1 - \lambda)L}(e_3 + e_1) \text{ at } x = (1 - \lambda - \beta_1)\,L$$

$$\frac{d^2 e}{dx^2} = -\frac{2}{L^2}\frac{(e_1 + e_3 - h_1)}{(1 - \lambda - \beta_1)^2} = -\frac{2}{L^2}\frac{(e_1 + e_3)}{(1 - \lambda - \beta)(1 - \lambda)} \qquad (6.16)$$

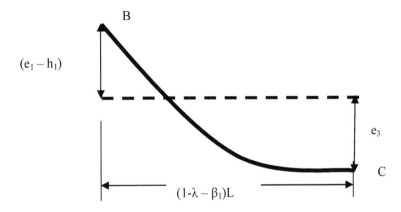

Fig. 6.9 Parabolic segment 2

Parabola 3: As shown in Fig. 6.10, the second segment parabola CD has a positive (i.e. below the centroidal axis) eccentricity of e_3 at C and a negative (i.e. above the centroidal axis) of $(e_2 - h_2)$ at D. The length of the segment is $(\lambda - \beta_2)\,L$, where L is the span. The equation to the parabola is given by

$$e = A_3 x^2 + B_3 x + C_3$$

where A_3, B_3 and C_3 are constants.

The boundary conditions are:

At C: $x = 0$, $e = e_3$ and $de/dx = 0$. This condition ensures continuity at the junction of parabolas 2 and 3.

At D: $x = (\lambda - \beta_2)\,L$, $e = -(e_2 - h_2)$

The constants are

$$C_3 = e_3, \ B_3 = 0, \ A_3 = -\frac{(e_3 + e_2 - h_2)}{(\lambda - \beta_2)^2 L^2}$$

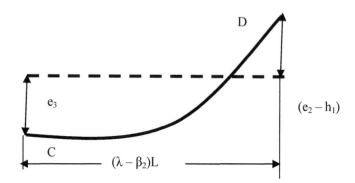

Fig. 6.10 Parabolic segment 3

Substituting for the constants,

$$e = e_3 - \frac{(e_3 + e_2 - h_2)}{(\lambda - \beta_2)^2}(\frac{x}{L})^2 \ , 0 \le \ x \le \ (\lambda - \beta_2) \ L$$

$$\frac{d^2 e}{dx^2} = -\frac{2}{L^2}\frac{(e_3 + e_2 - h_2)}{(\lambda - \beta_2)^2}$$

From parabola 3, the slope at $x = (\lambda - \beta_2) \ L$ is

$$\frac{de}{dx} = -\frac{2}{L}\frac{(e_3 + e_2 - h_2)}{(\lambda - \beta_2)} \tag{6.17}$$

Parabola 4: As shown in Fig. 6.11, the third segment parabola DE has a negative (i.e. above the centroidal axis) eccentricity of $(e_2 - h_2)$ at D and e_2 at E   The length of the segment is β L where L is the span.   The equation of the parabola is given by

$$e = A_4 x^2 + B_4 x + C_4$$

where A_4, B_4 and C_4 are constants.   The boundary conditions are:

At D: $x = 0$, $e = -(e_2 - h_2)$ and $\dfrac{de}{dx} = -\dfrac{2}{L}\dfrac{(e_3 + e_2 - h_2)}{(\lambda - \beta_2)}$.   This ensures continuity of parabolas 3 and 4 at their junction.

At D: $x = \beta_2 L$, $e = -e_2$ and $de/dx = 0$

There are thus four conditions to be satisfied, leading to

$$C_4 = -(e_2 - h_2), \quad B_4 = -2\frac{h_2}{\beta_2 L}, \quad A_4 = \frac{h_2}{(\beta_2 L)^2}$$

$$e = -(e_2 - h_2) - 2h_2 \frac{x}{\beta_2 L} + h_2 \left(\frac{x}{\beta_2 L}\right)^2$$

$$\text{Slope at } D = \frac{de}{dx} = -\frac{2}{L}\frac{h_2}{\beta_2}, \quad \frac{d^2 e}{dx^2} = 2\frac{h_2}{(\beta_2 L)^2}$$

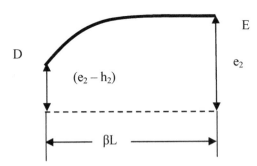

Fig. 6.11 Parabolic segment 4

For continuity of slope at D, the slopes of parabolic segments 3 and 4 must be the same:

$$\frac{de}{dx} = -\frac{2}{L}\frac{(e_3 + e_2 - h_2)}{(\lambda - \beta_2)} = -\frac{2}{L}\frac{h_2}{\beta_2}$$

$$h_2 = (e_3 + e_2)\frac{\beta_2}{\lambda} \tag{6.18}$$

Substituting for h_2, $\dfrac{de}{dx} = -\dfrac{2(e_3 + e_2)}{\lambda L}$ at $x = D$

Substituting for $h_1 = (e_3 + e_1)\dfrac{\beta_1}{(1-\lambda)}$, $h_2 = (e_3 + e_2)\dfrac{\beta_2}{\lambda}$, the curvatures are

given by

Segment 1: $\dfrac{d^2 e}{dx^2} = \dfrac{2}{L^2}\dfrac{(e_1 + e_3)}{(1-\lambda)\beta_1}$ $\qquad (6.19)$

Segment 2: $\dfrac{d^2 e}{dx^2} = -\dfrac{2}{L^2}\dfrac{(e_1 + e_3)}{(1-\lambda-\beta_1)(1-\lambda)}$ $\qquad (6.20)$

Segment 3: $\dfrac{d^2e}{dx^2} = -\dfrac{2}{L^2}\dfrac{(e_3+e_2)}{(\lambda-\beta_2)\lambda}$ (6.21)

Segment 4: $\dfrac{d^2e}{dx^2} = \dfrac{2}{L^2}\dfrac{(e_3+e_2)}{\lambda\beta_2}$ (6.22)

Similarly, ignoring the signs, the equations for the slopes are

Segment 1: $\dfrac{de}{dx}=0$ at A and $\dfrac{de}{dx}=2\dfrac{(e_1+e_3)}{(1-\lambda)L}$ at B (6.23)

Segment 2: $\dfrac{de}{dx}=2\dfrac{(e_1+e_3)}{(1-\lambda)L}$ at B and $\dfrac{de}{dx}=0$ at C (6.24)

Segment 3: $\dfrac{de}{dx}=0$ at C and $\dfrac{de}{dx}=\dfrac{2}{\lambda L}(e_3+e_2)$ at D (6.25)

Segment 4: $\dfrac{de}{dx}=\dfrac{2}{\lambda L}(e_3+e_2)$ at D and $\dfrac{de}{dx}=0$ at E (6.26)

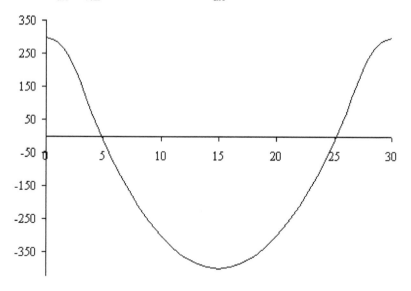

Fig. 6.12 Cable profile made up of four parabolic segments.

Example: L = 30 m, the lengths of the segments are respectively 3 m, 12 m, 12 m and 3 m. $e_1 = 300$ mm, $e_2 = 300$ mm, $e_3 = 400$ mm.

$\beta_1 \times 30 = 3$, $\beta_1 = 0.1$, $(1 - \lambda - \beta_1) \times 30 = 12$, $\lambda = 0.5$

$\beta_2 \times 30 = 3$, $\beta_2 = 0.1$

Fig. 6.12 shows a plot of the profile of the cable.

The curvatures and the radius curvature of the four segments are:

Segment 1: 3.11×10^{-2} m^{-1}, 32.1 m (Convex)

Segment 2: 7.78×10^{-3} m^{-1}, 129.0 m (Concave)

Segment 3: 7.78×10^{-3} m^{-1}, 129.0 m (Concave)

Segment 4: 3.11×10-2 m-1, 32.1 m (Convex)

The slopes are:
Segment 1: At A: de/dx = 0, At B: de/dx = 93.33×10^{-3}
Segment 2: At B: de/dx = 93.33×10^{-3}, At C: de/dx = 0

It is worth pointing out that a cable with a *convex* profile will induce upward load on the cable but ***downward load on the concrete***. Similarly a cable with a *concave* profile will induce downward load on the cable but ***upward load on the concrete***.

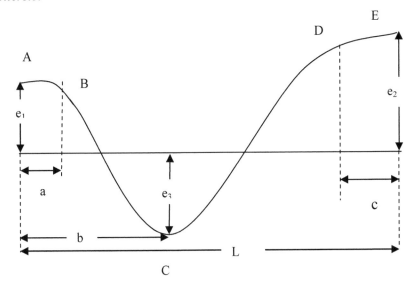

Fig. 6.13 A three parabola configuration

6.2.3.1 Alternative Profile Consisting of Three Parabolas Instead of Four Parabolas

In the profile considered in section 6.2.3, the parameter λ dictated the position of zero slope in the concave part of the profile. It is possible to combine parabolas 2 and 3 into a single parabola and for prefixed values of e_1 and e_2 to let the cable profile determine the position of the zero slope on the concave side.

Fig. 6.13 shows the profile considered. Parabolas AB and DE are convex and parabola BCD is concave with zero slope at distance b from the left-hand support.

Taking the origin at the left-hand support, with abscissa x, positive to the right and ordinate y positive above, the equations of the three parabolas can be written as follows:

Parabola AB: $y = e_1 + A_1 x^2, 0 \le x \le a$

Parabola BCD: $y = -e_3 + A_2 (b-x)^2, a < x \le (L-c)$

Note that the slope is zero when x = b.

Parabola DE: $y = e_2 + A_3 (L-x)^2, (L-c) \le x \le L$

Continuity of both slope and deflection are maintained at B the junction of parabolas AB and BCD and also at D where the parabolas BCD and DE meet. Substituting x = a in parabolas AB and BCD for deflection and slope continuity at B,

$$e_1 + A_1 a^2 = -e_3 + A_2 (b-a)^2$$
$$2 A_1 a = -2A_2 (b-a)$$

Solving for A_1 and A_2,

$$A_1 = -\frac{(e_1 + e_3)}{ab}, \quad A_2 = \frac{(e_1 + e_3)}{b(b-a)} \tag{6.27}$$

Substituting x = (L−c), in parabolas BCD and DE for deflection and slope continuity at D,

$$e_2 + A_3 c^2 = -e_3 + A_2 (b - L + c)^2$$
$$-2 A_3 c = -2A_2 (b - L + c)$$

Solving for A_2 and A_3,

$$A_3 = -\frac{(e_2 + e_3)}{c(L-b)}, \quad A_2 = -\frac{(e_2 + e_3)}{(L-b)(b-L+a)} \tag{6.28}$$

Since the two expressions for A_2 from equations (6.27) and (6.28) must be equal,

$$\frac{(e_1 + e_3)}{b(b-a)} = -\frac{(e_2 + e_3)}{(L-b)(b-L+a)}$$

$$(b - L + c)(L - b) = -[(e_2 + e_3)/(e_1 + e_3)](b^2 - b\,a) \tag{6.29}$$

Multiplying through,
$$(b L - L^2 + L c - b^2 + b L - b c) = -[(e_2 + e_3)/(e_1 + e_3)](b^2 - b\,a)$$

Dividing throughout by L^2 and gathering terms in $(b/L)^2$ and b/L,

$$\{1 - \alpha\}(\frac{b}{L})^2 - \frac{b}{L}(2 - \alpha\frac{a}{L} - \frac{c}{L}) + (1 - \frac{c}{L}) = 0 \tag{6.30}$$

where $\alpha = \frac{(e_2 + e_3)}{(e_1 + e_3)}\}, A = 1 - \alpha, B = (2 - \alpha\frac{a}{L} - \frac{c}{L}), C = (1 - \frac{c}{L})$ \tag{6.31}

Substituting for A, B and C, (6.30) can be simplified as

$$A(\frac{b}{L})^2 - B\frac{b}{L} + C = 0 \tag{6.32}$$

Solving the quadratic equation (6.32),

$$\frac{b}{L} = \frac{B - \sqrt{(B^2 - 4\,A\,C)}}{2A} \tag{6.33}$$

For given values of $\frac{(e_2 + e_3)}{(e_1 + e_3)}$, a/L and c/L, equation (6.33) gives the value of b/L.

The ordinates at B and D are given by
$$y_B = e_1 - (e_1 + e_3)(a/b), \quad y_D = e_2 - (e_2 + e_3)[c/(L-b)]$$
The curvatures of the three parabolas are:

Parabola AB:
$$y = e_1 + A_1 x^2$$
Substituting for A_1 from (6.27) and simplifying,
$$y = e_1 - \frac{(e_1 + e_3)}{ab} x^2, \quad 0 \le x \le a$$

Slope at B: $x = a$, $\dfrac{dy}{dx} = -2\dfrac{(e_1 + e_3)}{b}$, $\dfrac{d^2 y}{dx^2} = -2\dfrac{(e_1 + e_3)}{ab}$ $\qquad(6.34)$

Parabola BCD:
$$y = -e_3 + A_2(b-x)^2, \quad a < x \le (L-c)$$
Substituting for A_2 from (6.27) and simplifying
$$y = -e_3 + \frac{(e_1 + e_3)}{b(b-a)}(b-x)^2$$
$$\frac{dy}{dx} = -2\frac{(e_1 + e_3)}{b(b-a)}(b-x)$$

At B, $x = a$, $\dfrac{dy}{dx} = -2\dfrac{(e_1 + e_3)}{b}$

At D, $x = (L-c)$, $\dfrac{dy}{dx} = -2\dfrac{(e_1 + e_3)}{b(b-a)}(b-L+c)$

$$\frac{d^2 y}{dx^2} = 2\frac{(e_1 + e_3)}{b(b-a)} \qquad(6.35)$$

Alternatively, substituting for A_2 from (6.28) and simplifying
$$y = -e_3 - \frac{(e_2 + e_3)}{(L-b)(b-L+a)}(b-x)^2$$
$$\frac{dy}{dx} = 2\frac{(e_2 + e_3)}{(L-b)(b-L+a)}(b-x)$$

At D, $x = (L-c)$, $\dfrac{dy}{dx} = -2\dfrac{(e_1 + e_3)}{b(b-a)}(b-L+c)$

Sincey the two expressions for A_2 from equations (6.27) and (6.28) must be equal,
$$\frac{(e_1 + e_3)}{b(b-a)} = -\frac{(e_2 + e_3)}{(L-b)(b-L+a)}$$
$$\frac{(e_1 + e_3)}{b(b-a)}(b-L+a) = -\frac{(e_2 + e_3)}{(L-b)}$$

At D, $x = (L-c)$, $\dfrac{dy}{dx} = -2\dfrac{(e_1 + e_3)}{b(b-a)}(b-L+c) = 2\dfrac{(e_2 + e_3)}{(L-b)}$

Parabola DE:
$$y = e_2 + A_3 (L - x)^2, \qquad (L-c) \le x \le L$$
Substituting for A_3 from (6.28) and simplifying
$$y = e_2 - \frac{(e_2 + e_3)}{c(L-b)}(L-x)^2$$

$$\frac{dy}{dx} = 2\frac{(e_2 + e_3)}{c(L-b)}(L-x)$$

At D, $x = (L-c)$, $\dfrac{dy}{dx} = 2\dfrac{(e_2 + e_3)}{(L-b)}$

$$\frac{d^2 y}{dx^2} = -2\frac{(e_2 + e_3)}{c(L-b)} \tag{6.36}$$

Example: $a/L = 0.1$, $c/L = 0.1$, $e_1 = 150$ mm, $e_2 = 350$ mm, $e_3 = 200$ mm
$\alpha = (350 + 200)/(150 + 200) = 1.5714$
$A = 1 - \alpha = -0.5714$, $B = 2 - \alpha (a/L) - c/L = 1.7429$, $C = 1 - c/L = 0.9$
$-0.5714 (b/L)^2 - 1.7429 (b/L) + 0.9 = 0$
Solviong the quadratic equation, $b/L = 0.45$

6.2.3.2 Alternative Profile Consisting of Two Parabolas Instead of Three
Parabolas

As happens in the case of end spans, there is a convex parabola at one end only as shown in Fig. 6.14. Assuming that it is at the right-hand support, substituting a = 0 in the equations of the previous section,

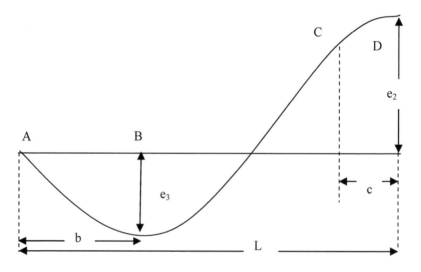

Fig. 6.14 A three-parabola configuration

Parabola ABC: $y = -e_3 + A_2(b-x)^2$, $0 < x \le (L-c)$
Note that the slope is zero when $x = b$.
When $x = 0$, $y = 0$.

$$A_2 = e_3/b^2 \tag{6.37}$$

Parabola CD: $y = e_2 + A_3(L-x)^2$, $(L-c) \le x \le L$
Continuity of both slope and deflection are maintained at D where the parabolas ABC and CD meet.
Substituting $x = (L-c)$, for continuity at D,
$e_2 + A_3 c^2 = -e_3 + A_2(b-L+c)^2$ for equal deflections.
$-2 A_3 c = -2A_2(b-L+c)$, for equal slopes.
Solving for A_2 and A_3,

$$A_2 = -\frac{(e_2+e_3)}{(b-L+c)(L-b)} \quad , \quad A_3 = -\frac{(e_2+e_3)}{(L-b)c} \tag{6.38}$$

Equating the two expressions for A_2 from (6.37) and (6.38),

$$\frac{e_3}{b^2} = -\frac{(e_2+e_3)}{(b-L+c)(L-b)}$$

$$(b-L+c)(L-b) = -[(e_2+e_3)/e_3](b^2)$$

Dividing throughout by L^2 and gathering terms in $(b/L)^2$ and b/L,

$$\{1-\alpha\}(\frac{b}{L})^2 - \frac{b}{L}(2-\frac{c}{L}) + (1-\frac{c}{L}) = 0 \tag{6.39}$$

Set: $\alpha = \dfrac{(e_2+e_3)}{e_3}$, $A = 1-\alpha$, $B = (2-\dfrac{c}{L})$, $C = (1-\dfrac{c}{L})$ $\tag{6.40}$

Substituting for A, B and C, equation (6.39) can be simplified as

$$A(\frac{b}{L})^2 - B\frac{b}{L} + C = 0 \tag{6.41}$$

Solving the quadratic equation (6.41),

$$\frac{b}{L} = \frac{B - \sqrt{(B^2 - 4AC)}}{2A} \tag{6.42}$$

For given values of $\dfrac{(e_2+e_3)}{e_3}$ and c/L, equation (6.40) gives the value of b/L.

Parabola ABC:

$y = -e_3 + A_2(b-x)^2$, $0 \le x \le (L-c)$
Substituting for A_2 from (6.37),

$$y = -e_3 + e_3\frac{(b-x)^2}{b^2}$$

$$\frac{dy}{dx} = -2e_3\frac{(b-x)}{b^2}$$

At A, $x = 0$, $\dfrac{dy}{dx} = -2\dfrac{e_3}{b}$

$$\frac{d^2y}{dx^2} = 2\frac{e_3}{b^2} \tag{6.43}$$

Parabola CD:

$y = e_2 + A_3 (L - x)^2, \quad (L - c) \le x \le L$

Substituting for A_3 from (6.38), $y = e_2 - \dfrac{(e_2 + e_3)}{(L-b)c}(L-x)^2$

$\dfrac{dy}{dx} = 2 \dfrac{(e_2 + e_3)}{(L-b)c}(L-x)$

At C, $x = (L-c)$, $\dfrac{dy}{dx} = 2 \dfrac{(e_2 + e_3)}{(L-b)}$

$\dfrac{d^2 y}{dx^2} = -2 \dfrac{(e_2 + e_3)}{(L-b)c}$ \hfill (6.43)

The ordinates at B and D are given by
$y_B = e_1 - (e_1 + e_3) (a/b)$, $y_D = e_2 - (e_2 + e_3) [c/ (L-b)]$
The curvatures of the two parabolas are:

Example: $c/L = 0.1$, $e_2 = 350$ mm, $e_3 = 200$ mm
$\alpha = (350 + 200)/ 200 = 2.75$
$A = 1 - \alpha = -1.75$, $B = 2 - c/L = 1.9$, $C = 1 - c/L = 0.9$

$-1.75 (b/L)^2 - 1.9 (b/L) + 0.9 = 0$

$\dfrac{b}{L} = \dfrac{B - \sqrt{(B^2 - 4AC)}}{2A} = \dfrac{1.9 - \sqrt{(1.9^2 - 4 \times (-1.75) \times 0.9)}}{2 \times (-1.75)} = 0.36$

6.2.4 Prestress Loss and Equivalent Loads

Loss of prestress due to friction and draw-in of the anchor causes variation of the prestress along the length of the cable as will be discussed in Chapter 12. This effect can be included in calculating the equivalent loads by multiplying the curvature by an average value of the prestress over different segments of the profile.

6.3 FIXED END MOMENTS

In order to determine the moments due to prestress, it is necessary to analyse the structure under the equivalent loads due to prestress. Because the cable profile consists of a series of parabolic segments with positive or negative curvature, the equivalent loads on an element also consist of segments of distributed loads which can be positive or negative. Manual method of structural analysis such as the moment distribution method requires the calculation of fixed forces on an element. The necessary equations are summarised below.

Fig. 6.15 shows the uniformly distributed patch load q loading on a beam of span L. The fixed end moments (clockwise positive) are given by

$$M_A = -q\frac{L^2}{12}[(1-\alpha_1)^3(1+3\alpha_1) - \alpha_2^3(4-3\alpha_2)]$$

$$M_B = q\frac{L^2}{12}[(1-\alpha_2)^3(1+3\alpha_2) - \alpha_1^3(4-3\alpha_1)] \qquad (6.44)$$

Note: For clamped beams, where the load is towardsthe left or right support only, the following simplified equations can be used.
Case 1: Load towards the left-hand support only: $\alpha_1 = 0$, $\alpha_2 > 0$,

$$M_A = -q\frac{L^2}{12}[1 - \alpha_2^3(4-3\alpha_2)], \quad M_B = q\frac{L^2}{12}[(1-\alpha_2)^3(1+3\alpha_2)] \qquad (6.45)$$

Case 2: Load towards the right-hand support only: $\alpha_1 > 0$, $\alpha_2 = 0$,

$$M_A = -q\frac{L^2}{12}[(1-\alpha_1)^3(1+3\alpha_1)], \quad M_B = q\frac{L^2}{12}[1 - \alpha_1^3(4-3\alpha_1)] \qquad (6.46)$$

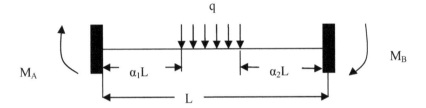

Fig. 6.15 Beam with partial loading in the middle

In the case of propped cantilevers, where the left end is simply supported, the following simplified equations can be used.
Case 1: Load towards the fixed support B only: $\alpha_1 > 0$, $\alpha_2 = 0$,

$$M_A = 0, \quad M_B = q\frac{L^2}{8}(1-\alpha_1^2)^2 \qquad (6.47)$$

Case 2: Load towards the simple .support A only: $\alpha_1 = 0$, $\alpha_2 > 0$,

$$M_A = 0, \quad M_B = q\frac{L^2}{8}(1-\alpha_1)^2\{2-(1-\alpha_1)^2\} \qquad (6.48)$$

6.3.1 Fixed End Moment for the Three Parabola Cable Profile

For the cable profile considered in section 6.2.2, the fixed end moment can be calculated as shown in Table 6.1. Taking L = 30 m
$M_A = -0.312$ P kNm, $M_B = -0.395$ P kNm
where P = Prestressing force in kN.

Table 6.1 Fixed end moments for span 1-2

Parabolic segment No.	q/P + down	$\alpha_1 L$	$\alpha_2 L$	Loaded length	M_A/P	M_B/P
1	-0.25×10^{-2}	0	15	15	1.520×10^{-1}	-0.880×10^{-1}
2	-1.30×10^{-2}	15	3	12	1.706×10^{-1}	-4.594×10^{-1}
3	3.89×10^{-2}	27	0	3	-0.108×10^{-1}	1.525×10^{-1}
SUM				30	0.312	−0.395

6.3.2 Fixed End Moment for the Four Parabola Cable Profile

For the cable profile considered in section 6.2.3, the fixed end moment can be
calculated as shown in Table 6.2.   L = 30 m

Table 6.2 Fixed end moments for span 2-3

Parabolic segment No.	q/P + down	$\alpha_1 L$	$\alpha_2 L$	Loaded length	M_A	M_B
1	3.11×10^{-2}	0	27	3	-1.220×10^{-1}	-0.009×10^{-1}
2	-0.778×10^{-2}	3	15	12	3.705×10^{-1}	-1.801×10^{-1}
3	-0.778×10^{-2}	15	3	12	1.801×10^{-1}	-3.705×10^{-1}
4	3.11×10^{-2}	27	0	3	-0.009×10^{-1}	1.220×10^{-1}
SUM				30	0.42	−0.42

$M_A = 0.42$ P kNm, $M_B = -0.42$ P kNm
where P = Prestressing force in kN.

6.4 ANALYSIS OF A CONTINUOUS BEAM FOR MOMENT DISTRIBUTION DUE TO PRESTRESS

Fig. 6.16 shows a three span continuous beam of uniform flexural rigidity EI and
constant span L.

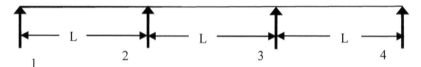

Fig. 6.16 Three span continuous beam

If the analysis is carried out by any convenient frame analysis software using the uniformly distributed loads calculated in Tables 6.1 and 6.2, then there is no need to calculate the fixed end moments.

From the analysis the final moments at the ends of spans are

$M_{12} = 0$, $M_{21} = -0.472P$,

$M_{23} = 0.472P$, $M_{32} = -0.472P$,

$M_{34} = 0.472P$, $M_{43} = 0$.

Fig. 6.17 and Fig. 6.18 show respectively the loadings on span 1-2 and span 2-3.

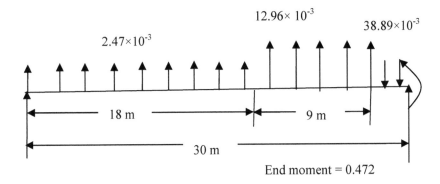

Fig. 6.17 Loads on span 1-2.

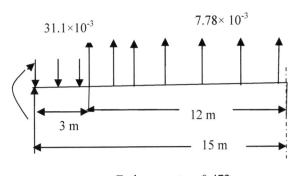

End moments = 0.472

Fig. 6.18 Loads on a symmetrical half of span 2-3

In span 1, the reactions are

$V_1 = \{-2.47 \times 10^{-3} \times 18 \times (18/2 + 12) - 12.96 \times 10^{-3} \times 9 \times (9/2 + 3)$
$\qquad + 38.89 \times 10^{-3} \times 3 \times (3/2) + 0.472/30\} P = -38.71 \times 10^{-3} P$

$V_2 = \{-2.47 \times 10^{-3} \times 18 \times (18/2) - 12.96 \times 10^{-3} \times 9 \times (9/2 + 18)$
$\qquad + 38.89 \times 10^{-3} \times 3 \times (3/2 + 27) - 0.472/30\} P = -5.72 \times 10^{-3} P$

In span 2, reactions are

$V_2 = V_3 = \{31.1 \times 10^{-3} \times 3 - 7.78 \times 10^{-3} \times 12\} P = -0.06 \times 10^{-3} P$

In the continuous beam the net reactions are:

$V_1 = V_4 = -38.71 \times 10^{-3}$ P, $V_2 = V_3 = \{-5.72 - 0.06\} \times 10^{-3}$ P $= -5.78 \times 10^{-3}$ P

If the intermediate supports are removed, then the end reactions in the simply supported beam are

$V_1 = V_4 = \{-2.47 \times 10^{-3} \times 18 - 12.96 \times 10^{-3} \times 9 + 38.89 \times 10^{-3} \times 3 + 31.1 \times 10^{-3} \times 3$
$- 7.78 \times 10^{-3} \times 12\}P = -44.49 \times 10^{-3}$ P

$V_2 = V_3 = 0$

The reactions induced by preventing displacements at the supports 2 and 3 are:

$V_1 = V_4 = (-38.71 + 44.49) \times 10^{-3}$ P $= 5.78 \times 10^{-3}$ P

$V_2 = V_3 = (-5.78 - 0) \times 10^{-3}$ P $= -5.78 \times 10^{-3}$ P

The secondary moments induced by the reactions preventing displacements at the intermediate supports are

$V_1 \times 30 = 0.174$ P

Fig. 6.19 shows the moment distribution due to secondary moments.

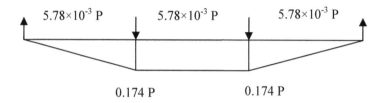

Fig. 6.19 Secondary moment induced by reactions

Having obtained the support moments, the bending moment distribution in each of the spans can be determined. Since the final bending moments arise from the prestressing force acting at an *'effective eccentricity'*, the effective eccentricity has the same variation along the span as the bending moment. Fig. 6.20 and Fig. 6.21 show respectively the actual and effective eccentricities for spans 1-2 and span 2-3. The corresponding bending moments are equal to the product of prestressing force and the effective eccentricity with a negative sign.

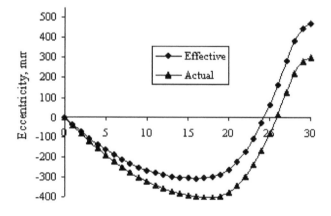

Fig. 6.20 Actual and effective eccentricities for span 1-2

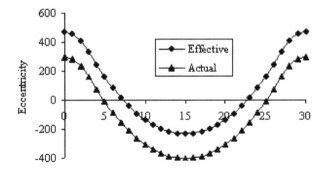

Fig. 6.21 Actual and effective eccentricities for span 2-3

It can be seen that at *mid-span* the effective eccentricity is *smaller* than the eccentricity as provided by the cable position in the spans. On the other hand, at *supports* the effective eccentricity is *larger* than the eccentricity as provided by the cable position in the spans. This is due to the secondary moments induced by support reactions as shown in Fig. 6.19.

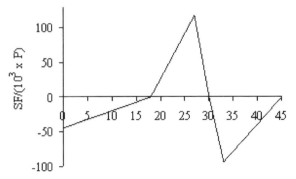

Fig. 6.22 Shear force distribution for symmetric half.

6.4.1 Distribution of Shear Force

The shear force distribution in the simply supported beam of $30 \times 3 = 90$ m span due to equivalent loads is shown in Fig. 6.22.

From the values of slope calculated in section 6.2.2 for span 1 and in section 6.2.3 for span 2, it can be seen that the shear force at any section due to prestress in the statically determinate beam is equal to the vertical component of the shear force at the section. It can therefore be stated that in a statically determinate structure, for any section, bending moment due to prestress is equal to the product of the prestress and eccentricity at that section. Similarly, the shear force at a section is equal to the vertical component of the prestress at the section. In the case of a statically indeterminate structure, both bending moment and shear for are modified by the parasitic moments.

6.5 CABLE PROFILE CONSISTING OF LINEAR VARIATION BETWEEN SUPPORTS

Consider the cable profile shown in Fig. 6.23. It consists of *zero eccentricity at the end supports* but otherwise it varies linearly between supports. In order to maintain the cable in the given form, equivalent loads will be concentrated loads applied to the cable in line with the supports. Evidently the load on the structure will therefore consist of concentrated loads acting only at the supports and no load being applied in the span. Therefore this form of cable profile will affect the axial stresses in the beam and also the load on the supports. These forces at the supports will influence the design of the supports.

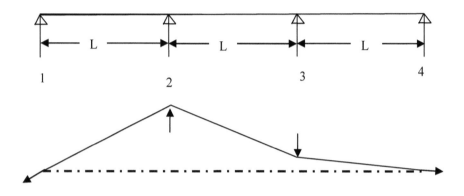

Fig. 6.23 Cable profile with linear variation between supports

The fact that linear variation of eccentricities between supports will not affect the bending stresses is a very useful observation. For example if an assumed cable profile is found to be satisfactory in every respect except that the profile cannot be easily fitted in the structure, the cable profile can be adjusted by varying the position of the cable at *internal supports only* without affecting the calculated bending stresses.

6.6 DETERMINATION OF PRESTRESSING FORCE AND CABLE PROFILE: EXAMPLE OF A CONTINUOUS BRIDGE BEAM

Fig. 6.24 shows the cross section of a closed box with side cantilevers. It is used in a four span continuous beam shown in Fig. 6.25.
The cross sectional properties of the beam are
Area = 5.75×10^6 mm^2, Centroidal distance from soffit, y_b = 1495 mm,
$y_t = 2500 - y_b = 1005$ mm
Second moment of area = 5.09×10^{12} mm^2
Section moduli: $Z_t = 5.06 \times 10^9$ mm^2, $Z_b = 3.40 \times 10^9$ mm^2

The beam is used in a four span continuous bridge structure shown in Fig. 6.25. The end spans are approximately 10% shorter than the interior spans. The structure is subjected to the following loads:

Unit weight of concrete = 25 kN/m³

Self weight of the box beam = 143.75 kN/m

Super dead load (assuming 150 mm thick surfacing) = 37.5 kN/m

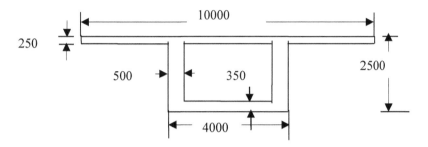

Fig. 6.24 Box beam with side cantilevers (dimensions in mm)

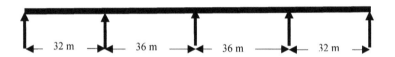

Fig. 6.25 Four span continuous beam bridge

Live load = 100 kN/m

It is assumed that the live load can occupy any part of the bridge.

Load factors on prestress: $\gamma_{superior} = 1.1$, $\gamma_{Inferior} = 0.90$

Loss of prestress = 10% at transfer and 25% at service.

Ratio $P_s/P_t = (1-25\%)/ (1-10\%) = 0.83$

6.6.1 Analysis of the Bridge

The bridge is analysed by the Matrix Stiffness method. The self weight and super dead load cover all the spans. However the live loads need to occupy only part of the structure. From the influence diagrams it is well known that

- For maximum moment in span it is necessary to apply maximum load only to that span and to every alternate span.
- For maximum moment at support it is necessary to apply maximum load on spans on both sides of a support and to every alternate span.

The following five live load cases are considered.

Case 1: Maximum span moments in spans 1-2 and 3-4. Live load covers only these two spans.

Case 2: Maximum span moments in spans 2-3 and 4-5. Live load covers only these
two spans.

Case 3: Maximum support moment over support 2. Live loads cover spans 1-2,
2-3 and 4-5.

Case 4: Maximum support moment over support 3. Live loads cover spans 2-3
and 3-4 only.

Case 5: Maximum support moment over support 4. Live loads cover spans 1-2,
3-4 and 4-5.

Table 6.3 Support moments for various load cases

M_{12}	M_{21} & M_{23}	M_{32} & M_{34}	M_{43} & M_{45}	M_{54}	
0	$1.708 \times 10^{+04}$	$1.475 \times 10^{+04}$	$1.708 \times 10^{+04}$	0	DEAD
0	$4.457 \times 10^{+03}$	$3.847 \times 10^{+03}$	$4.457 \times 10^{+03}$	0	SDL
0	$4.666 \times 10^{+03}$	$5.129 \times 10^{+03}$	$7.219 \times 10^{+03}$	0	Case 1
0	$7.219 \times 10^{+03}$	$5.129 \times 10^{+03}$	$4.666 \times 10^{+03}$	0	Case 2
0	$1.370 \times 10^{+04}$	$3.393 \times 10^{+03}$	$5.125 \times 10^{+03}$	0	Case 3
0	$4.942 \times 10^{+03}$	$1.373 \times 10^{+04}$	$4.942 \times 10^{+03}$	0	Case 4
0	$5.125 \times 10^{+03}$	$3.393 \times 10^{+03}$	$1.370 \times 10^{+04}$	0	Case 5

Fig. 6.26 to Fig. 6.29 show the distribution of bending moment along the span and
Fig. 6.30 and Fig. 6.31 show the moment envelopes considering self weight, super
dead load and live loads for spans 1-2 and 2-3 respectively.

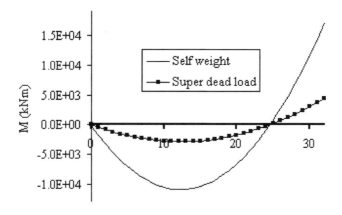

Fig. 6.26 Bending moment due to self weight and super dead load in span 1-2

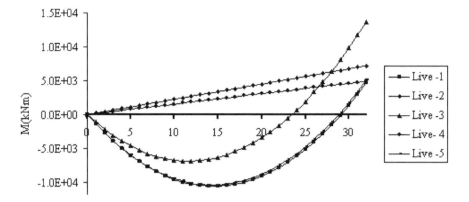

Fig. 6.27 Bending moment due to live load in span 1-2

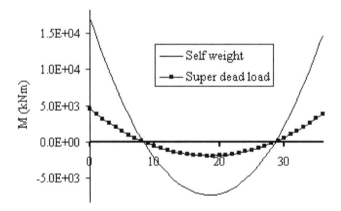

Fig. 6.28 Bending moment due to self weight and super dead load in span 2-3

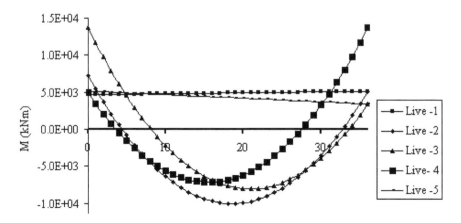

Fig. 6.29 Bending moment due to live loads in span 2-3

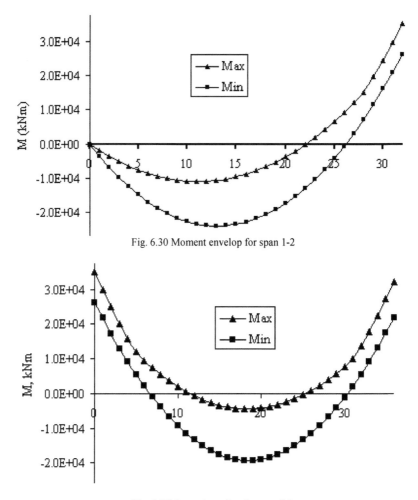

Fig. 6.30 Moment envelop for span 1-2

Fig. 6.31Moment envelop for span 2-3

6.6.2 Determination of Prestress and Eccentricity

The main difficulty in determining the prestress and eccentricity is that the parasitic moments affect the ***effective eccentricity***. The cable profile is concave in the span and convex over the supports. It is worth pointing out that in most continuous structures, the parasitic moments increase the magnitude of the positive post-tensioning moments at interior supports and reduce the negative post-tensioning moments between supports. The determination of prestress and eccentricity at a section is therefore a trial and error process.

Assuming that the centroid of the cable is at say 200 mm from the top and bottom fibres, then

$e_{top} = y_t - 200 = 1005 - 200 = 805$ mm

$e_{bottom} = y_b - 200 = 1495 - 200 = 1295$ mm

These are the provided eccentricities. The parasitic moment will alter these valuesand the effective eccentricities will be different from the above values. The parasitic reactions will increase the effective eccentricity over the support, with a consequent reduction in effective eccentricity in the span. Let the provided eccentricities over the support and in the span be 800 mm and 1300 mm respectively. The eccentricities at the end supports are taken as zero.

For an initial estimation of prestress, assume single parabola profile discussed in section 6.2.1.

Equivalent uniformly distributed load, q acting upwards = 8 × drape × (P/L^2)

For span 1-2: span = 32 m, $e_1 = 0$, $e_2 = -800$ mm, $e_3 = 1300$ mm at mid-span.

$$\text{Drape} = e_3 - (e_1+e_2)/2 = 1.698 \text{ m}$$

$q = 13.26 \times 10^{-3}$ P kN/m

For span 2-3: span = 36 m, $e_1 = -800$, $e_2 = -800$ mm, $e_3 = 1300$mm at mid-span.

$$\text{Drape} = e_3 - (e_1+e_2)/2 = 2.1 \text{ m}$$

$q = 12.96 \times 10^{-3}$ P kN/m

Moments at the supports caused by equivalent loads can be shown to be:

$M_{12} = 0$, $M_{21} = -M_{23} = -1.562$ P, $M_{32} = -M_{34} = -1.319$,

$M_{45} = -M_{54} = -1.562$, $M_{54} = 0$

The mid-span moments in span 1-2 and 4-5 = 0.917 P kNm (tension at top face)

The mid-span moments in span 2-3 and 3-4 = 0.6595 P kNm (tension at top face).

Generally the stress state causing tension at the bottom fibre is critical. Assuming P_s is in kN, the value of prestress P_s in order to limit the tensile stress at mid-span and support sections in different spans can be determined as follows. In the following $\gamma_{Inferior} = 0.90$ and $f_{st} = 3.5$ MPa.

i. Mid-span in Span 1-2: The maximum moment at mid-span from self weight, super dead load and live load case 1 = $(9.858 + 2.572 + 10.467 = 22.97) \times 10^3$ kNm causing tension at the bottom. The moment due to prestress is 0.917 P_s kNm.

$$(-\frac{P_s}{A} - \frac{P_s\, e}{Z_b})\gamma_{inf} + \frac{M_{max}}{Z_b} \leq f_{st}$$

$$(-\frac{P_s}{5.75\times10^6} - \frac{P_s \times 917}{3.40\times10^9})\times10^3 \times 0.90 + \frac{22970\times10^6}{3.40\times10^9} \leq 3.5$$

$$-0.3992\times10^{-3} \times P_s + 6.76 \leq 3.5$$

$$P_s \geq 8098\text{kN}$$

ii. First interior support: The maximum moment from self weight, super dead load and live load case 3 = $(17.084 + 4.467 + 13.702 = 34.253) \times10^3$ kNm causing tension at the top. The moment due to prestress is 1.562 P_s kNm

$$(-\frac{P_s}{A} + \frac{P_s\, e}{Z_t})\gamma_{inf} + \frac{M_{max}}{Z_t} \leq f_{st}$$

$$(-\frac{P_s}{5.75\times10^6} - \frac{P_s \times 1562}{5.06\times10^9})\times10^3 \times 0.90 + \frac{34253\times10^6}{5.06\times10^9} \leq 3.5$$

$$-0.4290\times10^{-3} \times P_s + 6.77 \leq 3.5$$

$$P_s \geq 7621 \text{ kN}$$

iii. Mid-span in span 2-3: The maximum moment at mid-span from self weight, super dead load and live load case 1 = (7.373 + 1.923 + 10.026 = 19.32) × 10^3 kNm causing tension at the bottom. The moment due to prestress is 0.6595 P_s kNm.

$$(-\frac{P_s}{A} - \frac{P_s\,e}{Z_b})\gamma_{inf} + \frac{M_{max}}{Z_b} \le f_{st}$$

$$(-\frac{P_s}{5.75\times10^6} - \frac{P_s\times659.5}{3.40\times10^9})\times10^3 \times0.90 + \frac{19322\times10^6}{3.40\times10^9} \le 3.5$$

$$-0.3311\times10^{-3}\times P_s + 5.68 \le 3.5$$

$$P_s \ge 6593 \text{ kN}$$

iv. Second interior support, the maximum moment from self weight, super dead load and live load case 3 = (14.745 + 3.847 + 13.729 = 32.321) × 10^3 kNm causing tension at the top. The moment due to prestress is 1.319 P_s kNm

$$(-\frac{P_s}{A} + \frac{P_s\,e}{Z_t})\gamma_{inf} + \frac{M_{max}}{Z_t} \le f_{st}$$

$$(-\frac{P_s}{5.75\times10^6} - \frac{P_s\times1319}{5.06\times10^9})\times10^3 \times0.90 + \frac{32321\times10^6}{5.06\times10^9} \le 3.5$$

$$-0.3912\times10^{-3}\times P_s + 6.39 \le 3.5, \quad P_s \ge 7382 \text{ kN}$$

Clearly P_s will be approximately 8000 kN. The approximate value can be refined by doing a correct analysis by taking into account the actual profile of the cable in the spans.

6.6.3 Refined Analysis due to Equivalent Loads

Having estimated the required prestress using the approximate analysis, the next step is to do a refined analysis taking into account the exact profile of the cable. It is also necessary to include loss of prestress in the analysis but in this example this part is ignored for simplicity at this stage.

Table 6.4 Fixed end moment calculation for span 1-2

Parabolic segment No.	q/P + down	$\alpha_1 L$	$\alpha_2 L$	Loaded length	M_A	M_B
1	-1.3265×10^{-2}	0	18	14	0.666	−0.255
2	-1.5766×10^{-2}	14	3.2	14.8	0.549	−0.972
3	7.2917×10^{-2}	28.8	0	3.2	−0.023	0.325
SUM				32	1.192	−0.902

M_A = 1.192 P kNm, M_B = −0.902 P kNm, P = Prestressing force in kN.

6.6.3.1 Fixed End Moment for the Three- parabola Cable Profile

For the end spans using a three-parabola cable profile, with $e_1 = 0$, $e_2 = 800$ mm, $e_3 = 1300$ mm, $(1-\lambda) L = 14$ m, $\beta L = 3.2$ m, $L = 32$ m, using the equations developed in section 6.3, the fixed end moment can be calculated as shown in Table 6.4.

6.6.3.2 Fixed End Moment for the Four-parabola Cable Profile

For the end spans using a three-parabola cable profile, with $e_1 = 600$, $e_2 = 600$ mm, $e_3 = 1300$ mm, $(1-\lambda) L = 18$ m, $\beta L = 3.6$ m, $L = 36$ m, using the equations developed in section 6.2.2, the fixed end moment can be calculated as calculated as shown in Table 6.5. $L = 36$ m

Table 6.5 Fixed end moment calculation for span 2-3

Parabolic segment No.	q/P + down	$\alpha_1 L$	$\alpha_2 L$	Loaded length	M_A	M_B
1	6.4815×10^{-2}	0	32.4	3.6	−0.366	0.026
2	-1.6204×10^{-2}	3.6	18.0	14.4	1.112	−0.540
3	-1.6204×10^{-2}	18.0	3.6	14.4	0.540	−1.112
4	6.4815×10^{-2}	32.4	0	3.6	−0.026	0.366
SUM				36	1.26	−1.26

$M_A = 1.26$ P kNm, $M_B = -1.26$ P kNm, P = Prestressing force in kN.

6.6.3.3 Moments at Supports for the Cable Profile

From the Matrix analysis of the structure, the end moments are given by
$M_{12} = 0$, $M_{21} = -1.389$ P, $M_{23} = 1.389$ P, $M_{32} = -1.1955$ P,
$M_{34} = 1.1955$ P, $M_{43} = -1.389$ P, $M_{45} = 1.389$ P, $M_{54} = 0$.

6.6.3.4 Choice of Prestress at Service

Guided by the approximate analysis, assume the prestress P_s, is 8500 kN. P_t, the prestress at transfer is $P_t = P_s/0.83 = 10200$ kN. The effective moment provided by the prestressing force at service is obtained as the product of P_s and the effective eccentricity at the section and are shown in Fig. 6.32 and Fig. 6.33 for spans 1-2 and 2-3 respectively.

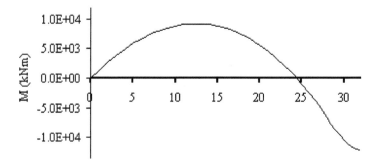

Fig. 6.32 Moment due to prestress in span 1-2

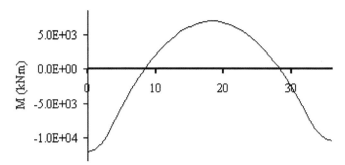

Fig. 6.33 Moment due to prestress in span 2-3

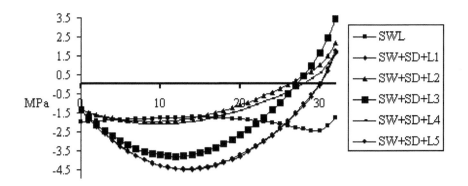

Fig. 6.34 Stress at top fibre in span 1-2

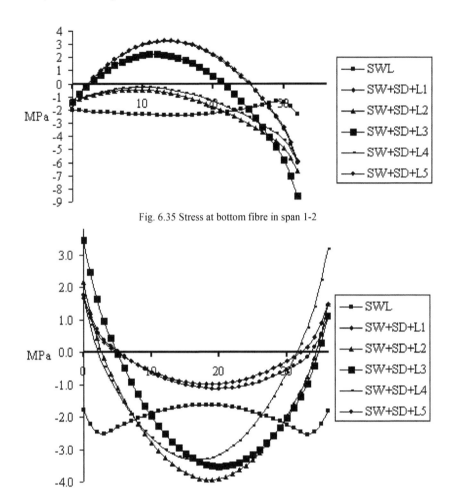

Fig. 6.35 Stress at bottom fibre in span 1-2

Fig. 6.36 Stress at top fibre in span 2-3

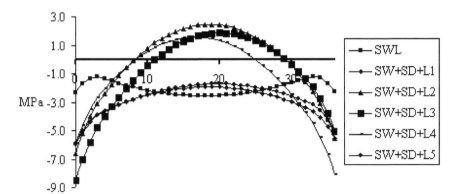

Fig. 6.37 Stress at bottom fibre in span 2-3

6.6.3.5 Stress Check at Transfer and Service

Once the effective prestress and eccentricity at a section are known, then the stress distribution in the cross section can be determined using the equations (4.1) and (4.2) at transfer and equations (4.7) and (4.8) at service.
In these equations $P_t = 10440$ kN, $P_s = 8700$ kN, $\gamma_{superior} = 1.1$, $\gamma_{Inferior} = 0.90$
$f_{tt} = 2.6$ N/mm^2, $f_{tc} = -15.0$ N/mm^2, $f_{st} = 3.5$ N/mm^2, $f_{sc} = -24.0$ N/mm^2

The calculated stress distributions are shown in Figs. 6.34 to 6.37. Note that when only self weight is acting, the prestress value corresponds to P_t with an appropriate load factor on the prestress and the state of stress is at transfer. On the other hand, when live loads are acting the prestress force corresponds to P_s with appropriate an load factor on the restress and the state of stress corresponds to that at service. The stress values are satisfactory.

6.7 CONCORDANT CABLE PROFILE

It is clear from the calculations in the previous sections that the cable profile in general results in parasitic moments and hence the effective eccentricity will be different from the one provided. However if the cable profile matches that of the bending moment distribution in a structure due to any loading, then it can be shown that the secondary or parasitic moments will be zero. Such a cable profile is known as a Concordant profile. The main attraction of the concordant profile is that design of continuous structures can be carried out in the same way as for a statically determinate structure because the actual and effective eccentricities are identical. Using modern methods of analysis, this advantage is irrelevant. In addition, it has to be emphasised that these so-called concordant profiles in general lead to uneconomic prestressing steel.

6.8 CHOICE OF TENDON SIZE AND LOCATION OF TENDONS

In case of bridge decks, for example using VSL the anchorages, number of 7-wire strands of nominal diameter of 15.2 mm or 15.7 mm can vary from 1 to 55. The maximum force that can be applied can vary from 206 kN to 11336 kN.
In choosing the number of cables, as suggested by Robert Benaim, a few important points should be kept in mind.
- Tendons are normally evenly distributed among the webs in a cross section.
- Very large cables require large diameter ducts to accommodate them. For example with say 19 tendon cable the external diameter of the duct can be about 90 mm. This might dictate the minimum width of webs. Webs of large width increase the self weight of the structure.
- Large cables need large jacks to stress them and it is difficult to manoeuvre the heavy jacks in a confined space.

- As an upper limit, it is generally useful to consider 19 strands of 15.2 mm or 15.7 mm with a maximum load at stressing of 3916 kN and 4297 kN respectively as most suitable.

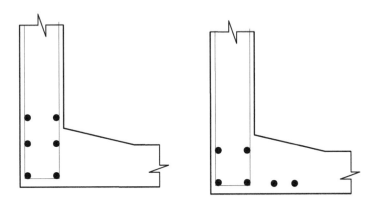

Fig. 6.38 a Alternative location of tendons in span section of a box girder

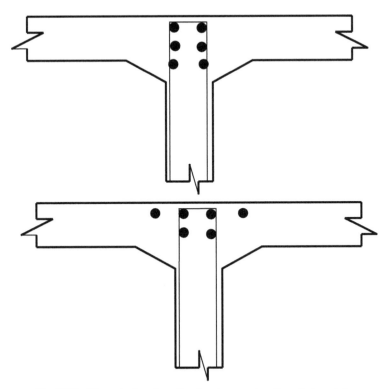

Fig. 6.38 b Alternative location of tendons in support section of a box girder

Tendons can be located inside the web reinforcement or outside it as shown in Fig. 6.38. Generally it is possible to maximise the eccentricity by locating the cables partly inside and partly outside the web stirrups.

6.9 EQUIVALENT LOADS AND SHIFT IN THE CENTROIDAL AXIS

Fig. 6.39 shows a two span continuous beam with a step change in depth at the interior support. It is prestressed with a force of 400 kN. The eccentricity of the cable is zero at the ends. Ignoring the reverse curvature over the interior support, the profile of the cable consists of two parabolas.

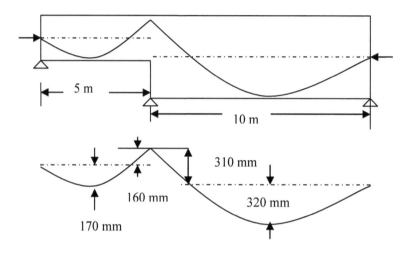

Fig. 6.39 Continuous beam with step change in depth

The dimensions of the two beams are as follows:

Beam AB: Span = 5 m, depth = 400 mm, eccentricity at mid-span is 170 mm and at the right hand support 160 mm.

Taking the origin at the centroid at the left-hand end, equation of the cable profile is $y = (x/5)^2 - 0.84 (x/5)$

$dy/dx = 2(x/5) - 0.84/5$

Slope at left hand end = $-0.84/5 = -0.168$

Slope at right hand end = $(2 - 0.84)/5 = 0.232$

Horizontal and vertical components of P at the left and right hand ends are:

At A : 400 and 400 × 0.168 = 67.2 kN, At B : 400 and 400 × 0.232 = 92.8 kN

Moment at right hand end = Pe = 400 × 160 × 10^{-3} = 64 kNm (Anticlockwise)

Curvature = $2/5^2$ = 0.08 m^{-1}

Equivalent uniformly distributed load q_1 = P × curvature
$$= 400 \times 0.08 = 32 \text{ kN/m}$$

Beam BC: Span = 10 m, depth = 700 mm, eccentricity at mid-span is 320 mm and at the left-hand support 310 mm.

Taking the origin at the centroid at the right-hand end, the equation of the cable profile is $y = 1.9(x/10)^2 - 1.59 (x/10)$
$dy/dx = 3.8(x/10) - 1.59/10$
Slope at the left hand end $= (3.8 - 1.59)/10 = 0.221$
Slope at the right hand end $= -1.59/10 = -0.159$
Horizontal and vertical components of P at the left and right hand ends are:
At B $=400$ & $400 \times 0.221 = 88.4$ kN, At C $=400$ & $400 \times 0.159 = 63.6$ kN
Moment at the left hand end $= Pe = 400 \times 310 \times 10^{-3} = 124$ kNm (clockwise)
Curvature $= 2 \times 1.9/10^2 = 0.038$ m^{-1}
Equivalent uniformly distributed load $q_2 = P \times$ curvature
$$= 400 \times 0.038 = 15.2 \text{ kN/m}$$
As shown in Fig. 6.40, the continuous beam needs to be analysed for an upward uniformly distributed load of 32 kN/m in span AB and 15.2 kN/mm in span BC. In addition at support B there is a clockwise moment of $(-64 + 124) = 60$ kNm. Note that the concentrated couple at B is equal to the product of the prestress P and the difference in the eccentricities at B in span AB and span BC:
$M_B = P (e_{BC} - e_{AB})$
Assuming uniform width and taking the second moment of area as proportional to the cube of the depth, the ratio $I_{BC}: I_{AB} = 1: 5.36$.
Fig. 6.41 shows the results of the analysis in two parts.
Part 1: Due to a moment at B of 60 kNm
Part 2: Due to uniformly distributed loads.
Part 3: Sum of Part 1 and Part 2.
Note the discontinuity of moment at the support.

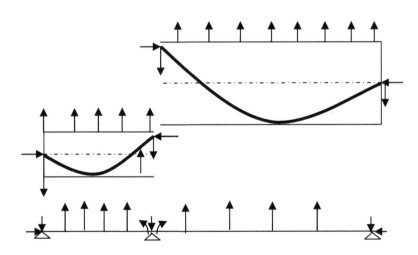

Fig. 6.40 Loads on the continuous beam due to prestress

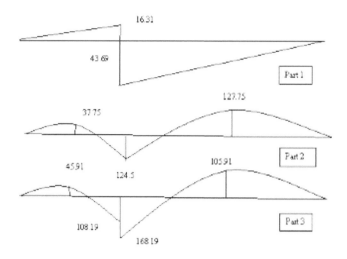

Fig. 6.41 Bending moment diagram due to prestress.

6.9.1 Shift in the Centroidal Axis in Box Girders

The problem described in section 6.10 commonly occurs in box girders where for example the thickness of one of the flanges is increased to accommodate large stresses. Fig. 6.42 shows a box girder with a straight cable in the top flange but the bottom flange is thickened near the central support.

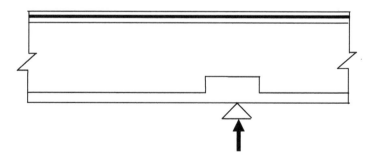

Fig. 6.42 Shift in centroidal axis in box girders

6.10 EQUIVALENT LOADS AND VARIABLE SECOND MOMENT OF AREA

Fig. 6.43 shows a symmetrical half of a beam with variable second moment of area. The depth of the beam is h at the ends and 4h/3 at the centre. It is prestressed with a cable of parabolic profile with eccentricity h/4 at the ends and –h/3 at the centre.

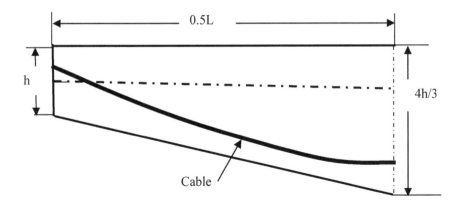

Fig. 6.43 Beam with variable second moment of area

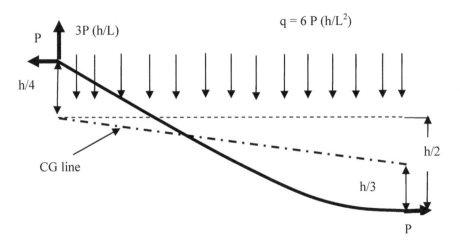

Fig. 6.44 Forces acting on the cable

Taking the origin at the centroid at the left hand end (y + up, x + right), the equation of the parabola is

$$y_{Parabola} = \frac{h}{4}\left[1 - 12\,\frac{x}{L}\left(1 - \frac{x}{L}\right)\right]$$

$$x = 0,\ y = h/4 \text{ and } x = 0.5L,\ y = -h/2$$

$$\frac{dy}{dx} = -3\frac{h}{L}\left(1 - 2\frac{x}{L}\right)$$

At x = 0, $\dfrac{dy}{dx} = -3\dfrac{h}{L}$, At $\dfrac{x}{L} = 0.5$, $\dfrac{dy}{dx} = 0$

$$\frac{d^2y}{dx^2} = 6\frac{h}{L^2}$$

Assuming the slopes are small, sin θ ≈ tan θ and cos θ ≈ 1, the forces acting on the cable are as shown in Fig. 6.44.

Lateral load $q = P \times$ curvature $= P \times 6h/L^2$

The horizontal and vertical components respectively of the prestress at the ends are:

At x = 0, P and P dy/dx = −3Ph/L and at x = 0.5L, P and 0.

Transferring the reversed loads on the cable on to the beam, the loads acting on the beam are as shown in Fig. 6.44.

The equation of the line defining the centroid is

$$y_{CG} = -\frac{h}{3}\frac{x}{L}$$

The equation for the eccentricity is

$$e = y_{Parabola} - y_{CG} = \frac{h}{4}\{1 - 12\frac{x}{L}(1 - \frac{x}{L})\} + \frac{h}{3}\frac{x}{L}$$

$$e = \frac{h}{4}[1 - \frac{32}{3}\frac{x}{L} + 12(\frac{x}{L})^2]$$

At x = 0, e = h/4 and at x = 0.5 L, e = −h/3.

The moments acting at the ends are

At x = 0, M = Ph/4 and at x/L = 0.5, M = −Ph/3.

The beam needs to be analysed for the forces shown in Fig. 6.45. In a frame analysis program, if an element to simulate the variation of depth shown is available, then it can be used.

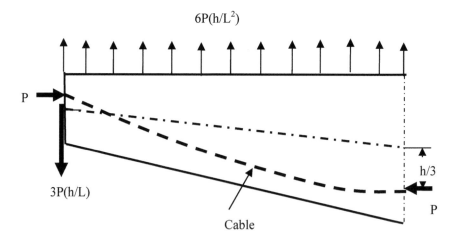

Fig. 6.45a Forces acting on the beam

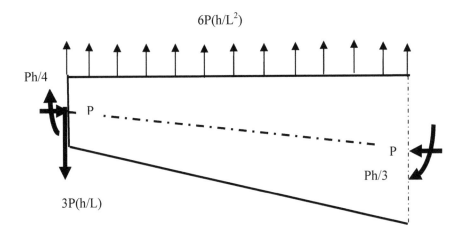

Fig. 6.45b Equivalent forces on the beam for analysis

6.11 THERMAL STRESS ANALYSIS IN CONTINUOUS STRUCTURES

Section 4.7, Chapter 4, explained in detail the calculation of stresses due to thermal gradients in the cross section of a statically determinate beam. In the case of a statically determinate beam, only self-equilibrating stresses are produced but no net axial force on moment at a section. This is not true in the case of statically indeterminate beam. An example using the section shown in Fig. 6.24 will illustrate this point.

6.11.1 Thermal Stress Calculation: Heating

$h = 2500$ mm, $0.3h = 750$ mm, $h_1 = 0.3h \leq 150$ mm,
$h_2 = 0.3h$ but ≥ 100 mm and ≤ 250 mm,
$h_3 = 0.3h$ but $\leq (100$ mm + surfacing depth in mm)
Therefore take: $h_1 = 150$ mm, $h_2 = 250$ mm, $h_3 = 200$ mm
From Table 4.2, $\Delta T_1 = 13.0°C$, $\Delta T_2 = 3.0°C$, $\Delta T_3 = 2.5°C$
Corresponding restraining stresses are: $\sigma = -\alpha_T \times E_c \times \Delta T$
Assume: Coefficient of thermal expansion $\alpha_T = 10 \times 10^{-6}/°C$,
　　　　Young's modulus, $E_c = 34$ GPa
$\alpha_T \times E_c = 0.34$ MPa
$\sigma = -0.34 \times \Delta T = (-4.42, -1.02, -0.85)$ MPa
Table 6.6 shows the restraining stresses at various levels in the cross section.
　　Having obtained the restraining stresses, the next step is to calculate the forces F in different segments of the cross section using the formula
$$F = \text{Average stress} \times \text{Area}$$
The values of the forces are shown in Table 6.7.

Table 6.6 Restraining stresses in the cross section: Heating

Level from top, mm	σ, MPa
0	−4.42
150	−1.02
250	−0.61
400	0
2300	0
2500	−0.85

Table 6.7 Restraining forces in the cross section: Heating

σ at top, MPa	σ at bottom, MPa	Area	F, kN
−4.42	−1.02	10000 × 150	−4080.00
−1.02	−0.61	10000 × 100	−815.00
−0.61	0	2 × 500 × 150	−45.75
0	−0.85	4000 × 200	−340.00

Having calculated the forces, the next step is to calculate the moment due to the restraining forces about the neutral axis of the beam. Table 6.8 shows the details of the calculation of the moment due to force F. Note that if the stress distribution is as shown in Fig. 4.16, the resultant is at a distance C from the top given by equation (4.24):

$$C = \frac{D}{3} \times \frac{(2\sigma_{Bottom} + \sigma_{Top})}{(\sigma_{Bottom} + \sigma_{Top})}$$

Table 6.8 Calculation of moment: Heating

F, kN	D mm	C mm	Lever arm mm	Moment kNm
−4080.00	150	59	1005 − 59 = 936	−3818.88
−815.00	100	46	1005−150−46 = 809	−659.34
−45.75	150	50	1005−250−50 = 705	−32.25
−340.00	200	133	−(1495−200 + 133) = −1428	485.52
Σ=−5280.75				Σ= −4024.95

Note: y_b = 1495 mm, y_t = 1005 mm

In each span, there is a restraining compression axial force equal to 5280.75 kN and a moment equal to 4024.95 kNm causing compression at the top fibre. Fig. 6.46 shows the restraining axial forces and moments on each span. In order to obtain the stresses caused in the continuous beam by thermal gradients, the axial

force is released by applying a tensile force of 5280.75 kN. As there is no restraint
to rotation at the ends of the beam, the restraining moment M equal to 4024.95
kNm at the end of the beam is released resulting in the moments of 0.2M and
0.1 M over the supports as shown in Fig. 6.46.

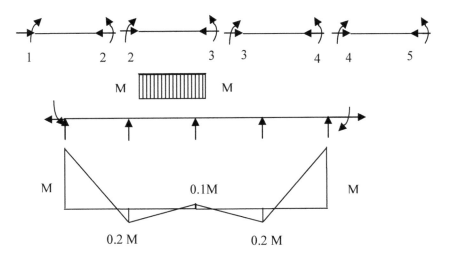

Fig. 6.46 Restraining release forces

The final stress is the sum of restraining stresses plus the stresses due to released
forces.

Table 6.9 Thermal stress calculation: Heating

y mm	σ MPa	− F/A	End support		Penultimate support	
			−(M/I) y	Final stress MPa	(0.2 M/I)y	Final stress MPa
1005	−4.4 2	0.92	0.80	−2.7	−0.16	−3.66
855	−1.0 2	0.92	0.68	0.58	−0.14	−0.24
755	−0.6 1	0.92	0.60	0.91	−0.12	0.19
605	0	0.92	0.48	1.40	−0.10	0.82
0	0	0.92	0	0.92	0	0.92
−1295	0	0.92	−1.03	−0.11	0.21	1.13
−1495	−0.8 5	0.92	−1.19	−1.12	0.24	0.31

Note: y_t = 1005 mm, y_b = 1495 mm, A = 5.75 × 10^6 mm², I = 5.05× 10^{12} mm⁴
y is measured from the neutral axis.

Table 6.9 shows the details of the stress calculation. It has to be noted that unlike in a statically determinate beam, in general a net moment acts at a section, although there is no net axial force. At the end support the state of stress will be a set of purely self-equilibrating stresses. However this is not the case at the penultimate support where although there is no net axial force, there is a net moment equal to 1.2M, where M is the restraining moment. Fig. 6.47 shows the stress distribution at two sections.

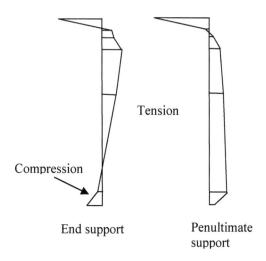

Fig. 6.47 Stress distribution: Heating

Table 6.10 Restraining stresses in the cross section: Cooling

Level from top	σ
0	2.86
250	0.17
500	0
1250	0
1875	0.34
2150	1.17*
2500	2.21

* By interpolation

6.11.2 Thermal Stress Calculation: Cooling

The procedure to be adopted is similar to the calculation of stresses due to heating:
$h = 2500$ mm, $0.2h = 500$ mm, $h_1 = h_4 = 0.2h \leq 250$ mm
$0.25h = 625$ mm, $h_2 = h_3 = 0.25h$ but ≥ 200 mm
Therefore: $h_1 = h_4 = 250$ mm, $h_2 = h_3 = 625$ mm
From Table 4.3, $\Delta T_1 = -8.4°C$, $\Delta T_2 = -0.50°C$, $\Delta T_3 = -1.0°C$, $\Delta T_4 = -6.5°C$

The corresponding restraining stresses are:

$\sigma = -\alpha_T \times E_c \times \Delta T = -0.34 \times \Delta T = (2.86, 0.17, 0.34, 2.21)$ MPa

Table 6.10 shows the restraining stresses in the cross section and the corresponding forces and moments are shown in Tables 6.11 and 6.12 respectively. Table 6.13 shows the final stresses due to temperature differentials due to cooling. Fig. 6.48 shows the stress distribution resulting from cooling at two different cross sections.

Table 6.11 Restraining forces in the cross section: Cooling

σ at top MPa	σ at bottom MPa	Area	F, kN
2.86	0.17	10000 × 250	3787.5
0.17	0	2 × 500× 250	21.25
0	0.34	2 × 500× 625	106.25
0.34	1.17	2 × 500× 275	207.63
1.17	2.21	4000 × 350	2366.0

Table 6.12 Calculation of moment: Cooling

F, kN	D mm	C mm	Lever arm mm	Moment, kNm
3787.5	250	88	1005 − 88 = 917	3473.14
21.25	250	83	1005−250−83 = 672	14.28
106.25	625	417	−(1495−350−275−625+ 417) = −662	−70.34
207.63	275	163	−(1495−350−275+163) = −1033	−214.48
2366.0	350	193	−(1495−350 + 190) = −1338	−3165.59
Σ = 6488.63				Σ= 37.01

Table 6.13 Thermal stress calculation: Cooling

			End support		Penultimate support	
y mm	σ MPa	−Σ F/A MPa	−(Σ M/I)y	Final stress MPa	−0.2 Σ (M/I)y	Final stress MPa
1005	2.86	−1.13	−0.007	1.72	≈ 0	1.73
755	0.17	−1.13	−0.006	−1.84		−0.96
505	0	−1.13	−0.004	−1.13		−1.13
0	0	−1.13	0	−1.13		−1.13
−245	0	−1.13	0	−1.13		−1.13
−870	0.34	−1.13	0.006	−0.78		−0.79
−1145	1.17*	−1.13	0.008	0.05		0.04
−1495	2.21	−1.13	0.011	1.09		1.08

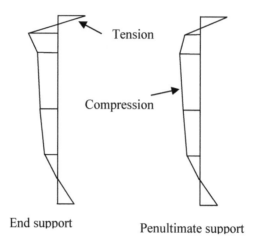

End support   Penultimate support

Fig. 6.48 Stress distribution: Cooling

6.12 REDUCTION OF MOMENT OVER A SUPPORT IN CONTINUOUS

BEAMS

In the example considered in section 6.6, it was assumed that the supports are point supports. This leads to the conclusion that the moment over a support has a sharp kink. In practice the support has a finite width and this leads to a smoothed-out bending moment diagram over the support. In addition it also leads to a reduction of moment over the support.

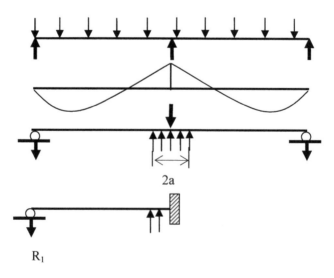

Fig. 6.49 Two span continuous beam

As an example consider a two span continuous beam of two equal spans of length L and subjected to a uniformly distributed moment of q per unit length as shown in Fig. 6.49. The moment over the support is $qL^2/8$ and the mid-span support reaction is $5qL/5$.

In order to calculate the effect of spreading the central reaction R over a width of 2a, consider the two span continuous beam subjected to a concentrated force R over the central support and the same force spread over a width 2a in the opposite direction. The value of the distributed load per unit length is q_1, where $q_1 = 0.5 R/a$.

The reaction R_1 in the propped cantilever is given by equating the deflection at the support caused by reaction R_1 and by q_1 to zero:

$$R_1 \frac{L^3}{3} = q_1 [\frac{a^4}{8} + \frac{a^3}{6}(L-a)]$$

$$R_1 = \frac{1}{8} \frac{a^3}{L^3} q_1 (4L - a)$$

The reduction of moment at the support due to spreading of the central support reaction is M, where

$$M = R_1 L - q_1 \frac{a^2}{2}$$

$$M = -q_1 a^2 \frac{1}{2} [1 - \frac{1}{4} \frac{a}{L} (4 - \frac{a}{L})]$$

$$= -\frac{1}{8} 2a R \ [1 - \frac{a}{L} + \frac{1}{4}(\frac{a}{L})^2 \ ]$$

The terms inside the square brackets sum to near unity. For example, if $a/L = 0.01$ and 0.02 then the value of the terms inside the square brackets are 0.99 and 0.98 respectively. The reduction of moment at the support due to the reduction caused by the central reaction spread over a width of 2a is approximately

Reduction of moment = Reaction at support × Width of support / 8.

6.13 REFERENCE TO EUROCODE 2 CLAUSE

Reduction of moment over continuous support: 5.3.2.2 (4)

CHAPTER 7

ULTIMATE BENDING STRENGTH CALCULATIONS

7.1 INTRODUCTION

In prestressed concrete design, structures are generally designed at the serviceability limit state (SLS). However, it is equally important to ensure that the designed structure has sufficient moment and shear capacity at the ultimate limit state (ULS). In this chapter, checking for bending capacity at the ULS is discussed.

7.2 STRESS DISTRIBUTION AT DIFFERENT STAGES OF LOADING

Fig. 7.1 shows the load-deflection behaviour of a prestressed beam. Fig. 7.2 shows the stress distribution in concrete at different stages of loading.

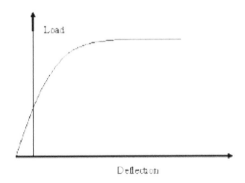

Fig. 7.1 Load-deflection relationship.

When the beam is prestressed, the large value of the prestress makes the beam hog up. Assuming that deflection is positive downward, at zero load the deflection is negative. At this stage, the stress distribution in concrete is compressive at the bottom fibre and tensile at the top fibre as shown in Fig. 7.2(a). As the load is increased, the compressive stress at the soffit begins to decrease so that as shown in Fig. 7.2(b), at a certain stage it becomes zero. However, the opposite happens at the top face where the tensile stress reverses and becomes compressive. On further increase in load, at some stage the stress at the soffit becomes tensile and the compressive stress at the top face continues to increase as shown in Fig. 7.2 (c). On increasing the load further, cracks appear at the soffit and the stress for a certain distance from the soffit becomes zero as shown in Fig. 7.2(d). However, the compressive stress and strain at the top face continue to increase. With further increase in load, cracks at the soffit continue to travel towards the top face. The compressive strain at the top face keeps increasing and the stress distribution in the compressive zone becomes non-linear as shown in Fig. 7.2(e). On further increase in load, the compressive strain in the top face continues to increase till it reaches the maximum value permissible, equal to ε_{cu3}. The maximum value of the permissible strain is smaller for higher strength concrete compared with the 'normal' strength concrete. At this stage, concrete has reached its maximum resistance and the concrete begins to crumble. This represents the ultimate capacity of the section in bending.

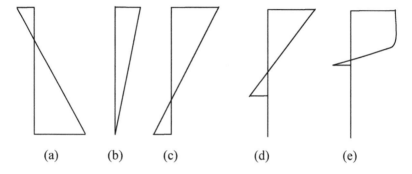

(a)   (b)   (c)   (d)   (e)

Fig. 7.2 Stress distribution in concrete at various stages of loading

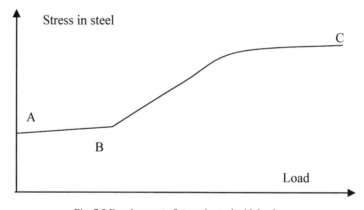

Fig. 7.3 Development of stress in steel with load

As far as the stress in steel is concerned, when jacked the stress in steel will be equal to the jacking stress but on release from the abutments or from the jack, it will reduce by about 10% due to immediate elastic compression of concrete. Over a period of time it continues to decrease due to creep and shrinkage of concrete and relaxation of steel to approximately 75% of the jacking stress. On application of load, until cracks start at the soffit, because of the full bond between steel and concrete, the stress in steel increases very gradually as shown in region AB in Fig. 7.3. However once the concrete cracks, stress increases steadily because of the rapid increase in strain in steel due to bending deformation of the cross section. Depending on the strain at the time when concrete at the top face reaches the maximum strain, steel might or might not reach yield stress as at C in Fig. 7.3. Thus when a prestressed member is subjected to a monotonically increasing bending moment, the stress and strain in the cross section change from the elastic state to a non-linear stage and the changes are quite large.

7.3 STRESS-STRAIN RELATIONSHIP FOR CONCRETE

The stress-strain relationship for concrete was discussed in section 3.4 in Chapter 3. It is usually assumed to be parabolic-rectangular form as shown in Fig. 3.5. However this relationship is somewhat complicated for routine design work. It is therefore simplified, without too much loss of accuracy to the bilinear form shown in Fig. 3.6.

7.4 RECTANGULAR STRESS BLOCK IN BENDING CALCULATIONS

At the ultimate stage, the strain distribution in the cross section is linear with the strain at the compressive face being the maximum value equal to ε_{cu3}. The stress distribution in concrete is rectangular-parabolic or the simplified bilinear form. However, this can be simplified still further by assuming that the stress in the compression region is uniform equal to $\eta\, f_{cd}$ over a depth of λ times the neutral axis depth as shown in Fig. 7.4.

$$\lambda = 0.8, \qquad\qquad\qquad f_{ck} \leq 50 \text{ MPa}$$
$$\lambda = 0.8 - \frac{(f_{ck} - 50)}{400}, \qquad 50 < f_{ck} \leq 90 \text{ MPa} \tag{7.1}$$

$$\eta = 1.0, \qquad\qquad\qquad f_{ck} \leq 50 \text{ MPa}$$
$$\eta = 1.0 - \frac{(f_{ck} - 50)}{200}, \qquad 50 < f_{ck} \leq 90 \text{ MPa} \tag{7.2}$$

$$f_{cd} = \frac{f_{ck}}{\gamma_m}, \ \gamma_m = 1.5 \tag{7.3}$$

$$\varepsilon_{cu3} = 3.5 \times 10^{-3}, \qquad\qquad\qquad \sigma_c \leq 50 \text{MPa}$$
$$= [2.6 + 35 \times \{\frac{90 - f_{ck}}{100}\}^4] \times 10^{-3}, \qquad \sigma_c > 50 \text{MPa} \tag{7.4}$$

Note: If the width of the compression zone decreases in the direction of the extreme compression fibre, the value of the average stress $\eta\, f_{cd}$ in the stress block should be reduced by 10 %

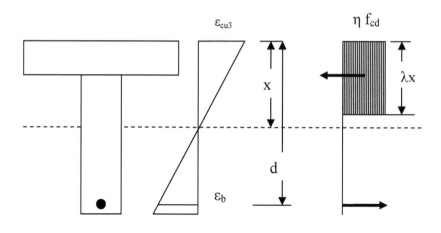

Fig. 7.4 Rectangular stress block

7.5 STRESS-STRAIN RELATIONSHIP FOR STEEL

The stress-strain relationship for steel was discussed in section 3.6 in Chapter 3. The accurate relationship shown in Fig. 3.7 is somewhat too complicated for routine design work. It is therefore simplified, without too much loss of accuracy, to the bilinear form shown in Fig. 3.10.

7.6 STRAIN AND STRESS IN STEEL

Unlike reinforced concrete where the steel is initially unstressed, in prestressed concrete the steel is stressed even before an external load begins to act. Once the load acts on the structure, additional strains are induced in the steel due to bending. Therefore the total strain in steel is the sum of prestrain and strain due to bending.

7.6.1 Prestress and Prestrain in Steel

If the prestress at the section at the serviceability limit state is P_s and A_{ps} is the *area of all stressed steel*, then provided all steel is stressed to the same level, prestress σ_{pe} and prestrain ε_{pe} in steel are

$$\sigma_{pe} = \frac{P_s}{A_{ps}} \, , \, \varepsilon_{pe} = \frac{\sigma_{pe}}{E_s} \tag{7.5}$$

where E_s = Young's modulus for steel. This is normally taken as equal to 195 GPa for a 7-wire strand.

7.6.2 Strain due to Bending in Steel

At the ultimate limit state, as shown in Fig. 7.4, the strain ε_b in steel at a depth d from the compression face due to bending is given by

$$\varepsilon_b = \varepsilon_{cu3} \frac{(d-x)}{x} \tag{7.6}$$

7.6.3 Total Strain and Stress in Steel

The total strain ε_s in steel at distance from the compression face is given by

$$\varepsilon_s = \varepsilon_{pe} + \varepsilon_b = \frac{P_s}{A_{ps} E_s} + \varepsilon_{cu3} \frac{(d-x)}{x} \tag{7.7}$$

From the stress-strain diagram shown in Fig. 3.10, the stress in steel σ_s is related to the total strain ε_s in steel by

$$f_{pd} = f_{p0.1\,k} / \gamma_s \approx 0.85\, f_{pk}/1.15 \approx 0.74\, f_{pk.}, \quad \sigma_s = \varepsilon_s E_s \le f_{pd} \tag{7.8}$$

7.7 STRAIN COMPATIBILITY METHOD

In order to determine the ultimate moment capacity of a given section which has been designed to the serviceability limit state of bending, it is necessary to determine the position of the neutral axis at the ultimate limit state. This is a trial and error procedure. Initially a value for the neutral axis depth x is assumed. For the assumed depth, compressive forces in various parts of concrete and tensile force in all steel reinforcement are calculated. For equilibrium, the algebraic sum of total tensile and compressive forces must be zero. If the sum is not zero, calculations are repeated for a new value of neural axis depth and the calculations are continued till the correct value of x is found.

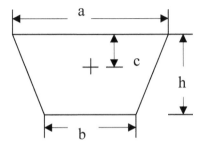

Fig. 7.5 A trapezium

7.8 PROPERTIES OF A TRAPEZIUM

Very often in ultimate moment calculations, certain properties of a trapezium are required. Fig. 7.5 shows a trapezium. The area and the distance c to the centroid from the top face are as follows:

$$\text{Area} = \frac{h}{2}(a+b), \quad c = \frac{h}{3}[1 + \frac{b}{(a+b)}] \tag{7.9}$$

7.9 ULTIMATE MOMENT CALCULATION OF A BRIDGE BEAM

Fig. 7.6 shows a bridge beam.   The details of material properties, prestress details, etc. are follows.

Concrete: Concrete grade 40/50, f_{ck} = 40 MPa, material safety factor, γ_m = 1.50,

$f_{cd} = f_{ck}/ \gamma_m$ = 26.7 MPa

Using a rectangular stress block: λ = 0.8, η = 1.0, $f_{ck} \leq$ 50 MPa, $\eta\, f_{cd}$ = 26.7 MPa

Maximum strain in compression ε_{cu3} = 3.5×10^{-3}, $\sigma_c \leq$ 50 MPa

Steel: 12.7 mm nominal diameter 7-wire strand with an effective cross-sectional area of 112 mm^2.

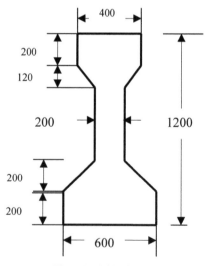

Fig. 7.6 A bridge beam

f_{pk} = 1860 MPa, $f_{p0.1k} \approx$ 0.88 $\times$ f_{pk} = 1637 MPa, γ_m = 1.15

$f_{pd} = f_{p0.1k} /\gamma_m$ = 1424 MPa, E_s = 195 GPa

Prestressing details: 20 strands are located at the following distances from the soffit: six each at 50, 100 and 200 mm and two at 1100 mm from the soffit. The cables are spaced horizontally at 50 mm c/c.

Total prestress at service P_s = 2037 kN

Stress σ_{pe} due to prestress force in the strand = P_s/Area of 20 stressed strands

$$= 2037 \times 10^3 / (20 \times 112) = 909 \text{ MPa}$$

Prestrain $\varepsilon_{pe} = \sigma_{pe} / E_s = 909 / (195 \times 10^3) = 4.66 \times 10^{-3}$

The calculations consist of the following steps.

Step 1: Estimate a value for neutral axis depth:

i. Total tensile force: Assume all 18 strands in the *bottom three rows only* 'yield' i.e. stress is maximum of f_{pd} equal to 1424 MPa. The area of each strand = 112 mm^2.

Therefore total tensile force T = $18 \times 112 \times 1424 \times 10^{-3} = 2871$ kN

ii. Total compressive force: Assuming a rectangular stress block, the uniform compressive stress = $\eta\, f_{cd} = 26.7$ MPa, $\lambda = 0.8$

(a). Total compressive force in top flange = $26.7 \times 200 \times 400 \times 10^{-3} = 2136$ kN

(b). Total compressive force in the top trapezium

$$= 26.7 \times 0.5 \times (200 + 400) \times 120 \times 10^{-3} = 961.2 \text{ kN}$$

Total compressive force C = 2136.0 + 961.2 = 3097.2 kN

C > T. Therefore stress block $\lambda x < (200 + 120)$ or x< 400 mm

Step 2: For an assumed value of neutral axis depth x, calculate the total tensile and compressive forces and check for equilibrium.

Iteration 1: Assume x = 350 mm.

Rectangular stress block depth s = $\lambda x = 0.8 \times 350 = 280$ mm.

The uniform compressive stress = $\eta\, f_{cd} = 26.7$ MPa

ii. Total compressive force:

(a) Compressive force in top flange = 2136.0 kN

(b) Total compressive force in part of the trapezium:

Depth d_x of stress block inside the trapezium = s – depth of top flange

$$= 280 - 200 = 80 \text{ mm}$$

Width B_x of the trapezium at the bottom edge of stress block:

B_x = width of web + (width of top flange – width of web) × $(1-d_x$ /depth of trapezium)

$B_x = 200 + (400 - 200) \times (1 - 80 / 120) = 267$ mm

Compressive force in the trapezium = $26.7 \times 0.5 \times (400 + 267) \times 80 \times 10^{-3}$

$$= 712.4 \text{ kN}$$

Total compressive force C = 2136.0 + 712.4 = 2848.4 kN

iii. Total tensile force:

Bending strain ε_b in strand at d from the compression face = $\varepsilon_{cu3} \times (d - x)/x$

Calculations are shown in Table 7.1.

T = 2965.3 kN

The difference between the total tensile force and compressive force

$$= 2965 - 2848 = 117 \text{ kN}$$

The compressive force is too small, indicating that the neutral axis depth is larger than 350 mm.

Table 7.1 Calculation of total tensile force

Distance from soffit, mm	d mm	ε_b $\times 10^3$	$\varepsilon_s =(\varepsilon_b+ \varepsilon_{pe})$ $\times 10^3$	$\sigma_s,$ MPa	No. of strands	T, kN
50	1150	8.0	12.66	1424	6	957
100	1100	7.5	12.16	1424	6	957
200	1000	6.5	11.16	1424	6	957
1100	100	−2.5	2.16	421	2	94
SUM					20	2965

Note: $\varepsilon_{cu3} = 3.5 \times 10^{-3}$, $\varepsilon_{pe} = 4.66 \times 10^{-3}$

Step 3: Assume x = 450 mm. Stress block depth s = λx = 0.8 × 450 = 360 mm

Iteration 2:

ii. Total compressive force:

(a). Total compressive force in top flange = $26.7 \times 200 \times 400 \times 10^{-3} = 2136$ kN

(b). Total compressive force in the top trapezium:

Depth d_x of stress block inside the trapezium = s – depth of top flange

$$= 360 - 200 = 160 \text{ mm} > 120 \text{ mm}$$

The entire trapezium is in compression.

Compressive force in the trapezium $= 26.7 \times 0.5 \times (400 + 200) \times 120 \times 10^{-3}$

$$= 961.2 \text{ kN}$$

(c). Compressive force in the web:

Depth of web in compression $= s - 200 - 120 = 40$ mm

Compressive force in web $= 26.7 \times 40 \times 200 \times 10^{-3} = 213.6$ kN

Total compressive force $C = 2136.0 + 961.2 + 213.6 = 3310.8$ kN

iii. Total tensile force:

Bending strain ε_b in strand at d from the compression face $= \varepsilon_{cu3} \times (d - x)/x$

Calculations are shown in Table 7.2

$T = 2955.6$ kN

The difference between the total tensile force and compressive force

$$= 2956 - 3311 = -355 \text{ kN}$$

Table 7.2 Calculation of total tensile force

Distance from soffit, mm	d mm	ε_b $\times 10^{-3}$	$\varepsilon_s = (\varepsilon_b + \varepsilon_{pe})$ $\times 10^{-3}$	σ_s, MPa	No. of strands	T, kN
50	1150	5.44	10.10	1424	6	957
100	1100	5.06	9.72	1424	6	957
200	1000	4.28	8.94	1424	6	957
1100	100	−2.72	1.94	378.0	2	85
SUM					20	2956

Note: $\varepsilon_{cu3} = 3.5 \times 10^{-3}$, $\varepsilon_{pe} = 4.66 \times 10^{-3}$

Step 4:

Interpolation: From the two values of x and the corresponding difference between the total tensile and compressive forces, the value of x for which the difference is zero can be determined by interpolation.

x	T−C
350	117
450	−355

$x = 350 + (450 - 350) \times 117/ [117 - (- 355)] = 375$ mm

Step 5: Final calculation:

Assume $x = 375$ mm. Stress block depth $s = \lambda x = 0.8 \times 375 = 300$ mm

ii. Total compressive force:

(a). Total compressive force in top flange $= 26.7 \times 200 \times 400 \times 10^{-3} = 2136$ kN

(b). Total compressive force in the top trapezium:

Depth d_x of stress block inside the trapezium $= s -$ depth of top flange

$$= 300 - 200 = 100 \text{ mm} \quad < 120 \text{ mm}$$

Width B_x of the trapezium at the bottom edge of stress block $=$ width of web $+$

(width of top flange – width of web) $\times$ $(1-d_x/\text{depth of trapezium})$

$= 200 + (400 - 200) \times (1-100/120) = 233$ mm

Compressive force in the trapezium $= 26.7 \times 0.5 \times (400 + 233) \times 100 \times 10^{-3}$

$$= 845.5 \text{ kN}$$

Total compressive force $C = 2136.0 + 845.5 = 2981.5$ kN

iii. Total tensile force:

Calculations are shown in Table 7.3.

$T = 2962$ kN

Table 7.3 Calculation of total tensile force

Distance from soffit, mm	d mm	ε_b $\times 10^3$	ε_s $\times 10^3$	$\sigma_s,$ MPa	No. of strands	T, kN
50	1150	7.23	11.89	1424	6	957
100	1100	6.77	11.43	1424	6	957
200	1000	5.83	10.49	1424	6	957
1100	100	−2.57	2.09	408.0	2	91
sum					20	2962

Note: $\varepsilon_{cu3} = 3.5 \times 10^{-3}$, $\varepsilon_{pe} = 4.66 \times 10^{-3}$

The difference between the total tensile force and compressive force

$$= 2962 - 2982 = -20 \text{ kN}$$

This is small enough to be ignored. In fact the correct value of $x = 372$ mm. The 'error' is because of linear interpolation.

Step 6:

Ultimate moment: The ultimate moment is obtained by calculating the moment of all forces, tensile as well as compressive about the soffit.

(a). Compressive force in top flange = 2136 kN

  Lever arm from the soffit = 1200 – 200/2 = 1100 mm

(b). Compressive force in the top trapezium = 845.5 kN

Depth of stress block inside the trapezium d_x = 100 mm

Width of the trapezium at the bottom edge of stress block B_x = 233 mm

Lever arm = $1200 – 200 – (d_x/3) \times [1 + B_x / (400 + B_x)]$ = 954 mm

iii. Tensile force forces at various level act at their position from the soffit as shown in Table. 7.3.

The ultimate bending moment M_u is given by

$M_u = [2136 \times 1100 + 845.5 \times 954$ -957 $\times$ 50 – 957 $\times$ 100 – 957 $\times$ 200

$\qquad$ – 91.4 $\times$ 1100] $\times$ 10^{-3} = 2721.1 kNm

7.10 ULTIMATE MOMENT CALCULATION OF A COMPOSITE BRIDGE
BEAM

Fig. 7.7 shows a pretensioned Y5 beam which has been made composite with an in-situ reinforced concrete slab. The details of material properties, prestress details, etc. are follows.

Y-5 beam: The beam has an overall depth of 1100 mm. The key at the top is 313 $\times$ 50 mm deep. A 20 mm thick formwork is used for casting the in-situ concrete slab.

Concrete and steel properties are as in section 7.9.

λ = 0.8, η f_{cd} = 26.7 MPa, ε_{cu3} = 3.5 $\times$ 10^{-3}

f_{pd} = 1424 MPa, E_s = 195 GPa

Steel: 15.9 mm nominal diameter 7-wire strand with an effective cross-sectional area of 140 mm^2.

Prestressing details: twenty two 7-wire strands are located at the following distances from the soffit: 10 each at 60 mm and 110 mm and two at 1000 mm from the soffit. The cables are spaced horizontally at 50 mm c/c.

Total prestress at service P_s = 3226 kN

Stress σ_{pe} due to prestress force in the strand = P_s/Area of 22 stressed strands

$$= 3226 \times 10^3 / (22 \times 140) = 1047 \text{ MPa}$$

Prestrain $\varepsilon_{pe} = \sigma_{pe}/E_s = 1047/(195 \times 10^3) = 5.37 \times 10^{-3}$

Reinforced concrete slab:

Concrete: Concrete grade: 30/37, $f_{ck} = 30$ MPa (N/mm^2),

Material safety factor, $\gamma_m = 1.50$, $f_{cd} = f_{ck}/\gamma_m = 20.0$ MPa

Using rectangular stress block: $\lambda = 0.8$, $\eta = 1.0$, $f_{ck} \leq 50$ MPa, $\eta\, f_{cd} = 20.0$ MPa

Maximum strain in compression $\varepsilon_{cu3} = 3.5 \times 10^{-3}$, $\sigma_c \leq 50$ MPa

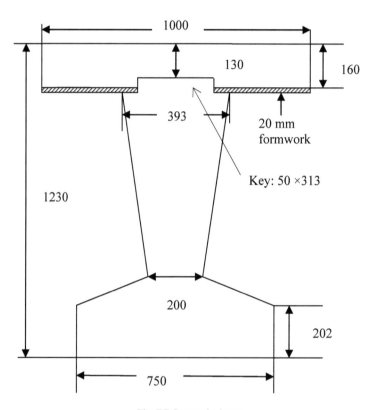

Fig. 7.7 Composite beam

Step 1: Estimate a value for neutral axis depth:

i. Total tensile force: Assume all 20 strands in the bottom three rows 'yield' i.e. stress is maximum of f_{pd} equal to 1424 MPa. The area of each strand = 140 mm^2.

Therefore total tensile force T = 20 ×140 ×1424 × 10^{-3} = 3985 kN

ii. Total compressive force: Assuming rectangular stress block, the uniform compressive stress $= \eta_{fcd} = 26.7$ MPa

(a). Total compressive force in the top 130 mm of slab $= 20.0 \times 1000 \times 130 \times 10^{-3}$

$$= 2600 \text{ kN}$$

(b). Total compressive force in the bottom 30 mm of slab

$$= 20.0 \times (1000 - 313) \times 30 \times 10^{-3} = 412.2 \text{ kN}$$

(c). Total compressive force in the key of the beam

$$= 26.67 \times 313 \times 50 \times 10^{-3} = 417.3 \text{ kN}$$

Total compressive force from (a) + (b) + (c) = 3430 kN

The above is smaller than the tensile force T from (i).   The stress block will extend into the web of the Y-5 beam.

Taking an average width of web $= (393 + 200)/2 =$ say 300 mm, the depth a of the web in compression will be

$a = (3985 - 3430) \times 10^3 / (26.67 \times 300) =$ say 70 mm

Taking $\lambda = 0.8$,

$\lambda x > (160 + 20 + 70)$ or $x > 313$ mm

Table 7.4 Calculation of total tensile force

Distance from soffit, mm	d, mm	ε_b $\times 10^{-3}$	ε_s $\times 10^{-3}$	σ_s, MPa	No. of strands	T, kN
60	1170	9.71	15.08	1424	10	1994
110	1120	9.15	14.52	1424	10	1994
1000	230	−0.90	4.47	872	2	244
SUM					22	4232

Note: $\varepsilon_{cu3} = 3.5 \times 10^{-3}$, $\varepsilon_{pe} = 4.66 \times 10^{-3}$

Step 2: Iteration 1: Assume x = 310 mm

i. Compressive force in web:

Stress block depth $s = \lambda x = 0.8 \times 310 = 248$ mm

Depth d_x of web in compression $= 248 - 160 - 20 = 68$ mm

Width B_x of web at the bottom of the stress block:

$B_x = 200 + (393 - 200) \times (1 - 68/671) = 373$ mm

Compressive force in web $= 26.67 \times (393 + 373)/2 \times 68 \times 10^{-3} = 694.6$ kN

Total compressive force C from (Slab + Key + Web) $= 3430 + 694.6 = 4124.0$ kN

ii. Total tensile force:

Bending strain ε_b in strand at d from the compression face $= \varepsilon_{cu3} \times (d - x)/x$

Depth of composite beam $= 1230$ mm

The tensile force calculation is shown in Table 7.4.

The difference between the total tensile force and compressive force

$$= 4232 - 4124 = 108 \text{ kN}$$

The compressive force is too small, indicating that the neutral axis depth is larger than 310 mm.

T $= 4232$ kN

Step 3: Iteration 2: Assume $x = 340$ mm

i. Compressive force in web:

Stress block depth $s = \lambda x = 0.8 \times 340 = 272$ mm

Depth d_x of web in compression $= 272 - 160 - 20 = 92$ mm

Width B_x of web at the bottom of the stress block:

$B_x = 200 + (393 - 200) \times (1 - 92/671) = 367$ mm

Compressive force in web $= 26.67 \times (393 + 367)/2 \times 92 \times 10^{-3} = 931.8$ kN

Total compressive force C from Slab + Key + Web $= 3430 + 931.8 = 4362.0$ kN

Table 7.5 Calculation of total tensile force

Distance from soffit, mm	d mm	ε_b $\times 10^{-3}$	ε_s $\times 10^{-3}$	σ_s MPa	No. of strands	T kN
60	1170	8.54	13.91	1424	10	1994
110	1120	8.03	13.40	1424	10	1994
1000	230	−1.13	4.23	826	2	231
SUM					22	4218

Note: $\varepsilon_{cu3} = 3.5 \times 10^{-3}$, $\varepsilon_{pe} = 4.66 \times 10^{-3}$

ii. Total tensile force:

Bending strain ε_b in strand at d from the compression face $= \varepsilon_{cu3} \times (d - x)/x$

Depth of composite beam = 1230 mm

Detailed calculations are shown in Table 7.5.

T = 4218 kN

The difference between the total tensile force and compressive force

$$= 4218 - 4362 = -144 \text{ kN}$$

The compressive force is too large, indicating that the neutral axis depth is smaller than 310 mm.

Step 4: Interpolation: From the two values of and the corresponding difference between the total tensile and compressive forces, the value of x for which the difference is zero can be determined by interpolation.

x	T–C
310	108
340	-140

x = 310 + (340 – 310) × 108/ (108 + 140) = 323 mm

Table 7.6 Calculation of total tensile force

Distance from soffit, mm	d mm	ε_b $\times 10^{-3}$	ε_s $\times 10^{-3}$	σ_s MPa	No. of strands	T kN
60	1170	9.18	14.55	1424	10	1994
110	1120	8.64	14.01	1424	10	1994
1000	230	-1.01	4.36	851	2	238
SUM					22	4226

Note: $\varepsilon_{cu3} = 3.5 \times 10^{-3}$, $\varepsilon_{pe} = 4.66 \times 10^{-3}$

Step 5: Final calculation:

Assume x = 323 mm

i. Compressive force in web:

Stress block depth s = λx = 0.8 × 323 = 278 mm

Depth d_x of web in compression = 258 – 160 – 20 = 78 mm

Width B_x of web at the bottom of the stress block

B_x = 200 + (393 – 200) × (1 – 78/671) = 371 mm

Compressive force in web = $26.67 \times (393 + 371)/2 \times 78 \times 10^{-3} = 794.1$ kN

Total compressive force C from Slab + Key + Web = $3430 + 794 = 4224.0$ kN

ii. Total tensile force:

Bending strain ε_b in strand at d from the compression face = $\varepsilon_{cu3} \times (d - x)/x$

Depth of composite beam = 1230 mm

Detailed calculations are shown in Table 7.6.

T = 4226 kN

The difference between the total tensile force and compressive force

$$= 4226 - 4224 = 2.0 \text{ kN}$$

This is small enough to be ignored. In fact the correct value of x = 372 mm. The 'error' is because on linear interpolation.

Step 6: Ultimate moment: The ultimate moment is obtained by calculating the moment of all forces, tensile as well as compressive, about the soffit.

(a). Total compressive force in top 130 mm of slab = 2600 kN

Lever arm from soffit = $1230 - 130/2 = 1165$ mm

(b). Total compressive force in the bottom 30 mm of slab = 412.2 kN

Lever arm from soffit = $1230 - 130 - 30/2 = 1085$ mm

(c). Total compressive force in the key of the beam = 417.3 kN

Lever arm from the soffit = $1100 - 50/2 = 1075$ mm

(d). Compressive force in the web = 794 kN

Depth of stress block in the web $d_x = 78$ mm

Width of the web at the bottom edge of stress block $B_x = 371$ mm

Lever arm = $1100 - 50 - (d_x/3) \times [1 + B_x / (393 + B_x)] = 1011$ mm

iii. Tensile force forces at various level act at their position from the soffit as shown in Table 7.6.

The ultimate bending moment M_u is given by

$M_u = [2600 \times 1165 + 412 \times 1085 + 417 \times 1075 + 794 \times 1011 - 1994 \times 60$

$\qquad - 1994 \times 110 - 238 \times 1100] \times 10^{-3} = 4150$ kNm

7.11   USE OF ADDITIONAL UNSTRESSED STEEL

In the design of partially prestressed concrete structures, all the tension is not resisted by prestressing steel. It is common to use ordinary unstressed steel in addition to prestressed steel. The advantage is that apart from economy, the structures will be more ductile and generally cracks tend to be well distributed. The ultimate moment can be calculated using the strain compatibility method. The only difference is that while the stressed steel has a prestrain, for unstressed steel the prestrain is zero. In addition, the stress-strain relationships for the two steel will also be different.

7.12 STRESS-STRAIN RELATIONSHIP FOR UNSTRESSED REINFORCING STEEL

Eurocode 2 allows for two different bilinear stress-strain relationships for use in design as shown in Fig. 7.8. In the relationship marked as (1) in Fig. 7.8, although the maximum stress is higher than that in (2), there is a limitation on the maximum permitted strain. However, in (2) the maximum permitted stress is lower than that in (1) but there is no limitation on the maximum permitted strain. From the computation point of view, the stress-strain curve (2) is simpler to use than (1).

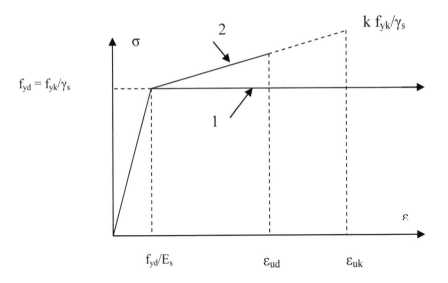

Fig. 7.8 Stress-strain relationship for reinforcing steel

Table 7.7 shows the value of k and ε_{uk} for different classes of reinforcing steel. $f_{0.2k}$ represents the stress at 0.2% strain. ε_{ud} is approximately 0.9 ε_{uk}. The value of Young's modulus E_s is taken as 200 GPa.

Table 7.7 Properties of reinforcing steel

Class	A	B	C
Minimum value of k	$k \geq 1.05$	$k \geq 1.08$	$1.35 > k \geq 1.15$
ε_{uk}	$\geq 2.5 \times 10^{-2}$	$\geq 5.0 \times 10^{-2}$	$\geq 7.5 \times 10^{-2}$
f_{yk} or $f_{0.2k}$	400 – 600 MPa		

7.13 EXAMPLE OF ULTIMATE MOMENT CALCULATION WITH STRESSED AND UNSTRESSED STEELS

Calculate the ultimate moment capacity of the bridge beam considered in section 7.9 when four 25 mm reinforcing bars are added to the bottom row at 50 mm from the soffit. Take $f_{yk} = 460$ MPa.

Reinforcing steel: Area of 25 mm bar = 491 mm², $f_{pd} = 460/1.15 = 400$ MPa

$\varepsilon_y = f_{pd}/E = 400/ (200 \times 10^3) = 2.0 \times 10^{-3}$

The calculations consist of the following steps.

Step 1: Estimate a value for neutral axis depth:

i. Total tensile force: Assume all 18 strands in the bottom three rows and also the four reinforcing bars 'yield'.

Therefore total tensile force T = $(18 \times 112 \times 1424 + 4 \times 491 \times 400) \times 10^{-3} = 3656$ kN

ii. Total compressive force: Assuming a rectangular stress block, the uniform compressive stress = $\eta_{fcd} = 26.7$ MPa

From section 7.9, compressive force in top (flange + trapezium) = 3097.2 kN

Total compressive force is smaller than the tensile force T. Therefore the stress block extends into the web.

Compressive force in web = $26.7 \times 200 \times (\lambda x - 200 - 120) \times 10^{-3}$

$$= 5.34(\lambda x - 320) \text{ kN}$$

Equating tensile and compressive forces, $\lambda x = 425$mm, x > 531 mm

Step 2: For an assumed value of neutral axis depth x, calculate the total tensile and compressive forces and check for equilibrium.

Iteration 1: Assume x = 520 mm.

Rectangular stress block depth s = $\lambda x = 0.8 \times 520 = 416$ mm, $\eta_{fcd} = 26.7$ MPa

ii. Total compressive force:

From section 7.9, compressive force in top (flange + trapezium) = 3097.2 kN

Compressive force in web = $26.7 \times (\lambda x - 200 - 120) \times 200 \times 10^{-3} = 512.6$ kN

Total compressive force C = $3097.2 + 512.6 = 3609.8$ kN

iii. Total tensile force: Detailed calculations are shown in Table 7.7.

Bending strain ε_b in strand at d from the compression face = $\varepsilon_{cu3} \times (d - x)/x$

Table 7.7 Calculation of total tensile force

Distance from soffit, mm	d mm	ε_b $\times 10^{-3}$	ε_s $\times 10^{-3}$	σ_s MPa	No. of strands	T kN
50	1150	4.24	8.90	1424	6	957
100	1100	3.90	8.56	1424	6	957
200	1000	3.23	7.89	1424	6	957
1100	100	-3.43	1.23	358.0	2	80.2
Reinforcing steel (prestrain is zero)						
50	1150	4.24	4.24	400	4	785.4
SUM					20	3736.6

The difference between the total tensile force and the compressive force

$$= 3736.6 - 3609.8 = 126.8 \text{ kN}$$

The compressive force is too small, indicating that the neutral axis depth is larger than 520 mm.

Step 3: Assume x = 570 mm. Stress block depth s = $\lambda x = 0.8 \times 570 = 456$ mm

Iteration 2:

ii. Total compressive force:

From section 7.9, compressive force in top (flange + trapezium) = 3097.2 kN

Compressive force in web = $26.7 \times (456 - 200 - 120) \times 200 \times 10^{-3} = 726.2$ kN

Total compressive force C = $3097.2 + 726.2 = 3823.4$ kN

iii. Total tensile force: Detailed calculations are shown in Table 7.8.

Bending strain ε_b in strand at d from the compression face = $\varepsilon_{cu3} \times (d - x)/x$

The difference between the total tensile force and the compressive force

$$= 3733.7 - 3823.4 = -89.7 \text{ kN}$$

Table 7.8 Calculation of total tensile force

Distance from soffit, mm	d mm	ε_b × 10^{-3}	ε_s × 10^{-3}	σ_s, MPa	No. of strands	T, kN
50	1150	3.56	8.2	1424	6	957
100	1100	3.25	7.91	1424	6	957
200	1000	2.64	7.30	1424	6	957
1100	100	-2.89	1.77	345	2	77.3
Reinforcing steel (prestrain is zero)						
50	1150	3.56	3.56	400	4	785.4
SUM					20	3733.7

Step 4:

Interpolation: From the two values of x and the corresponding difference between the total tensile and compressive forces, the value of x for which the difference is zero can be determined by interpolation.

x	T–C
520	126.8
570	-89.7

x = 520 + (570 – 520) × 126.8/ [126.8– (– 89.7)] = 549 mm

Step 5: Final calculation:

Assume x = 549 mm. Stress block depth s = λx = 0.8 × 549 = 439 mm

ii. Total compressive force:

From section 7.9, compressive force in top (flange + trapezium) = 3097.2 kN

 Compressive force in web= 26.7 × (λx – 200 – 120) × 200 × 10^{-3} = 635.5 kN

Total compressive force C = 3097.2 + 635.5 = 3732.7 kN

iii. Total tensile force: Detailed calculations are shown in Table 7.9.

The difference between the total tensile force and compressive force

$$= 3735.0– 3732.7 = 2.3 \text{ kN}$$

This is small enough to be ignored. In fact the correct value of $x = 550$ mm. The 'error' is because of linear interpolation.

Table 7.9 Calculation of total tensile force

Distance from soffit, mm	d mm	ε_b $\times 10^{-3}$	ε_s $\times 10^{-3}$	σ_s MPa	No. of strands	T kN
50	1150	3.83	8.49	1424	6	957
100	1100	3.51	8.17	1424	6	957
200	1000	2.88	7.54	1424	6	957
1100	100	−2.86	1.80	351	2	78.6
Reinforcing steel (prestrain is zero)						
50	1150	3.83	3.83	400	4	785.4
SUM					20	3735.0

Ultimate moment: The ultimate moment is obtained by calculating the moment of all forces, tensile as well as compressive about the soffit.

(a). Compressive force in top flange = 2136 kN

   Lever arm from the soffit = $1200 - 200/2 = 1100$ mm

(b). Compressive force in the top trapezium = 961.2 kN

Lever arm = $1200 - 200 - (120/3) \times [1 + 200 / (400 + 200)] = 946$ mm

(c). Compressive force in web: = 635.5 kN

Lever arm = $1200 - 200 - 120 - 0.5 \times (439 - 200 - 120) = 821$ mm

iii. Tensile force forces at various level act at their position from the soffit as shown in Table 7.9

The ultimate bending moment M_u is given by

$$M_u = [2136 \times 1100 + 961.2 \times 946 + 635.5 \times 821 - (957 + 785.4) \times 50 - 957 \times 100$$
$$957 \times 200 - 78.6 \times 1100] \times 10^{-3} = 3320.0 \text{ kNm}$$

7.14 CALCULATION OF M_U USING TABULAR VALUES

Fig. 7.9 shows a rectangular beam with prestressing steel of area A_{ps} concentrated at a level d from the compression face. From section 7.4, for the rectangular stress block, if $f_{ck} \leq 50$ MPa, then $\lambda = 0.8$ and $\eta = 1.0$. Taking the material safety factor for concrete as $\gamma_m = 1.5$, the total compressive force is given by

$$C = b \times \lambda x \times \eta \times f_{ck}/\gamma_m = 0.533 \times b \times d \times f_{ck} \times (x/d)$$

The total strain ε_s in steel is

$$\varepsilon_s = \frac{f_{pe}}{E_s} + \varepsilon_{cu3} \times (\frac{d-x}{x})$$

The total stress σ_s in steel is

$$\sigma_s = \varepsilon_s E_s = f_{pe} + \varepsilon_{cu3} \times [\frac{d-x}{x}] \times E_s$$

Taking $\varepsilon_{cu3} = 3.5 \times 10^{-3}$, Young's modulus for steel $E_s = 195$ GPa and maximum stress in steel from section 7.6.4 is $f_{pd} = 0.74\, f_{pk}$,

$$\sigma_s = f_{pe} + 682.5 \times [\frac{d-x}{x}] \leq 0.74\, f_{pk}$$

$$\frac{\sigma_s}{f_{pk}} = \frac{f_{pe}}{f_{pk}} + \frac{682.5}{f_{pk}}[\frac{d}{x} - 1] \leq 0.74$$

Total tensile force T is

$$T = A_{ps}\, \sigma_s = A_{ps}\, f_{pk} \times (\sigma_s/f_{pk})$$

For fixed values of the parameters $\dfrac{f_{pk}\, A_{ps}}{bd\, f_{ck}}$ and $\dfrac{f_{pe}}{f_{pk}}$, the values of x/d is

calculated such that total tensile and compression forces are equal. Table 7.10 shows the results of such calculations where it has been assumed that $f_{ck} \leq 50$ MPa and $f_{pk} = 1860$ MPa.

The ultimate moment capacity M_u is given by

$$M_u = A_{ps}\, f_{pk}\, d[1 - 0.4\frac{x}{d}]\frac{\sigma_s}{f_{pk}}$$

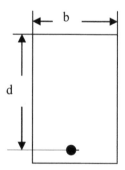

Fig. 7.9 A rectangular beam

Although f_{pk} is taken as 1860 MPa, the tabulated values will be valid for any other value of f_{pk}. Although the calculations have been done for a rectangular beam, the tabular values are valid even for T-beams provided $\lambda x/d \le h_f/d$, where h_f = depth of top flange.

Table 7.10 Values of x/d and σ_s/f_{pk}

$\dfrac{f_{pk}\,A_{ps}}{f_{ck}\,b\,d}$	$\dfrac{\sigma_s}{f_{pk}}$			$\dfrac{x}{d}$		
	$\dfrac{f_{pe}}{f_{pk}}=$ 0.6	$\dfrac{f_{pe}}{f_{pk}}=$ 0.5	$\dfrac{f_{pe}}{f_{pk}}=$ 0.4	$\dfrac{f_{pe}}{f_{pk}}=$ 0.6	$\dfrac{f_{pe}}{f_{pk}}=$ 0.5	$\dfrac{f_{pe}}{f_{pk}}=$ 0.4
0.04	0.74	0.74	0.74	0.06	0.06	0.06
0.08	0.74	0.74	0.74	0.11	0.11	0.11
0.12	0.74	0.74	0.74	0.17	0.17	0.17
0.16	0.74	0.74	0.74	0.22	0.22	0.22
0.20	0.74	0.74	0.74	0.28	0.28	0.28
0.24	0.74	0.74	0.74	0.33	0.33	0.33
0.28	0.74	0.74	0.74	0.39	0.39	0.39
0.32	0.74	0.74	0.74	0.44	0.44	0.44
0.36	0.74	0.74	0.74	0.50	0.50	0.50
0.40	0.74	0.74	0.72	0.56	0.56	0.54
0.44	0.74	0.74	0.68	0.61	0.61	0.56

7.15 CALCULATION OF M_U FOR STATICALLY INDETERMINATE BEAMS

In the case of statically determinate structures, at the ultimate limit state the applied moment at a section is due to factored external loads only. However, in the case of a statically indeterminate structure, the total moment is the sum of the moment due to factored loads and a fraction of the parasitic moment. If at collapse a sufficient number of plastic hinges form to convert it into a statically determinate structure, then the parasitic moments will be reduced to zero. For this to happen, there should be sufficient rotation capacity at positions where plastic hinges can form.

Unfortunately, prestressed concrete structures are not as a rule as ductile as reinforced concrete structures. It is therefore possible that at collapse the structure is not converted into a statically determinate structure and therefore the parasitic moments do not disappear. It is therefore suggested that in checking the required ultimate moment capacity at a section, the total moment is taken as the sum of the factored external load and the unfactored parasitic moment.

7.16 REFERENCES TO EUROCODE 2 CLAUSES

In this chapter, reference has been made to the following clauses in Eurocode 2.

Design assumptions: reinforcing steel: 3.2.7

Design assumptions: prestressing steel: 3.3.6

Rectangular stress distribution: 3.1.7(3)

CHAPTER 8

ANALYSIS OF CRACKED SECTIONS

8.1 INTRODUCTION

In Chapters 4 and 5 the design at the serviceability limit state was carried out for sections where the permissible tensile stress criterion for concrete was not violated either at the transfer stage or at the service stage. However, especially in the case of partially prestressed structures, sections do crack under the serviceability limit state. In this case a cracked section analysis is required not only to ensure safe design in terms of the stress in the steel but also to keep the width and spacing of cracks within acceptable limits.

8.2 CRACKED SECTION ANALYSIS

In the case of prestressed members with bonded tendons, if under frequent load combination, the tensile stress at the soffit exceeds the allowable tensile stress f_{ctm}, then the section should be analysed as a cracked section to determine that the widths of the cracks do not exceed the permitted limits. The analysis of cracked sections is similar to the analysis of sections at ultimate limit state to calculate their moment capacity discussed in Chapter 7. However, the analysis is more complex for two reasons. In the case of cracked section analysis, for *a given applied moment* both the maximum compression strain and the neutral axis depth are unknown. Therefore the analysis has to proceed as follows.

Step 1: Assume a value for compression strain

Step 2: Calculate the depth of neutral axis to balance total tension and compression using the strain compatibility method as explained in section 7.7.

Step 3: Calculate the moment capacity.

Step 4: If the calculated moment capacity is less than the applied moment, increase the compression strain and repeat the calculations in steps 1 to 3.

Step 5: Calculate the maximum crack width for a moment value corresponding to the frequent load combination.

Step 6: If the crack width exceeds the acceptable limit, redesign the beam.

It has to be appreciated that there is no simple procedure to design a beam to satisfy a prescribed maximum crack width criterion at a moment value corresponding to frequent load combination. It is best to simply calculate the moment capacity for a series of assumed compression strains and then to interpolate the required compressive strain for a given value of the moment corresponding to the frequent load combination and then calculating the maximum crack width. The procedure is demonstrated by an example.

8.3 CRACKED SECTION ANALYSIS OF DOUBLE T-BEAM

Example: Fig 4.1 shows the cross section of a double T-beam used as a simply supported prestressed beam of 12 m span to support a live load of 8.0 kN/m². From section 4.2.2, Chapter 4, beam cross-sectional properties and permissible stresses are:

$A = 435.0 \times 10^3$ mm², $y_{bar} = 532$ mm

$I = 2.219 \times 10^{10}$ mm⁴, $Z_t = 101.8 \times 10^6$ mm³, $Z_b = 41.73 \times 10^6$ mm³

Assumed that short-term loss is 10% and long-term loss is 25% of P_{jack}.

γ $S_{uperior} = 1.05$, γ $_{Inferior} = 0.95$, $\eta = 0.83$

$f_{tt} = 2.6$ MPa, $f_{tc} = -15.0$ MPa

$f_{st} = 3.5$ MPa, $f_{sc} = -24.0$ MPa

Solution: In order to have a cracked section under frequent load combination, when drawing the Magnel diagram at mid-span, assume $f_{st} = 4.5$ MPa, which is larger than the permissible tensile stress of 3.5 MPa. Determine the required prestress and eccentricity.

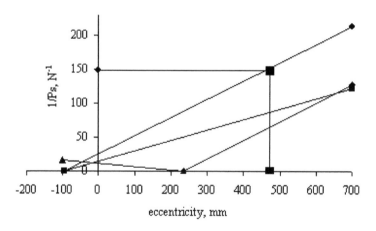

Fig. 8.1 Magnel diagram for cracked section

Magnel Equations:

Transfer: Using equations 4.5b and 4.6b at top and bottom faces respectively from section 4.3.5, Chapter 4,

At top: $(-229.885 + 0.982 \text{ e}) \leq 3.56 (10^{-8} /P_s)$

At bottom: $(-229.885 - 2.396 \text{ e}) \geq -16.05 (10^8/P_s)$

Service: Using equations 4.9a from section 4.3.5, Chapter 4,

At top: $(-229.885 + 0.982 \text{ e}) \geq -19.68 (10^8/P_s)$

Using equations 4.10a from section 4.3.5, Chapter 4, except replacing $f_{st} = 4.5$ instead of 3.5,

At bottom: $(-229.885 - 2.396\ e) \geq -8.95\ (10^8/P_s)$

Fig. 8.1 shows the Magnel diagram for the mid-span section.

Assume a 15.7 mm 7-wire strand with f_{pk} equal to 1860 MPa and a cross-sectional area of 150 mm². $f_{p0.1k} = 0.88\ f_{pk}$. The maximum permitted load P_{max} at jacking is

$P_{max} = 0.9 \times 0.88 \times 1860 \times 150 \times 10^{-3} = 221$ kN

Taking $10^8/P_s = 149$, $P_s = (10^8/149) \times 10^{-3} = 671.1$ kN

Prestressing force at jacking: assuming loss of prestress at service stage as 25% of jacking force, then force at jacking

$P_{jack} = P_s/0.75 = 671.1/0.75 = 894.9$ kN

Maximum number of strands required = 894.9/221 = 4.05, say total of four strands i.e. two strands per web. Assuming that strands can be provided in the web at two strands per layer and the layers are spaced at 50 mm intervals with the first layer at 60 mm from the soffit, then the value of eccentricity is

$e = y_b - 60 = 532 - 60 = 472$ mm

From Magnel diagram, for e = 472 mm, $10^8/P_s = 149$, $P_s = 671.1$ kN is satisfactory.

The stress in the cables = $671.1 \times 10^3/ (4 \times 150) = 1119$ MPa

The stress distributions at transfer and service stages are as follows.

Transfer: $P_t = 805.4$ kN, e = 472 mm. Stresses at top and bottom faces are given by equations (4.1) and (4.2) from Chapter 4:

$\sigma_{Top} = [-1.85 + 3.73] \times 1.05 - 1.9 = \{0.1 \leq f_{tt}\}$

$\sigma_{Bottom} = [-1.85 - 9.10] \times 1.05 + 4.7 = \{-6.8 \geq f_{tc}\}$

Service: $P_s = 671.1$ kN, e = 472 mm. Stress as at top and bottom faces are given by equations (4.7) and (4.8).

$\sigma_{Top} = [-1.49 + 3.11] \times 0.95 - 5.3 = \{-3.8 \geq f_{sc}\}$

$\sigma_{Bottom} = [-1.49 - 7.59] \times 0.95 + 13.0 = \{4.4 \leq f_{st}\}$

Clearly the tensile stress at the soffit is 4.4 MPa which is larger than the actual permitted tensile stress of 3.5 MPa. Therefore under the service load, the section is cracked and requires a cracked section analysis to determine the actual stress distribution.

8.3.1 Stress-Strain Relationship: Concrete

Fig. 3.6 in Chapter 3 shows the bilinear stress-strain diagram for concrete. The elastic Young's modulus E_c is given by

$$E_c = f_{cd}/\varepsilon_{c3}$$
$$f_{cd} = f_{ck}/\gamma_m = 40/1.5 = 26.67 \text{ MPa}$$
$$\varepsilon_{c3} = 1.75 \times 10^{-3}$$
$$E_c = 15.24 \text{ GPa}$$

8.3.2 Stress-Strain Relationship for Steel

Fig. 3.10 in Chapter 3 shows the bilinear stress-strain diagram for steel. The elastic Young's modulus is taken as 195 GPa.

$f_{pk} = 1870$ MPa, $f_{p0.1k} = 0.88 \times 1870 = 1646$ MPa
$\gamma_m = 1.15$, $f_{pd} = 1646/1.15 = 1431$ MPa, $E_s = 195$ GPa, $\varepsilon_{yield} = 7.39 \times 10^{-3}$

8.3.3 Cracked Section Analysis

Using the double T- beam section and material property values, a series of analyses were carried out taking into account the tensile force contribution of uncracked concrete area. Table 8.1 shows the results of such an analysis. In order to keep the scale of plotting reasonable, Figs. 8.2 to 8.4 show only some of the results. The stress distribution at various stages can be summarised as follows.

(i). *Stress distribution due to prestress at service*:

$P_s = 671$ kN, $e = 472$ mm, $\gamma_{\text{Inferior}} = 0.95$

$$[-\frac{P_s}{A} + \frac{P_s e}{Z_t}]\gamma_{\text{Inferior}} = [-1.49 + 3.11] \times 0.95 = 1.5 \text{ MPa}$$

$$[-\frac{P_s}{A} - \frac{P_s e}{Z_b}]\gamma_{\text{Inferior}} = [-1.49 - 7.59] \times 0.95 = -8.6 \text{ MPa}$$

(ii). *Stress distribution due to prestress and moment to cause cracking:*

Assuming that the permissible tensile stress is 3.5 MPa, the moment required to cause a tensile stress at the soffit equal to 3.5 MPa is given by

$$[-\frac{P_s}{A} - \frac{P_s e}{Z_b}]\gamma_{\text{Inferior}} + \frac{M_{\text{crack}}}{Z_b} = [-1.49 - 7.59] \times 0.95 + \frac{M_{\text{crack}}}{Z_b} = 3.5 \text{ MPa}$$

$M_{\text{crack}} = 506.0$ kNm

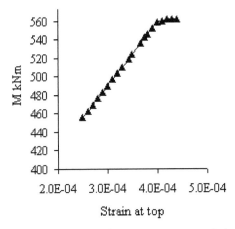

Fig. 8.2 Variation of strain at top vs. moment applied

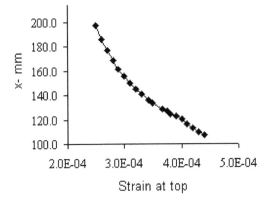

Fig. 8.3 Variation of neutral axis depth vs. strain at top

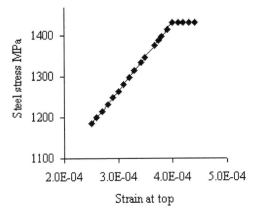

Fig. 8.4 Variation of stress in steel vs. strain at top

Table 8.1 Results of cracked section analysis

ε_c-top	X, mm	teel stress, MPa	M, kNm	Remarks
1.40×10^{-04}	690	1063	369.0	Zero compression at steel level
2.5×10^{-04}	197	1185	455.9	
2.6×10^{-04}	186	1200	462.6	
2.7×10^{-04}	177	1216	469.3	
2.8×10^{-04}	168	1232	476.1	
2.9×10^{-04}	161	1248	482.9	
3.0×10^{-04}	155	1264	489.7	
3.1×10^{-04}	150	1281	496.6	Cracking
3.2×10^{-04}	145	1297	503.5	
3.3×10^{-04}	141	1314	510.4	
3.4×10^{-04}	136	1334	518.9	
3.5×10^{-04}	133	1347	524.3	
3.7×10^{-04}	128	1375	535.8	
3.7×10^{-04}	127	1387	541.8	Service
3.8×10^{-04}	125	1398	545.2	
3.9×10^{-04}	123	1415	552.2	
4.0×10^{-04}	120	1431	558.8	
4.1×10^{-04}	116	1431	559.5	
4.2×10^{-04}	113	1431	561.3	
4.3×10^{-04}	110	1431	561.7	
4.4×10^{-04}	107	1431	561.4	
3.5×10^{-03}	17	1431	586.0	Ultimate

Calculations from accurate strain-compatibility method:

Assume x = 380 mm, strain ε_c at top = $0.235 \times 10^{-3} < \varepsilon_{c3}$. Therefore concrete is in an elastic state.

Compression forces:

Flange: Compression at top = $\varepsilon_c \times E_c$ = 3.6 MPa,

Compression at bottom = 2.6 MPa

$C_{flange} = 0.5 \times (3.58 + 2.64) \times 2400 \times 100 \times 10^{-3} = 746.4$ kN

Lever arm from soffit = $[750 - (100/3) \times \{1 + 2.6/(2.6 + 3.6)\}] \times 10^{-3} = 0.70$ m

$C_{Web} = 0.5 \times 2.6 \times (2 \times 150) \times (x - 100) \times 10^{-3} = 110.9$ kN

Lever arm from soffit = $[750 - 100 - (x - 100)/3] \times 10^{-3} = 0.56$ m

$C_{total} = 746.4 + 110.9 = 857.3$ kN

Tension forces:

$P_s \gamma_{inferior} = 671.1 \times 0.95 = 637.6$ kN, Area A_{ps} of cables $= 4 \times 150 = 600$ mm^2

Young's modulus E_s for steel $= 195$ GPa

Stress σ_{pe} due to prestress $= P_s \gamma_{inferior} / A_{ps} = 1063$ MPa

Strain ε_{pe} due to prestress $= \sigma_{pe} / E_s = 5.45 \times 10^{-3}$

Strain ε_{pb} due to bending $= \varepsilon_c \times (750 - 60 - x)/x = 0.192 \times 10^{-3}$

Total strain ε_{total} in steel $= \varepsilon_{pe} + \varepsilon_{pb} = 5.64 \times 10^{-3}$

Stress σ_s in steel $= \varepsilon_{total} \times E_s = 1100$ MPa

Tensile force $T_{steel} = \sigma_s \times A_{ps} = 659.9$ kN acts at 0.06 m from the soffit

Tensile strain ε_t at the soffit $= \varepsilon_c \times (750 - x)/x = 0.229 \times 10^{-3}$

Tensile stress σ_t at the soffit $= \varepsilon_t \times E_c = 3.5$ MPa; the tensile stress is at the limit of cracking

Tensile force T_{web} in web $= 0.5 \times \sigma_t \times (750 - x) \times (2 \times 300) \times 10^{-3} = 194.25$ kN

Lever arm from the soffit $= [(2/3) \times (750 - x)] \times 10^{-3} = 0.25$ m

$T_{total} = T_{steel} + T_{web} = 659.9 + 194.25 = 854.2$ kN

$T - C = 854.2 - 857.3 = -3.1$ kN ≈ 0 to ensure equilibrium

$M = 746.4 \times 0.70 + 110.9 \times 0.56 - 659.9 \times 0.06 - 194.25 \times 0.25 = 496.4$ kNm

The accurate value of $M_{crack} = 496.4$ is about 2% less than the approximate value of 506.0 kNm.

(iii). *Stress distribution due to prestress and serviceability loads*:

Applied moment $= 541.8$ kNm

$$[-\frac{P_s}{A} + \frac{P_s e}{Z_t}]\gamma_{Inferior} - \frac{M_{service}}{Z_t} = [-1.49 + 3.11] \times 0.95 - 5.32 = -3.8 \text{ MPa}$$

$$[-\frac{P_s}{A} - \frac{P_s e}{Z_b}]\gamma_{Inferior} + \frac{M_{service}}{Z_b} = [-1.49 - 7.59] \times 0.95 + 12.98 = 4.4 \text{ MPa}$$

The beam cracks under this load. Analysis must be carried out as a cracked section.

Calculations from accurate strain-compatibility method:

Assume $x = 127$ mm, strain ε_c at top $= 0.374 \times 10^{-3} < \varepsilon_{c3}$. Therefore concrete is in an elastic state.

Compression forces:

Flange: Compression at top $= \varepsilon_c \times E_c = 5.7$ MPa

Compression at bottom $= 1.2$ MPa

$C_{flange} = 0.5 \times (5.70 + 1.20) \times 2400 \times 100 \times 10^{-3} = 828.0$ kN

Lever arm from soffit $= [750 - (100/3) \times \{1 + 1.20/(1.20 + 5.70)\}] \times 10^{-3} = 0.71$ m

$C_{Web} = 0.5 \times 1.20 \times (2 \times 150) \times (x - 100) \times 10^{-3} = 4.9$ kN

Lever arm from soffit $= [750 - 100 - (x-100)/3] \times 10^{-3} = 0.64$ m

$C_{total} = 828.0 + 4.9 = 832.9$ kN

Tension forces:

$P_s \gamma_{inferior} = 671.1 \times 0.95 = 637.6$ kN, Area A_{ps} of cables $= 4 \times 150 = 600$ mm^2

Young's modulus E_s for steel $= 195$ GPa

Stress σ_{pe} due to prestress $= P_s \gamma_{inferior} / A_{ps} = 1063$ MPa

Strain ε_{pe} due to prestress $= \sigma_{pe} / E_s = 5.45 \times 10^{-3}$

Strain ε_{pb} due to bending $= \varepsilon_c \times (750 - 60 - x)/x = 1.658 \times 10^{-3}$

Total strain ε_{total} in steel $= \varepsilon_{pe} + \varepsilon_{pb} = 7.11 \times 10^{-3}$

Stress σ_s in steel $= \varepsilon_{total} \times E_s = 1386$ MPa.

Tensile force $T_{steel} = \sigma_s \times A_{ps} = 831.6$ kN acts at $= 0.06$ m from the soffit.

Tensile strain ε_t at the soffit $= \varepsilon_c \times (750 - x)/x = 1.835 \times 10^{-3}$

Concrete cracks at this stage. Ignoring any small tensile force contribution from the web,

$T_{total} = T_{steel} + T_{Web} = 831.6 + 0 = 831.6$ kN

$T - C = 831.6 - 832.9 = -1.3$ kN ≈ 0 to ensure equilibrium

$M = 828.0 \times 0.71 + 4.9 \times 0.64 - 831.6 \times 0.06 = 541.1$ kNm

 (iv). **Compressive strain at steel level is zero**: Applied moment $= 369.02$ kNm. The compressive strain and stress at the top fibre are respectively 0.140×10^{-3} and 2.1 MPa. $E_{concrete} = 15.24$ GPa, $x = 690$ mm. The stress in the steel $= 1063$ MPa.

Calculations from accurate strain-compatibility method:

Compression forces:

Flange: Compression at top $= \varepsilon_c \times E_c = 2.14$ MPa

Compression at bottom $= 1.83$ MPa

$C_{flange} = 0.5 \times (2.14 + 1.83) \times 2400 \times 100 \times 10^{-3} = 476.4$ kN

Lever arm from soffit $= [750 - (100/3) \times \{1 + 1.83/(1.83 + 2.14)\}] \times 10^{-3} = 0.70$ m

$C_{Web} = 0.5 \times 1.83 \times (2 \times 150) \times (x - 100) \times 10^{-3} = 162.0$ kN

Lever arm from soffit $= [750 - 100 - (x - 100)/3] \times 10^{-3} = 0.45$ m

$C_{total} = 476.4 + 162.0 = 638.4$ kN

Tension forces:

P_s $\gamma_{inferior}$ = 671.1 × 0.95 = 637.6 kN, Area A_{ps} of cables = 4 × 150 = 600 mm^2

Young's modulus E_s for steel = 195 GPa

Stress σ_{pe} due to prestress = P_s $\gamma_{inferior}$ / A_{ps} = 1063 MPa

Strain ε_{pe} due to prestress = σ_{pe} /E_s = 5.45 × 10^{-3}

Strain ε_{pb} due to bending = 0

Total strain ε_{total} in steel = ε_{pe}

Stress σ_s in steel = ε_{total} × E_s = 1063 MPa

Tensile force T_{steel} = σ_s × A_{ps} = 637.8 kN acts at = 0.06 m from the soffit

T – C = 637.8 – 638.4 = –0.6 kN ≈ 0 to ensure equilibrium

M = 476.4 × 0.70 + 162.0 × 0.45 – 637.8 × 0.06 = 368.1 kNm

Table 8.2 Results from cracked section analysis

ε at top	x mm	σ_s MPa	M kNm	Notes
0.140 × 10^{-3}	690	1063	368.1	Zero compression at steel level
0.235 × 10^{-3}	380	1100	496.4	Concrete cracks
0.374 × 10^{-3}	127	1386	541.1	Serviceability moment
3.500 × 10^{-3}	17	1431	586.0	Ultimate moment

(v). **Ultimate moment capacity**: Applied moment = 586.0 kNm. The compressive strain and stress at the top fibre are respectively 3.5 ×10^{-3} and 26.7 MPa. The neutral axis depth = 17 mm. The stress in the steel = 1431 MPa.

(vi). **Stress differences in steel**: Table 8.2 shows a summary of the results from the last five analyses. This is required in determining the width of cracks as explained in Chapter 10.

(a). The stress difference Δσ in the cables from the state when the compressive strain at steel level was zero and the serviceability moment is given by

Δσ = 1386 – 1063 = 323 MPa

(b). The stress difference Δσ in the cables with the serviceability moment acting and the prestress of 671.0 kN applied by four cables each of 150 mm^2 is given by

Δσ = 1386 – 671.0 × 10^3 / (4 × 150) = 1386 – 1118 = 268 MPa (c). The stress difference $\Delta\sigma$ in the cables from the state when the compressive strain at steel level was zero and the cracking moment is given by

$\Delta\sigma = 1100 - 1063 = 37$ MPa

Note that the ultimate moment capacity of this section is only 586.7 kNm which might mean that when load factors are applied to the serviceability loads, the required ultimate moment might very well be larger than 586.7 kNm.

8.4 PARTIALLY PRESTRESSED BEAM

Partially prestressed beams are where there is a mix of prestressed strands and unstressed reinforcement. As an example to demonstrate the effect of providing unstressed steel, consider the double T-beam in section 8.3 modified by the addition of a 12 mm high yield steel reinforcing bar. For the high yield steel, the following properties are assumed:

Yield stress = 460 MPa, material safety factor $\gamma_m = 1.15$,

Young's modulus = 200 GPa.

The results from cracked section analysis can be summarised as follows.

(i). *Stress distribution due to prestress and moment to cause cracking.*

Assume x = 382 mm. Strain ε_c at top = 0.235×10^{-3}

Compression forces:

Flange: Compression at top = $\varepsilon_c \times E_c = 3.58$ MPa,

Compression at bottom = 2.64 MPa

$C_{flange} = 0.5 \times (3.58 + 2.64) \times 2400 \times 100 \times 10^{-3} = 746.4$ kN

Lever arm from soffit = $[750 - (100/3) \times \{1+ 2.64/ (2.64 + 3.58)\}] \times 10^{-3} = 0.70$ m

$C_{Web} = 0.5 \times 2.64 \times (2 \times 150) \times (x - 100) \times 10^{-3} = 111.7$ kN

Lever arm from soffit = $[750 - 100 - (x-100)/3] \times 10^{-3} = 0.556$ m

$C_{total} = 746.4 + 111.7 = 858.1$ kN

Tension forces:

(a). Prestressing strands:

Strain ε_{pe} due to prestress = 5.45×10^{-3}

Strain ε_{pb} due to bending = $\varepsilon_c \times (750 - 60 - x)/x = 0.19 \times 10^{-3}$

 Total strain ε_{total} in steel = $\varepsilon_{pe} + \varepsilon_{pb} = 5.64 \times 10^{-3}$

stress σ_s in steel = $\varepsilon_{total} \times E_s = 1100$ MPa.

Tensile force $T_{prestressed\ steel} = \sigma_s \times A_{ps} = 659.9$ kN acts at = 0.06 m from the soffit

(b). Unstressed steel:

A_s = Area of two 12 mm bars = 226 mm^2

Strain ε_{pb} due to bending = $\varepsilon_c \times (750 - 60 - x)/x = 0.19 \times 10^{-3}$

Stress σ_s in steel = $\varepsilon_{pb} \times E_s = 38$ MPa

Tensile force $T_{unstressed\ steel} = \sigma_s \times A_s = 8.6$ kN acts at = 0.06 m from the soffit.

(c). Tensile force in web:

Tensile strain ε_t at the soffit = $\varepsilon_c \times (750 - x)/x = 0.226 \times 10^{-3}$

Tensile stress σ_t at the soffit = $\varepsilon_t \times E_c = 3.5$ MPa

Tensile force T_{web} in web = $0.5 \times \sigma_t \times (750 - x) \times (2 \times 150) \times 10^{-3} = 191.1$ kN

Lever arm from the soffit = $[(2/3) \times (750 - x)] \times 10^{-3} = 0.25$ m

$T_{total} = T_{prestressing\ steel} + T_{unstressed\ steel} + T_{Web} = 659.9 + 8.6 + 191.1 = 859.6$ kN

$T - C = 859.6 - 858.1 = 1.5$ kN ≈ 0 to ensure equilibrium.

$M = 746.4 \times 0.70 + 111.7 \times 0.56 - 659.9 \times 0.06 - 8.6 \times 0.06 - 191.1 \times 0.25$

= 497.2 kNm

The accurate value of $M_{crack} = 497.2$ is slightly (about 2%) less than the approximate value of 506.0 kNm. As can be seen, because of the fact the concrete has not cracked, the strain in unstressed steel is very small and it has very little effect on the cracking moment.

(ii). *Stress distribution due to prestress and serviceability loads*:

Applied moment = 541.8 kNm.

$$[-\frac{P_s}{A} + \frac{P_s e}{Z_t}] \gamma_{Inferior} - \frac{M_{service}}{Z_t} = [-1.49 + 3.11] \times 0.95 - 5.32 = -3.78 \text{ MPa}$$

$$[-\frac{P_s}{A} - \frac{P_s e}{Z_b}] \gamma_{Inferior} + \frac{M_{service}}{Z_b} = [-1.49 - 7.59] \times 0.95 + 12.98 = 4.35 \text{ MPa}$$

The beam cracks under this load. Analysis must be carried out as a cracked section. The results from such an analysis are:

Calculations from accurate strain-compatibility method:

Neutral axis depth = 150 mm. Strain ε_c at top = 0.338×10^{-3}

Compression forces:

Flange: Compression at top = $\varepsilon_c \times E_c = 5.16$ MPa

Compression at bottom = 1.70 MPa

$C_{flange} = 0.5 \times (5.16 + 1.70) \times 2400 \times 100 \times 10^{-3} = 823.2$ kN

Lever arm from soffit = $[750 - (100/3) \times \{1 + 1.70/(1.70 + 5.16)\}] \times 10^{-3}$

= 0.708 m

$C_{Web} = 0.5 \times 1.70 \times (2 \times 150) \times (x - 100) \times 10^{-3} = 12.8$ kN

Lever arm from soffit $= [750 - 100 - (x-100)/3] \times 10^{-3} = 0.63$ m

$C_{total} = 823.2 + 12.8 = 836.0$ kN

Tension forces:

(a). Prestressing strands:

Strain ε_{pe} due to prestress $= 5.45 \times 10^{-3}$

Strain ε_{pb} due to bending $= \varepsilon_c \times (750 - 60 - x)/x = 1.217 \times 10^{-3}$

Total strain ε_{total} in steel $= \varepsilon_{pe} + \varepsilon_{pb} = 6.67 \times 10^{-3}$

Stress σ_s in steel $= \varepsilon_{total} \times E_s = 1301$ MPa.

Tensile force $T_{prestressed\ steel} = \sigma_s \times A_{ps} = 780.6$ kN acts at $= 0.06$ m from the soffit.

(b). Unstressed steel:

$A_s =$ Area of two 12 mm bars $= 226$ mm^2

Strain

ε_{pb} due to bending $= \varepsilon_c \times (750 - 60 - x)/x = 1.217 \times 10^{-3}$

Stress σ_s in steel $= \varepsilon_{pb} \times E_s = 245$ MPa.

Tensile force $T_{unstressed\ steel} = \sigma_s \times A_s = 55.0$ kN acts at $= 0.06$ m from the soffit

Tensile strain ε_t at the soffit $= \varepsilon_c \times (750 - x)/x = 1.35 \times 10^{-3}$

Concrete cracks at this stage. Ignoring any small tensile force contribution from the web

$T_{total} = T_{prestressing\ steel} + T_{Unstressed\ steel} = 780.6 + 55.0 = 835.6$ kN

$T - C = 835.6 - 836.0 = -0.4$ kN ≈ 0 to ensure equilibrium.

$M = 823.2 \times 0.708 + 12.8 \times 0.63 - 835.6 \times 0.06 = 540.8$ kNm

(iii). **Compressive strain at steel level is zero**: The results will be as for the beam without unstressed steel because the strain in the unstressed steel is zero.

(iv). **Ultimate moment capacity**: Applied moment $= 586.0$ kNm. The compressive strain and stress at the top fibre are respectively 3.5×10^{-3} and 26.7 MPa. The neutral axis depth $= 18.5$ mm. The stress in the steel strands $= 1431$ MPa, the stress in unstressed steel $= 400$ MPa, Ultimate moment $= 647.0$ kNm. Comparing the ultimate moments for beams with and without unstressed steel, one can see that the ultimate moment has increased from 586.0 kNm to 647.0 kNm, about a 10% increase in capacity due to unstressed steel.

(v). **Stress differences in steel**: Table 8.3 shows a summary of the results from the cracked section analyses.

(a). The stress difference $\Delta\sigma$ in the cables due from the state when the compressive strain at steel level was zero and the serviceability moment is given by

$\Delta\sigma = 1301 - 1063 = 238$ MPa.

(b). The stress difference $\Delta\sigma$ in the cables with the serviceability moment acting and the prestress of 671.0 kN applied by four cables each of 150 mm² is given by

$\Delta\sigma = 1301 - 671.0 \times 10^3 / (4 \times 150) = 1301 - 1118 = 183$ MPa.

(c). The stress difference $\Delta\sigma$ in the cables from the state when the compressive strain at steel level was zero and the cracking moment is given by

$\Delta\sigma = 1100 - 1063 = 37$ MPa

Note that the ultimate moment capacity of this section is only 586.7 kNm which might mean that when load factors are applied to the serviceability loads, the required ultimate moment might very well be larger than 586.7 kNm.

Table 8.3 Results from cracked section analysis

ε at top	x mm	σ_s MPa	M kNm	Notes
0.140×10^{-3}	690	1063	368.1	Zero compression at steel level
0.235×10^{-3}	382	1100	499.8	Concrete cracks
0.338×10^{-3}	150	1301	541.1	Serviceability moment
3.500×10^{-3}	19	1431	647.0	Ultimate moment

8.5 COMPOSITE BEAM

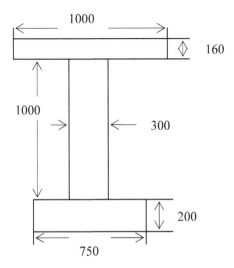

Fig. 8.5 Composite beam

Fig 8.5 shows a composite beam. The precast inverted T-beam 1200 mm deep with a flange 750 × 200 mm and a web thickness of 200 mm is compositely attached to a cast in-situ slab 1000 × 160 mm.

8.5.1 Magnel Diagram for Composite Beam

Draw the Magnel diagram for the composite beam shown in Fig. 8.5. It is used to carry an 18 kN/m live load over a simply supported beam span of 24 m. Use the following data.

a. Precast beam properties:

Area of cross section, $A_{beam} = 450.0 \times 10^3$ mm^2

$y_b = 500$ mm, $y_t = 700$ mm

$I = 6.15 \times 10^{10}$ mm^4

$Z_b = 123.0 \times 10^6$ mm^3, $Z_t = 87.86 \times 10^6$ mm^3

$1/A = 1/ (450 \times 10^3) = 222.22 \times 10^{-8}$, $1/Z_b = 1/ (123.0 \times 10^6) = 0.813 \times 10^{-8}$

$1/Z_t = 1/ (87.86 \times 10^6) = 1.138 \times 10^{-8}$.

The beam is made from C40/50 concrete: $f_{ck} = 40$ MPa, $f_{cm} = 48$ MPa.

Secant modulus from (3.4), $E_c = 35.22$ GPa.

b. Composite beam properties:

Slab = 1000 ×160 mm

Slab is made from C25/30 concrete: $f_{ck} = 25$ MPa, $f_{cm} = 33$ MPa.

Secant modulus from (3.4), $E_c = 31.48$ GPa.

Modular ratio = $E_{c\ Slab}/E_{C\ Beam} = 0.894$

Effective width = 1000 × Modular ratio = 894 mm

$A_{comp} = 593.04 \times 10^3$ mm^2

$y_{b\ comp} = 688$ mm

y_t to top of precast = 1200 – 688 = 512 mm

$I_{Comp} = 12.78 \times 10^{10}$ mm^4

Z_{Comp} to top of precast = 249.69 × 10^6 mm^3, $Z_{b\ comp} = 185.77 \times 10^6$ mm^3

$Z_{t\ Comp} = 190.24 \times 10^6$ mm^3

c. Concrete strength: Beam

The permissible stresses at transfer: $f_{ck} = 25/ 30$ MPa. From equation (3.21)

$f_{tt} = f_{ctm} = 0.30 \times 25^{2/3} = 2.6$ MPa , $f_{tc} = -0.6 \times 25 = -15.0$ MPa

The permissible stresses at service: $f_{ck} = 40/50$ MPa. From (3.21)

$f_{st} = f_{ctm} = 0.30 \times 40^{2/3} = 3.5$ MPa , $f_{sc} = -0.6 \times 40 = -24.0$ MPa

From (3.4), secant modulus = $E_c = 22 \times (0.1 \times \dfrac{f_{ck}}{\gamma_m})^{0.3} = 29.5$ GPa

In order to ensure that the beam is cracked under serviceability limit state loads, assume that $f_{st} = 4.5$ MPa.

d. Load factors for prestress: $\gamma_{Superior} = 1.05$, $\gamma_{Inferior} = 0.95$

e. Loss of prestress: loss of 10% at transfer and 25% at service,

$$\eta = P_s/P_t = 0.75/0.9 = 0.83$$

f. Shrinkage stresses:

Assume $\varepsilon_{Shrinkage} = 213 \times 10^{-6}$ and $E_c = 31.48$ GPa

Stress σ due to restraining free shrinkage = $\varepsilon_{Shrinkage} \times E_c = 6.7$ MPa

Total restraining force F = $\sigma \times$ (Area of slab = 160×10^3) = 1072 .7 kN

This force F acts at the centre of the slab.

The eccentricity ℓ of force with respect to the centroidal axis of the composite beam is given by

ℓ = [(Total depth of beam – Slab thickness /2 = 1360 – 160/2) – $y_{b\ comp}$]

$$= 1280 - 688 = 592 \text{ mm}$$

F $\times \ell$ = 635.04 kNm

Shrinkage stresses in the precast beam and slab are as follows:

Shrinkage stresses in the slab: From equations (4.20) and (4.21),

$$\sigma_{top} = \dfrac{F}{A_{slab}} - [\dfrac{F}{A_{comp.\ beam}} + \dfrac{F\ell}{Z_{top\ of\ comp.\ beam}}] \times \dfrac{E_{slab}}{E_{beam}}$$
$$= 6.70 - [1.80 + 3.34] \times 0.894 = 2.11$$

$$\sigma_{bottom} = \dfrac{F}{A_{slab}} - [\dfrac{F}{A_{comp.\ beam}} + \dfrac{F\ell}{Z_{slab\ bottom\ of\ comp.\ beam}}] \times \dfrac{E_{slab}}{E_{beam}}$$
$$= 6.70 - [1.80 + 2.54] \times 0.894 = 2.82$$

Shrinkage stresses in the precast beam: From equations (4.22) and (4.23),

$$\sigma_{top} = -\dfrac{F}{A_{comp.\ beam}} - \dfrac{F\ell}{Z_{Comp.\ beam\ to\ top\ of\ precast\ beam}}$$
$$= -1.80 - 2.54 = -4.34$$

$$\sigma_{bottom} = -\frac{F}{A_{comp.beam}} + \frac{F\ell}{Z_{bottom\,of\,comp.beam}}$$

$$= -1.80 + 3.42 = 1.62$$

It is worth mentioning that the stresses due to shrinkage form a self-equilibrating system leading to zero resultant axial force and moment.

Magnel diagram for mid-span:

Unit weight of concrete = 25 kN/m³

$q_{self\;weight}$ = Area of beam (= 4.50×10^5) $\times 10^{-6} \times 25 = 11.25$ kN/m

$M_{self\;weight} = 11.25 \times 24^2/8 = 810$ kNm

$\sigma_{top} = 810 \times 10^6/Z_t = 9.22$ MPa, $\sigma_{bottom} = 837 \times 10^6/Z_b = 6.59$ MPa

q_{slab} = Area of slab (= 1.6×10^5) $\times 10^{-6} \times 25 = 4.0$ kN/m

Allow 10% additional load for formwork.

$q_{slab + formwork} = 1.1 \times 4.0 = 4.4$ kN/m

$M_{slab + formwork} = 4.4 \times 24^2/8 = 316.8$ kNm

$\sigma_{top} = 316.8 \times 10^6/Z_t = 3.61$ MPa, $\sigma_{bottom} = 316.8 \times 10^6/Z_b = 2.58$ MPa

$q_{live} = 18.0$ kN/m, $M_{live} = 18.0 \times 24^2/8 = 1296.0$ kN/m

$\sigma_{top} = 1296 \times 10^6/Z_{t\;comp}$ to top of precast = 5.20 MPa,

$\sigma_{bottom} = 1296 \times 10^6/Z_{b\;comp} = 6.98$ MPa

$\sigma_{top\;composite} = 1296 \times 10^6/Z_{t\;comp} = 6.81$ MPa

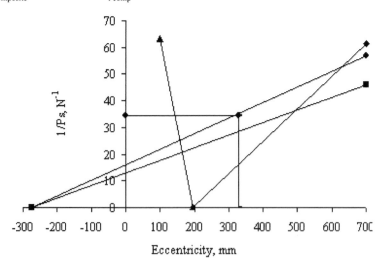

Fig. 8.6 Magnel diagram for mid-span section

Substituting in the formulae (4.5)and (4.6)

At transfer:

$$-222.23 + 1.138\,e \leq [\frac{0.83}{1.05}\langle 9.22 + 2.6 \rangle = 9.34] \frac{10^8}{P_s}$$

$$-222.23 - 0.813\,e \geq [\frac{0.83}{1.05}\langle -6.59 - 15.0 \rangle = -17.07] \frac{10^8}{P_s}$$

Substituting in the formulae (4.18) and (419)

At service:

$$-222.225 + 1.138\,e \geq [\frac{1}{0.95}(9.22 + 3.61 + 5.20 + 4.35 - 24.0) = -1.71] \frac{10^8}{P_s}$$

$$-222.225 - 0.813\,e \geq [\frac{1}{0.95}(-6.59 - 2.58 - 6.98 - 1.61 + 4.5) = -13.96] \frac{10^8}{P_s}$$

Fig. 8.6 shows the Magnel diagram for the mid-span section. From the Magnel diagram we can choose $10^8/P_s = 34.5$, $P_s = 2898.6$ kN.

In this example assume that the maximum breaking load per 12.7 mm 7-wire strand with f_{pk} equal to 1860 MPa and a cross sectional area of 112 mm^2 is 208 kN. From (3.24), the maximum permitted load P_{max} at jacking :

$P_{max} = 0.77 \times 208 = 160.4$ kN.

Prestressing force at jacking: assuming loss of prestress at service stage as 25% of jacking force, then force at jacking

$P_{jack} = P_s/0.75 = 2898.6/0.75 = 3864.7$ kN

Maximum number of strands required = 3864.7/160.4 = 24.1, say a total of 24 strands. The strands can be arranged as follows:

10 of 60mm, 12 of 110mm and 2 of 1100 mm from the soffit. This gives an eccentricity e of

$e = (y_b = 500) - (10 \times 60 + 12 \times 110 + 2 \times 1100)/24 = 328$ mm

The prestress in the strands $\sigma_{pe} = P_s / (24 \times 112) = 1078$ MPa

The prestrain in the strands = $\sigma_{pe}/ E_s = 5.53 \times 10^{-3}$

Stress distribution in the precast beam:
At Transfer: Using equations (4.1) and (4.2) but replacing P_t by P_s/η

$$[-\frac{1}{A} + \frac{e}{Z_t}]P_s\gamma_{sup}\frac{1}{\eta} - \frac{M_{self\,weight}}{Z_t} \leq f_{tt}$$

$$[-8.12 + 13.65 - 9.22 = -3.7] < 2.6$$

$$[-\frac{1}{A}-\frac{e}{Z_b}]P_s\,\gamma_{sup}\,\frac{1}{\eta}+\frac{M_{self\,weight}}{Z_b}\geq f_{tc}$$

$$[-8.12-9.75+6.59=-11.28]>-15.0$$

At service: Using equations (4.18) and (4.19),

$$[-\frac{1}{A}+\frac{e}{Z_t}]P_s\,\gamma_{Inf}-\{\frac{(M_{self\,weight}+M_{slab})}{Z_t}+\frac{M_{live}}{Comp.Z_{top\,of\,precast}}$$

$$+\,Shrinkage\,stress\}\leq f_{sc}$$

$$[-6.12+10.29-9.22-3.61-5.19-4.35=-18.20]>-24.0$$

$$[-\frac{1}{A}-\frac{e}{Z_b}]P_s\,\gamma_{Inf}+\{\frac{(M_{self\,weight}+M_{slab})}{Z_b}+\frac{M_{live}}{Comp.Z_b}$$

$$+\,Shrinkage\,stress\}\leq f_{st}$$

$$[-6.12-7.35+6.59+2.58+6.98+1.61=4.29]>3.5$$

Clearly the beam cracks at serviceability load. Therefore stress calculations done using uncracked section is inaccurate.

Stress distribution in the cast in-situ slab:
The stresses in the slab are due to shrinkage and the live load only.
$\sigma_{top}=2.11-6.81\times0.894=-4.0$ MPa, $\sigma_{bottom}=2.82-5.19\times0.894=-1.8$ MPa
$0.894=$ modular ratio

Cracking moment:
If the tensile stress at the soffit of the precast beam is limited to 3.5 MPa, then

$$[-\frac{1}{A}-\frac{e}{Z_b}]P_s\,\gamma_{Inf}+\{\frac{(M_{self\,weight}+M_{slab})}{Z_b}+\frac{M_{From\,live\,loads}}{Comp.Z_b}$$

$$+\,Shrinkage\,stress\}=f_{st}$$

$$-6.12-7.35+6.59+2.58+\frac{M_{From\,live\,loads}}{Comp.Z_b}+1.61=3.5$$

$M_{From\,live\,loads}=1147.5$ kNm

$M_{crack}=M_{dead}+M_{slab+Formwork}+M_{From\,live\,loads}=810.0+316.8+1147.5$
      $=2274.3$ kNm

Stress distribution when the compressive strain at the bottom steel level is zero:

$$\sigma=[-\frac{1}{A}-\frac{e\times y_1}{I_{beam}}]P_s\,\gamma_{Inf}+\{\frac{\{M_{self\,weight}+M_{slab}\}\times y_1}{I_{beam}}+\frac{M_{From\,live\,loads}\times y_2}{I_{Comp}}$$

$$-\frac{F_{shr}}{A_{comp}}+\frac{M_{Shr}\times y_2}{I_{comp}}\}$$

The stress distribution in layers of steel at 60 mm, 110 mm and 1100 mm from the soffit can be calculated by substituting the values of y_1 and y_2 respectively {(440, 628), (390, 578), (−600, −412)}.

At 60 mm from soffit, for zero stress in concrete

$[-6.12 - 6.47] + \{5.80 + 2.27\} + M_{\text{From live loads}} \times 4.91 \times 10^{-3} + \{-1.80 + 3.12\} = 0$

$M_{\text{From live loads}} = 652$ kNm

$M_{\text{zero compression}} = M_{\text{dead}} + M_{\text{slab + Formwork}} + M_{\text{From Live loads}} = 810.0 + 316.8 + 651.0$
$= 1777.8$ kNm

The stresses in concrete at layer of steel at 60 mm, 110 mm and 1100 mm are

- $(2.4, 1.55, -15.86)$ MPa at cracking moment $= 2274.3$ kN
- $(0, -0.7$ and $-14.3)$ MPa at zero compression moment $= 1777.8$ kNm

The strains in steel at 60 mm, 110 mm and 1100 mm are

- $(1.57, 1.02, -10.41) \times 10^{-4}$ at cracking moment $= 2274.3$ kN
- $(0, -0.46, -9.38) \times 10^{-4}$ at zero compression moment $= 1777.8$ kNm

Adding the prestrain of 5.53×10^{-3} and taking the Young's modulus E_s for steel as $E_s = 195$ GPa, the stress in steel at 60mm, 110 mm and 1100 mm are

- $(1109, 1098, 875)$ MPa at cracking moment $= 2274.3$ kN
- $(1078, 1069, 895)$ MPa at zero compression moment $= 1777.8$ kNm

Once the precast beam starts to crack, stress redistribution will take place between the beam and the slab. Calculations then proceed as follows.

- Calculate the stresses in the slab and precast beam due to self weight, slab load, shrinkage and prestress.
- Calculate the corresponding strains using the appropriate value of Young's modulus.

The results are shown in Table 8.4.

Table 8.4 Stress and strain distribution in the precast beam and slab

Position	Stresses				Strain×10^4
	sw+slab	Prestress	Shrinkage	Total	
Slab, top			2.1	2.1	2.20
Slab, bot.			2.8	2.8	2.94
Web, top	−12.8	4.2	−4.4	−13.0	−8.53
Web, bot.	5.5	−10.5	0.6	−4.4	−2.89
Soffit	9.2	−13.5	1.6	−2.7	−1.77

Tabular values were calculated using the following data:
$M_{\text{sw + slab}} = 810.0 + 316.8 = 1126.8$ kNm,
Prestress $P_s = 2898.6$ kN, $\gamma_{\text{inf}} = 0.95$, e = 328 mm
$I = 6.15 \times 10^{10}$ mm^4, $y_b = 500$ mm
$\gamma_m = 1.5$, $\varepsilon_{c3} = 1.75 \times 10^{-3}$
f_{ck} beam = 40 MPa, f_{cd} beam = 40/ γ_m = 26.67
f_{ck} slab = 25 MPa, f_{cd} slab = 25/ γ_m = 16.67
E_c beam = f_{cd} beam/ε_{c3} = 15.24 GPa
E_c slab = f_{cd} slab/ε_{c3} = 9.53 GPa

- Assume a strain distribution due to the moment applied by the live load moment at the serviceability limit state. This is done by assuming a compressive strain at the top of the slab and a value of the neutral axis depth.

- Calculate the total strain at any section and the corresponding value of stress. Wherever the tensile stress exceeds the permitted value, the stress is reduced to zero.
- Keep varying the strain at the top and the neutral axis depth until the required total moment is obtained.
- Calculate the total value of the compressive force and check if it is equal to the applied prestressing force.
- Calculate the strains and stresses in steel due to bending strains.
- Take the moment of all forces about the point of application of the prestressing force and check if the value of the moment is equal to the total value of the moment due to self weight, slab weight and the SLS live load moment.
- If the calculated values of the normal force and moment match with the applied values, then the assumed values of compressive strain and neutral axis depth are correct. Otherwise calculations are continued using a new value for the strain distribution until calculated and applied values of the force and moment are in agreement.

Calculations at serviceability moment:

SLS moment = $M_{Sw + slab}$ + M_{live} at SLS = $810.0 + 316.8 + 1296 = 2423$ kNm,
As a trial, assume an additional compressive strain at the top of the composite beam of 6.0×10^{-4} and a neutral axis depth of 650 mm. Adding the additional strain to the strains shown in Table 8.4, the strains and stresses in the cross section are as shown in Table 8.5.

Table 8.5 Final stress and strain distribution in the precast beam and slab at SLS moment

Position	Strain×10^4 from Table 8.4	Additional Strain ×10^4	Total Strain×10^4	Stress MPa
Slab, top	2.20	−6.0	−3.8	−3.6
Slab, bot.	2.94	−4.52	−1.58	−1.5
Web, top	−8.53	−4.52	−13.05	−19.9
Web, bot.	−2.89	4.71	1.82	2.8
Soffit	−1.77	6.55	4.78	0*

*Note that for the strain of 4.78×10^{-4}, the stress at the soffit exceeds the permitted tensile stress of 3.5 MPa and is therefore is taken as zero.
Note the discontinuity in stress at the slab-web junction.
The stress value of 3.5 MPa occurs at 31 mm below the bottom flange-web junction. Below this value, the bottom flange is fully cracked.
Compression force in slab = $0.5 \times (3.6 + 1.5) \times 160 \times 1000 \times 10^{-3} = 408$ kN
Position of the resultant from top of slab = $\dfrac{160}{3} \times \dfrac{(2 \times 1.5 + 3.6)}{(1.5 + 3.6)} = 69$ mm

Compression force in web = $0.5 \times (19.9 - 2.8) \times 300 \times 1000 \times 10^{-3} = 2565$ kN
Position of the resultant from top of slab =
$$160 + \dfrac{1000}{3} \times \dfrac{\{2 \times (-2.80) + 19.9\}}{(-2.8 + 19.9)} = 160 + 279 = 439 \ mm$$

Tension force in the bottom flange = $0.5 \times (2.8 + 3.5) \times 31 \times 750 \times 10^{-3}$
$$= 73 \text{ kN}$$
Position of the resultant from top of slab =
$$1160 + \frac{31}{3} \times \frac{\{2 \times 2.8 + 3.5\}}{(2.8 + 3.5)} = 1160 + 15 = 1175 \text{ mm}$$
Table 8.6 shows the strains and stresses in steel due to the bending strains assumed.

Table 8.6 Strains due to bending and forces in steel

Position from soffit	Strain$\times 10^4$	Number	Force, kN
1100	−11.56	2	50.5
110	3.15	12	−82.6
60	3.89	10	−85.0

Tabular values calculated assuming $E_s = 195$ GPa, Area per strand = 112 mm^2

Table 8.7 shows a summary of the calculations.

Table 8.7 Calculated force and moments

Position	Force, kN	Position from top, mm	Position from Prestressing force, mm	Moment, kNm
Slab	408	69	1119	456.6
Web	2565	439	749	1921.2
Bot. flange	−73	1175	13	−0.95
Steel at 1100	50.5	260	928	46.9
Steel at 110	−82.6	1250	−62	5.1
Steel at 69	−85.0	1300	−112	9.5
Sum	2783			2437.8

Note: Position from prestressing force = 1188 − position from top

Applied prestressing force = $\{P_s = 2898.6\} \times (\gamma_{inf} = 0.95) = 2754$ kN
Calculated compressive force /Applied force = 2783/ 2754 = 1.01
Applied moment = $M_{sw + slab} + M_{live}$ at SLS
$$= 810.0 + 316.8 + 1296 = 2423 \text{ kNm,}$$
Calculated moment = 2437.8 kNm
Calculated/Applied moment = 2437.8 / 2423.0 = 1.006
Note that if the prestrain of $5.53 \times 10^{-3} \times (\gamma_{inf} = 0.95) = 5.25 \times 10^{-3}$ is added to the strains due to bending, the final values in the strands at the three levels are
  • strains $(4.10, 5.57, 5.64) \times 10^{-3}$
  • stresses (800, 1086, 1100) MPa
  • forces (179, 1460, 1232) kN
Total force T in steel = 2871 kN
Total compressive force C in concrete = 408 + 2565 − 73 = 2900 kN
C − T = 2900 − 2871 = 29 kN, which is a small error
Therefore the assumed values of additional strain and neutral axis depth are correct.

Ultimate limit state:
Taking $\lambda = 0.8$ and $\eta = 1.0$, $\varepsilon_{cu3} = 3.5 \times 10^{-3}$, $f_{cd\ beam} = 26.67$ MPa,
$f_{cd\ slab} = 16.67$, f_{pd} strands $= 1424$ MPa, $E_s = 195$ GPa,
Area of a strand $= 112$ mm^2.
It can be shown that the neutral axis depth at ULS is 363 mm. Compression force
in slab and beam are respectively 2667 kN and 1040 kN.
 The bending strains in the steel at three levels are $(-0.99, 8.57, 9.05) \times 10^{-3}$.
Adding the prestrain of 5.53×10^{-3}, the final stresses in the steel are (885, 1424,
1424) MPa
The corresponding tensile forces in the steel are (198, 1914, 1595) kN.
The ultimate moment is 4070 kNm.

Summary of results: Table 8.8 shows the results of analysis.
(a). The maximum stress difference $\Delta\sigma$ in the cables from the state when the
compressive strain at steel level was zero and the serviceability moment is given
by

$\Delta\sigma = 1154 - 1109 = 45$ MPa.

(b). The maximum stress difference $\Delta\sigma$ in the cables between the serviceability
moment acting and the prestress of 2898.6 kN applied by 24 cables each of

112 mm^2 is given by

$\Delta\sigma = 1154 - 2898.6 \times 10^3 / (24 \times 112) = 1154 - 1078 = 76$ MPa.

(c). The maximum stress difference $\Delta\sigma$ in the cables due between the state when
the compressive strain at steel level was zero and the cracking moment is given by

$\Delta\sigma = 1178 - 1109 = 69$ MPa

Table 8.8 Results from cracked section analysis

σ_s ,MPa	M, kNm	Notes
(875, 1098, 1109)	1778	Zero compression at steel level
(895, 1069, 1078)	2274	Concrete cracks
(852, 1141, 1154)	2423	Serviceability moment
(885, 1424, 1424)	4070	Ultimate moment

CHAPTER 9

ULTIMATE SHEAR AND TORSIONAL STRENGTH CALCULATIONS

9.1 INTRODUCTION

In general, when a beam is subjected to a loading inducing shear force and bending moment, the behaviour depends on the distribution of bending moment M and shear force V along the span. In the regions where M is large compared with V, vertical flexural cracks appear and the resistance of the beam corresponds to the ultimate bending strength as discussed in Chapter 7. In regions where both M and V are large, an interaction between M and V takes place. Fig. 9.1 shows the typical regions in the case of a simply supported beam under two-point loading. The interaction between moment and shear force is very complex and is not amenable to simple analysis.

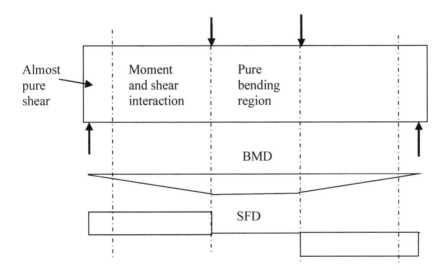

Fig. 9.1 Bean under two point loading

Tests on beams show many different types of shear failure. Some common types of failure are discussed below.

Shear compression: Cracks which start as vertical cracks due to flexural stress, because of the effect of shear stresses, become inclined as they penetrate deeper into the beam. If the compression area above the inclined crack is inadequate to resist the compression resulting from bending moment, concrete in the compression zone crushes. This type of failure is called *shear compression failure*. Fig. 9.2 shows a typical shear compression failure

.

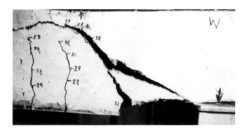

Fig. 9.2 Shear compression failure

Web crushing: In regions where shear force is large compared to bending moment, inclined shear cracks known as *web shear cracks* appear. In beams with thin webs, the compression force in the direction of the cracks lead to the crushing of concrete. Fig. 9.3 shows a typical *web crushing type of failure*.

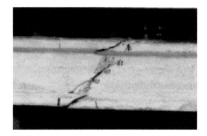

Fig. 9.3 Web crushing failure

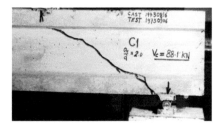

Fig. 9.4 Diagonal cracking failure

Diagonal cracking: If the web is thick enough to resist crushing, the web suffers inclined cracking due to shear stresses and this leads to failure. This type of failure is known as *diagonal cracking failure*. Fig. 9.4 shows a typical diagonal cracking failure.

In design, the calculation of shear capacity is divided into two basic categories.

a. Regions largely uncracked in flexure (meaning high V, small M). This type of failure occurs near the simply supported ends of beams. The criterion for the determination of shear capacity is the limit on the maximum tensile stress permitted in the web. The maximum tensile stress is the principal tensile stress caused by the interaction between the compressive stress caused by prestress (bending stress is ignored as the bending moment is likely to be small) and the shear stress caused by the shear force. The stresses are calculated on the elastic basis.

b. Regions cracked in flexure (meaning moderate V and moderate M). This type of failure takes place at internal supports of continuous beams and also in areas away from support regions in simply supported beams. For regions cracked in flexure, the shear capacity is calculated using an empirical formula developed on the basis of tests. A section is judged to be cracked in flexure if the flexural tensile stress is greater than $f_{ctk, \, 0.05}/ \gamma_c$.

9.2 SHEAR CAPACITY OF A SECTION WITHOUT SHEAR REINFORCEMENT AND UNCRACKED IN FLEXURE

Calculation of shear capacity involves elastic stress calculation. The stresses involved are:

a. Prestress: The stress due to prestress N_{Ed} is calculated using the usual combination of axial and bending stress given by

$$\sigma_{cp} = \frac{N_{Ed}}{A} - \frac{N_{Ed} \times e \times y}{I} \qquad (9.1)$$

y = distance from neutral axis (positive up)

$N_{Ed} = P_s$, prestressing force at service, e = eccentricity of prestressing force.

b. Shear Stress

The shear stress due to shear force V is calculated using the formula

$$\tau = V \times \frac{S}{I \times b_w} \qquad (9.2)$$

S = first moment of area about the neutral axis of the area of cross section above

the level at which τ is being determined

b_w = width of the section at the position where the shear stress is being calculated.

See Fig. 9.5 for the notation. The area for calculating S is shown hatched.

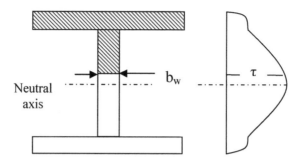

Fig. 9.5 Shear stress distribution in a cross section.

c. Principal tensile stress

From the normal stress σ_{cp} and shear stress τ, the principle tensile stress σ_1 can be calculated:

$$\sigma_1 = -0.5\ \sigma_{cp} + \sqrt{\{(-0.5\sigma_{cp})^2 + \tau^2\}}$$

$$\tau^2 = (\sigma_1 + 0.5\ \sigma_{cp})^2 - (0.5\sigma_{cp})^2 = (\sigma_1^2 + \sigma_1\ \sigma_{cp})$$

$$\tau = \sqrt{(\sigma_1^2 + \sigma_1\ \sigma_{cp})}$$

Equating σ_1 to the permissible tensile stress f_{ctd},

$$\tau = \sqrt{(f_{ctd}^2 + \sigma_{cp}\ f_{ctd})} \tag{9.3}$$

d. Calculation of permissible shear force

If the shear force $V = V_{RD,\ c}$ the shear force that concrete alone can resist at a section without the maximum tensile stress exceeding f_{ctd} is

$$\tau = V_{RD,\ c} \times S/\ (I \times b_w) = \sqrt{(f_{ctd}^2 + \sigma_{cp}\ f_{ctd})}$$

$$V_{RD,c} = \frac{I \times b_w}{S}\sqrt{[f_{ctd}^2 + \sigma_{cp}\ f_{ctd}]} \tag{9.4}$$

Since the full σ_{cp} might not be available at a section, depending on the transmission length, the above equation is modified as follows:

$$V_{RD,c} = \frac{I \times b_w}{S}\sqrt{[f_{ctd}^2 + \alpha_\ell\ \sigma_{cp}\ f_{ctd}]} \tag{9.5}$$

$\alpha_\ell = \ell_x/\ \ell_{pt2} \leq 1.0$ for pretensioned tendons

  = 1.0 for other types of prestressing

ℓ_x = distance of the section from the starting point of transmission length.

$\ell_{pt2} = 1.2\ell_{pt}$ which is the upper bound of the transmission length. ℓ_{pt} is given by equation (2.1).

9.2.1 Example of Shear Capacity of Sections without Shear Reinforcement and Uncracked in Bending

Calculate the shear capacity of the double T-beam considered in Chapter 4 and shown in Fig. 4.1.

The relevant section properties are

$A = 435.0 \times 10^3$ mm^2, $y_b = 532$ mm, $y_t = 218$ mm, $I = 2.219 \times 10^{10}$ mm^4

$Z_t = 101.8 \times 10^6$ mm^3, $Z_b = 41.73 \times 10^6$ mm^3

Concrete grade is C40/50: $f_{ck} = 40$ MPa

From (3.2), $f_{ctm} = 0.30 \times f_{ck}^{0.667} = 3.5$ MPa,

From (3.3), $f_{ctk, \, 0.05} = 0.7 \times f_{ctm} = 2.5$ MPa, $f_{ctd} = 1.6$ MPa

The beam is prestressed with four strands per web. $P_s = 833.3$ kN, $e = 447$ mm.

The beam is simply supported over a span L =12 m.

Characteristic loads are: Self weight = 10.875 kN/m, Live load = 19.2 kN/m

Using load factors for self weight of 1.35 and for live load of 1.5, at ultimate limit state the uniformly distributed load on the beam is

q = 10.875 × 1.35 + 19.2 × 1.5 = 43.48 kNm

Shear force V at support = q × L/2 = 260.9 kN

Bending moment M at mid-span = q× L^2/ 8 = 782.6 kNm

Fig. 9.6 shows the shear force and bending moment diagrams.

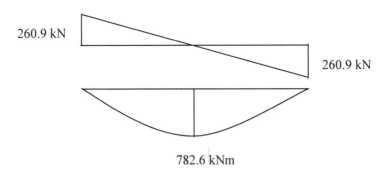

260.9 kN

260.9 kN

782.6 kNm

Fig. 9.6 Shear force and bending moment distribution

At the neutral axis, $\sigma_{cp} = N_{Ed}/A = P_s/A = 833.3 \times 10^3/ (435.0 \times 10^3) = 1.9$ MPa

Assume $\alpha_1 = 0.5$

$$\tau = \sqrt{(f_{ctd}^2 + \alpha_1 \sigma_{cp} f_{ctd})} = \sqrt{(1.6^2 + 0.5 \times 1.9 \times 1.6} = 2.0 \, \text{MPa}$$

S = 100 × 2400 × (750 – 532 – 100/2) for top flange
   + 2× 150× (750 – 532 -100)^2/2 for two webs
   = (40.32 + 2.09) × 10^6 = 42.41× 10^6 mm^3
b_w = width of two webs = 2 × 150 = 300 mm

$$V_{Rd,c} = \frac{I\,b_w}{S}\sqrt{(f_{ctd}^2 + \alpha_1 \sigma_{cp}\, f_{ctd})} = \frac{2.219\times10^{10}\times300}{42.41\times10^6}\times2.0\times10^{-3} = 314.9 \text{ kN}$$

9.3 CHECKING FOR START OF CRACKED SECTION

Example: For the beam considered in section 9.2, calculate the start of the section from where the beam should be considered as cracked in bending.

   If the flexural stress is greater than f_{ctd} = 2.5 MPa, the section should be considered as cracked.

$$1.6 = -\frac{P_s}{A} - \frac{P_s \times e}{Z_b} + \frac{\frac{q}{2}\times(Lx - x^2)}{Z_b}$$

$$1.6 = -\frac{833.3\times10^3}{435.0\times10^3} - \frac{833.3\times10^3 \times447}{41.73\times10^6} + \frac{43.48\times0.5\times(12x - x^2)\times10^6}{41.73\times10^6}$$

$$1.6 = -1.92 - 8.92 + 0.521\times(12x - x^2)$$

$$x^2 - 12x + 23.88 = 0$$

x = 2.5 m and 9.5 m

As shown in Fig.9.8, the beam should be considered as cracked beyond 2.5 m from the supports.

9.4 SHEAR CAPACITY OF SECTIONS WITHOUT SHEAR
REINFORCEMENT AND CRACKED IN BENDING

In the case of sections where the flexural tensile stress is greater than f_{ctd} the shear capacity of the section is given by the following semi-empirical formula.:

$$V_{Rd,c} = [C_{Rd,c}\, k(100\rho_t\, f_{ck})^{\frac{1}{3}} + k_1 \sigma_{cp}]b_w d \geq [v_{min} + k_1 \sigma_{cp}]b_w d \qquad (9.6)$$

γ_c = 1.5, $C_{Rd,c}$ = 0.18/γ_c = 0.18/1.5 = 0.12

$$k = 1 + \sqrt{\frac{200}{d}} \leq 2.0,\, d = \text{effective depth to steel in tensile zone}$$

k_1 = 0.15
σ_{cp} = N_{Ed}/A = P_s/A $\leq 0.2\, f_{cd}$

Average stress in stress block, $f_{cd} = 0.85 \times f_{ck}/ (\gamma_m = 1.5) = 0.57 f_{ck}$

$$v_{min} = 0.035 k^{\frac{3}{2}} f_{ck}^{\frac{1}{2}}$$

A_{sl} =area of tensile steel which extends a distance at least equal to $(\ell_{bd} + d)$ beyond the section considered.

ℓ_{bd} = anchorage length

b_w = width of web

$\rho_t = A_{sl}/ (b_w \times d) \leq 0.02$

9.4.1 Example of Shear Capacity of Sections without Shear Reinforcement and Cracked in Bending

Example: For the beam considered in section 9.2, calculate the shear capacity at a section which is cracked in bending. The tension steel consists in each web of four 7-wire 15.2 mm diameter tendons with an effective cross sectional area of 112 mm². The eccentricity of the steel is 447 mm. The beam is made from concrete C40/50.

$C_{Rd, c} = 0.12$

At the neutral axis, $\sigma_{cp} = N_{Ed}/A = P_s/A = 833.3 \times 10^3/ (435.0 \times 10^3) = 1.9$ MPa

$f_{ck} = 40$ MPa, $f_{cd} = 0.57 f_{ck} = 22.7$ MPa, $0.2 f_{cd} = 4.5$ MPa

$\sigma_{cp} = 1.9 < (0.2 f_{cd} = 4.5)$

$k_1 = 0.15$

$d = h - y_{bar} + e = 750 - 533 + 447 = 664$ mm

$$k = 1 + \sqrt{\frac{200}{d}} = 1 + \sqrt{\frac{200}{664}} = 1.55 \leq 2.0$$

$$v_{min} = 0.035 k^{\frac{3}{2}} f_{ck}^{\frac{1}{2}} = 0.035 \times 1.55^{\frac{3}{2}} \times 40^{\frac{1}{2}} = 0.43 \, \text{MPa}$$

A_{sl} = Eight 7-wire tendons each 112 mm² = 8 × 112 = 896 mm²

b_w = 2 webs × 150 = 300 mm, d = 664 mm

$\rho_t = 896/ (300 \times 664) = 0.0045 < 0.02$

Substituting the values into equation (9.6)

$$V_{Rd,c} = [C_{Rd,c} k(100 \rho_t f_{ck})^{\frac{1}{3}} + k_1 \sigma_{cp}] b_w d \geq [v_{min} + k_1 \sigma_{cp}] b_w d$$

$$V_{Rd,c} = [0.12 \times 1.55 \times (100 \times 0.0045 \times 40)^{\frac{1}{3}} + 0.15 \times 1.9](300) \times 664 \times 10^{-3}$$

$$\geq [0.49 + 0.15 \times 1.9](300) \times 664 \times 10^{-3}$$

$$= (0.488 + 0.285) \times 300 \times 664 \times 10^{-3} \geq (0.42 + 0.285) \times 300 \times 664 \times 10^{-3}$$

$$= 154.0 \geq 140.4$$

$V_{Rd, c} = 154.0$ kN

Fig. 9.7 shows the shear capacities in the cracked and uncracked sections. Clearly the beam has sufficient shear capacity in both uncracked and cracked regions.

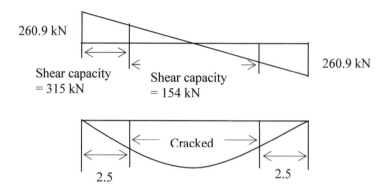

Fig. 9.7 Shear capacities

9.5 DESIGN OF SHEAR REINFORCEMENT

Eurocode 2 assumes that if the design shear force is larger than the capacity of the section without shear reinforcement, then the entire shear force is resisted by shear reinforcement. There is no allowance made for shear force resisted by concrete alone. However, the maximum shear force cannot exceed $V_{Rd, max}$ given below based on the maximum compressive stress in the struts. If the applied shear force is larger than $V_{Rd, max}$ then the section will have to be redesigned.

$$V_{Rd,max} = \sigma_c b_w z \frac{1}{(\cot\theta + \tan\theta)} \tag{9.7}$$

- The maximum stress in the compression struts is σ_c, where

$$\sigma_c = \alpha_{cw} v_1 f_{cd} \tag{9.8}$$

- Parameter α_{cw} allows for any applied compression to the member.

$$\begin{aligned}\alpha_{cw} &= 1 + r, & 0 < r \le 0.25 \\ &= 1.25, & 0.25 < r \le 0.50 \\ &= 2.5(1 - r), & 0.50 < r \le 1.0\end{aligned}$$

$$r = \frac{\sigma_{cp}}{f_{cd}} \tag{9.9}$$

- efficiency factor parameter v_1 allows for the actual distribution of stresses in the strut:

$$v_1 = 0.6[1 - \frac{f_{ck}}{250}] \tag{9.10}$$

- The value of $\cot\theta$ should be within the limits $1 \le \cot\theta \le 2.5$

The shear force resisted by vertical links is given by

$$V_{RD,s} = \frac{A_{sw}}{s} z f_{ywd} \cot\theta \tag{9.11}$$

f_{ywd} = design yield strength of shear reinforcement

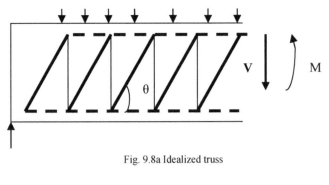

Fig. 9.8a Idealized truss

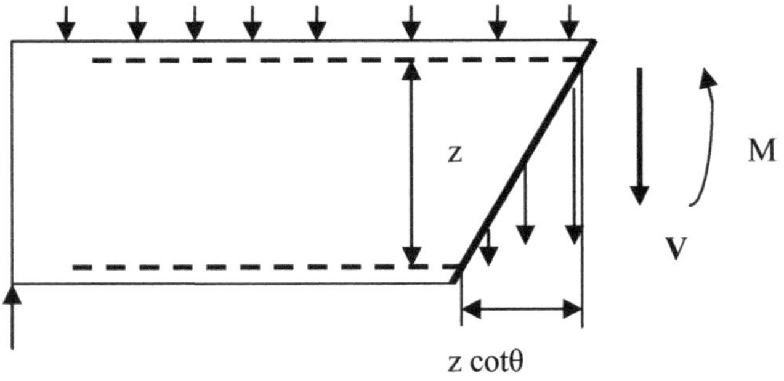

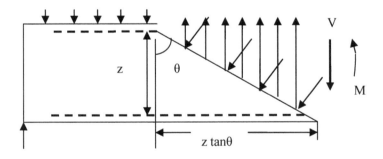

Fig. 9.8b A section parallel to compression struts

Fig. 9.8c A section perpendicular to compression struts

9.5.1 Derivation of Shear Design Equations (9.7) and (9.11)

Fig. 9.8a shows an idealized truss. It consists of tension and compression chords along the resultant compression and tension forces to balance the forces generated due to bending moment. In addition, it has concrete compression struts inclined at an angle θ to the x-axis and shear links as vertical web members to resist shear forces. z is the lever arm of forces in the chords and b_w the width of the web.

Taking a cut parallel to the compression struts as shown in Fig. 9.8b and ignoring any shear force resisted by inter-granular friction, the total shear force V at the section is resisted by tension in the shear reinforcement. The horizontal projection of the cut section is z cot θ. If A_{sw} is the area of one link, s is the spacing of the links and f_{yw} is the stress in the links, the number of links in the distance z cot θ is
z cot θ /s. The shear force V is therefore given by

$$V = A_{sw} \frac{z \cot \theta}{s} f_{yw} = A_{sw} \frac{z}{s} f_{yw} \cot \theta \qquad (9.12)$$

$$A_{sw} \frac{z}{s} f_{yw} = V \tan \theta \qquad (9.13)$$

Taking a cut parallel to the struts as shown in Fig. 9.8c, the length of the cut section is z /cos θ and the horizontal projection of the cut section is z tan θ. The number of links in the cut section is z tan θ /s. If σ_c is the stress in the concrete compression struts, the total compressive force F_c acting normal to the cut section is

$$F_c = \sigma_c b_w \frac{z}{\cos \theta} \qquad (9.14)$$

The total vertical tensile force F_s due to the shear reinforcement at the cut section is

$$F_s = A_{sw} \frac{z \tan \theta}{s} f_{yw} \qquad (9.15)$$

Using (9.13), replacing $A_{sw} \frac{z}{s} f_{yw}$ by $V \tan \theta$ in (9.15)

$$F_s = V \tan^2 \theta \qquad (9.16)$$

Equilibrium of vertical forces gives

$$V = F_c \sin \theta - F_s = F_c \sin \theta - V \tan^2 \theta$$
$$V = \frac{\sin \theta}{(1 + \tan^2 \theta)} F_c \qquad (9.17)$$

Using equation (9.14),

$$V = \frac{\sin \theta}{(1 + \tan^2 \theta)} \sigma_c b_w \frac{z}{\cos \theta} = \sigma_c b_w z \frac{\tan \theta}{(1 + \tan^2 \theta)}$$

$$= \sigma_c b_w z \frac{1}{(\cot \theta + \tan \theta)} \qquad (9.18)$$

The maximum shear force that can be resisted without crushing the compression struts is therefore given by (9.18) which is the same as equation (9.7) in Eurocode 2. Similarly the shear force resisted by links is given by (9.12) which is same as (9.11) in Eurocode 2.

9.5.2 Procedure for Shear Link Design

The design procedure for shear reinforcement calculation is as follows.

i. Calculate the design shear force V. This is calculated at a section an effective depth d from the face of the support. However check that the shear force at the support does not exceed $V_{RD, max}$.

ii. Calculate the shear capacity of concrete alone, $V_{RD, c}$, using (9.5) or (9.6) depending on whether the section is uncracked or cracked in flexure.

iii. If $V_{RD, c} > V$, then theoretically no shear reinforcement is required but a minimum amount will always be provided, given by

$$\frac{A_{sw}}{s \times b_w} \ge 0.08 \frac{\sqrt{f_{ck}}}{f_{yk}} \tag{9.19}$$

iv. If $V_{RD, c} < V$, then shear reinforcement is required and the whole of shear force V is resisted by shear links.

v. Equating the required shear force V to be supported by shear links to the maximum value of the shear force $V_{RD, max}$ that the section can resist given by (9.7), calculate the value of θ and hence cot θ. Check that this value is within the limits $1 \le \cot \theta \le 2.5$. If it is outside the limits, fix a value to lie within the permitted limits.

vi. Assume a value for the total area of a shear link.

vii. Using the value of cot θ calculated in (v), calculate the spacing of the links such that $V \le V_{RD, s}$.

viii. Check that the spacing meets maximum spacing requirements.

9.5.3 Design of a Beam Not Needing Design Shear Reinforcement

Design necessary shear reinforcement for the double T-beam considered in sections 9.3 to 9.5.

The shear force at the support is $43.48 \times 12/2 = 260.9$ kN.

The shear capacity of a section uncracked in flexure = 321.8 kN.

The shear capacity of a section cracked in flexure = 154.0 kN.

The beam is cracked beyond 2.5 m from supports.

In the length uncracked in flexure, the shear capacity is 321.8 kN, which is greater than the maximum applied shear force of 260.9 kN.

At 2.5 m, the applied shear force is

V = 260.9 – 43.48 × 2.5 = 152.2 kN

This shear force is less than 154.0 kN which is the shear capacity of a section cracked in flexure. Therefore in this beam, there is no need for any designed shear links. Only nominal links will have to be provided.

For a two-leg 6 mm diameter links, the area of a shear link, $A_{sw} = 57$ mm^2.

$f_{ck} = 40$ MPa, assume $f_{yk} = 456$ MPa, b_w = width of one web = 150 mm.

Substituting in equation (9.19),

$$[\frac{A_{sw}}{s \times b_w} = \frac{57}{s \times 150}] \geq 0.08[\frac{\sqrt{f_{ck}}}{f_{yk}} = \frac{\sqrt{40}}{456}], \; s \leq 343 \, mm$$

Maximum spacing $s \leq (0.75 \; d = 0.75 \times 664 = 500$ mm
Provide 2-leg 6 mm diameter links in each web at a spacing of 340 mm c/c.

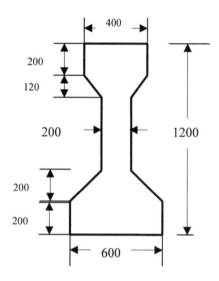

Fig. 9.9 A Bridge beam

9.5.4 Design of a Beam Needing Design Shear Reinforcement

Design shear reinforcement for the bridge I-beam shown in Fig. 9.9. The beam is simply supported over a span of 20 m. The relevant section properties are

$A = 412 \times 10^3$ mm^2, $y_b = 530$ mm, $y_t = 670$ mm, $I = 6.17 \times 10^{10}$ mm^4

$Z_b = 116.6 \times 10^6$ mm^3, $Z_t = 92.1 \times 10^6$ mm^3

At ultimate limit state the uniformly distributed load on the beam is q = 53.0 kNm

The beam is prestressed with twenty 7-wire strands spaced as follows:

6 at 50 mm, 6 at 100 mm, 6 at 200 mm and 2 at 1100 mm from the soffit giving

$P_s = 2037$ kN, e = 315 mm

Ignoring the cable at 100 mm from the soffit, effective depth d:

d $= 1200 - (6 \times 50 + 6 \times 100 + 6 \times 2000) / 18 = 1083$ mm

The beam is made from concrete grade C40/50.

$f_{ck} = 40$ MPa, $f_{ctm} = 0.30 \times f_{ck}^{0.667} = 3.5$ MPa, $f_{ctd} = 1.6$ MPa

Step 1: Calculate the shear capacity of concrete alone at an uncracked section

At the neutral axis, $\sigma_{cp} = N_{Ed}/A = P_s/A = 2037.0 \times 10^3/ (412.0 \times 10^3) = 4.94$ MPa

$f_{ctd} = 1.64$ MPa from section 9.2.1, assume $\alpha_1 = 0.5$

$$\sqrt{(f_{ctd}^2 + \alpha_1 \sigma_{cp} f_{ctd})} = \sqrt{(1.64^2 + 0.5 \times 4.94 \times 1.64)} = 2.60 \text{ MPa}$$

$S = 400 \times 200 \times (1200 - y_{bar} - 200/2)$ for top flange

$\qquad + 0.5 \times (400 - 200) \times 120 \times (1200 - y_{bar} - 200 - 120/3)$ for top triangle

$\qquad + 200 \times (1200 - y_{bar} - 200)^2/2$ for the web

$S = (45.6 + 5.16 + 22.09) \times 10^6 = 72.85 \times 10^6 \text{ mm}^3$

$b_w = 200$ mm

Substituting in (9.5)

$$V_{Rd,c} = \frac{I b_w}{S} \sqrt{(f_{ctd}^2 + \alpha_1 \sigma_{cp} f_{ctd})}$$

$$= \frac{6.17 \times 10^{10} \times 200}{72.85 \times 10^6} \times 2.60 \times 10^{-3} = 440.4 \text{.kN}$$

Step 2: Check the start of the cracked section

If the flexural stress is greater than $f_{ctd} = 1.6$ MPa, the section should be considered as cracked.

$$1.6 = -\frac{P_s}{A} - \frac{P_s \times e}{Z_b} + \frac{q \times 0.5 \times (Lx - x^2)}{Z_b}$$

$$1.6 = -\frac{2037 \times 10^3}{412.0 \times 10^3} - \frac{2037 \times 10^3 \times 315}{116.6 \times 10^6} + \frac{53.0 \times 0.5 \times (20x - x^2) \times 10^6}{116.6 \times 10^6}$$

$$1.6 = -4.94 - 5.50 + 0.227 \times (20x - x^2)$$

$$x^2 - 20x + 53.0 = 0$$

$$x = 3.1 \text{ and } 16.9$$

The beam should be considered as cracked beyond 3.1 m from the supports.

Step 3: Calculate the shear capacity of concrete alone at a cracked section using (9.6)

$C_{Rd,c} = 0.12$

At the neutral axis, $\sigma_{cp} = N_{Ed}/A = P_s/A = 2037.0 \times 10^3/ (412.0 \times 10^3) = 4.94$ MPa

Average stress in stress block, $f_{cd} = 22.7$ MPa

$0.2 f_{cd} = 4.5$ MPa

$\sigma_{cp} = 4.94 > (0.2 f_{cd} = 4.5)$

Take $\sigma_{cp} = 4.5$ MPa in further calculations.

$k_1 = 0.15$

Note: Only the steel areas in the 'tension zone' should be included. The two cables at 1100 mm from the soffit are therefore not included. $d = 1083$ mm

$$k = 1 + \sqrt{\frac{200}{d}} = 1 + \sqrt{\frac{200}{1083}} = 1.43 \leq 2.0$$

$$v_{min} = 0.035 k^{\frac{3}{2}} f_{ck}^{\frac{1}{2}} = 0.035 \times 1.43^{\frac{3}{2}} \times 40^{\frac{1}{2}} = 0.38 \, \text{MPa}$$

A_{sl} = 18 number 7-wire tendons each 112 mm^2 =18 × 112 = 2016 mm^2

b_w = 200 mm, d = 1083 mm

ρ_t = 2016/ (200 × 1083) = 0.009 < 0.02

Substituting the values into (9.6),

$$V_{Rd,c} = [C_{Rd,c}\, k (100 \rho_t\, f_{ck})^{\frac{1}{3}} + k_1 \sigma_{cp}] b_w d \geq (v_{min} + k_1 \sigma_{cp}) b_w d$$

$$V_{Rd,c} = [0.12 \times 1.43 \times (100 \times 0.009 \times 40)^{\frac{1}{3}} + 0.15 \times 4.5](200) \times 1083 \times 10^{-3}$$

$$\geq (0.38 + 0.15 \times 4.5] \times 200 \times 1083 \times 10^{-3}$$

$$= 268.9 \geq 228.5 \, \text{kN}$$

$V_{Rd, c}$ = 268.9 kN

Step 4: Design necessary shear reinforcement

The shear force at the support is 53.0 × 20/2 = 530.0 kN.

The shear capacity of a section cracked in flexure = 268.9 kN

The shear capacity of a section uncracked in flexure = 440.4 kN

The beam is cracked beyond 3.1 m from supports.

The applied shear force is equal to cracked section shear force capacity at

268.9 = 530 − 53x, x = 4.9 m

Shear reinforcement is required for a distance of 4.9 m from the supports. In the rest of the span only nominal steel is required.

Shear V at d from support = 530.0 − q d = 530 − 53 × 1083× 10^{-3} = 472.6 kN

From section 9.4.1, f_{cd} = 22.7 MPa, σ_{cp} = 4.5 MPa, $r = \dfrac{\sigma_{cp}}{f_{cd}} = 0.20 < 0.25$

α_{cw} = 1+r = 1.20, $v_1 = 0.6[1 - \dfrac{f_{ck}}{250}]$ = 0.6 [1- 40/250] = 0.504

$\sigma_c = \alpha_{cw}\, v_1 f_{cd}$ = 1.20 × 0.504 × 22.7 = 13.72 MPa, b_w = 200 mm

z = lever arm

It can be verified from section 7.9 that at the ultimate limit state the compression force acts at 1060 mm from the soffit and the tension force at 148 mm from soffit

z = 1060 − 148 = 911 mm

$\sigma_c \times b_w \times z \times 10^{-3}$ = 13.72 × 200 × 911 × 10^{-3} = 2499.8 kN

Equating V to $V_{RD, max}$

$$472.6 = \sigma_c b_w\, z\, \frac{1}{(\cot\theta + \tan\theta)} = 2499.8 \times \frac{1}{(\cot\theta + \tan\theta)}$$

$$\frac{1}{(\cot\theta + \tan\theta)} = 0.1891$$

One can solve for θ as follows:

$$(\cot\theta + \tan\theta) = \frac{\cos\theta}{\sin\theta} + \frac{\sin\theta}{\cos\theta} = \frac{\cos^2\theta + \sin^2\theta}{\sin\theta\,\cos\theta} = \frac{1}{\sin\theta\,\cos\theta} = \frac{2}{\sin 2\theta}$$

$$\frac{1}{(\cot\theta + \tan\theta)} = 0.5\sin 2\theta = 0.1891$$

$\sin 2\theta = 0.3781, \theta = 11.11°$, $\cot\theta = 5.09 > 2.5$

Restricting $\cot\theta = 2.5$, $\tan\theta = 0.4$,

$V_{Rd,\,max} = 2499.8/(2.5 + 0.4) = 862.0 \text{ kN} > (V = 472.6 \text{ kN})$

Design of shear reinforcement:

The shear force equal to 472.6 kN is resisted by vertical links

Assume: $f_{ywd} = 460/(\gamma_s = 1.15) = 400$ MPa, $b_w = 200$ mm, $z = 911$ mm

Area A_{sw} of one 2-legged 8 mm link in the web = $2 \times 50.27 = 101$ mm^2

Substituting in (9.11),

$$V_{RD,s} = 472.6 \times 10^3 = \frac{101}{s} \times 911 \times 400 \times (\cot\theta = 2.5)$$

$s = 195$ mm

Maximum spacing = $0.75\, d = 0.75 \times 1083 = 812$ mm

Minimum shear links: Using 2- leg 8 mm diameter links, area of a shear link,

Substituting in (9.19),

$$[\frac{A_{sw}}{s \times b_w} = \frac{101}{s \times 200}] \geq 0.08[\frac{\sqrt{f_{ck}}}{f_{yk}} = \frac{\sqrt{40}}{400}], \, s \leq 400\,mm$$

Provide 2-leg 8 mm diameter links in the web at a spacing of 400 mm c/c.

9.6 SHEAR CAPACITY OF A COMPOSITE BEAM

In composite beams dead loads are resisted by the precast beam and live loads by the composite beam. This fact has to be taken into account when doing shear capacity calculations. This will lead to slightly different procedures from what has been explained in the previous sections. In this section, an example is used to illustrate the differences.

Example: Calculate the uncracked shear capacity of a precast pretensioned inverted T-beam made composite with a 1000 mm wide and 160 mm thick cast in-situ reinforced concrete slab. It is used to carry at SLS an 18 kN/m live load over a simply supported beam span of 24 m. Use the following data from section 4.5.3.

Precast beam properties: Bottom flange: 750 mm wide × 300 mm thick,

Web: 300 mm thick, Overall depth: 1100 mm, y_{bar} from soffit: 434 mm

$A_{beam} = 465.0 \times 10^3$ mm^2, $I_{beam} = 4.96 \times 10^{10}$ mm^4, $Z_b = 114.29 \times 10^6$ mm^3.

Beam is made from grade C50/60. $f_{ck} = 50$ MPa, $E_{cm\,beam} = 37.28$ GPa.

Unit weight of concrete = 25 kN/m^3

$q_{self\,weight}$ = Area of beam (= 4.65×10^5) × 10^{-6} × 25 = 11.625 kN/m

In-situ concrete slab: 1000 mm wide × 160 mm thick

Slab is made from grade C25/30. $f_{ck} = 25$ MPa, $E_{cm\ slab} = 31.48$ GPa.

Modular ratio $= E_{cm\ slab}/E_{cm\ beam} = 31.48/37.28 = 0.844$

Effective width of slab $= 1000 ×$ modular ratio $= 844$ mm

Section properties of composite beam: $A_{comp} = 600.04 × 10^3$ mm^2

$y_{bar\ comp} = 602$ mm, $I_{comp} = 10.82 × 10^{10}$ mm^4, $Z_{b\ comp} = 179.73 × 10^6$ mm^3

$q_{slab} =$ Area of slab $(= 1.6 × 10^5) × 10^{-6} × 25 = 4.0$ kN/m

Allow 10% additional load for formwork.

$q_{slab\ +\ Formwork} = 1.1 × 4.0 = 4.4$ kN/m

$q_{live} = 18.0$ kN/m,

At mid-span: $P_s = 3226$ kN, $e = 338$ mm. Prestress is provided by 28 strands each of 112 mm^2 cross sectional area and arranged as 12 at 60, 12 at 110 and

4 at 160 mm from the soffit.

At support: $P_s = 921.7$ kN, $e = 299$ mm. Prestress is provided by 8 strands each of 112 mm^2 cross sectional area and arranged as 4 at 110 and 4 at 160 mm from the soffit.

Using load factors for self weight of 1.35 and for live load of 1.5, at the ultimate limit state, the uniformly distributed loads and shear forces due to dead load and live load at the support are

$q_{dead} = q_{self\ weight} + q_{slab\ +\ formwork} = 1.35 × (11.625 + 4.4) = 21.63$ kN/m,

$q_{live} = 1.5 × 18 = 27.0$ kN/m

At support: $V_{dead} = 21.63 × 24/2 = 259.6$ kN, $V_{live} = 27.0 × 24/2 = 324.0$ kN

(i). Calculation of $V_{Rd,c}$ at sections without shear reinforcement and uncracked in bending:

The value of $V_{Rd,c}$ will be calculated by limiting the principle elastic tensile stress at the neutral axis of the composite section to permissible tensile stress f_{ctd}. Calculate material strengths from equations (3.1) and (3.2) from Chapter 3.

$f_{ctm} = 0.3 × 50^{0.667} = 4.07$ MPa, $f_{ctk,0.05} = 0.7 f_{ctm} = 2.85$, $f_{ctd} = \dfrac{2.85}{1.5} = 1.9$ MPa

Calculate σ_{cp} due to prestress at the supports at the *neutral axis of the composite section*:

$$\sigma_{cp} = \frac{P_s}{A} - \frac{P_s\ e}{I} × (y_{comp} - y_{beam})$$

$$= \frac{921.7 × 10^3}{465 × 10^3} - \frac{921.7 × 10^3 × 299}{4.96 × 10^{10}} × (602 - 434) = 1.98 - 0.93 = 1.05\,\text{MPa}$$

Using (9.3) and assuming $\alpha_1 = 0.5$, calculate the permissible shear stress τ.

$$\tau = \sqrt{(f_{ctd}^2 + \alpha_1\, \sigma_{cp}\, f_{ctd})} = \sqrt{(1.9^2 + 0.5 \times 1.9 \times 1.05)} = 2.15 \text{ MPa}$$

In calculating the shear stress at the neutral axis of the composite section, the contribution to shear stresses from V_{dead} and V_{live} are calculated from the formulae similar to (9.5):

$$\tau_{dead} = V_{Dead} \times \frac{S_{beam}}{I_{beam} \times b_w}, \quad \tau_{live} = V_{Live} \times \frac{S_{Comp\,beam}}{I_{Comp\,beam} \times b_w}$$

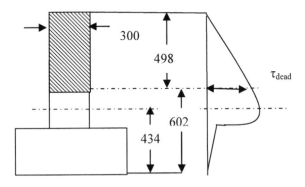

Fig. 9.10 Calculation of S for τdead

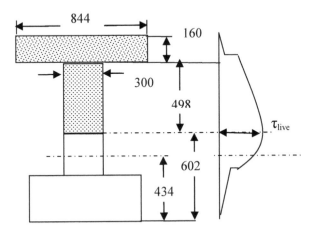

Fig. 9.11 Calculation of S for τ_{live}

The value of S_{beam} is the first moment of area of the precast beam above the neutral axis of the composite beam as shown hatched in Fig. 9.10 taken about the *neutral axis of the precast beam.*

$S_{beam} = 300 \times 498 \times (602 - 434 + 498/2)$ for web $= 62.3 \times 10^6$ mm^3

$$\tau_{dead} = 259.6 \times 10^3 \times \frac{62.3 \times 10^6}{4.96 \times 10^{10} \times 300} = 1.09 \, \text{MPa}$$

The value of $S_{comp. \, beam}$ is the first moment of area of the composite beam above the neutral axis of the composite beam as shown 'dotted' in Fig. 9.11 taken about the *neutral axis of the composite beam.*

$S_{comp. \, beam} = 844 \times 160 \times (498 + 160/2)$ for slab $+ 300 \times 498 \times (498/2)$ for web
$$= 115.3 \times 10^6 \, \text{mm}^3$$

$$\tau_{live} = 388.8 \times 10^3 \times \frac{115.3 \times 10^6}{10.82 \times 10^{10} \times 300} = 1.38 \, \text{MPa}$$

Calculate the portion α of τ_{live} together with τ_{dead} equal to the maximum permissible shear stress τ:

$\tau = 2.15 = 1.09 + \alpha \times 1.38$, $\alpha = 0.77$

$V_{Rd,c} = V_{dead} + \alpha \times V_{live} = 259.6 + 0.77 \times 388.8 = 558.2 \, \text{kN}$

The above is an 'accurate' method for the calculation of $V_{Rd, \, c}$.

An approximate method is to assume that all shear force at the section is resisted by the composite cross section. In this case, $V_{Rd, \, c}$ is given by

$$V_{Rd,c} = \frac{I_{comp} \, b_w}{S_{comp}} \sqrt{(f_{ctd}^2 + \alpha_1 \, \sigma_{cp} \, f_{ctd})} = \frac{10.82 \times 10^{10} \times 300}{115.3 \times 10^6} \times 2.15 \times 10^{-3} = 605.3. \, \text{kN}$$

This value is an 8.4% over-estimation of the actual shear capacity.

The actual applied shear force $V = V_{dead} + V_{live} = 259.6 + 324.0 = 583.6$

Clearly $V > V_{RD, \, c}$, indicating that shear reinforcement is necessary.

As all steel is in the tensile zone, effective depth d:
$d = h_{comp} - y_{bar \, beam} + e = (1100 + 160) - 434 + 299 = 1125$ mm
Shear force at a distance d from the support is

$V = 583.6 - (q_{dead} + q_{live}) \times (d = 1.125) = 528.9 \, \text{kN} < V_{Rd,c} = 558.2 \, \text{kN}$
   Therefore only nominal shear reinforcement is required. The minimum value is given by (9.19)

(ii). Calculation of the start of the cracked section: In the absence of more information about debonding, assuming that the prestress at mid-span will be present at sections liable to crack,

$$f_{ctd} = -\frac{P_s}{A} - \frac{P_s \times e}{Z_b} + [\frac{q_{dead}}{Z_b} + \frac{q_{live}}{Z_{bcomp}}] \times 0.5 \times (Lx - x^2)$$

$$1.9 = -\frac{3226 \times 10^3}{465.0 \times 10^3} - \frac{3226 \times 10^3 \times 338}{114.29 \times 10^6}$$

$$+ \left[\frac{21.63}{114.29 \times 10^6} + \frac{27.0}{164.44 \times 10^6}\right] \times 0.5 \times (24x - x^2) \times 10^6$$

$$1.9 = -6.94 - 9.54 + 0.177 \times (24x - x^2)$$

$$x^2 - 12x + 103.84 = 0$$

x = 5.7 and 18.3 m

The section should be treated as cracked from 5.7 m from the supports.

(iii). Calculation of $V_{Rd,c}$ at sections without shear reinforcement and cracked in bending

From (9.6), $C_{Rd,c} = 0.12$.
Assuming that the prestress values at mid-span are valid,

$$\sigma_{cp} = \frac{P_s}{A} - \frac{P_s\ e}{I} \times (y_{comp} - y_{beam})$$

$$= \frac{3226 \times 10^3}{465 \times 10^3} - \frac{3226 \times 10^3 \times 338}{4.96 \times 10^{10}} \times (602 - 434) = 6.94 - 3.69 = 3.25\ \text{MPa}$$

$f_{ck} = 50$ MPa, $f_{cd} = 0.85 \times 50/1.5 = 28.3$ MPa
$0.2\ f_{cd} = 5.7$ MPa
$\sigma_{cp} = 3.25 < (0.2\ f_{cd} = 5.7)$
$k_1 = 0.15$
As all steel is in the tensile zone,
$d = h_{comp} - y_{bar\ beam} + e = (1100 + 160) - 434 + 338 = 1164$ mm
$A_{sl} = 28$ number 7-wire tendons each 112 mm$^2 = 28 \times 112 = 3136$ mm^2

$$k = 1 + \sqrt{\frac{200}{d}} = 1 + \sqrt{\frac{200}{1164}} = 1.42 \le 2.0$$

$$v_{min} = 0.035\,k^{\frac{3}{2}}\ f_{ck}^{\frac{1}{2}} = 0.035 \times 1.42^{\frac{3}{2}} \times 50^{\frac{1}{2}} = 0.42\ \text{MPa}$$

$b_w = 300$ mm, $d = 1164$ mm
$\rho_t = 3136/(300 \times 1164) = 0.009 < 0.02$
Substituting the values into (9.6),

$$V_{Rd,c} = [C_{Rd,c}\,k\,(100\,\rho_t\,f_{ck})^{\frac{1}{3}} + k_1\,\sigma_{cp}]b_w d \ge (v_{min} + k_1\,\sigma_{cp})b_w d$$

$$V_{Rd,c} = [0.12 \times 1.42 \times (100 \times 0.009 \times 50)^{\frac{1}{3}} + 0.15 \times 3.25](300) \times 1164 \times 10^{-3}$$

$$\geq (0.42 + 0.15 \times 3.25](300) \times 1164 \times 10^{-3}$$

$$= (0.61 + 0.49) \times 300 \times 1164 \times 10^{-3} \geq (0.42 + 0.49) \times 300 \times 1164 \times 10^{-3}$$

$$= 384.1 \geq 317.8$$

$V_{Rd,c}$ = 384.1 kN, this value occurs at 4.1 m from the supports.

Fig. 9.12 shows the shear force distribution and shear capacities without shear reinforcement. The beam requires only nominal reinforcement as given by (9.19).

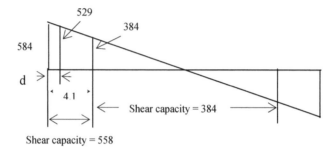

Fig. 9.12 Shear force and shear capacity distribution

9.7 EFFECTIVE WEB WIDTH IN THE PRESENCE OF DUCTS

When calculating shear capacity of a section, the effective width b_w of the web should account for the presence of ducts in post-tensioned beams by using $b_{w.\,nom}$ instead of b_w as follows:

- If the web contains grouted ducts with an outer diameter $\varphi > b_w/8$, $b_{w.\,nom} = b_w - 0.5\Sigma\varphi$
  $\Sigma\varphi$ is determined for the most unfavourable level.
- If the web contains grouted metal ducts with an outer diameter $\varphi \leq b_w/8$, $\quad b_{w.\,nom} = b_w$
- If the web contains grouted plastic ducts, non-grouted ducts or unbonded tendons, $\quad b_{w.\,nom} = b_w - 1.2\,\Sigma\varphi$

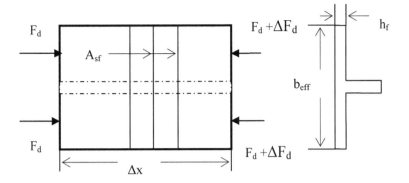

Fig. 9.13 Shear stress at flange-web junction

9.8 INTERFACE SHEAR BETWEEN WEB AND FLANGE IN T-SECTIONS

The shear strength of the web-flange junction is calculated in the usual way by modelling the flange as a truss with concrete struts and steel reinforcement. Using the notation shown in Fig. 9.13, the interface shear stress v_{ED} is given by

$$v_{ED} = \frac{\Delta F_d}{h_f \, \Delta x} \qquad (9.20)$$

where ΔF_d is the change in compressive force in one half of the flange over a distance Δx. $h_f =$ thickness of flange. Euroode 2 permits design of reinforcement to be based on an average shear stress by taking Δx equal to one half of the distance between the section where the moment is a maximum and the section where the moment is zero. In order to prevent the crushing of the struts in the flange, the limit on v_{ED} is given by

$$v_{ED} \leq \{ v f_{cd} \sin \theta_f \cos \theta_f = 0.5 v f_{cd} \sin 2\theta_f \}, \ v = 0.6[1 - \frac{f_{ck}}{250}] \qquad (9.21)$$

For compression flanges: $1.0 \leq \cot \theta_f \leq 2.0$ or $1.0 \leq \sin 2\theta_f \leq 0.80$
For tension flanges: $1.0 \leq \cot \theta_f \leq 1.25$ or $1.0 \leq \sin 2\theta_f \leq 0.98$
The required shear reinforcement A_{sf} at a spacing s_f should satisfy the equation

$$\frac{A_{sf}}{s_f} f_{yd} \geq \frac{v_{ED} \, h_f}{\cot \theta_f} \qquad (9.22)$$

In the case of shear between web and flange combined with transverse flexure, the total area of reinforcement A_{sf} should be greater than the value calculated for shear alone or half that calculated for shear plus that required for transverse bending. If the value of $v_{ED} \leq 0.4 \, f_{ctd}$, then only reinforcement needed for flexure need be provided.

9.8.1 Example of Reinforcement Calculation for Interface Shear between Web and Flange

Fig. 9.14 shows the cross section of a box girder with side cantilevers. It is used to span a two span continuous beam of two equal spans of 40 m. Full details are given in Chapter 16, Section 16.13.

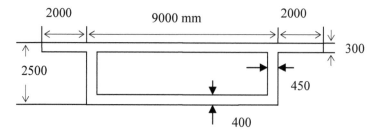

Fig. 9.14 Cross section of the box girder

Table 9.1 gives the details of neutral axis depth, strain at top and stress at top and bottom of the flange in the longitudinal direction obtained from strain compatibility analysis.

Table 9.1 Stresses and strains in the flange

Section	M	x, mm	ε at top	σ at top	σ at bottom
Midspan	52046	606	0.52×10^{-3}	7.9	4.0
Quarter span	47070	830	0.425×10^{-3}	6.5	4.1

Step 1: Compressive forces in the flange at mid-span and quarter spans are:

$$F_d = \frac{(6.5+4.1)}{2} \times 2000 \times 300 \times 10^{-3} = 3180 \, kN$$

$$F_d + \Delta F_d = \frac{(7.9+4.0)}{2} \times 2000 \times 300 \times 10^{-3} = 3570 \, kN$$

$\Delta F_d = 3570 - 3180 = 390$ kN

Step 2: Calculate v_{ED} from (9.20)

$h_f = 300$ mm, $\Delta x = $ Span/4 = 10 m

$$v_{ED} = \frac{\Delta F_d}{h_f \, \Delta x} = \frac{390 \times 10^3}{300 \times 10 \times 10^3} = 0.13 \, MPa$$

Step 3: Check compression struts will not be crushed using (9.21)

Assume $\sin 2\theta_f = 0.9$, $\theta_f = 32°$, $f_{ck} = 40$ MPa, $f_{cd} = 26.7$ MPa

$$v = 0.6[1 - \frac{f_{ck}}{250}] = 0.84 \,, \; 0.5 \, v \, f_{cd} \sin 2\theta_f = 10.1 \text{ MPa} \,, \; v_{ED} = 0.13 < 10.1$$

Step 4: Check if $v_{ED} < k \times f_{ctd}$, $k = 0.4$, $f_{ctd} = 2.3$ MPa

$v_{ED} = 0.13 < (0.4 \times 2.3 = 0.9 \text{ MPa})$

Therefore no additional reinforcement other than that required for cantilever bending need be provided.

Step 5: If required, then calculate transverse reinforcement using (9.22).
Take $f_{yd} = 400$ MPa, $\cot \theta_f = \cot 32° = 1.6$

$$\frac{A_{sf}}{s_f} \geq [\frac{v_{ED} \, h_f}{\cot \theta_f} \frac{1}{f_{yd}} = \frac{0.13 \times 300}{1.6 \times 400}] = 0.061$$

Assuming 12 mm diameter bars, $A_{sf} = 113$ mm^2, $s_f \leq 1853$ mm

9.9 INTERFACE SHEAR BETWEEN PRECAST BEAM AND CAST IN-SITU SLAB

It was stated in section 4.5 that for composite action to be present, it is necessary to prevent slip between the precast beam and the cast in-situ slab. It is therefore necessary to check that at the interface between the cast in-situ slab and the precast beam, the shear resistance is sufficient to resist the applied shear force.

The design shear resistance at the interface $v_{Rd, i}$ is a combination of frictional and cohesional resistance given by

$$v_{Rdi} = c \, f_{ctd} + \mu \, \sigma_n + \rho \, f_{yd} (\mu \sin \alpha + \cos \alpha) \leq 0.5 \, v \, f_{cd} \qquad (9.23)$$

where c and μ are 'cohesion' and friction factors depending on the interface surface characteristics. $c = 0.45$ and $\mu = 0.7$ for surfaces with exposed aggregate.

σ_n = normal stress per unit area (+ve for compression and –ve for tension) caused by the minimum normal force across the interface that can act simultaneously with the shear force.

$$\rho = A_s/A_i \qquad (9.24)$$

A_s = area of reinforcement crossing the joint including ordinary shear reinforcement with adequate anchorage on both sides of the interface.

A_i = area of the joint.

In $\rho \, f_{yd} (\mu \sin \alpha + \cos \alpha)$, the first term is the contribution to frictional resistance and the second term is the along the joint resistance due to tension in the shear links.

f_{yd} = design tensile strength of reinforcement crossing the joint

$$v = 0.6[1 - \frac{f_{ck}}{250}] \,, \; 45° \leq \alpha \leq 90°$$

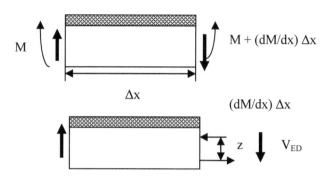

Fig. 9.15 Interface shear stress

The shear stress $v_{Rd\,i}$ caused by loads can be calculated as follows. As shown in Fig. 9.15, consider two sections Δx apart. The differential moment $(dM/dx)\,\Delta x$ can be replaced by tension and compressive forces acting at a lever arm z. The total compressive force C is given by

$$C = \frac{dM}{dx}\frac{1}{z}\Delta x, \qquad \frac{dM}{dx} = V_{Ed}, \qquad C = V_{Ed}\frac{1}{z}\Delta x \qquad (9.25)$$

where V_{ED} is the shear force at the section.

The compressive force acting on the slab only is only a fraction β of the total compressive force C. If the contact width between the slab and the beam is b_i, the applied interfacial shear stress v_{Edi} is given by

$$\beta C = \beta V_{Ed}\frac{1}{z}\Delta x = v_{Edi}\,b_i\,\Delta x, \qquad v_{Edi} = \beta\frac{V_{Ed}}{z\,b_i} \qquad (9.26)$$

Note that the shear force V_{Ed} and the lever arm z refer to the section where the interface shear stress v_{Edi} is being calculated. However, at the support the moment is zero and therefore the lever arm z cannot be calculated. It is thought reasonable to use the lever arm calculated at mid-span for sections other than that at mid-span.

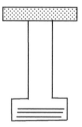

Fig. 9.16 Composite beam cross section

Example: Fig. 9.16 shows a composite beam considered in section 9.6. It is used as a simply supported beam over a 24 m span. It is prestressed by 7-wire tendons of 112 mm² cross sectional area. The tendons are arranged as follows.

12 each at 60 mm and 110 mm from the soffit and four at 160 mm from the soffit. At mid-span the prestressing force $P_s = 3226$ kN and $e = 338$ mm.

Slab: 1000×160 mm, $f_{ck} = 25$ MPa, $\gamma_m = 1.5$.

Beam: Total depth: 1100 mm, Web thickness = 300 mm,

bottom flange = 750×300 mm, $f_{ck} = 50$ MPa, $\gamma_m = 1.5$

(i). Calculate the interfacial shear stress:

At ultimate limit, the uniformly distributed loads on the beam are

$q_{dead} = 21.6$ kN/m, $q_{live} = 27.0$ kN/m

At the ultimate limit state it can be shown that the neutral axis depth is 430 mm.

$0.8 x = 344$ mm. The stress in the slab is $f_{cd} = 25/1.5 = 16.67$ MPa

The stress in the beam is $f_{cd} = 50/1.5 = 33.3$ MPa.

The force in the slab is

$C_{slab} = 16.67 \times 160 \times 1000 \times 10^{-3} = 2667$ kN

C acts at $(1260 - 160/2) = 1180$ mm from the soffit.

The compressive force in the beam is

$C_{beam} = 33.33 \times (0.8 x - 160) \times 300 \times 10^{-3} = 1840$ kN

C_{beam} acts at $(1260 - 160 - (0.8x - 160)/2) = 1008$ mm from the soffit.

$C_{concrete} = 2667 + 1840 = 4507$ kN

The resultant compressive force acts at

$\ell_{concrete} = (2667 \times 1180 + 1840 \times 1008)/ 4507 = 1110$ mm from the soffit

All three layers of steel yield with a stress of 1432 MPa and the forces in the three layers are

$T_{60} = 12 \times 112 \times 1432 \times 10^{-3} = 1925$ kN

$T_{110} = 12 \times 112 \times 1432 \times 10^{-3} = 1925$ kN

$T_{160} = 4 \times 112 \times 1432 \times 10^{-3} = 642$ kN

$T_{steel} = 1925 + 1925 + 642 = 4492$ kN

The resultant tension force acts at

$\ell_{steel} = (1925 \times 60 + 1925 \times 110 + 642 \times 160)/4492 = 96$ mm from the soffit

Lever arm $z = 1110 - 96 = 1014$ mm

$\beta = C_{slab}/C_{concrete} = 2667/4507 = 0.59$

V_{Ed} = shear force at support = $(q_{dead} + q_{live}) \times 24.0/2 = 583.6$ kN

b_i = contact width = 300 mm

$$v_{Edi} = \beta \frac{V_{Ed}}{z\,b_i} = 0.59 \times \frac{583.6 \times 10^3}{1014 \times 300} = 1.13\,\text{MPa}$$

(ii). Calculate the interfacial shear strength:

Slab concrete: $f_{ck} = 25$ MPa, $f_{cd} = 16.7$ MPa, from (3.3), $f_{ctd} = 1.2$ MPa

$v = 0.6 \times (1 - 25/250) = 0.54$

For shear links, $f_{yk} = 500$ MPa, $\gamma_m = 1.15$, $f_{yd} = 435$ MPa

$\sigma_n = (q_{dead} + q_{live}) \times 10^3/\{1000 \times (\text{contact width} = \text{web width} = 300\text{ mm})\}$

   $= 0.16$ MPa

$v_{Rdi} = 0.45 \times 1.2 + 0.6 \times 0.16 + \rho \times 435 \times (0.6 \times 1 + 0)$

(iii). Design reinforcement:

$v_{Rdi} = 0.45 \times 1.2 + 0.6 \times 0.16 + \rho \times 435 \times (0.6 \times 1 + 0) \geq 1.13$

$\rho \geq 0.19\%$

Two-legged 12 mm diameter links giving $A_{sw} = 226$ mm^2 at a spacing not exceeding approximately 0.75 d, with d = 1125 mm give the maximum spacing $s_{max} = 844$ mm

b_i = contact width= 300 mm.

$\rho = A_{sw}/(s \times b_i) = 226/(s \times 300) \geq 0.19\%$

$s \leq 398$ mm, say s = 400 mm.

Provide two-legged 12 mm diameter shear links at 400 mm spacing.

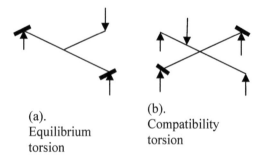

(a).
Equilibrium
torsion

(b).
Compatibility
torsion

Fig. 9.17 Equilibrium and compatibility torsion

9.10 DESIGN FOR TORSION

In practice there are two types of twisting moment, called equilibrium torsion and compatibility torsion. Fig. 9.17 shows simple examples of how these arise. In

Fig. 9.17a, the twisting moment induced in the longitudinal beam is required to maintain equilibrium. On the other hand, the twisting moment induced in the transverse beam in Fig. 9.17b arises purely in order to maintain compatibility of rotations between the longitudinal and transverse beams. In the case of compatibility torsion only minimum reinforcement need be provided. On the other hand, in the case of equilibrium torsion, properly designed reinforcement need to be provided.

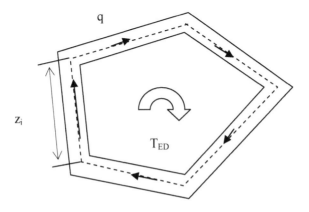

Fig. 9.18 Hollow tube subjected to twisting moment

Fig. 9.18 shows a thin-walled hollow tube subjected to a torsional moment T_{ED}. The shear flow q, which is the product of shear stress in the wall and its thickness, is constant in the walls and is given by

$$q = \tau_{t,i} \times t_{ef,i} = \frac{T_{ED}}{2A_k} \tag{9.27}$$

where $\tau_{t,i}$ is the torsional shear stress and $t_{ef,i}$ is the effective thickness of the ith wall.

The shear force $V_{ED,i}$ in the i_{th} wall is given by $V_{ED,i} = q\, z_i$

where z_i is the side length of the ith wall, equal to the distance between the intersection points with the adjacent walls.

$t_{ef,i}$ is the effective wall thickness. For hollow sections, it is equal to the actual thickness. However, it can be taken as the ratio A/u, where A is the total area of cross section within the outer circumference including any hollow areas and u is the outer circumference of the cross section. It should not be taken as less than the twice the distance between the edge and the centre line of the longitudinal reinforcement.

The shear force $V_{ED, i}$ in the walls causes inclined shear cracks as shown in Fig. 9.19 and is resisted by longitudinal reinforcement and links as shown in Fig. 9.20.

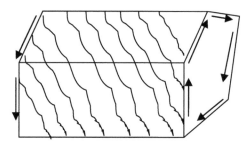

Fig. 9.19 Inclined shear cracks in the walls

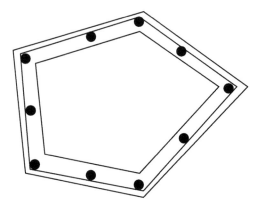

Fig. 9.20 Reinforcement in a hollow tube subjected to a twisting moment

The total longitudinal reinforcement $\Sigma A_{s\ell}$ is given by

$$\Sigma A_{s\ell} = \frac{T_{ED}}{2A_k}\frac{u_k}{f_{yd}}\cot\theta \tag{9.28}$$

where u_k = perimeter of area A_k

f_{yd} = design yield strength of longitudinal reinforcement

θ = angle of inclination of compression struts

Note: When torsion is combined with bending, the area of longitudinal reinforcement can be modified as follows. In compressive chords, the area of longitudinal reinforcement can be reduced in proportion to the available compressive force and in tension chords the reinforcement due to bending is added to the reinforcement required for torsion. The longitudinal reinforcement should be distributed over the side length z_i.

Example 1: Fig. 9.21 shows a rectangular hollow section. It is subjected to an ultimate twisting moment T_{ED} = 350 kNm. Determine the shear stresses in the walls of the tube and design the required reinforcement.

The centre-line dimensions are

Width b = 1000 – 2 × 200/2 = 800 mm,

Depth h = 1500 – 150/2 – 100/2 = 1375 mm   Area $A_k = b × h = 1.1 × 10^6$ mm^2

From (9.27), shear flow q = T_{ED}/ (2 × A_k) = 350 × 10^6/ (2 × 1.1 × 10^6)

$$= 159 \text{ N/mm}$$

Shear stress τ_i in the ith wall = q/ t_i

τ in webs = 159/ 200 = 0.8 N/mm^2 and in the upper and lower flanges equal to 1.1 and 1.6 N/mm^2 respectively.

Shear force in the webs = q × h × 10^{-3} = 218.6 kN

Shear force in the flanges = q × b × 10^{-3} = 127.2 kN

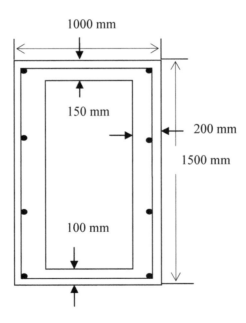

Fig. 9.21 Rectangular hollow section

i. Design longitudinal reinforcement:

Taking cot $\theta = 1$, $f_{yb} = f_{yk}/\gamma_s = 500/1.15 = 435$ MPa

$u_k = 2\,(b + h) = 4350$ mm. Substituting in (9.28),

$$\Sigma A_{s\ell} = \frac{T_{ED}}{2\,A_k}\frac{u_k}{f_{yd}}\cot\theta = \frac{350\times10^6}{2\times1.1\times10^6}\times\frac{4350}{435}\times1.0 = 1591 \ \ mm^2$$

Using 16 mm diameter bars, area of one bar = 201 mm². Number of bars required = 1591/ 201 = 8.

ii. Design link reinforcement:

Link reinforcement and corresponding spacing are designed for shear force in the web. Taking cover = 30 mm, link reinforcement as 12 mm and the diameter of longitudinal bar as 16 mm, the effective depth d for the bottom bar is d =1500 – 30 – 12 – 16/2 = 1450 mm. Taking the spacing of bars vertically as 475 mm and taking the bottom two bars as tension steel, the effective depth d is d = {1450 + 1450 - 475}/ 2 = 1213 mm.

(a). Calculate $V_{RD, c}$:

$C_{Rd,\,c} = 0.12$

$$k = 1 + \sqrt{\frac{200}{d}} = 1 + \sqrt{\frac{200}{1213}} = 1.41 \le 2.0$$

$$v_{min} = 0.035\,k^{\frac{3}{2}}\,f_{ck}^{\frac{1}{2}} = 0.035\times1.41^{\frac{3}{2}}\times40^{\frac{1}{2}} = 0.37\,\text{MPa}$$

A_{sl} = Two 16 mm bars each 201 mm² = 402 mm²
b_w = 200 mm, d = 1213 mm
$100\,\rho_t$ =100 × [402/ (200 × 1213)] = 0.17 < 2.0
σ_{cp} = 0 as there is no axial force
Substituting the values into the formula for $V_{RD,\,c}$ given by (9.6).

$$V_{Rd,c} = [C_{Rd,c}\,k\,(100\,\rho_t\,f_{ck})^{\frac{1}{3}} + k_1\,\sigma_{cp}]\,b_w d \ge (v_{min} + k_1\,\sigma_{cp})\,b_w d$$

$$V_{Rd,c} = [0.12\times1.41\times(100\times0.0017\times40)^{\frac{1}{3}} + 0](2\,00)\times1213\times10^{-3}$$

$$= 77.1 \ge [(v_{min}+0)\times(2\,00)\times1213\times10^{-3} = 89.8]$$

$V_{Rd,C}$ =89.8kN

$V_{Rd,\,c}$ = 89.9 kN < Applied shear force in web = 218.6 kN

(b). Check concrete struts do not crush

Take f_{ck} for concrete =40 MPa, f_{cd} = 26.7 MPa, σ_{cp} = 0
α_{cw} = 1, v_1 = 0.6
$\sigma_c = \alpha_{cw}\,v_1\,f_{cd}$ = 1.0 × 0.6 × 26.7 = 16.0 MPa, b_w = 200 mm
$z \approx 0.9d = 0.9 \times 1213 = 1092$ mm
$\sigma_c \times b_w \times z \times 10^{-3}$ = 16.0 × 200 × 1092 × 10⁻³ = 3494 kN
Substituting in equation (9.7),

$$V_{Rd,max} = V = 218.6 = \sigma_c b_w \, z \, \frac{1}{(\cot\theta + \tan\theta)} = 3494 \times \frac{1}{(\cot\theta + \tan\theta)}$$

$$\frac{1}{(\cot\theta + \tan\theta)} = 0.0623$$

$\cot\theta = 0.063$ or 15.94. Restricting $\cot\theta = \tan\theta = 1.0$,
$V_{Rd,\, max} = 3494/\,(1.0 + 1.0) = 1747 \text{ kN} > (V = 218.6 \text{ kN})$

(c). Design necessary shear reinforcement:

f_{ywd} = design yield strength of shear reinforcement.
Assuming that $f_{ywd} = 500/1.15 = 435$ MPa, $b_w = 200$ mm, $z = 1092$ mm,
Assuming one 12 mm link in the web, $A_{sw} = 113 \text{ mm}^2$
Substituting in equation (9.12),

$$V_{RD,s} = \frac{A_{sw}}{s} z f_{ywd} \cot\theta$$

$$V_{RD,s} \times 10^3 = 218.6 \times 10^3 = \frac{113}{s} \times 1092 \times 435 \times (\cot\theta = 1.0)$$

$s = 246$ *mm*

Provide 12 mm links at 240 mm c/c.

Example 2: Fig. 9.22 shows a trapezoidal hollow section. It is subjected to an ultimate twisting moment $T_{ED} = 350$ kNm. Determine the shear stresses in the walls of the tube.

The centre line dimensions are:

$b_{top} = 1000 - 150/\sin\theta = 849$ mm

$b_{bottom} = 600 - 150/\sin\theta = 449$ mm

Depth $h = 1500 - 150/2 - 100/2 = 1375$ mm, $h/\sin\theta = 1387$ mm

Area $A_k = 0.5 \times (b_{top} + b_{bottom}) \times h = 0.89 \times 10^6 \text{ mm}^2$

Perimeter $u_k = (b_{top} + b_{bottom}) + 2 \times h/\sin\theta = 4072$ mm

Shear flow $q = T_{ED}/\,(2 \times A_k) = 350 \times 10^6/\,(2 \times 0.89 \times 10^6)$

$$= 197 \text{ N/ mm}$$

Shear stress τ_i in the ith wall $= q/\,t_i$

τ in webs $= 197/\,150 = 1.3 \text{ N/mm}^2$ and in the upper and lower flanges equal to 1.3 and 2.0 N/mm² respectively.

Shear force in the webs $= q \times h/\sin\theta \times 10^{-3} = 273.3$ kN

i. Design longitudinal reinforcement:

Taking $\cot\theta = 1$, $f_{yb} = f_{yk}/\gamma = 500/1.15 = 435$ MPa. Substituting in (9.28),

$$\Sigma A_{s\ell} = \frac{T_{ED}}{2\,A_k}\frac{u_k}{f_{yd}}\cot\theta = \frac{350\times10^6}{2\times0.89\times10^6}\times\frac{4072}{435}\times1.0 = 1841\,mm^2$$

Using 16 mm diameter bars, area of one bar = 201 mm^2.

Number of bars required = 1591/ 201 = 9.2 say 11 bars. Provide five bars in each web and an additional bar at the middle of the top flange.

(ii). Design link reinforcement and spacing for shear force in the web

Taking cover as 30 mm, link reinforcement as 12 mm and the diameter of longitudinal bar as 16 mm, the total depth h for bottom bar is h =1500 – 30 – 12 – 16/2 = 1450 mm. Taking the spacing of bars vertically as 375 mm and taking the bottom three bars as tension steel, effective depth h is given by

h = {1450 + 1450 – 375 + 1450 – (375 + 375)}/ 3 = 1075 mm.

d = h/ sin θ = 1085 mm

(a). Calculate $V_{RD,\,c}$:

$$C_{Rd,\,c} = 0.12, \quad k = 1 + \sqrt{\frac{200}{d}} = 1 + \sqrt{\frac{200}{1085}} = 1.43 \le 2.0$$

$$v_{min} = 0.035 k^{\frac{3}{2}}\, f_{ck}^{\frac{1}{2}} = 0.035\times1.43^{\frac{3}{2}}\times40^{\frac{1}{2}} = 0.38\,MPa$$
A_{sl} = three 16 mm bars each 201 mm^2 = 603 mm^2
b_w = 200 mm, d = 1075 mm
100 ρ_t =100 × [603/ (200 × 1075)] = 0.28 < 2.0
Substituting the values into the formula for $V_{RD,\,c}$ given by (9.6).

$$V_{Rd,c} = [C_{Rd,c}\, k\,(100\,\rho_t\, f_{ck})^{\frac{1}{3}} + k_1\,\sigma_{cp}]\,b_w d \ge (v_{min} + k_1\,\sigma_{cp})\,b_w d$$

$$V_{Rd,c} = [0.12\times1.43\times(0.28\times40)^{\frac{1}{3}}]\,(2\,00)\times1085\times10^{-3}$$

$$= 83.3 \ge [0.38\times(2\,00)\times1085\times10^{-3} = 82.5]$$

$V_{Rd,C} = 83.3\,kN$

$V_{Rd,\,c}$ = 89.9 kN < Applied shear force in web = 273.3 kN

(b). Check concrete struts do not crush:

$\sigma_c = \alpha_{cw}\, v_1\, f_{cd} = 16.0\,MPa$, b_w = 200 mm
$z \approx 0.9d = 0.9 \times 1085 = 977$ mm
$\sigma_c \times b_w \times z \times 10^{-3} = 16.0 \times 200 \times 977 \times 10^{-3} = 3126$ kN

$$V_{Rd,max} = V = 273.3 = \sigma_c b_w\, z\,\frac{1}{(\cot\theta+\tan\theta)} = 3126\times\frac{1}{(\cot\theta+\tan\theta)}$$

$$\frac{1}{(\cot\theta+\tan\theta)} = 0.0874$$

cot θ = 0.088 or 11.35. Restricting cot θ = tan θ = 1.0,
$V_{Rd, max}$ = 3126/ (1.0 + 1.0) = 1563 kN > (V = 273.3 kN)

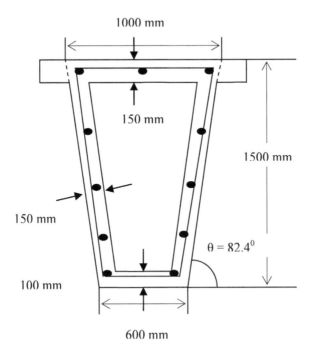

Fig. 9.22 a Trapezoidal hollow section

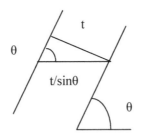

Fig 9.22 b Horizontal width of web

(c). Design of shear reinforcement:
Substitute in (9.12),

$$V_{RD,s} = \frac{A_{sw}}{s} z f_{ywd} \cot θ$$

$$V_{RD,s} \times 10^3 = 273.3 \times 10^3 = \frac{113}{s} \times 977 \times 435 \times (\cot\theta = 1.0)$$

$s = 176\ mm$

Provide 12 mm links at 175 mm c/c.

Example 3: Fig. 9.23 shows a solid rectangular section. It is subjected to an ultimate twisting moment T_{ED} = 350 kNm. Determine the shear stresses in the cross section.

Assuming 30 mm cover, link diameter of 10 mm and diameter of longitudinal bars 25 mm, the centre line dimensions are

Width, b = 1000 – 2 × (30 + 10) –25 = 895 mm,

Depth, h = 1500 – 2 × (30 + 10) –25 = 1395 mm

Area A_k = b × h = 1.25 × 10⁶ mm²

Shear flow q = T_{ED}/ (2 × A_k) = 350 × 10⁶/ (2 ×1.25 × 10⁶) = 140 N/mm

Area, A enclosed by the outer circumference = 1000 × 1500 = 1.5 × 10⁶ mm²

Outer circumference, u of the cross section = 2 × (1000 + 1500) = 5× 10³ mm

A/u = 300 mm. This is considered as the thickness of the equivalent hollow section. Shear stress τ in the cross section = q/ t = 140/ 300 = 0.47 N/mm²

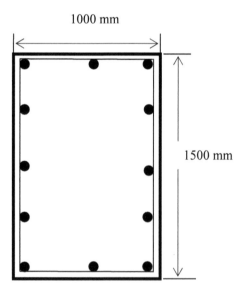

Fig. 9.23 Solid rectangular section

9.10.1 SPACING OF TORSION REINFORCEMENT

The longitudinal spacing of torsion links should not exceed u/8 , or 0.75d or the lesser dimension of the beam cross section. The longitudinal bars should be so arranged such that there is at least one bar at each corner and the others are distributed evenly around the inner periphery of the links. The spacing should not exceed 350 mm.

9.11 DESIGN FOR COMBINED SHEAR FORCE AND TORSION

Fig. 9.24 shows a box section subjected to shear force and torsion. As can be seen, in certain parts the shear stresses due to shear force and torsion are additive. It is important to ensure that the concrete struts are not over-stressed. This is ensured by the following inequality:

$$\frac{T_{ED}}{T_{Rd,\,max}} + \frac{V_{ED}}{V_{Rd,\,max}} \le 1.0 \tag{9.29}$$

$$T_{Rd,\,max} = 2\,v\,\alpha_{cw}\,f_{cd}\,A_k\,t_{ef,\,i}\,\sin\theta\cos\theta$$

$$v = 0.6[1 - \frac{f_{ck}}{250}], \ r = \frac{\sigma_{cp}}{f_{cd}}$$

$$\begin{aligned}
\alpha_{cw} &= 1 + r, & 0 < r \le 0.25 \\
&= 1.25, & 0.25 < r \le 0.50 \\
&= 2.5(1 - r), & 0.50 < r \le 1.0
\end{aligned} \tag{9.30}$$

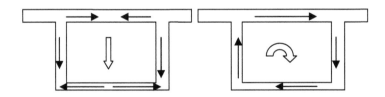

Fig. 9.24 Box section under shear force and torsion.

In the case of approximately solid sections only minimum reinforcement needs to be provided if

$$\frac{T_{ED}}{T_{Rd,\,c}} + \frac{V_{ED}}{V_{Rd,\,c}} \le 1.0,$$

$$T_{ED,c} = 2\,A_k \times f_{ctd} \times t_{ef} \quad , \; t_{ef} = \frac{A}{u} \tag{9.31}$$

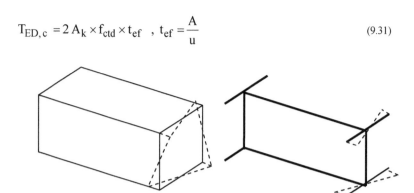

Fig. 9.25 Warping displacements in closed and open sections.

9.12 WARPING TORSION

When sections are subjected to torsion, the cross section rotates as a rigid body but in addition there are displacements in the axial direction. If these displacements are restrained, then additional stresses are set up. Fig. 9.25 shows the warping displacements in a closed rectangular tube and an open I-section. In a closed section, warping displacements are generally small and can be ignored. However this is not the case in the case of open sections. Warping displacements set up bending and shear stresses in the walls of the section and might needs to be investigated using the well-established Theory of Thin walled section.

9.13 REFERENCES TO EUROCODE 2 CLAUSES

In this chapter, the following clauses of Eurocode 2 have been referred to.

Shear:

General verification procedure for shear: 6.2.1

Members not requiring design shear reinforcement: 6.2.2

Members requiring design shear reinforcement: 6.2.3

Shear between web and flanges of T-sections: 6.2.4

Shear at the interface between concrete cast at different times: 6.2.5

Minimum shear reinforcement: 9.2.2(5) and 9.2.2(6)

Torsion:

Design procedure: 6.3.2, Warping torsion: 6.3.3

Detailing: 9.2.3

CHAPTER 10

CALCULATION OF CRACK WIDTHS

10.1 INTRODUCTION

In the case of prestressed members with bonded tendons, if under frequent load combination the tensile stress at the soffit exceeds the allowable tensile stress fctm, then the section should be analysed as a cracked section to determine that the width of the crack does not exceed the permitted limits.

10.2 EXPOSURE CLASSES

Eurocode 2 has six classes of exposure. They are

1. X0: No risk of corrosion or attack
2. XC1–XC4: Corrosion induced by carbonation
3. XD1–XD3: Corrosion induced by chlorides
4. XS1–XS3: Corrosion induced by chlorides from sea water
5. XF1–XF4: Freeze–Thaw attack
6. XA1–XA3: Chemical attack

From the point of view of limiting crack width, only the first four exposure classes are considered as shown in Table 10.1.

10.3 RECOMMENDED VALUES OF MAXIMUM CRACK WIDTH

Table 10.2 shows the permitted maximum crack width for various exposure classes

Table 10.2 Recommended values of maximum crack width

Exposure class	Frequent load combination	Quasi permanent load combination
X0, XC1	0.2 mm	
XC2 – XC4	0.2 mm	Check decompression
XD1, XD2, XS1–XS3	Check decompression	

Table 10.1 Exposure classes related to environmental conditions

Class designation	Description of the environment	Examples of where exposure classes might occur
No risk of corrosion or attack		
X0	For concrete without reinforcement or with reinforcement in very dry environment	Concrete inside buildings with very low humidity
Corrosion induced by carbonation		
XC1	Dry or permanently wet	Concrete inside buildings with very low humidity. Concrete permanently submerged in water
XC2	Wet, rarely dry	Concrete surfaces subjected to long-term water contact. Many foundations
XC3	Moderate humidity	Concrete inside buildings with moderate or high air humidity. External concrete sheltered from rain
XC4	Cyclic wet and dry	Concrete surfaces subjected to water contact, not within exposure class XC2
Corrosion induced by chlorides.		
XD1	Moderate humidity	Concrete surfaces exposed to airborne chlorides
XD2	Wet, rarely dry	Swimming pools Concrete components exposed to industrial waters containing chlorides
Corrosion induced by chlorides from seawater		
XS1	Exposed to airborne salt but not in direct contact with seawater	Structures near to or on the coast
XS2	Permanently submerged	Parts of marine structures
XS3	Tidal, splash and spray zones	Parts of marine structures

Note that

- In the case of exposure classes X0 and XC1, the crack width limitation is meant to ensure acceptable appearance rather than influence durability.

- Decompression means that all tendons must lie at least 25 mm with in the compression zone.

10.4 MINIUMUM STEEL AREAS

The amount of bonded reinforcement required to control cracking may be estimated from the equilibrium between the tensile force in concrete just before cracking and the tensile force in reinforcement at yielding. The minimum area of steel should be calculated for individual parts such as webs and flanges of a profiled cross section such as T-beams, box girders, etc.

$$A_{s,min}\, \sigma_s + \xi_1\, A_p'\, \Delta\sigma_p \geq k_c\, k\, f_{ct,eff}\, A_{ct} \tag{10.1}$$

Where

$A_{s,\,min}$ = minimum area of reinforcing steel in the tensile zone.

A_{ct} = area of concrete in the tensile zone. This is the area of the cross section in tension before the formation of the first crack.

σ_s is generally taken as the yield strength of reinforcement.

$f_{ct,\,eff}$ is generally taken as the tensile strength of concrete f_{ctm} unless if cracks are expected to occur earlier than 28 days, in which case f_{tcm} (t).

k is factor introduced to reduce the amount of minimum reinforcement required in sections where because of internal restraint, a higher tensile stress \will develop at the surface than can be predicted by linear stress distribution leading occurring at a lower load:

k = 1.0 for web depth $\leq$ 300 mm or flanges with width < 300 mm

k = 0.65 for web depth $\geq$ 800 mm or flanges with width > 800 mm. (10.2)

It is permissible to interpolate for intermediate values.

k_c is a factor accounting for different forms of stress distribution as follows. For bending combined with axial forces:

- For rectangular sections and webs of box and T-sections:

$$k_c = 0.4 \left[1 - \frac{\sigma_c}{k_1 \dfrac{h}{h^*} f_{ct,eff}} \right] \leq 1.0 \tag{10.3}$$

h^* = h for h < 1.0 m and h^* = 1.0 m for h $\geq$ 1.0 m

$$\sigma_c = \frac{N_{Ed}}{bh} \tag{10.4}$$

N_{Ed} = axial force (positive, compression) at SLS acting on the part of the cross section under consideration

$k_1 = 1.5$ if N_{ED} is compressive and $k_1 = 0.67\ h^*/h$ if N_{Ed} is tensile

For flanges of box and T-sections:

$$k_c = 0.9\frac{F_{cr}}{A_{ct}\ f_{ct,eff}} \geq 0.5 \tag{10.5}$$

where

F_{cr} = absolute value of tensile force within the flange immediately prior to cracking
     due to cracking moment

$A_p{}'$ = area of pre- or post-tensioned tendons within $A_{c,\ eff}$

$A_{c,\ eff}$ = effective area of concrete in tension surrounding the reinforcement or
     prestressing tendons of depth $h_{c,\ eff}$

$h_{c,\ eff}$ = MIN [2.5 (h–d), (h–x)/3, h/2]

ξ_1 = Adjusted bond strength taking into account the different bar diameters of
     prestressing and reinforcing steel

$$\xi_1 = \sqrt{[\frac{\phi_s}{\phi_p}\xi]} \tag{10.6}$$

ϕ_s = largest bar diameter of reinforcing steel

ϕ_p = equivalent diameter of tendon = 1.75 ϕ_{wire} for 7-wire strands made from wires
     of diameter ϕ_{wire}

$\xi = 0.6$ for strands in pre-tensioned systems

    = 0.5 for concrete grade ≤ C50/60 and 0.25 for concrete grade ≥ C70/85 in
     post-tensioned systems

If only prestressing steel is used then $\xi_1 = \sqrt{\xi}$

$\Delta\sigma_p$ = Stress variation in prestressing tendons from the state of zero strain of the concrete at the same level

Note: In prestressed members no minimum reinforcement is required in sections where under characteristic combinations of loads and prestress the concrete is compressed or the absolute value of the tensile stress in the concrete is below $f_{ct,eff}$.

10.4.1 Example of Minimum Steel Area Calculation

Using the given data, calculate the minimum steel reinforcement for the webs of the double T-beam shown in Fig. 4.1. A detailed analysis of the beam as a cracked section is given in section 8.3.3, Chapter 8. The beam is used as a simply supported beam of 12 m span supporting a live load of 8.0 kN/m² at SLS.

Beam properties: $A = 435.0 \times 10^3$ mm², $y_{bar} = 532$ mm

$I = 2.219 \times 10^{10}$ mm⁴, $Z_t = 101.8 \times 10^6$ mm³, $Z_b = 41.73 \times 10^6$ mm³

Concrete grade: C40/50, $f_{ck} = 40$, $f_{ctm} = 4.0$ MPa

Load factors for prestress: $\gamma_{superior} = 1.05$, $\gamma_{inferior} = 0.95$

Prestress, $P_s = 671.1$ kN provided by two strands per web located at 60 mm from the soffit giving a net eccentricity of e = 472 mm

σ_s = yield strength of unstressed steel = 460 MPa

h = depth of web = 750 mm

x = neutral axis depth at serviceability moment = 127 mm

d = effective depth = 750 – 60 = 690 mm

Step 1: Using (10.20), calculate the value of k for web a depth of 750 mm

$k = 0.65 + (1 – 0.65) \times (800 – 750)/ (800 – 300) = 0.685$

Step 2: Using (10.4), calculate σ_c

N_{Ed} = Portion of P_s acting on the web

$N_{ED} \approx P_s \times \gamma_{inferior} \times$ (Area of web/A)

$= 671.0 \times 0.95 \times (2 \times 150 \times 750)/ (435 \times 10^3) = 329.7$ kN

Substituting in (10.3),

$$\sigma_c = \frac{N_{Ed}}{bh} = \frac{329.7 \times 10^3}{2 \times 150 \times 750} = 1.47 MPa$$

Step 3: Using (10.3) calculate k_c

h* = h = 750 mm as h < 1 m

$k_1 = 1.5$ as N_{Ed} is compressive, $f_{ctm} = 4.0$ MPa

Substituting in (10.3),

$$k_c = 0.4[1 - \frac{\sigma_c}{k_1 \frac{h}{h^*} f_{ct,eff}}] = 0.4[1 - \frac{1.47}{1.5 \times \frac{750}{750} \times 4.0}] = 0.30 \le 1.0$$

Step 4: calculate ξ_1

$\xi = 0.6$ for strands in pre-tensioned systems

As only prestressing steel is used, $\xi_1 = \sqrt{\xi} = 0.78$

Step 5: calculate $\Delta\sigma_p$

$\Delta\sigma_p$ = Stress variation in prestressing tendons from the state of zero strain of concrete at the same level to the stress under cracking moment

Results from the cracked section analysis in Chapter 8 are given in summary form in Table 8.1. From Table 8.1,

Stress in steel at zero compression in concrete at steel level = 1063 MPa
Stress in steel at cracking moment = 1100 MPa
$\Delta\sigma_p = 1100 - 1063 = 37$ MPa

A_p' = area of two prestressing strands per web = (two webs × 2) × 150 = 600 mm^2

A_{ct} = area of beam in tension:

$A_{ct} = 2 \times 150 \times (h - x) = 2 \times 150 \times (750 - 380) = 111.0 \times 10^3$ mm^2

Substituting in (10.1),

$A_{s,min}\,\sigma_s + \xi_1\,A_p\,\Delta\sigma_p \quad k_c\,k\,f_{ct,eff}\,A_{ct} = A_{s,\,min} \times 460 + 0.78 \times 600 \times 37$

$$= A_{s,\,min} \times 460 + 17.3 \times 10^3$$

$k_c\,k\,f_{ct,eff}\,A_{ct} = 0.30 \times 0.69 \times 4.0 \times 111.0 \times 10^3 = 91.9 \times 10^3$

$A_{s,\,min} \times 460 + 17.3 \times 10^3 = 91.9 \times 10^3$

$A_{s,\,min} = 162$ mm^2, or say one 12 mm bar per web

10.5 CALCULATION OF CRACK WIDTHS w_k

Crack width may be calculated from the following expression

$$w_k = S_{r,max} \times [\dfrac{\sigma_s - k_t\,\dfrac{f_{ct,eff}}{\rho_{p,eff}}\,(1+\alpha_e\,\rho_{p,eff})}{E_s} \geq 0.6\dfrac{\sigma_s}{E_s}] \qquad (10.7)$$

where

σ_s = stress variation in prestressing tendons from the state of zero strain in the concrete at the same level.

α_e = modular ratio = E_s/E_{cm}

$\rho_{p,\,eff} = (A_s + \xi_1^2\,A_p')/A_{ct,\,eff}$

A_p', $A_{c,\,eff}$, ξ_1 = as defined earlier in section 10.4.

$k_t = 0.6$ for short-term loading and 0.4 for long-term loading

10.5.1 Crack Spacing $S_{r, max}$

(i). In situations where bonded reinforcement is fixed at spacing $\leq 5(\text{cover} + \text{bar diameter}/2)$, the maximum crack spacing may be calculated from

$$S_{r,max} = 3.4\,c + 0.425\,k_1\;k_2\;\frac{\phi}{\rho_{p,eff}} \tag{10.8}$$

where

$k_1 = 1.6$ for prestressing tendons, $k_2 = 0.5$ for bending

Φ = bar diameter or if it consists of bars of different numbers and diameters then the equivalent bar diameter Φ_{eq} defined by

$$\phi_{eq} = \frac{\Sigma\,\phi_i^2\,n_i}{\Sigma\,\phi_i\,n_i} \tag{10.9}$$

c = cover to reinforcement

(ii). In situations where bonded reinforcement is fixed at spacing $> 5(\text{cover} + \text{bar diameter}/2)$, the maximum crack spacing may be calculated from

$$S_{r, max} = 1.3\,(h - x) \tag{10.10}$$

where h = total dept, x = neutral axis depth

10.5.2 Example of Crack Width and Spacing Calculation

Calculate the crack width for the double T-beam considered in section 10.4.1.

From the results of the cracked section analysis shown in Table 10.3, stresses in prestressing tendons at the state of zero strain in concrete and at the serviceability limit state are respectively 1063 MPa and 1386 MPa. Therefore

$\sigma_s = 1386 - 1063 = 323$ MPa

E_s = Young's modulus for steel = 195 GPa

$E_{cm} = 22 \times (0.1\times f_{cm})^{0.3} = 22 \times (0.1\times 48)^{0.3} = 35.2$ GPa

α_e = modular ratio = E_s/E_{cm} = 195.0/35.2 = 5.5

From 10.4.1, $A_p^{'} = 600$ mm^2, h = 750 mm, x = 127 mm, $f_{ctm} = 4.0$

d = effective depth = 750 – 60 = 690 mm

$h_{c, eff}$ = MIN [2.5 (h–d), (h–x)/3, h/2]

   = MIN [2.5× (750 – 690), (750 – 127)/3, 750/2] = 150 mm

$A_{c, eff}$ = effective area of concrete of depth $h_{c, eff}$ in tension surrounding the
   reinforcement or prestressing tendon

$\qquad$ = two webs × width of web × $h_{c,\ eff}$ = 2 × 150 × 150 = 45 × 10^3 mm^2

A_s = 0 as no unstressed steel is used.

ξ = 0.6 for strands in pre-tensioned systems, $\xi_1^2 = \xi = 0.6$

$\rho_{p,\ eff} = (A_s + \xi_1^2\ A_p^{'})/A_{ct,\ eff} = (0 + 0.6 \times 600)/ (45 \times 10^3) = 0.008$ or 0.8%

k_t = 0.4 for long-term loading.

Substituting in (10.7),

$$\sigma_s - k_t\frac{f_{ct,eff}}{\rho_{p,eff}}(1+\alpha_e\,\rho_{p,eff}) = 323 - 0.4\times\frac{4.0}{0.008}(1+5.5 \times 0.008)$$

$$=114.0 \geq [0.6\,\sigma_s = 194]$$

Take $\sigma_s - k_t\dfrac{f_{ct,eff}}{\rho_{p,eff}}(1+\alpha_e\,\rho_{p,eff}) = 194.0$

As there is no bonded reinforcement, assuming that crack spacing $S_{r,max}$ is given by

$$(10.10)$$

$S_{r,max}$ = 1.3 (h - x) = 1.3 × (750 - 127) = 810 mm

$$w_k = S_{r,max} \times [\frac{\sigma_s - k_t\dfrac{f_{ct,eff}}{\rho_{p,eff}}(1+\alpha_e\,\rho_{p,eff})}{E_s} \geq 0.6\frac{\sigma_s}{E_s}] = 810\times\frac{194}{195\times10^3} = 0.8\,mm$$

Crack width is 0.8 mm which is unacceptably large. This indicates that the beam needs bonded un-tensioned steel to control the width of the crack.

10.6 EXAMPLE OF A PARTIALLY PRESTRESSED BEAM

Consider the same double T-beam from section 10.4 except that there is a 12 mm unstressed bar in each web. From the cracked section analysis, the results shown in Table 8.3 can be obtained.

10.6.1 Example of Minimum Steel Area

Data for (10.1):

x = neutral axis depth at serviceability moment = 150 mm

Using the data from section 10.4.1,

k = 0.685, N_{Ed} = 329.7 kN, k_1 = 1.5., σ_c = 1.47 MPa, k_c = 0.29, $A_p^{'}$ = 600 mm^2,

ξ_1 = 0.89

Stress in steel at zero compression in concrete at steel level $= 1063$ MPa

Stress in steel at cracking moment $= 1100$ MPa

$\Delta\sigma_p = 1100 - 1063 = 37$ MPa

Neutral axis depth at cracking moment, $x = 382$ mm

$A_{ct} = 2 \times 150 \times (h - x) = 2 \times 150 \times (750 - 382) = 110.4 \times 10^3$ mm^2

$A_{s,min}\, \sigma_s + \xi_1\, A_p\, \Delta\sigma_p = A_{s,\,min} \times 460 + 0.89 \times 600 \times 37$

$$= A_{s,\,min} \times 460 + 19.8 \times 10^3$$

$k_c\; k\; f_{ct,eff}\; A_{ct} = 0.29 \times 0.69 \times 4.0 \times 110.4 \times 10^3 = 88.4 \times 10^3$

$A_{s,\,min} \times 460 + 19.8 \times 10^3 = 88.4 \times 10^3$

$A_{s,\,min} = 149$ mm^2 or say one 12 mm bar per web.

10.6.2 Example of Width and Spacing of Cracks

From the results of the cracked section analysis shown in Table 8.3, stress in prestressing tendons at the state of zero strain in concrete and at the serviceability limit state are respectively 1063 MPa and 1301 MPa. Therefore

$\sigma_s = 1301 - 1063 = 238$ MPa

$x =$ depth of neutral axis at SLS $= 150$ mm

Using the results from section 10.5.2, $\alpha_e = 5.8$, $A_p' = 600$ mm^2, $h = 750$ mm,

$d = 690$ mm, $f_{ctm} = 4.0$ MPa

$h_{c,\,eff} = $ MIN[$2.5 \times (750 - 690)$, $(750 - 150)/3$, $750/2$] $= 150$ mm

$A_{c,\,eff} = 45 \times 10^3$ mm^2

$A_s =$ area of two 12 mm bars, one in each web $= 226$ mm^2.

From (10.6), $\xi_1^2 = \xi\, \dfrac{\phi_s}{\phi_p} = 0.6 \times \dfrac{12}{9.1} = 0.79$

$k_t = 0.4$ for long-term loading.

Substituting in (10.7)

$\rho_{p,\,eff} = (A_s + \xi_1^2\, A_p')/A_{ct,\,eff} = (226 + 0.79 \times 600)/(45 \times 10^3) = 0.016$ or 1.6%

$$\sigma_s - k_t\, \frac{f_{ct,eff}}{\rho_{p,eff}}\,(1 + \alpha_e\, \rho_{p,eff}) = 238 - 0.4 \times \frac{4.0}{0.016}\,(1 + 5.5 \times 0.016)$$

$$= 129.0 \geq [0.6\,\sigma_s = 194]$$

$$\frac{\sigma_s - k_t \dfrac{f_{ct,eff}}{\rho_{p,eff}}(1 + \alpha_e\,\rho_{p,eff})}{195 \times 10^3} = \frac{194}{195 \times 10^3} = 0.001$$

φ = As the combination is one bar of 12 mm diameter in each web and two prestressing tendons of 9.1 equivalent diameter,

$$\phi_{eq} = \frac{\Sigma\,\phi_i^2\,n_i}{\Sigma\,\phi_i\,n_i} = \frac{(12^2 + 2 \times 9.1^2)}{(12 + 2 \times 9.1)} = 10.3\,\text{mm}$$

Taking cover, c = 45 mm, k_1 = 1.6, k_2 = 0.5 and substituting in (10.6),

$$S_{r,max} = 3.4\,c + 0.425\,k_1\,k_2\,\frac{\phi}{\rho_{p,eff}} = 3.4 \times 45 + 0.425 \times 1.6 \times 0.5 \times \frac{10.3}{0.016} = 370\,\text{mm}$$

$$w_k = S_{r,max} \times \left[\frac{\sigma_s - k_t\dfrac{f_{ct,eff}}{\rho_{p,eff}}(1 + \alpha_e\,\rho_{p,eff})}{E_s} \geq 0.6\frac{\sigma_s}{E_s}\right] = 370 \times 0.001 = 0.37\,\text{mm}$$

10.7 CONTROL OF CRACKING WITHOUT DIRECT CALCULATION

Crack width can be controlled *either* by restricting the bar diameter as specified in Tables 10.3 *or* by controlling the spacing specified in Table 10.4.

Note that

- In the case of pre-tensioned concrete, where crack control is mainly provided by tendons with direct bond, the values of the stress in the tables refer to total stress minus prestress.

- In the case of post-tensioned concrete where crack control is mainly provided by ordinary reinforcement, the value of stress in the ordinary reinforcement is caused by loads as well as by the prestress should be used.

- When the member is subjected to bending moment with at least part of the section is in compression, the maximum bar diameter should be modified as follows:

$$\phi_s = \phi_s^* \frac{f_{ct,eff}}{2.9} \frac{k_c\,h_{cr}}{2(h - d)} \qquad (10.11)$$

φ_s = adjusted maximum bar diameter

Φ_s^* = diameter from Table 10.3

h = overall depth of section

d = effective depth to the centroid of the outer layer of reinforcement

h_{cr} = depth of the tensile zone immediately prior to cracking, considering the characteristic values of prestress under ***quasi-permanent combination of actions***

Table 10.3 Maximum bar diameter Φ_s^* for crack control

Steel stress, MPa	Maximum width of crack		
	0.4 mm	0.3 mm	0.2 mm
160	40	32	25
200	32	25	16
240	20	16	12
280	16	12	10
320	12	10	8
360	10	8	6
400	8	6	4
450	6	5	

Table 10.4 Maximum spacing for crack control

Steel stress, MPa	Maximum width of crack		
	0.4 mm	0.3 mm	0.2 mm
160	300	300	200
200	300	250	150
240	250	200	100
280	200	150	50
320	150	100	
360	100	50	

10.8 REFERENCES TO EUROCODE 2 CLAUSES

In this chapter, the following clauses in Eurocode 2 have been referred to:

Environmental conditions: 4.2

Exposure classes: Table 4.1

Crack control: 7.3

Recommended values of w_{max}: Table 7.1N

Minimum steel areas: 7.3.2

Control of cracking without direct calculation: 7.3.3

Calculation of crack widths: 7.3.4

Calculation of maximum crack spacing: 7.3.4.(3)

CHAPTER 11

LOSS OF PRESTRESS

11.1 INTRODUCTION

It was mentioned briefly in Chapters 2 and 3 that the final value of prestress remaining is only about 75% of the prestress at the time of jacking. This loss takes place in approximately two stages.

- Loss at transfer: Prestress applied at the time of jacking reduces immediately after jacking due to elastic contraction in the case of both pre- and post-tensioned beams. In the case of post-tensioned structures, during jacking due to friction between the duct and the cable, the prestress decreases from the jacking end to the anchored end. In addition, there is loss due to slip at anchors. The loss at transfer can be of the order of about 10% of the jacking force
- Long-term loss: Over a period of time there is further reduction in prestress mainly due to creep and shrinkage of concrete and relaxation of steel. The final loss can be of the order of 25% of the jacking force.

It is necessary to take account of these losses during design. In this chapter the calculation of loss due to several sources will be discussed in detail. It is important to emphasize that accurate assessment of losses is not possible because of many uncertainties involved in the parameters governing loss calculations.

11.2 IMMEDIATE LOSS OF PRESTRESS

The reason for the loss of prestress at transfer depends on whether the beam is pre-tensioned or post-tensioned.

11.2.1 Elastic Loss in Pre-tensioned Beams

In the case of pre-tensioned beams, the main reason is the elastic compression of concrete when the force is transferred from the abutments on to the beam. However, as the strands are tensioned on to the abutments, even before the beam is cast, certain losses can take place. Eurocode 2 suggests that the following losses should be considered even before the force is transferred from abutments on to concrete:

- Loss due to friction at the bends in the case of deflected (draped/harped) tendons and loss at the anchors due to wedge draw in.
- Loss due to relaxation of strands in the period between stressing the tendons and transferring the force on to concrete.

However, one can make an allowance for these losses by direct measurement of strain in the strands. The elastic loss can be calculated as follows.

Let the beam be prestressed with a force at transfer of P_t applied at an eccentricity of e. Ignoring the self weight of the beam, the compressive stress σ_c in concrete at the centroid of the steel is given by

$$\sigma_c = \frac{P_t}{A} + P_t\, e\, \frac{e}{I} \tag{11.1}$$

where A = area and I = second moment of area of the beam cross section.
The compressive strain ε_c in concrete is given by

$$\varepsilon_c = \frac{\sigma_c}{E_{cm}(t)} \tag{11.2}$$

where E_{cm} (t) = secant modulus of elasticity of concrete at the time of transfer of force given by equation (3.6).
Since steel is fully bonded to concrete, the compressive strain ε_s in steel is the same as the strain in concrete and the compressive stress σ_s in steel is given by

$$\varepsilon_s = \varepsilon_c, \quad \sigma_s = \varepsilon_s\, E_s = \sigma_c\, \frac{E_s}{E_{cm}(t)} = \frac{P_t}{A}(1 + \frac{Ae^2}{I}) \times \frac{E_s}{E_{cm}(t)} \tag{11.3}$$

The loss of force in the steel is given by

$$Loss = \sigma_s\, A_{ps} = \{\frac{P_t}{A} + \frac{P_t}{I}e^2\}\frac{E_s}{E_{cm}(t)}A_{ps}$$

$$= P_t\{1 + \frac{Ae^2}{I}\}\frac{E_s}{E_{cm}(t)}\frac{A_{ps}}{A} \tag{11.4}$$

where A_{ps} = total area of cross section of the prestressing strands.
If P_{jack} is the force at abutments at the time of transfer,
$P_t = P_{jack} - Loss$

$$P_t = \frac{P_{Jack}}{[1 + \{1 + \frac{Ae^2}{I}\}\frac{E_s}{E_{cm}(t)}\frac{A_{ps}}{A}]} \tag{11.5}$$

Substituting for P_t, loss of prestress can be expressed as

$$Loss = A_{ps}\frac{\frac{P_{Jack}}{A}\{1 + \frac{Ae^2}{I}\}\frac{E_s}{E_{cm}(t)}}{[1 + \{1 + \frac{Ae^2}{I}\}\frac{E_s}{E_{cm}(t)}\frac{A_{ps}}{A}]} \tag{11.6}$$

The numerator is the stress at the centroid of steel at the time of jacking,

$$Loss = A_{ps}\frac{\text{Compressive stress in steel at steel centroid due to } P_{Jack}}{[1 + \{1 + \frac{Ae^2}{I}\}\frac{E_s}{E_{cm}(t)}\frac{A_{ps}}{A}]}$$

$$\% loss = \frac{Loss}{P_{Jack}} \times 100 = [\frac{\{1 + \frac{Ae^2}{I}\}\frac{E_s}{E_{cm}(t)}\frac{A_{ps}}{A}}{1 + \{1 + \frac{Ae^2}{I}\}\frac{E_s}{E_{cm}(t)}\frac{A_{ps}}{A}}] \times 100 \tag{11.7}$$

11.2.1.1 Example of elastic loss calculation

A pre-tensioned double T-beam is shown in Fig. 4.1. The details are as follows.
$A = 435.0 \times 10^3$ mm², $y_{bar} = 532$ mm, $I = 2.219 \times 10^{10}$ mm⁴, Span = 12m
Concrete is of grade 40/50, $f_{ck} = 40$ MPa
$P_{jack} = 1200.0$ kN is applied by four strands per web. Eccentricity e = 447 mm.
The area of each strand is 112 mm², giving $A_{ps} = 896$ mm². $E_s = 195$ GPa
Both P and e are constant over the entire span.
$f_{cm} = f_{ck} + 8 = 48$ MPa. From (3.4), secant modulus, $E_{cm} = 37.1$ GPa
Assuming Class R cement, s = 0.20, t = 3 days, from (3.1),

$$f_{cm}(t) = \exp\{s[1-\sqrt{(\frac{28}{t})}]\}\ f_{cm} = 31.8 \text{ MPa}$$

Secant modulus at time of transfer, $E_{cm}(t) = [\frac{f_{cm}(t)}{f_{cm}}]^{0.3} \times E_{cm} = 32.8$ GPa

$$\frac{Ae^2}{I} = \frac{435 \times 10^3 \times 447^2}{2.219 \times 10^{10}} = 3.92, \quad \frac{E_s}{E_{cm}(t)} = \frac{195}{32.8} = 5.95, \quad \frac{A_{ps}}{A} = \frac{896}{435 \times 10^3} = 2.06 \times 10^{-3}$$

Substituting in (11.7), % loss $= [\dfrac{(1+3.92) \times 5.95 \times 2.06 \times 10^{-3}}{[1+(1+3.92) \times 5.95 \times 2.06 \times 10^{-3}]}] \times 100 = 5.7\%$

If it is thought important to take the tensile stress due to self weight into
account, the compressive stress σ_c in concrete at the centroid of the steel can be
included as follows.

Note that only 2/3 of the self weight is taken into account because the bending
moment distribution due to self weight is parabolic and the 'average' bending
moment is 2/3 of the maximum value at mid-span.
In this example, assuming unit weight of concrete as 25 kN/m³,
Self weight q_{sw} per unit length $= 25 \times 435 \times 10^3 \times 10^{-6} = 10.88$ kN/m
Bending moment at mid-span, $M_{sw} = q_{sw} \times 12^2/8 = 192.75$ kNm,
Average stress due to self weight at steel centroid,

$$\sigma_{sw} = (2/3) \times M_{sw} \times e/ I = (2/3) \times 192.75 \times 10^6 \times 447/ (2.219 \times 10^{10}) = 2.6 \text{ MPa}$$

The total compressive at the steel centroid, σ_c is

$$\sigma_c = \frac{P_t}{A} + P_t e\frac{e}{I} - \frac{2}{3}\frac{M_{sw}}{I}e = P_t \times 10^3 \times 1.13 \times 10^{-5} - 2.6$$

where P_t is in kN

Loss of stress in steel, $\sigma_s = \sigma_c \times 195/32.8 = [0.067\ P_t - 15.1]$ MPa
Prestress loss $= [0.067\ P_t - 15.1] \times (A_{ps} = 896) \times 10^{-3}$
$\qquad\qquad = [0.06\ P_t - 13.5]$ kN
$P_t = P_{jack} -$ Loss $= 1200 - [0.06\ Pt - 13.5]$,
$(1 + 0.06)\ P_t = 1213.5$, $Pt = 1145.0$ kN
% loss $= [1 - 1145/1200] \times 100 = 4.6\%$

In a pre-tensioned beam, including the stress due to self weight decreases slightly the loss of prestress at transfer.

11.2.2 Elastic Loss in Post-tensioned Beams

In the case of post-tensioned beams, if all the cables are stressed simultaneously, then there is no loss due to elastic shortening because the force in the jack automatically compensates for the elastic compression of the beam. However, in general cables are stressed successively, so that when a new set of strands are prestressed, the resulting elastic compression will cause a loss of prestress in all the *previously* stressed cables.

Assume that jacking is done in N equal stages. Let $\Delta\sigma_c$ be the total compressive stress at the level of tendons in the corresponding pre-tensioned beam and A_{ps} is the total area of the tendons. The total loss of prestress in the pre-tensioned beam is $\Delta\sigma_s$ where

$$\Delta\sigma_s = \frac{E_s}{E_{cm}(t)}\,\Delta\sigma_c \tag{11.8}$$

The area of tendons stressed at each stage is A_{ps}/N.
The compressive stresses induced in successively stressed cables will be as follows.–
Cables stressed in the first step will suffer a compressive stress = $[(N-1)/N]\,\Delta\sigma_s$
Cables stressed in the second step will suffer a compressive stress = $[(N-2)/N]\,\Delta\sigma_s$
Cables stressed in the third step will suffer a compressive stress = $[(N-3)/N]\,\Delta\sigma_s$
This process continues, until cables stressed in the last but one step will suffer a compressive stress = $[1/N]\,\Delta\sigma_s$
Cables stressed in the last step will suffer a compressive stress $[0/N]\,\Delta\sigma_s$.
The total loss of prestress is therefore

$$\text{Loss} = \{\frac{N-1}{N}+\frac{N-2}{N}+\frac{N-3}{N}+....+\frac{1}{N}\}\times\Delta\sigma_s\times\frac{A_{ps}}{N}$$

$$= \{(N-1)+(N-2)+(N-3)+....+1\}\frac{\Delta\sigma_s\,A_{ps}}{N^2}$$

The terms inside {...} form an arithmetic series of (N−1) terms. Their sum is given by

$$\text{Sum} = \frac{\text{No. of terms}}{2}[\text{First term }+\text{ Last term}] = \frac{N-1}{2}\times[N-1+1] = \frac{N(N-1)}{2}.$$

The total loss of prestress is therefore

$$\text{Loss} = \frac{N(N-1)}{2}\frac{\Delta\sigma_s\,A_{ps}}{N^2} = \frac{(N-1)}{2N}\frac{E_s}{E_{cm}(t)}\Delta\sigma_c\,A_{ps} \tag{11.9}$$

Table 11.1 Loss of prestress with the number of stages

N	(N−1)/(2N)
1	0
2	0.25
3	0.33
4	0.375
10	0.45
20	0.475

As the number of stages increase, the total loss increases as shown in Table 11.1. In the absence of data at the preliminary design stage, (N−1)/ (2N) can be taken as equal to 0.50. The *maximum* loss of prestress in a post-tensioned beam is only 50% of that in an equivalent pre-tensioned beam.

11.2.3 Loss of Prestress due to Friction and Wobble

In the case of post-tensioned beams, because of the fact that the cable will be in contact with the duct during stressing, friction between the cable and the duct will cause loss of prestress. This is called curvature friction. Similarly because the duct can never be held rigidly in place, there is a certain amount of deviation (wobble) from the intended profile. The loss of prestress due to unintentional contact between the cable and the duct is called parasite or wobble friction and this will also cause loss of prestress.

11.2.3.1 Derivation of an equation for loss of prestress due to friction

Consider a segment of the tendon which forms part of a circular arc of radius R and an included angle of $d\theta$ as shown in Fig. 11.1. The total length of the segment is R $d\theta$. Let the prestress force at end 1 be P and that at end 2 be P + ΔP. The difference between the two forces is due to frictional force between the cable and the duct. ΔP is given by

$$\Delta P = \frac{dP}{d\theta} d\theta \qquad (11.10)$$

The tensile forces at the ends are inclined to the horizontal by $d\theta$/2. The total vertical component N of forces at the two ends is equal to

$$N = P \sin\frac{d\theta}{2} + (P + \Delta P)\sin\frac{d\theta}{2} \qquad (11.11)$$

Since $d\theta$ is small, replace sin ($d\theta$/2) $\approx$ $d\theta$/2 and ΔP by (11.10):

$$N = (P + P + \frac{dP}{d\theta} d\theta)\frac{d\theta}{2}$$

Ignoring second order terms in $(d\theta)^2$ and simplifying,

$$N = P \, d\theta \qquad (11.12)$$

The frictional force = friction coefficient $\times$ normal force = μ N
where μ = coefficient of friction between the cable and the duct

Considering the equilibrium in the horizontal direction, ΔP is equal to the normal force N multiplied by the frictional coefficient μ:

$$\Delta P = -\mu N = -\mu\, P\, d\theta \tag{11.13}$$

Equating ΔP from (11.1) and (11.13),

$$\frac{dP}{d\theta}\, d\theta = -\mu\, P\, d\theta \tag{11.14}$$

Cancelling $d\theta$ and dividing by P

$$\frac{dP}{P} = -\mu\, d\theta \tag{11.15}$$

Integrating (11.15)

$$\int_{P_1}^{P_2}\frac{dP}{P} = -\mu\int_{\theta_1}^{\theta_2} d\theta\,, \quad \ell n\, P_2 - \ell n\, P_1 = -\mu\,(\theta_2 - \theta_1)\,, \quad \ell n\,\frac{P_2}{P_1} = -\mu\,(\theta_2 - \theta_1)$$

$$\frac{P_2}{P_1} = e^{-\mu(\theta_2 - \theta_1)} \tag{11.16a}$$

Note that $ds = R\, d\theta$, where R is the radius of curvature. As the cable profile is quite flat, it is reasonable to assume that $dx \approx ds$, $dx \approx R\, d\theta$. Making this substitution, the variation of the prestressing force can be expressed as

$$\frac{P_2}{P_1} = e^{-\mu\frac{x_2 - x_1}{R}} \tag{11.16b}$$

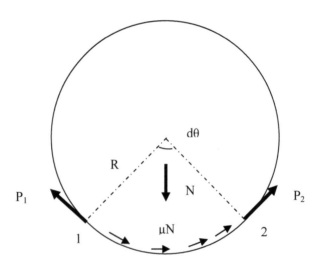

Fig. 11.1   A segment of cable under tension and frictional forces

Assuming that the tendon fills about half of the duct as shown in Fig. 5.3 in order that the tendons can be pushed or pulled into the empty ducts easily, the value of the friction coefficient μ for internal tendons is about 0.19. It is worth noting that in the case of unbonded tendons greased inside an extruded plastic sheath as shown in Fig. 2.23, the value of μ is smaller (μ ≈ 0.06) than for a tendon inside plastic or metal ducts. However the unbonded tendons are much less rigid than the tendons inside metal ducts and hence are liable to have a larger deviation from the intended profile, leading to a higher loss of prestress due to wobble friction.

Assuming that the effect of the wobble is to simply increase the value of friction, the total effect of friction and wobble can be summed and the variation of the force between ends 1 and 2 can be expressed as

$$\frac{P_2}{P_1} = e^{-\mu\{(\theta_2 - \theta_1) + k(x_2 - x_1)\}} \tag{11.16c}$$

The constant k is called the wobble coefficient and has a value in the range $0.005 < k < 0.01$ per metre.

Note that the angle θ is in radians and it is the sum irrespective of signs or directions of angular displacements, over the length x.

If $\mu\,(\theta + kx)$ is less than 0.2, $\quad e^{-\mu\{\theta + kx\}} \approx 1 - \mu(\theta + kx)$

$$P_1 \approx P_2\{1 - \mu(\theta + kx)\}\,,\;(P_1 - P_2) \approx P_1\,\mu\,(\theta + kx)$$

Note that where the profile of the tendon consists of a series of curves, the angle θ is the sum of all angles taken as positive.

11.2.3.2 Example of calculation of loss of prestress due to friction and wobble

Calculate the loss of prestress due to friction and wobble in a post-tensioned beam of 20 m span. The profile of the cable is parabolic with a central dip of 300 mm. The cable is jacked from one end with a force of 1400 kN. Calculate the change in prestress from the jacking end to the anchored end. Assume $\mu = 0.19$, $k = 0.008$ /m Consider a symmetric parabola with a central dip equal to Δ as shown in Fig. 11.2.

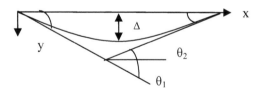

Fig. 11.2 Cable with a parabolic profile

Taking the origin at one end, the equation for the parabola is

$$y = \frac{4\,\Delta}{L^2} x\,(L - x),\quad \frac{dy}{dx} = \tan\theta = \frac{4\,\Delta}{L^2}(L - 2x)$$

Since θ is small, $\quad \tan\theta \approx \theta \approx \pm 4\dfrac{\Delta}{L}$ at $x = 0$ and $x = L$

The change in angle from $x = 0$ to $x = L$ is

$$\theta = \theta_2 - \theta_1 = 4\frac{\Delta}{L} - (-4\frac{\Delta}{L}) = 8\frac{\Delta}{L} = 8\frac{0.300}{20} = 0.12 \text{ radians}$$

$$x_2 - x_1 = L = 20 \text{ m}$$
$$\mu\,(\theta + kx) = 0.19 \times (0.12 + 0.008 \times 20) = 0.053$$
$$P_2 - P_1 = 0.053 \times 1400 = 74 \text{ kN}$$

11.2.3.3 Calculation of θ for different profiles

In sections 6.2.2 and 6.2.3 of Chapter 6, equations for cable profiles which were a combination of three and four parabolic segments with continuity of slope and deflection at their junctions were described in detail. The three-parabola profile is intended for end spans of continuous beams and the four-parabola profile is intended for interior spans of continuous beams. Fig. 11.3 and Fig. 11.4 show the profiles and the slopes at key points in the profile.

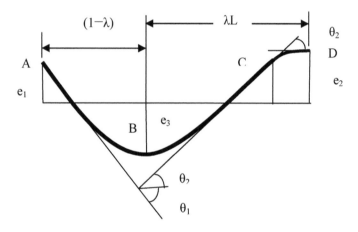

Fig. 11.3 Three-segment parabolic profile

Three-parabola profile: The three-parabola profile shown in Fig 11.3 consists of three segments as follows:

(i). Parabola AB: Extends over a length $(1-\lambda)\,L$ with ordinates $-e_1$ at A and $+e_3$ at B. The slope of the parabola at A to the x-axis (positive left) is $\theta_1 = \dfrac{2(e_1 + e_3)}{(1 - \lambda)L}$ and is zero at B.

(ii). Parabola BC: Extends over a length $(\lambda - \beta)\,L$ with ordinates $+e_3$ at B and $-(e_2 - h_2)$ at C. The slope of the parabola at C to the x-axis (positive right) is $\theta_2 = \dfrac{2(e_2 + e_3)}{\lambda L}$ and is zero at B.

(iii). Parabola CD: Extends over a length βL with ordinates $-e_2$ at D and $-(e_2 - h_2)$ at C. The slope of the parabola at C to the x-axis (Positive right) is

$$\theta_2 = \frac{2(e_2 + e_3)}{\lambda L} \text{ and is zero at D.}$$

The total angular change is therefore $\theta = \theta_1 + \theta_2 + \theta_2$:

$$\theta = \frac{2}{L} \times \{\frac{(e_1 + e_3)}{(1-\lambda)} + 2 \frac{(e_2 + e_3)}{\lambda}\} \qquad (11.17)$$

Example: Calculate the total angular change for the following data:
Span L = 30 m, the length of the first segment is 18 m, the length of the third segment is 3 m. $e_1 = 0$ mm, $e_2 = 300$ mm, $e_3 = 400$ mm.
$(1 - \lambda) \times 30 = 18$, $\lambda = 0.4$. $\beta \times 30 = 3$, $\beta = 0.1$
$\theta_1 = 2 \times (0 + 0.4)/18 = 0.044$ radians, $\theta_2 = 2 \times (-0.3 + 0.4)/12 = 0.017$ radians
$\theta = \theta_1 + \theta_2 + \theta_2 = 0.044 + 0.017 + 0.017 = 0.078$ radians

Four-parabola profile: The four-parabola profile shown in Fig 11.4 consists of four segments as follows:
(i). Parabola AB: Extends over a length $\beta_1 L$ with ordinates $-e_1$ at A and $-(e_1 - h_1)$ at B. The slope of the parabola at A to the x-axis (positive right) is zero and at B

is $\theta_1 = \dfrac{2(e_1 + e_3)}{(1-\lambda)L}$.

(ii). Parabola BC: Extends over a length $(1-\lambda - \beta_1) L$ with ordinates $-(e_1 - h_1)$ at B and $+e_3$ at B. The slope of the parabola at B to the x-axis (positive left) is

$\theta_1 = \dfrac{2(e_1 + e_3)}{(1-\lambda)L}$ and is zero at C.

(iii). Parabola CD: Extends over a length $(\lambda - \beta_2) L$ with ordinates $+e_3$ at C and $-(e_2 - h_2)$ at D. The slope of the parabola at C to the x-axis (positive right) is zero

and at D is $\theta_2 = \dfrac{2(e_2 + e_3)}{\lambda L}$.

(iv). Parabola DE: Extends over a length $\beta_2 L$ with ordinates $-e_2$ at E and $-(e_2 - h_2)$ at D. The slope of the parabola at D to the x-axis (positive left) is

$\theta_2 = \dfrac{2(e_2 + e_3)}{\lambda L}$ and is zero at E.

The total angular change is therefore $\theta = \theta_1 + \theta_1 + \theta_2 + \theta_2$:

$$\theta = \frac{4}{L} \times \{\frac{(e_1 + e_3)}{(1-\lambda)} + \frac{(e_2 + e_3)}{\lambda}\} \qquad (11.18)$$

Example: Calculate the total angular change for the following data:
Span L = 30 m, the lengths of the first and fourth segments are 3 m, the lengths of second and third segments are 12 m. Let the eccentricities be $e_1 = 300$ mm, $e_2 = 300$ mm, $e_3 = 400$ mm. Then
$(1 - \lambda) \times 30 = 15$, $\lambda = 0.5$. $\beta_1 \times 30 = \beta_2 \times 30 = 3$, $\beta_1 = \beta_2 = 0.1$
$\theta_1 = \theta_2 = 2 \times (-0.3 + 0.4)/15 = 0.013$ radians
$\theta = 2(\theta_1 + \theta_2) = 0.052$ radians

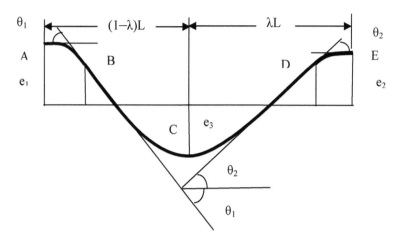

Fig. 11.4 Four -egment parabolic profile

Example: Calculate the total loss of prestress in a three span continuous beam with
each span 30 m long. The end spans have three-parabola cable profile and the
central span a four-parabola profile as discussed in the example above. The
structure is jacked from one end only. The total angular change θ is the sum of the
angular change in each of the three spans.
$\theta = 0.078 + 0.052 + 0.078 = 0.208$ radians $\approx 12°$, $x = 3 \times 30 = 90$ m
Assume $\mu = 0.19$, $k = 0.008$/m,
  $\mu\ (\theta + kx) = 0.19 \times (0.208 + 0.008 \times 90) = 0.176$
  $[1 - \mu\ (\theta + kx)] = 0.82$
$P_2 / P_1 = 0.82$, i.e. there is a large loss in prestress of 18 %.

Loss due to friction and wobble can be reduced using the following options:
  i.   Make the tendon profile as flat as possible. In many cases use of
              haunches at supports can lead to a flatter cable profile as shown in
              Fig. 11.5. The haunch increases the eccentricity over the support
              without having to increase the total angle turned through.

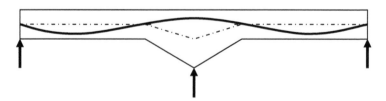

Fig. 11.5 Haunch at support to reduce friction loss

  ii.   Another possibility is to anchor the cables at intermediate positions as
              shown in Fig. 11.6. One can usea fixed anchorage at the top of the
              beam to avoid corrosion problems and stressing anchorages only at
              the ends.

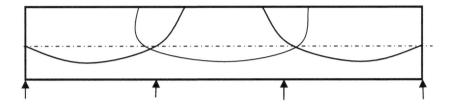

Fig. 11.6 Cable profile to reduce frictional losses

iii.   Provide additional short lengths of tendons in the spans far from the stressing end.

iv.   Stress the tendons from both ends, although this causes problems on site. Generally internal tendons can be up to 120 m long and external tendons can be up to 300 m long without the friction losses becoming excessive.

v.   Stress alternate tendons from the two ends. This will cause a more uniform prestressing force over the whole length of the structure.

vi.   In the case of slabs, it is possible to stress each bay as it is cast. This has the advantage of minimizing shrinkage strains. However, it requires extra hardware in the form of couplers.

vii.   Short lengths of cap cables can be used as shown in Fig. 11.7.

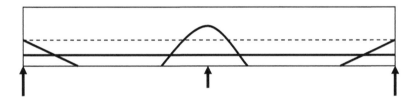

Fig. 11.7 Use of cap cables

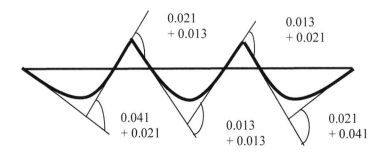

Fig. 11.8 Angular deviations in a three single-parabola system

Single-parabola approximation: As shown in Fig. 11.8, it is possible to ignore the reverse curvatures towards the supports and approximate the profile as a single parabola as discussed in section 6.2.1 in Chapter 6. For a single parabola, the expressions for the slopes are given by

$$\frac{de}{dx} = -\frac{(1-\lambda)^2 \ (e_1 - e_2) - (e_1 + e_3)}{\lambda \ (1-\lambda) L} \quad \text{at } x = 0$$

$$\frac{de}{dx} = \frac{(1-\lambda^2)(e_1 - e_2) - (e_1 + e_3)}{\lambda \ (1-\lambda) L} \quad \text{at } x = L$$

In the end spans, the eccentricities are $e_1 = 0$ mm, $e_2 = 300$ mm, $e_3 = 400$ mm then $(1 - \lambda) \times 30 = 18$, $\lambda = 0.4$.
$\theta_1 = 0.041$ radians, $\theta_2 = 0.021$ radians
In the middle span, the eccentricities are $e_1 = -300$ mm, $e_2 = 300$ mm, $e_3 = 400$ mm, $\lambda = 0.5$.
$\theta_1 = \theta_2 = 0.013$ radians
Thus considering all three parabolas, the total θ as shown in Fig. 11.8 is given by
$\theta = 2 \ (0.041 + 0.021) + 2(0.021 + 0.013 + 0.013) = 0.218$ radians compared with $\theta = 0.078 + 0.052 + 0.078 = 0.208$ radians when the short lengths of reverse curvature are taken into account. The error is only 4%.

11.2.4 Loss of Prestress due to Draw-in of Wedges

In post-tensioned systems, once the tendons have been stressed, before the pressure in the stressing jack is released, wedges are driven in so that the cables are held in tension when the force in the jack is reduced to zero. During this process before the wedges 'bite', a certain amount of slip between the cable and the wedges takes place. This is known as draw-in of the wedges and is generally of the order of 6 to 8 mm. During this stage, the cable tries to slip back but is resisted by the friction. During stressing friction resists the cable from stretching but during the draw-in friction prevents the cable from contracting. Fig. 11.9a shows the variation of force in the cable along the span. Along ABC the force variation is given by

$$P_x = P_A \ e^{-\mu\{\theta + kx\}}$$

If $\mu \ (\theta + kx)$ is less than 0.2, $e^{-\mu\{\theta + kx\}} \approx 1 - \mu(\theta + kx)$

$$P_x = P_A \ e^{-\mu\{\theta + kx\}} \approx P_A \{1 - \mu(\frac{x}{R} + kx)\}$$

where $\theta = x/R$, R = radius of curvature of the tendon profile
   Variation of P_x with x can be represented by a straight line as shown in Fig. 11.9b. ABC shows the variation of force if there was no draw-in. DBC shows the variation of prestress when there is draw-in. Loss of force from A to B must be the same as the gain of force from D to B because it is the same combination of friction and wobble which is causing the variation of the force. In other words the slopes of AB and DB are equal.

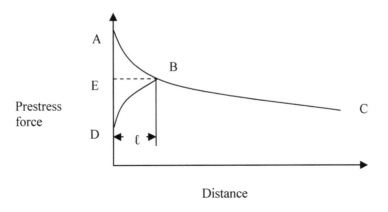

Fig. 11.9a Variation of prestress along the member

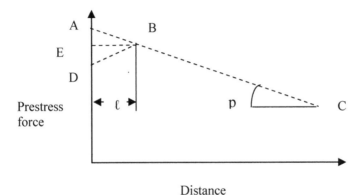

Distance

Fig. 11.9b Linear variation of prestress along the member

If the slope of line ABC is p, which is the loss of force due to friction and wobble per unit length, the loss of force due to anchorage draw-in at a distance x (measured from B and positive to left) = 2px
If A_{ps} is the area of the cable, the loss of stress σ_s in the cable and the corresponding strain ε_s are

$$\sigma_s = \frac{2px}{A_{ps}}, \ \varepsilon_s = \frac{\sigma_s}{E_s} = \frac{2px}{A_{ps} E_s}$$

The contraction of an element of length $dx = \varepsilon_s \, dx = \dfrac{2p}{A_{ps} E_s} \, dx$

The total contraction Δ is equal to 'draw-in' over the length ℓ and is given by

$$\Delta = \int_0^\ell \frac{2px}{A_{ps} E_s} \, dx = \frac{p \ell^2}{A_{ps} E_s}$$

$$\ell = \sqrt{\frac{\Delta A_{ps} E_s}{p}}$$

$$\ell = \sqrt{\frac{\Delta A_{ps} E_s}{\text{Loss of prestress per unit length}}} \qquad (11.20)$$

In the above discussion it was assumed that the bonded length is longer than the length ℓ so that there is no 'draw-in' loss at the 'fixed end'. In a short tendon, the whole length of the tendon will suffer the 'draw-in' loss. Fig. 11.10 shows the variation of prestress before and after 'draw-in' by AB and by CD respectively.

Note that in the case of unbonded tendons the draw-in loss extends over the entire length of the cable.

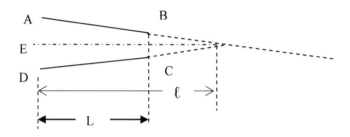

Fig. 11.10 Variation of prestress along a short member

11.2.4.1 Example of loss of prestress due to draw-in

Calculate the variation in cable force using the following data for the three-span continuous beam considered in section 11.2.3.3. The cable is stressed from one end with a force of $P_A = 7600$ kN. The cable is made up of thirty seven 7-wire strands of 15.2 mm diameter, giving $A_{ps} = 5180$ mm^2. $\mu = 0.19$, $k = 0.008$/m, $E_s = 195$ kN/mm^2, draw-in, $\Delta = 6$ mm.

As a first approximation, calculate the total loss of prestress in the first span of 30 m. The total change of angle is 0.078 radians. Over a length of 30 m, loss of prestress due to curvature and wobble is

Loss $= P_A - P_B = P_A \times \mu \, (\theta + k\ell) = 7600 \times \{0.19(0.078 + 0.008 \times 30)\} = 459.2$ kN
$p =$ loss per unit length $= 459.2/30 = 15.31$ kN

As a second approximation, calculate the loss over the first parabolic segment only then as shown in Fig. 11.3,
$\lambda = 0.4$, $\beta = 0.1$, $L = 30$ m, AB $= (1 - \lambda) L = 18$ m, $\theta_1 = 0.044$ radians.
Loss $= P_A \times \mu \, (\theta + k\ell) = 7600 \times \{0.19(0.044 + 0.008 \times 18)\} = 271.5$ kN
$p =$ loss per unit length $= 271.5/ 18 = 15.1$ kN
Both assumptions lead to almost identical values for p.

$$\ell = \sqrt{\frac{\Delta A_{ps}\, E_s}{\text{Loss of prestress per unit length}}} = \sqrt{\frac{(6\times10^{-3})\times5180\times195}{15.31}} = 19.90\,\text{m}$$

Loss due to wedge draw-in $\approx 2 \times 15.31 \times 19.90 = 609.4$ kN

$P_D = P_A - 609.4 = 6991$ kN

Fig. 11.3 shows the three-parabola cable profile in the first span. In the end span, $\lambda = 0.4$, $\beta = 0.1$, $L = 30$ m, $\theta_1 = 0.044$ radians.

The length of the first parabola, $AB = (1 - \lambda)\, L = 18$ m,

The length of 19.90 m over which draw-in has effect extends 1.9 m beyond B, the first parabola. The length of 1.9 m is in the second parabola, BC. The slope at $x = 1.9$ m in the second parabola BC and is given by

$$\theta_x = \frac{2x}{L^2} \frac{(e_3 + e_2)}{\lambda\,(\lambda - \beta)} = \frac{2 \times 1.90}{30^2} \frac{(300 + 400) \times 10^{-3}}{0.4 \times (0.4 - 0.1)} = 0.013$$

The total angle θ at 19.90 m from A is $\theta = (\theta_1 + \theta_x)$

$$= (0.044 + 0.013) = 0.057 \text{ radians.}$$

$P_{19.90} = 7660 \times \{1 - 0.19 \times (0.057 + 0.008 \times 19.90)\} = 7345$ kN

Table 11.1 shows the values of prestress at different sections along the beam.

Table 11.1 Variation of prestress along the beam

x- m	θ, Radians	$\mu(\theta + kx)$	$7600 \times e^{-\mu\,(\theta + kx)}$,kN
0	-	-	6991*
18.0	0.044	0.036	7311*
19.90	0.057	0.041	7345#
30	0.078	0.060	7154
45	0.104	0.088	6959
60	0.130	0.116	6768
72	0.164	0.141	6603
90	0.218	0.178	6359

* These values are in the region DB in Fig. 11. 9.

\# This value is at B in Fig. 11.9.

11.3 LOSS OF PRESTRESS DUE TO CREEP, SHRINKAGE AND RELAXATION

As discussed in Chapter 3, concrete under stress continues to deform over a period of time due to the property of creep and steel under strain continues to loose stress over a period of time due to the property of relaxation. In addition concrete shrinks due to loss of water as a result of drying and hydration. All these three effects lead to loss of prestress over time and this has to be taken account of in design.

The formula in Eurocode 2 can be derived as follows.

Let the beam be prestressed with a force at transfer of P_t applied at an eccentricity of e. Ignoring the self weight of the beam, the compressive stress σ_c in concrete at the centroid of the steel is given by (11.1) and the corresponding strain by (11.2) except that $E_{cm}(t)$ is replaced by the long-term Young's modulus allowing for creep deformation E_{cm}/φ, where φ is the creep coefficient.

(i). The loss of prestress due to creep is given by

$$\text{Loss} = \{\frac{P_t}{A} + \frac{P_t}{I} e^2\} \frac{E_s}{E_{cm}} \phi A_{ps} = P_t \chi \tag{11.21}$$

$$\chi = \{1 + \frac{A e^2}{I}\} \frac{E_s}{E_{cm}} \phi \frac{A_{ps}}{A} \tag{11.22}$$

(ii). Loss due to shrinkage $= \varepsilon_{cs} E_s A_{ps}$

where $\varepsilon_{cs} =$ strain due to shrinkage of concrete.

(iii). Loss due to relaxation $= \Delta\sigma_{pr} A_{ps}$

(iv). Loss due to creep strain due to quasi-permanent live and dead loads:

$$= -\sigma_{c,external} \frac{E_s}{E_{cm}} \phi$$

The negative sign is because the external load causes tensile stresses in concrete at the steel centroid.

The total loss due to (ii) to (iv) is

$$C = \{\varepsilon_{cs} E_s + \Delta\sigma_{pr} - \sigma_{c,externl} \frac{E_s}{E_{cm}} \phi\} A_{ps} \tag{11.23}$$

If P_{jack} is the force at abutments at the time of transfer, the prestress after all losses is given by

$$P_s = P_{Jack} - P_t \chi - C \tag{11.24}$$

Assuming $P_s \approx P_t$, and solving for P_t,

$$P_t = \frac{P_{Jack} - C}{[1 + \chi]}$$

$$\text{loss} = P_t \chi + C = \frac{(P_{Jack} - C)}{[1 + \chi]} \chi + C = \frac{(P_{Jack} \chi + C)}{[1 + \chi]} \tag{11.25}$$

The numerator can be written as

$$[\frac{P_{Jack}}{A} \{1 + \frac{A e^2}{I}\} - \sigma_{c,external}] \frac{E_s}{E_{cm}} \phi A_{ps} + [\varepsilon_{cs} E_s + \Delta\sigma_{pr}] A_{ps} \tag{11.26}$$

Setting, $\sigma_{c,QP} = \frac{P_{Jack}}{A} \{1 + \frac{A e^2}{I}\} - \sigma_{c,external} \tag{11.27}$

$$\text{Loss} = A_{ps} \frac{[\varepsilon_{cs} E_s + \Delta\sigma_{pr} + \sigma_{c,QP} \frac{E_s}{E_{cm}} \phi]}{[1 + \chi]} \tag{11.28}$$

Eurocode 2 makes a slight simplification to (11.28) and gives the loss as

$$\text{Loss} = A_{ps} \frac{\varepsilon_{cs}\, E_s + 0.8\,\Delta\sigma_{pr} + \dfrac{E_s}{E_{cm}}\,\phi\,\sigma_{c,QP}}{1 + \chi\{0.8 + \dfrac{1}{\phi}\}} \tag{11.29}$$

where $\sigma_{c,\,QP}$ includes the stress in concrete at the steel centroid due to quasi-permanent live loads, dead loads and also the stress due to P_{jack}. The factor 0.8 is introduced to account for the fact that loss of prestress due to relaxation of steel is influenced by the deformation of concrete due to creep and shrinkage.

11.3.1 Example of Final Loss Calculation

A simply supported rectangular beam 400 × 900 mm over a span of 12 m is prestressed with a force at jacking equal to 1570 kN using eight 7-wire strands giving $A_{ps} = 1120$ mm^2. The profile of the cable is parabolic with a net eccentricity at mid span of 350 mm and zero at supports. The beam is subjected to a dead load of 9.0 kN/m and a quasi-permanent live load of 32.0 kN/m. Calculate the loss of prestress in the long term. Concrete is of grade C40/50.
Solution:

(i). Shrinkage loss:
Following the details in section 3.4.3, Chapter 3, equation (3.11),
Specimen dimensions: 400 × 900, $A_c = 3.6 \times 10^5$ mm^2,
Assume only bottom and two sides are exposed: $u = 400 + 2 \times 900 = 2200$ mm
$h_0 = 2 \times A_c/ u = 327$ mm
Using Table 3.4 ,
$k_h = 0.75 - (0.75 - 0.70) \times (327 - 300)/ (500 - 300) = 0.74$
Using Class N cement for which from Table 3.3, $\alpha_{ds1} = 4$, $\alpha_{ds2} = 0.12$ and a relative humidity of say RH = 60%, substituting in (3.11)

$\varepsilon_{cd0} = 0.85[(220 + 110 \times 4) \times \exp\{-0.12 \times 0.1 \times 48\}] \times 1.55\{1 - (0.01 \times 60)^3\} \times 10^{-6}$
    $\varepsilon_{cd,0} = 0.38 \times 10^{-3}$

Substituting in $\varepsilon_{cs} = \varepsilon_{cd0} \times k_h + 2.5 (f_{ck} - 10) \times 10^{-6}$

$\varepsilon_{cs} = 0.38 \times 10^{-3} \times 0.74 + 2.5 \times (40 - 10) \times 10^{-6} = 0.355 \times 10^{-3}$
Total shrinkage strain, $\varepsilon_{cs} = 0.355 \times 10^{-3}$
$f_{cm} = f_{ck} + 8 = 48$ Mpa. From equation (3.4), $E_{cm} = 22$ (cm/ 10)$^{0.3} = 35.2$ GPa
Loss of prestress due to shrinkage is
$\varepsilon_{cs} \times E_s \times A_{ps} = 0.355 \times 10^{-3} \times 195 \times 10^3 \times 1120 \times 10^{-3} = 77.5$ kN

(ii). Relaxation loss:
Using Class 2, low-relaxation steel strands, 1000 hour loss = 2.5%
$f_{pk} = 1860$ MPa, $f_{p,\,0.1k} = 0.88\ f_{pk} = 1637$ MPa
$\sigma_{pi} = 0.9\ f_{p,\,0.1k} = 1473$ MPa
$\mu = \sigma_{pi}/f_{pk} = 1473/1860 = 0.79$
Relaxation after 50 years, $t = 365$ days × 50 = 18250 days:
Substituting for class 2 in (3.23),

$$\frac{\Delta\sigma_{pr}}{\sigma_{pi}} = 0.66 \times \rho_{1000} \times e^{9.1\,\mu} \times (\frac{t}{1000})^{0.75(1-\mu)} \times 10^{-5}$$

$$= 0.66 \times \rho_{1000} \times e^{9.1\times0.79} \times (\frac{18250}{1000})^{0.75(1-0.79)} \times 10^{-5}$$

$$= 0.66 \times \rho_{1000} \times 1325 \times 1.58 \times 10^{-5}$$

$$= 0.0138 \times 2.5 = 3.45\%$$

$$\Delta\sigma_{pr} = 1473 \times 3.45/100 = 51\,\text{MPa}$$

Loss of prestress due to relaxation is
$\Delta\sigma_{pr} \times (A_{ps} = 1120) \times 10^{-3} = 57.1\,\text{kN}$

(iii). Elastic and creep losses:
$E_{cm} = 22\,(f_{cm}/10)^{0.3} = 35.2\,\text{GPa}$, $E_s = 195\,\text{GPa}$
Calculation of creep coefficient φ:
Following the details in 3.4.2, Chapter 3,
RH = 60, $h_0 = 327$ mm, loaded at $t_0 = 3$ days, Class N cement, s = 0.25, α = 0,
$f_{cm} = 48$ MPa
Substituting in 3.8,

$$\alpha_1 = [\frac{35}{f_{cm}}]^{0.7} = [\frac{35}{48}]^{0.7} = 0.80, \quad \alpha_2 = [\frac{35}{f_{cm}}]^{0.2} = [\frac{35}{48}]^{0.2} = 0.94$$

$$\phi_{RH} = [1 + \frac{1-0.01\times RH}{0.1\times h_0^{0.333}}\alpha_1]\alpha_2 = [1 + \frac{1-0.6}{0.1\times327^{0.333}}\times0.80]\times0.94 = 1.38$$

$$t_0 = t_{0,T}[\frac{9}{2+t_{0,T}^{1.2}}+1]^{\alpha} = 3 \times [\frac{9}{2+3^{1.2}}+1]^0 = 3 \geq 0.5$$

$$\phi(\infty, t_0) = \phi_{RH} \times \frac{16.8}{\sqrt{f_{cm}}} \times \frac{1}{(0.1+t_0^{0.20})} = 1.38 \times \frac{16.8}{\sqrt{48}} \times \frac{1}{(0.1+3^{0.20})} = 2.49$$

$$f_{cm}(t_0) = \exp[s\{1-\sqrt{\frac{28}{t_0}}\}]f_{cm} = \exp[0.25\{1-\sqrt{\frac{28}{3}}\}]\times48 = 28.7$$

At transfer, $f_{cm}(t_0) = 28.7$ MPa, σ_c = compressive at transfer = 15 MPa,
$k_\sigma = \sigma_c/f_{cm}(t_0) = 0.52 > 0.45$
$\phi_k(\infty, t_0) = \phi(\infty, t_0) \times \text{Exp}[1.5\times(k_\sigma - 0.45)] = 2.49 \times 1.25 = 3.12$

A = 3.6 × 10⁵ mm², I = 2.43 × 10¹⁰ mm⁴, e = 350 mm,
$E_{cm} = 35.2$ GPa, $E_s = 195$ GPa, $A_{ps} = 1120$ mm², $\phi_k(\infty, t_0) = 3.12$
Substituting in (11.22),

$$\chi = \{1 + \frac{A\,e^2}{I}\}\frac{E_s}{E_{cm}}\,\phi\,\frac{A_{ps}}{A}$$

$$= \{1 + \frac{3.6\times10^5 \times 350^2}{2.43\times10^{10}}\}\frac{195}{35.2}\,3.12\,\frac{1120}{3.6\times10^5}$$

$$= \{1+1.82\}\times5.54\times3.12\times3.50\times10^{-3} = 0.17$$

$$\chi\,(0.8+\frac{1}{\phi}) = 0.19$$

$$\frac{P_{Jack}}{A} + \frac{P_{Jack}}{I}\,e^2 = \frac{1570\times10^3}{3.6\times10^5} + \frac{1570\times10^3}{2.43\times10^{10}}\,350^2$$

$$= 4.36 + 7.92 = 12.28\,\text{MPa}$$

$q_{external} = (q_{dead} = 9.0) + (q_{live} = 32) = 41\,\text{kN/m}$

$M_{external} = 41 \times 12^2/8 = 738\,\text{kNm}$, $M_{external} \times e/I = 10.63\,\text{MPa}$

Substituting in (11.27), at mid-span

$$\sigma_{c,QP} = \frac{P_{Jack}}{A}\{1 + \frac{A\,e^2}{I}\} - \sigma_{c,external} = 11.28 - 10.63 = 1.65\,\text{MPa}$$

At support: $\sigma_{c,\,QP} = 12.28 - 0 = 12.28\,\text{MPa}$

Average value $= (12.28 + 1.65\,)/\,2 = 7.0\,\text{MPa}$

$$\frac{E_s}{E_{cm}}\,\phi\,\sigma_{c,QP}\,A_{ps} = \frac{195.0}{35.2}\times3.12\times7.0\times1120\times10^{-3} = 135.4\,\text{kN}$$

Substituting in (11.28),

$$Loss = A_{ps}\,\frac{\varepsilon_{cs}\,E_s + 0.8\,\Delta\sigma_{pr} + \dfrac{E_s}{E_{cm}}\,\phi\,\sigma_{c,QP}}{1+\chi(0.8+\dfrac{1}{\phi})}$$

$$= \frac{77.5 + 0.8\times57.1 + 135.4}{1+0.19} = 217.3\,\text{kN}$$

% loss $= 217.3/\,1570.0 = 14\%$

11.4 REFERENCES TO EUROCODE 2 CLAUSES

In this chapter, the following clauses in Eurocode 2 have been referred to:

Immediate loss of prestress in pre-tensioning: 5.10.4

Immediate loss of prestress in post-tensioning: 5.10.5

Loss due to friction: 5.10.5.2, Time dependent loss of prestress: 5.10.6

CHAPTER 12

DESIGN OF SLABS

12.1 INTRODUCTION

It is generally felt that for spans greater than about 6.0 m, post-tensioned slabs start to become cost effective. The vast majority of post-tensioned slabs use unbonded tendons purely because they considerably reduce the construction cost. Some of the significant advantages of post-tensioned slabs compared with reinforced concrete slabs are:

- Longer spans with fewer columns leading to flexibility in the positioning of partitions.
- Thinner slabs lead to saving in construction costs and reduced height of the building.
- Especially in the case of car parks, the virtually crack-free slabs are a great advantage to limit damage due to seepage of water with de-icing salts from melting snow.

Table 12.1 Dimensions of different types of slabs

Slab type	Slab, span m	Slab depth, mm	Beam depth, mm	Span/Depth ratio, slab, beam
Solid flat slab	6.0	200		30
	8.0	250		32
Solid flat slab with drop panels	8.0	225		36
	12.0	300		40
One-way slab with band beams	6.0	150	300	40, 20
	8.0	200	375	40, 21
	12.0	300	550	40, 22
One-way slab with narrow beams	6.0	175	375	34, 16
	8.0	225	500	36, 16
	12.0	325	750	37, 16
Ribbed slab	6.0	300		27
	8.0	450		27
	12.0	575		26

12.2 TYPICAL SLAB AND BEAM DEPTHS

1. There are a large number of different types of post-tensioned slabs used in practice. Table 12.1 gives, for an assumed imposed load of 5 kN/m^2, typical slab and beam dimensions. The information is taken from:

Stevenson, A.M., 1994, *Post-tensioned concrete floors in multi-storey buildings*, (British Cement Association).

Table 12.2 gives span/depth ratios recommended by the Post-tensioning Institute of the USA. The design of the various types of slabs is discussed in the following sections.

Table 12.2 Span-to-depth ratios for different types of slabs

Type of slab	Span/depth ratio
One-way slab	48
Two-way slab	45
Two-way slab with drop panel (Minimum drop panel span/6 each way)	50
Two-way slab with two-way beams	55
Waffle (5 × 5 grid)	35
Beams b ≈ h/3	20
Beams b ≈ 3 h	30

Post-tensioned slabs can be used on their own or combined with reinforced concrete to provide a range of in-situ concrete floor options. The slabs can be one-way or two-way spanning depending on the circumstance. In general if the aspect ratio length/width ≤ 2, then one can assume two-way spanning. There are a number of different types available in practice and some of the common ones are discussed below.

12.2.1 Effective Span of Slabs for Different Support Conditions

The effective span ℓ_{eff} of slabs in Eurocode 2 is given by
ℓ_{eff} = clear span $\ell_n + a_1$ at end 1 + a_1 at end 2
where h = total depth of slab, t = width of the support, $a_1 = 0.5 \times min(h, t)$

12.3 ONE-WAY SPANNING SLABS

Fig. 12.1 shows a one-way spanning slab. This is the simplest form of slab. The slab spans in one direction only and is supported on beams or walls. For spans in the 8m to 18 m range, one can adopt the standard beam and slab system. The 'primary' beams can be band beams as shown in Fig. 12.2 or standard beams spanning between columns. A set of 'secondary' T-beams are used to span between the 'primary' beams. The 'slab' spans between the 'secondary' beams as shown in Fig. 12.1. The slab might be simply supported or continuous between secondary beams. The primary beams span between columns and again are designed as simply supported or continuous beams as appropriate. The prestressing cables in the slab run in the span direction between secondary beams. In the transverse direction, unstressed reinforcement is used to control cracking due to shrinkage and thermal cracking and also to distribute any local concentration of loads.

It is possible to use a combination of post-tensioned beams in conjunction with reinforced concrete using a system of wide, shallow band beams as shown in Fig. 12.2. The post-tensioned band beams span between the columns and support the thinner reinforced concrete slab spanning between the band beams. This form is suitable for spans for beams in the 6m – 20 m range and for slabs in the 7m – 10m range. The band strips are normally made equal to about 2.4 m which is the width of a standard plywood sheet.

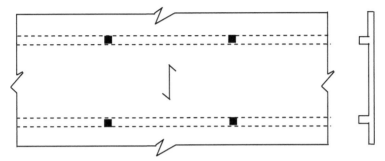

Fig. 12.1 One-way spanning slab.

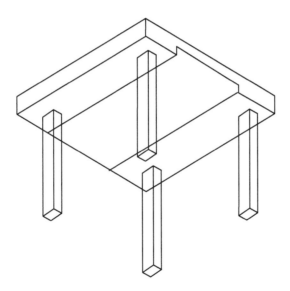

Fig. 12.2 Band beam and slab

12.3.1 Design of a One-way Spanning Slab

Design a three span continuous one-way slab spanning over integral beams. The spans are 8 m each.
Material properties: (see sections 3.3, 3.4 and 3.6 of Chapter 3)

Concrete: Assume C 30/37, f_{ck} = 30 MPa
At transfer: f_{ck} = 25 MPa
$f_{tt} = f_{ctm} = 0.30 \times 25^{2/3} = 2.6$ MPa , $f_{tc} = -0.6 \times 25 = -15.0$ MPa
At service: f_{ck} = 30 MPa
$f_{st} = f_{ctm} = 0.30 \times 30^{2/3} = 2.9$ MPa , $f_{sc} = -0.6 \times 30 = -18.0$ MPa
Secant modulus: E_{cm} = 22 × [(30 + 8)/10]$^{0.3}$ = 32.8 GPa
Tangent modulus ≈ 1.05 × 32.8 = 34.5 GPa
Steel: (see section 3.9, Chapter 3)
Monostrand, 13 mm (external 16 mm) diameter, f_{pk} = 1860 MPa,
$f_{p0.1k}$ = 0.88 f_{pk} = 1637 MPa, A_p = 100.5 mm^2, P_{max} = 0.9 A_p $f_{p0.1k}$ = 148.0 kN

Load factors for prestress in post-tensioning:
$\gamma_{Superior}$ = 1.1 at transfer and $\gamma_{Inferior}$ = 0.90 at service stage.

Loading:
Choose depth of slab: Span = 8 m, Assume a span/depth ratio = 35
Depth of slab = (8× 10^3)/35 = 229, say 250 mm
Unit weight of concrete = 25 kN/m^3
Self weight= 0.25 × 25 = 6.25 kN/m^2
Take partitions = 1.5 kN/m^2 and live load = 2.5 kN/m^2
Total load = 6.25 + 1.50 + 2.50 = 10.25 kN/m^2

12.3.2 Analysis for Applied Loading

Analysis is done for five load cases.
Case 1: Self weight of slab only on all spans, (Fig. 12.3a)

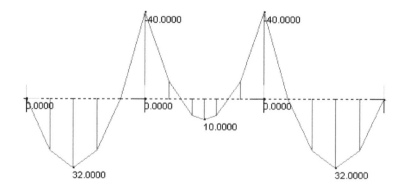

Fig. 12.3a Bending moment due to slab weight only

Case 2: Maximum load of 10.25 kN/m^2 on end spans and self weight of slab only on the middle span (Fig. 12.3b)

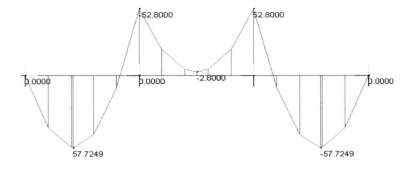

Fig. 12.3b Bending moment due to load pattern Max-Min-Max

Case 3: Maximum load of 10.25 kN/m² on middle span and self weight of slab
only on end spans (Fig. 12.3c)

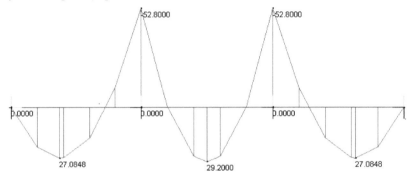

Fig. 12.3c Bending moment due to load pattern Min-Max-Min

Case 4: Maximum load of 10.25 kN/m² on first two spans and self weight of slab
only on the third span   (Fig. 12.3d)

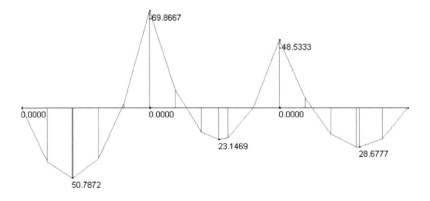

Fig. 12.3d Bending moment due to load pattern Max-Max-Min

Case 5: Maximum load of 10.25 kN/m² on second and third spans and self weight of slab only on the first span (Fig. 12.3e).

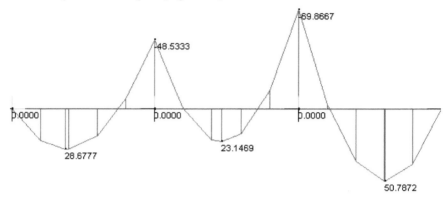

Fig. 12.3e Bending moment due to load pattern Min-Max-Max

Fig. 12.3f shows the moment envelope.

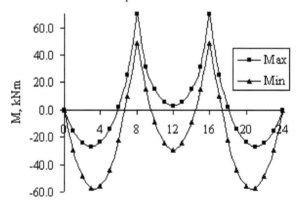

Fig. 12.3f Bending moment envelope due to applied loading

12.3.3 Choice of Prestress

Assume cover to steel = 30 mm
Diameter of strand = 16 mm
Strand at say (30 + 16/2) = 38 mm from top or bottom face.
Maximum eccentricity (above or below the neutral axis) available $\approx 250/2 - 38$
$$\approx 87 \text{ mm}$$
For initial sizing, ignoring reverse curvatures over supports and assuming parabolic profiles, maximum drapes available are:
Drapes:
End spans:
At end support, $e_1 = 0$, at penultimate support, $e_2 = 87$ mm

At mid-span, $e_3 = 87$ mm
Drape $\approx e_3 + (e_1 + e_2)/2 = 87 + (0 + 87)/2 = 131$ mm
Middle span:
At supports, $e_1 = e_2 = 87$ mm
At mid-span, $e_3 = 87$ mm.
Drape $= 87 + (87+87)/2 = 174$ mm
Load balancing:
Try balancing self weight + 10% superimposed loading.
Load to be balanced $= 6.25 + 0.1 \times (1.5 + 2.5) = 6.65$ kN/m^2
Ignore reverse curvatures over supports and calculate approximately the prestress
force required.
Prestress required: The relationship between equivalent load q and prestress for a
parabolic profile is given by (6.20):
$$q = 8 \times P \times \frac{drape}{L^2}$$
In end spans: $6.5 = 8 \times P \times$ drape/span$^2 = 8 \times P \times 131 \times 10^{-3}/8^2$, $P = 397$ kN/m
Keep this prestress as constant over all the three spans. In the middle span, the
equivalent load due to prestress is
$q = 8 \times 397 \times 174 \times 10^{-3}/8^2 = 8.64$ kN/m^2

Choose number of strands:
Maximum force at tensioning, $P_{max} = 148$ kN
Assuming 25% loss over long term, force per strand in service $= 0.75 \times 148$
$$= 110 \text{ kN}$$
Number of strands required $= 397/110 = 3.6$, say four strands/m
Provide four strands per meter. Prestress provided $= 110 \times 4 = 440$ kN/m

12.3.4 Calculation of Losses

(a). Loss due to curvature and wobble: Each strand is stressed from one end with
a force of $P_A = 148$ kN, $A_p = 100.5$ mm^2.
Assume: $\mu = 0.19$, $k = 0.008$/m, $E_s = 195$ kN/mm^2, draw-in, $\Delta = 6$ mm.

(i). End spans: From section 11.2.3.3, for the three-parabola profile in the end
spans, the total angle of deviation
$\theta = \theta_1 + \theta_2 + \theta_2$
where the eccentricity at the end support e_1 is zero and the eccentricities at mid-
span, e_3 and at first interior support e_2 are 87 mm each. The point of contra flexure
is at 800 mm from first interior support. Therefore
$\lambda = 0.5$, $\beta = 0.1$, $L = 8$ m
$$\theta_1 = \frac{2(e_1 + e_3)}{(1-\lambda)L} = \frac{2(0+87) \times 10^{-3}}{(1-0.5) \times 8} = 0.0435 \text{ radians}$$
$$\theta_2 = \frac{2(e_2 + e_3)}{\lambda L} = \frac{2(87+87) \times 10^{-3}}{0.5 \times 8} = 0.087 \text{ radians}$$
Total angle of deviation, $\theta = \theta_1 + \theta_2 + \theta_2 = 0.2175$ radians

(ii). Middle span: From section 11.2.3.3, for the four parabola profile shown in
Fig. 11.5, the total angular change is $\theta = \theta_1 + \theta_1 + \theta_2 + \theta_2$
where the eccentricity at the first interior support at left e_1 is 87 mm and the
eccentricity at mid-span e_3 is 87 mm and first interior support at right e_2 is 87 mm.
The point of contra flexure is at 800 mm from the first interior support. Therefore
$\lambda = 05$, $\beta = 0.1$, $L = 8$ m.

$$\theta_1 = \frac{2(e_1 + e_3)}{(1-\lambda)L} = \frac{2(87 + 87)\times 10^{-3}}{(1-0.5)\times 8} = 0.087 \text{ radians}$$

$$\theta_2 = \frac{2(e_2 + e_3)}{(1-\lambda)L} = \frac{2(87 + 87)\times 10^{-3}}{(1-0.5)\times 8} = 0.087 \text{ radians}$$

$\theta = \theta_1 + \theta_1 + \theta_2 + \theta_2 = 0.3480$ radians
The loss of prestress in different spans is calculated from the formula:
$$P_2 \approx P_1 \{1 - \mu (\theta + kx)\}$$
At first interior support: $P_B = 148 \times \{1 - 0.19 \times (0.2175 + 0.008 \times 8)\} = 140.1$ kN
At second interior support:
$P_C = 148 \times [1 - 0.19 \times \{(0.2175 + 0.3480) + 0.008 \times 16\}] = 128.5$ kN
At the right exterior support:
$P_D = 148 \times [1 - 0.19 \times \{0.2175 + 0.3480 + 0.2175 + 0.008 \times 24]$
    $= 120.6$ kN

(b). Loss of force due to draw-in:
Calculate the variation in cable force using the following data for the three span
continuous beam.
The cable is stressed from one end with a force of $P_A = 148$ kN. $A_{ps} = 100.5$ mm^2.
$\mu = 0.19$, $k = 0.008$/m, $E_s = 195$ kN/mm^2, draw-in length $\Delta = 6$ mm.
Average loss of prestress per unit length in the first span,
$p = (148 - 140.1)/8 = 1.0$ kN/m
Substituting in (11.20),

$$\ell = \sqrt{\frac{\Delta A_{ps} E_s}{\text{Loss of prestress per unit length}}} = \sqrt{\frac{(6\times 10^{-3})\times 100.5\times 195}{1.0}} = 10.8 \text{ m}$$

The wedge draw in length is larger than 8 m.
Loss of force due to wedge draw in $\approx 2p\ell = 2 \times 1.0 \times 10.8 = 21.6$ kN
At start after wedge draw-in, $P_A = 140.0 - 21.6 = 118.4$ kN
At B: At $x = 10.8$ m, $\theta = (0.2175 + 0.076) = 0.2935$ radians.
Note: The figure of 0.076 radians comes from slope of the second parabola at
$x = 10.8 - 8.0 = 2.8$ m given by (see section 6.2.2)

$$\theta_x = \frac{2x}{L^2} \frac{(e_3 + e_1)}{(1-\lambda)(1-\lambda-\beta)} = \frac{2\times 2.80}{8^2} \frac{(87+87)\times 10^{-3}}{0.5\times(1-0.5-0.1)} = 0.076$$

$P_{10.8} = 148 \times \{1 - 0.19 \times (0.2935 + 0.008 \times 10.8)\} = 137.3$ kN
Interpolating between 118.4 at $x = 0$ and 137.3 at $x = 10.8$, at $x = 8.0$, the force is
132.4
Table 12.3 shows the value of prestress at different sections along the beam.

.

Table 12.3 Variation of prestress along the beam

x- m	θ,Radians	$\mu(\theta + kx)$	$148 \times e^{-\mu(\theta + k\ell)}$, kN
0	-	-	122.1
8.0	-	-	136.7
10.8	0.2175	0.0545	138.9
16.0	0.8575	0.1870	120.3
24.0	1.1790	0.2610	109.4

12.3.5 Calculation of the Correct Equivalent Loads

Assuming 25% loss in the long term, the revised equivalent forces due to prestress for four cables are as follows:

Span 1:
Average P = 0.75 × 4 × (122.1+136.7)/2 = 388.2 kN/m
Segment 1:

$$P\frac{d^2e}{dx^2} = -P\frac{2}{L^2}\frac{(e_1+e_3)}{(1-\lambda)^2} = -388.2 \times \frac{2}{8^2}\frac{(0+87)\times 10^{-3}}{(1-0.5)^2} = -4.22 \text{ kN}/\text{m}^2$$

Segment 2:

$$P\frac{d^2e}{dx^2} = -P\frac{2}{L^2}\frac{(e_3+e_2)}{\lambda(\lambda-\beta)} = -388.2\times\frac{2}{8^2}\frac{(87+87)\times 10^{-3}}{0.5(0.5-0.1)} = = -10.55 \text{ kN}/\text{m}^2$$

Segment 3:

$$P\frac{d^2e}{dx^2} = P\frac{2}{L^2}\frac{(e_3+e_2)}{\lambda\beta} = 388.2\frac{2}{8^2}\frac{(87+87)\times 10^{-3}}{0.5\times 0.1} = 42.2 \text{ kN}/\text{m}^2$$

Span 2:
Average P = 0.75 × 4 × (136.7+120.3)/2 = 385.5 kN/m
Segment 1 and 4:

$$P\frac{d^2e}{dx^2} = P\frac{2}{L^2}\frac{(e_1+e_3)}{(1-\lambda)\beta_1} = 385.5\frac{2}{8^2}\frac{(87+87)\times 10^{-3}}{(1-0.5)\times 0.1} = 41.91 \text{ kN}/\text{m}^2$$

Segment 2 and 3:

$$P\frac{d^2e}{dx^2} = -P\frac{2}{L^2}\frac{(e_1+e_3)}{(1-\lambda-\beta_1)(1-\lambda)} = -385.5\frac{2}{8^2}\frac{(87+87)\times 10^{-3}}{(1-0.5-0.1)(1-0.5)} = -10.48 \text{ kN}/\text{m}^2$$

Span 3:
Average P = 0.75 × 4 × (120.3 + 109.4)/2 = 344.6 kN/m
Segment 1:

$$P\frac{d^2e}{dx^2} = -P\frac{2}{L^2}\frac{(e_1+e_3)}{(1-\lambda)^2} = -344.6\times\frac{2}{8^2}\frac{(0+87)\times 10^{-3}}{(1-0.5)^2} = -3.75 \text{ kN}/\text{m}^2$$

Segment 2:

$$P\frac{d^2e}{dx^2}=-P\frac{2}{L^2}\frac{(e_3+e_2)}{\lambda(\lambda-\beta)}=-344.6\times\frac{2}{8^2}\frac{(87+87)\times10^{-3}}{0.5(0.5-0.1)}==-9.37\,\text{kN/m}^2$$

Segment 3:

$$P\frac{d^2e}{dx^2}=P\frac{2}{L^2}\frac{(e_3+e_2)}{\lambda\beta}=344.6\frac{2}{8^2}\frac{(87+87)\times10^{-3}}{0.5\times0.1}=37.46\,\text{kN/m}^2$$

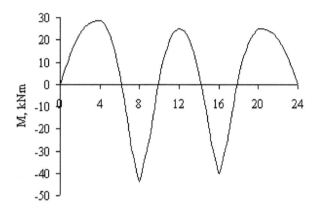

Fig. 12.4 Bending moment due to equivalent loads

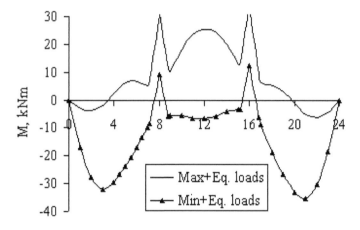

Fig. 12.5 Moment envelope at service due to prestress and applied loads

12.3.6 Calculation of moment distribution at service

Fig 12.4 shows the bending moment distribution due to equivalent loads.
Since at service the load factor on the prestressing force is 0.9, in calculating the
moment envelope and stresses, the prestressing force is multiplied by 0.9. Fig. 12.5

shows the final moment envelope. Using a load factor of 0.9, the prestressing forces in the three spans respectively are 349.4, 347.0, 310.1 kN/m².

12.3.7 Calculation of Stress Distribution at Service

The stress distribution in the slab is calculated using the following data:
Prestress P = 349.4, 347.0, 310.1 kN/m² respectively in the three spans
Slab: Area of cross section, A = 1000 × 250 = 2.5 × 10⁵ mm²,
section modulus, Z = 1000 × 250²/6 = 10.42 × 10⁶ mm³
σ_{top} = −P/A + {maximum or minimum moments due to (loads + prestress)}/Z
σ_{bottom} = −P/A − {maximum or minimum moments due to (loads + prestress)} Z

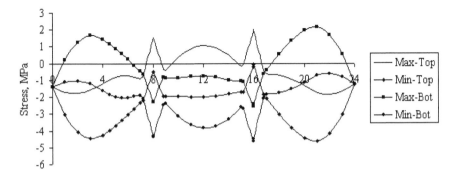

Fig. 12.6 Final stress distribution at serviceability limit state.

Fig. 12.6 shows the resulting stress distribution at the top and bottom of the slab. Comparing the stresses with the permissible values of 2.9 MPa in tension and −18.0 MPa in compression, it can be concluded that the stress distribution is satisfactory, although it is possible to reduce the thickness of the slab.

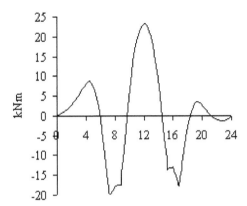

Fig. 12.7 Moment envelope at transfer due to prestress and applied loads

12.3.8 Calculation of Stress Distribution at Transfer

At transfer the only external load is the self weight of the slab. Assuming that at
transfer there is a loss of prestress of about 10%, the prestresses in the three spans
are obtained by multiplying the prestress at service by (0.9/0.75 = 1.2) because the
loss of prestress at service is assumed to be about 25%. The load factor at transfer
is 1.1. Fig. 12.7 shows the bending moment envelope and Fig 12.8 shows the
stress distribution. The state of stress is satisfactory as the compressive and tensile
stresses are below 15.0 MPA and 2.6 MPa.

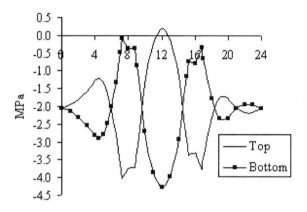

Fig. 12.8 Final stress distribution at transfer

12.4 EDGE-SUPPORTED TWO-WAY SPANNING SLABS

Fig. 12.9 shows an edge-supported two-way spanning slab. The slab is supported
on all four edges either by beams or walls. The slab spans in both directions and
the prestressing cables run in both directions.

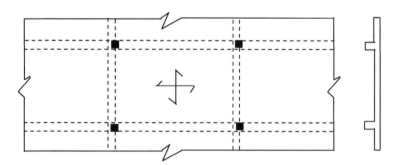

Fig. 12.9 Edge-supported two-way spanning slab

12.4.1 Design of a Two-Way Spanning Slab

Design a two-way spanning slab 8 m × 8 m supported on masonry walls providing no rotational restraint. The corners are assumed to be prevented from lifting. The slab has to support a live load of 4 kN/m² and superimposed dead load of 1.0 kN/m². Design using the data in section 12.3.1.

Choose a depth equal to span/45 = $(8 \times 10^3)/45 = 178$ mm, say 200 mm.

Self weight $= 0.2 \times 25 = 5$ kN/m²

Total load due to self weight, partitions and live load $= 5.0 + 1.0 + 4.0 = 10$ kN/m²

Prestressing force:

Balance approximately self weight + 10% of superimposed dead load.

Load to balance $= 5.0 + 0.1 \times 1.0 = 5.1$ kN/m²

Because of two-way action, half of this load will be balanced in each direction.

In the x-direction, maximum eccentricity e_x at mid-span = 200/2 – 38 = 62 mm

Assuming that the 13 mm diameter cables in the y-direction lie above those in the x-direction,

In the y-direction, maximum eccentricity e_y at mid-span = 200/2 – 38 –13 = 49 mm

Since the eccentricities at the ends of the parabola are zero, the drape is equal to the eccentricity at mid-span.

Prestress in x-direction: $0.5 \times 5.1 = (8 \times P_x \times 62 \times 10^{-3})/ 8^2$, $P_x = 329.0$ kN/m

Prestress in y-direction: $0.5 \times 5.1 = (8 \times P_x \times 49 \times 10^{-3})/ 8^2$, $P_y = 416.3$ kN/m

Maximum force at tensioning, $P_{max} = 148$ kN

Assuming 25% loss over the long term, force per strand in service $= 0.75 \times 148$
$$= 110 \text{ kN}$$

Number of strands required:

x-direction: $n_x = 329/110 = 3$ strands/m,

Prestress provided $= 110 \times 3 = 330$ kN/m

Load balanced, $q_x = (8 \times P_x \times 62 \times 10^{-3})/8^2 = 2.6$ kN/m²

y-direction: $n_y = 416.3/110 = 3.8$, say 4 strands /m

Prestress provided $= 110 \times 4 = 440$ kN/m

Load balanced, $q_y = (8 \times P_y \times 49 \times 10^{-3})/8^2 = 2.7$ kN/m²

Total load balanced $= q_x + q_y = 5.3$ kN/m²

Calculation of loss:

Loss due to curvature and wobble: Use $\mu = 0.19$, k = 0.008/m

Drape = mid-span eccentricity = 62 mm and 49 mm in the x- and y-directions, respectively.

From (6.1), angular deviation $\theta_x = 8 \times$ drape/span $= 8 \times 62 \times 10^{-3}/8 = 0.062$ radians

Angular deviation $\theta_y = 8 \times$ drape/span $= 8 \times 49 \times 10^{-3}/8 = 0.049$ radians

Net force $P_x = 330\{1 - 0.19(0.062 + 0.008 \times 8)\} = 322.1$ kN/m

Net force $P_y = 440\{1 - 0.19(0.049 + 0.008 \times 8)\} = 430.6$ kN/m

Loss due to anchorage slip:

The cable is stressed from one end with a force of $P_A = 148$ kN.

x-direction:

$A_{ps} = 3 \times 100.5 = 302$ mm², $E_s = 195$ kN/mm², draw-in length $\Delta = 6$ mm

Average loss of prestress p per unit length in the first span = (330 − 322)/8 = 1.0 kN/m

$$\ell = \sqrt{\frac{\Delta A_{ps}\, E_s}{\text{Loss of prestress per unit length}}} = \sqrt{\frac{(6\times10^{-3})\times302\times195}{1.0}} = 18.8\,\text{m} \ >> 8\ \text{m}$$

Calculate loss using the average strain $\varepsilon = (6 \times 10^{-3})\,/8 = 7.5 \times 10^{-4}$

Loss of force due to anchorage slip:

x-direction: for three cables = $(7.5 \times 10^{-4}) \times 195 \times 3 \times 100.5 = 44.2$ kN/m

y-direction: for four cables = $(7.5 \times 10^{-4}) \times 195 \times 4 \times 100.5 = 64.6$ kN/m

Note: Because of the short length of the cable, the loss due to anchorage slip is quite high.

Average force in x-direction= (330 + 322)/2 − 44.2 = 282.0 kN/m

Average force in y-direction= (440 + 430.6)/2 − 64.6 = 370.7 kN/m

Total load balanced using the revised forces in the -x and y-directions are

Load balanced, $q_x = (8 \times 282 \times 62 \times 10^{-3})/8^2 = 2.2$ kN/m^2

Load balanced, $q_y = (8 \times 370.7 \times 49 \times 10^{-3})/8^2 = 2.3$ kN/m^2

Total load balanced = $q_x + q_y = 4.5$ kN/m^2

Using a load factor on prestressing force at service of 0.9, net load at service is given by

q = Total load − balanced load = 10.0 − 4.5 × 0.9 = 5.95 kN/m^2

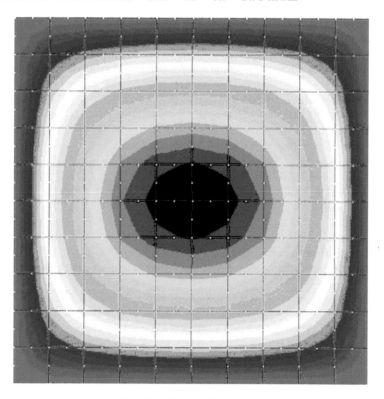

Fig. 12.10a Contour of moment M_x

Fig. 12.10a-d show the distribution of bending moment M_{xx} and twisting moment M_{xy} in a simply supported two-way slab subjected to uniformly distributed loading of 5.95 kN/m². Because of two-fold symmetry, the contours for the moment M_y will be identical to the contour for M_x. The analysis was done using an eight-node plate bending element. As can be seen, the bending moment is the maximum at mid-span and zero towards the supports. Because of the fact that the corners are held down, the twisting moment is maximum towards the corner and zero at the centre. The twisting moments need to be resisted by unstressed steel near the corners.

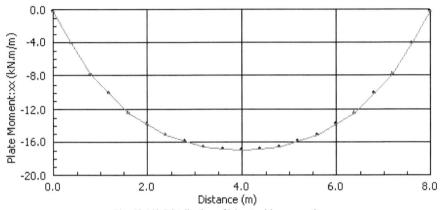

Fig. 12.10b Distribution of M_{xx} at mid-span section

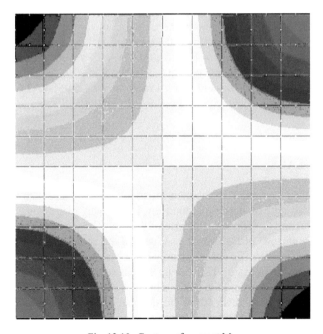

Fig. 12.10c Contour of moment M_{xy}

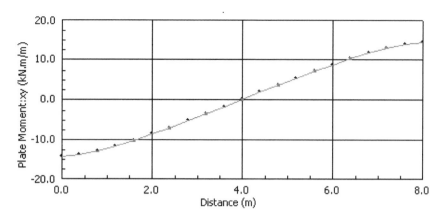

Fig. 12.10d Distribution of M_{xy} at support section

Stress check:
The maximum net bending moment at the mid-span section in both x- and y-directions is 17.02 kNm/m. The axial force is 282× 0.9 = 254 kN/m in the x-direction and 370.7× 0.9 = 334 kN/m in the y-direction.
Using A = 1000 × 200 = 2× 10^5 mm^2, Z = 1000 × $200^2/6$ = 6.67× 10^6 mm^3, calculated stresses using the formula
$\sigma_{top} = -P/A - M/Z$, $\sigma_{Bottom} = -P/A + M/Z$
 The stresses at top and bottom of the slab are (−3.8, 1.3) MPa in the x-direction and (−4.2, 0.9) MPa in the y-direction.
At transfer the prestress in is 0.9/0.75 of that at service. Using a load factor of 1.1, the axial force P is
282 × (0.9/0.75) × 1.1 = 372 kN/m in the x-direction
370.7 × (0.9/0.75 × 1.1 = 489 kN/m in the y-direction.
The load balanced by prestress is 4.5 × (0.9/0.75) × 1.1 = 5.94 kN/m^2
The net load acting is 5.94 – self weight = − 0.94 kN/m^2
The maximum moment at the centre is = 17.02 × (−0.94/5.95) = −2.67 kNm/m
The stresses are (−2.3, −1.5) MPa and (−2.9, −2.1) MPa in the x- and y-directions respectively.

12.5 FLAT SLABS

Fig. 12.11 shows a typical flat slab also called a beamless slab. The main attraction of this type of slab is that the formwork is very simple and it leads to an uncluttered flat soffit. However because of the fact the loads are transferred to the column over a narrow area, shear stresses around the column tend to be very high and can lead to failure. The shear capacity can be increased either by providing a column head as shown in Fig. 12.12 or by providing a drop panel, which is local thickening of the slab around the column head as shown in Fig. 12.13. Sometimes a steel shear head within the depth of the slab can be incorporated over columns to increase shear capacity and this has the advantage of eliminating the complexity introduced by column heads or drop panels. This form is generally suitable for spans in the range of 6 to 13 m.

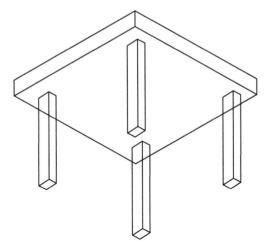

Fig. 12.11 Flat slab

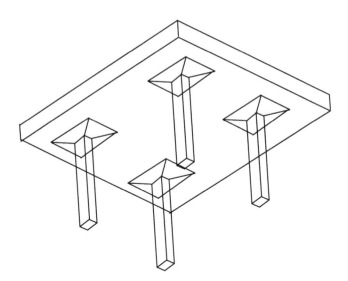

Fig. 12.12 Solid flat slab with column head

In order to reduce the dead weight of the slab, one can use a 'waffle slab' floor supported by columns as shown in Fig. 12.14. Waffle slabs are formed by using a grid work of fibre glass dome forms. This creates a grid work of deep beams between the 'domes' while the floor area between the beams is much thinner compared to the beams. From the underside, the slab resembles a waffle.
 Although the slab consists of a grid work of beams in two directions, they are normally analysed as a slab provided the following conditions are satisfied:
Rib spacing ≤ 1.5 m, depth of rib below the flange ≤ 4 × rib width
Depth of flange ≥ 50 mm or clear distance between rib /10

Transverse rib spacing $\leq$ 10 × depth of slab

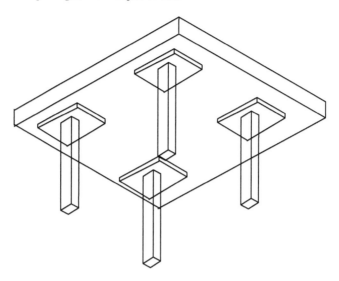

Fig. 12.13 Solid flat slab with drop panel

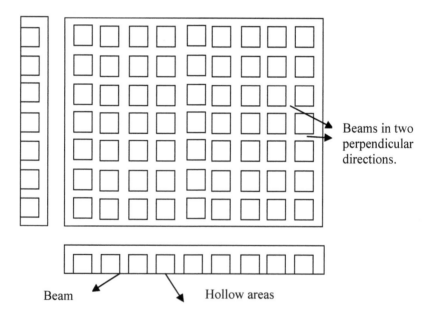

Beams in two perpendicular directions.

Beam   Hollow areas

Fig. 12.14 Underside of a 'waffle' slab

In order to increase the shear capacity near the columns, the slab is made solid around the columns. They are used in buildings subjected to heavy loading and spans are similar in both directions. Spans in the range of 10 to 20 m are common. In spite of the structural advantages, due to the perceived complexity of construction, there has been a falling out of favour of this form of construction. The ribbed one-way slab described in section 12.4 is a particular form of waffle slab where spans are predominantly in one direction.

12.6 METHODS OF ANALYSIS OF FLAT SLABS

Analysis of flat slabs can be done by a number of methods, some of which are less accurate than others. In all methods the analysis is carried out assuming elastic properties. The object is to obtain a set of stresses in equilibrium with external loads to be used in a lower bound design method. Some of the analysis methods used are

- Simple frame: In this method, the slab is divided into design strips and the slab along with the columns are analysed as a two-dimensional frame. The width of the slab is the width between the lines of zero shear.
- Equivalent frame: There are many types of idealizations, all known as equivalent frames. In the method as described in the ACI codes, the stiffnesses of slab and columns are adjusted to account for biaxial plate bending. It is generally thought as being more accurate than the simple frame model. However in the method given in Eurocode 2, described in greater detail in the next section, gross areas of cross sections are used in the analysis.
- Grillage analysis: The slab is divided into a set of beams in two directions and analysed using a grillage program. This could give more accurate results than a frame program, especially when irregular column layout needs to be catered for. However, the problem of allocating loads acting on the slab to beams in two different directions can cause some uncertainties.

The interaction of the slab with the columns can be allowed for by including a rotational spring to account for the stiffness of columns as shown in Fig. 12.15. The stiffness K of the rotational spring is equal to K_{fixed} or K_{pinned} depending on whether the far ends of the columns are fixed or pinned.

$$K_{fixed} = 4\{\frac{EI_1}{L_1} + \frac{EI_2}{L_2}\}$$

$$K_{pinned} = 3\{\frac{EI_1}{L_1} + \frac{EI_2}{L_2}\}$$

Using the appropriate values of the second moments of area I_1 and I_2 to account for rectangular shaped columns, spring stiffness is provided about the x- and y-axes.

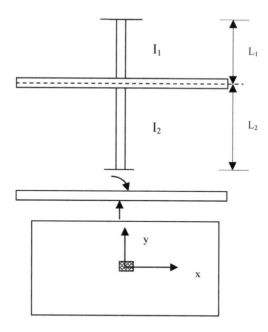

Fig. 12.15 Modelling for column stiffness

- Finite element method: This is the most accurate method of slab analysis and can cater for all complex situations. However, the output of stresses is not in the most convenient form for design purposes.

One of the main problems with this method is that the solutions are based on elastic analysis. This leads to high peaked moments at supports. It is generally believed that at the serviceability limit state these high moments do not exist in practice because of cracking and redistribution of stresses which lead to larger moments in span and smaller moments at supports. It is possible to adjust the stiffness of slab elements to reduce the concentration of moments but this complicates the analysis.

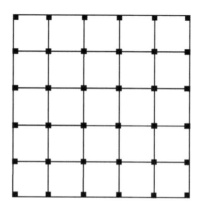

Fig. 12.16a Floor plan

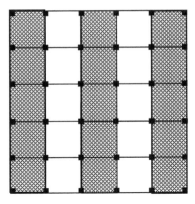

Fig. 12.16b Live load pattern 1

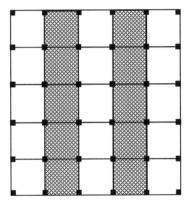

Fig. 12.16c Live load pattern 2

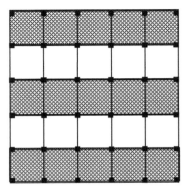

Fig. 12.16d Live load pattern 3

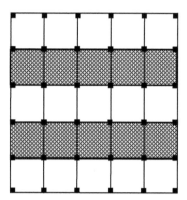

Fig. 12.16e Live load pattern 4

Regarding the load patterns to be considered, full factored dead load on the whole slab together with factored live loads on alternate bays should be examined. It is not necessary to use chequer-board pattern loading. For example for the slab shown in Fig. 12.16a, only the four live load patterns shown in Fig. 12.16 b-e need to be considered.

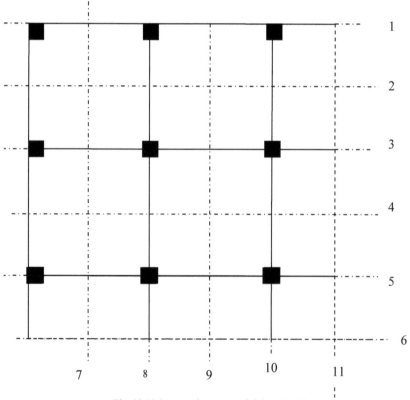

Fig. 12.17 Symmetric quarter of slab analysed

12.7 EXAMPLE OF THE DESIGN OF A FLAT SLAB

Fig. 12.16a shows the floor plan of a post-tensioned slab. It is assumed that the spans in both directions are 6.0 m. It is subjected to a live load of 2.5 kN/m^2 and the super dead load due to screed, partitions, ceiling, etc. is 1.5 kN/m^2. The internal columns are assumed to be 400 mm square and edge columns 300 mm square. The storey height is 3 m. There are rigid walls outside the area shown to resist horizontal loads.

Solution: Assume a span/depth ratio = 30
Depth ≈ 6000/30 = 200 mm

Loading:
Assuming unit weight at 25 kN/m^3, self weight = 0.2 × 25 = 5.0 kN/m^2
Total imposed load = 1.5 (SDL) + 2.5 (Live) = 4.0 kN/m^2

Material properties: Assume as in section 12.3.1.

Load factors for prestress in post-tensioning:
$\gamma_{Superior}$ = 1.1 at transfer and $\gamma_{Inferior}$ = 0.90 at service stage.

12.8 FINITE ELEMENT ANALYSIS OF FLAT SLAB

In order to get a feel for the distribution of bending moment in a flat slab, a finite element analysis program was used to analyse a symmetric quarter of the slab as shown in Fig. 12.17. The slab was modelled as 225 eight-node isotropic plate elements, each 1.0 × 1.0 m. Because of symmetry, it was assumed that there was no rotation about the axes of symmetry. The columns were modelled as point supports and the vertical displacement was restrained at the columns.
The stiffness of columns was represented as rotational springs acting about both x- and y- axes as explained in section 12.6.
Column height = 3 m,
Interior columns: 0.4 m × 0.4m, I = 2.133 × 10^{-3} m^4
Exterior columns: 0.3 m × 0.3m, I = 0.675 × 10^{-3} m^4
E = 32.8 GPa = 32.8 × 10^6 kN/m^2
Rotational stiffness K of columns: Calculate the stiffness assuming that the columns are fixed at the ends.
Interior columns; K = 2 × 4 × 32.8 × 10^6 × 2.133 × 10^{-3}/3 = 1.87 × 10^5 kNm/radian
Exterior columns: K = 2 × 4 × 32.8 × 10^6 × 0.675 × 10^{-3}/3 = 0.59 × 10^5 kNm/radian
Because of symmetry, only a quarter of slab needs to be analysed. The dead load is 5 kN/m^2 and live load is 4 kN/m^2.
Note: Because of symmetry, in all cases, M_{xx} along sections 7-11 will correspond respectively to M_{yy} for sections 2-6.

12.8.1 Results of Analysis for Dead Load

Fig. 12.18 shows the contours for the moment about the x-axis. Fig. 12.19a to k shows the distribution of moments at eleven sections. Fig. 12.20 shows the contours for the moment about the y-axis.

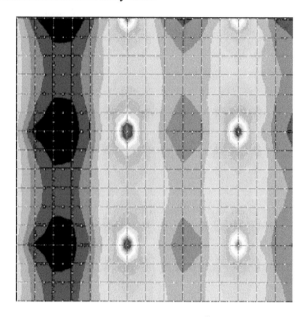

12.18 Contour of M_{xx} for dead load of 5 kN/m² over the whole slab.

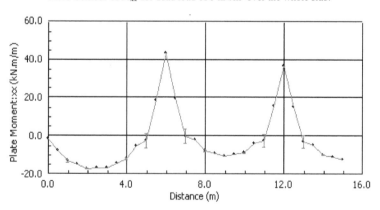

12.19a M_{xx} for dead load at section 1-1

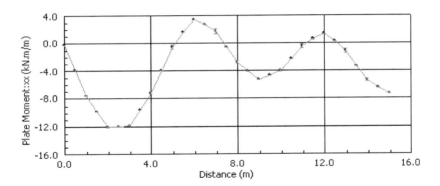

12.19b M_{xx} for dead load at section 2-2

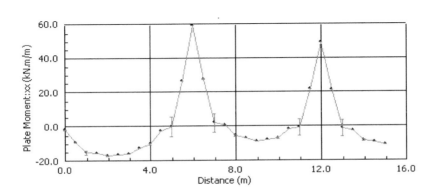

12.19c M_{xx} for dead load at section 3-3

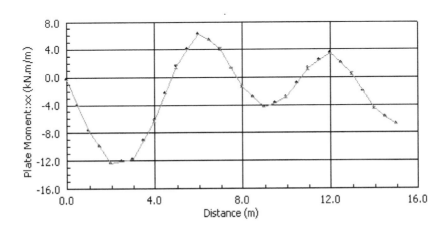

12.19d M_{xx} for dead load at section 4-4

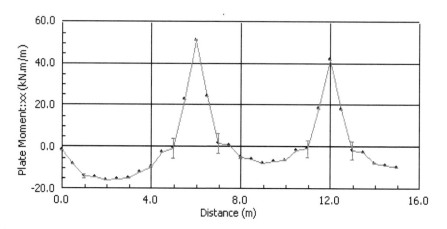

12.19e M_{xx} for dead load at section 5-5

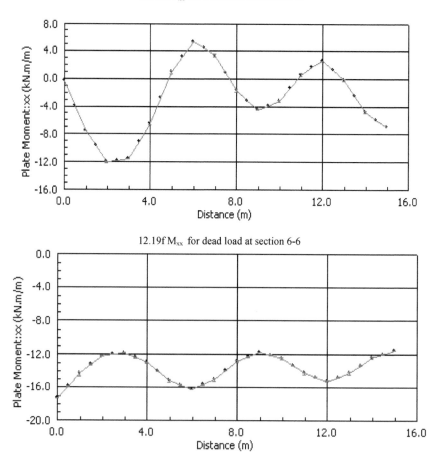

12.19f M_{xx} for dead load at section 6-6

12.19g M_{xx} for dead load at section 7-7

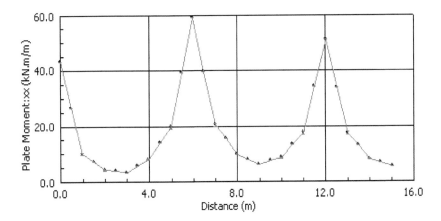

12.19h M$_{xx}$ for dead load at section 8-8

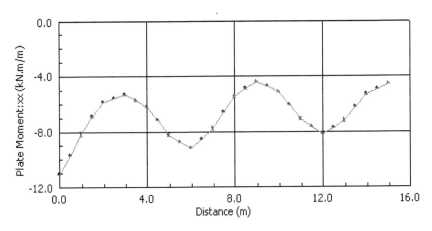

12.19i M$_{xx}$ for dead load at section 9-9

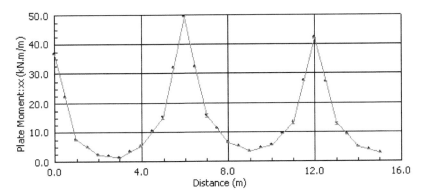

12.19j M$_{xx}$ for dead load at section 10-10

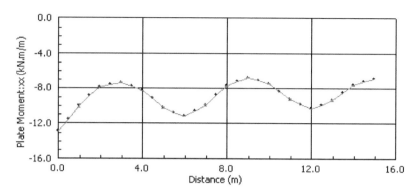

12.19k M_{xx} for dead load at section 11-11

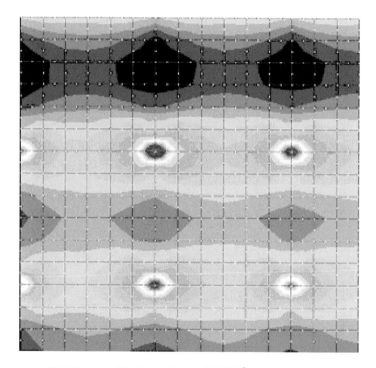

12.20 Contour of M_{yy} for dead load of 5 kN/m^2 over the whole slab

12.8.2 Results of Analysis for Dead plus Live Load Pattern 1

Fig. 12.21 shows the contours for the moment about the x-axis. Fig. 12.22 a to k show the distribution of moments at eleven sections. Fig. 12.23 shows the contours for the moment about the y-axis.

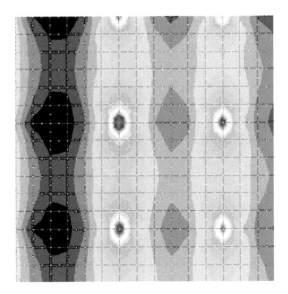

Fig. 12.21 Contour of M_{xx} for dead and live load pattern 1

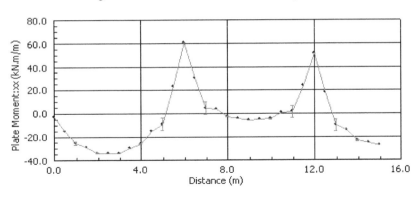

Fig. 12.22a M_{xx} at section 1-1 for dead and live load pattern 1

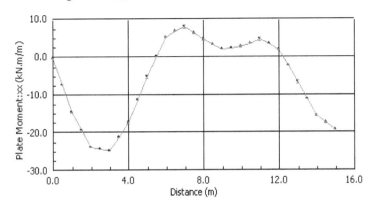

Fig. 12.22b M_{xx} at section 2-2 for dead and live load pattern 1

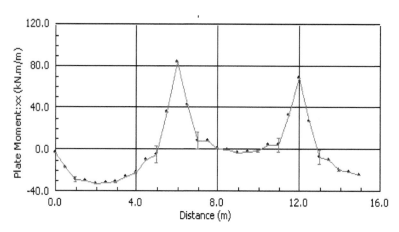

Fig. 12.22c M_{xx} at section 3-3 for dead and live load pattern 1

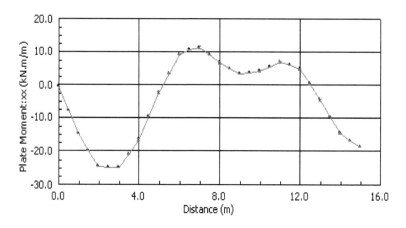

Fig. 12.22d M_{xx} at section 4-4 for dead and live load pattern 1

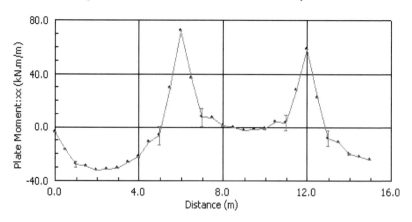

Fig. 12.22e M_{xx} at section 5-5 for dead and live load pattern 1

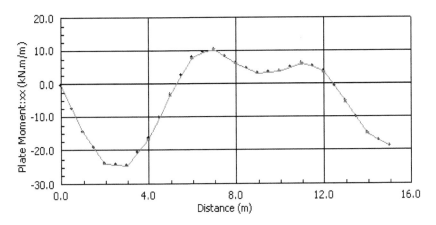

Fig. 12.22f M_{xx} at section 6-6 for dead and live load pattern 1

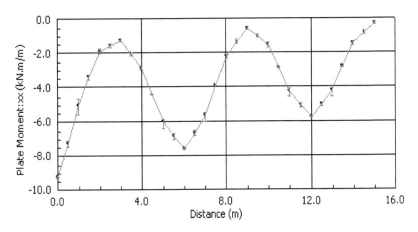

Fig. 12.22g M_{xx} at section 7-7 for dead and live load pattern 1

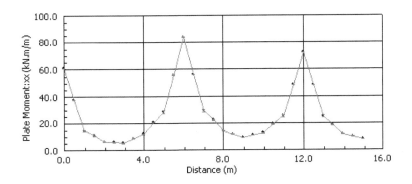

Fig. 12.22h M_{xx} at section 8-8 for dead and live load pattern 1

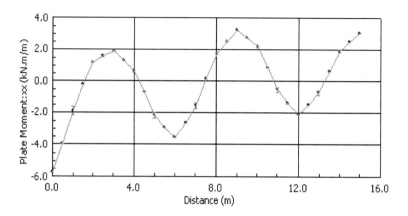

Fig. 12.22i M_{xx} at section 9-9 for dead and live load pattern 1

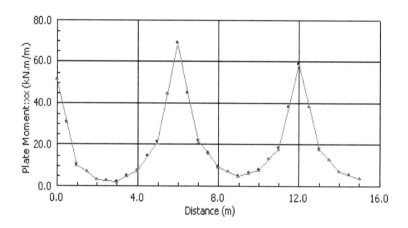

Fig. 12.22j M_{xx} at section 10-10 for dead and live load pattern 1

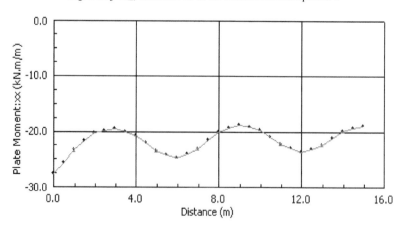

Fig. 12.22k M_{xx} at section 11-11 for dead and live load pattern 1

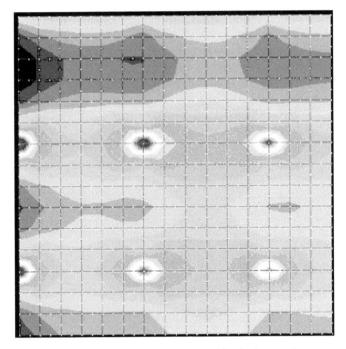

Fig. 12.23 Contour of M_{yy} for dead and live load pattern 1

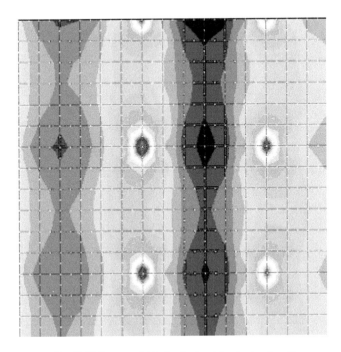

Fig. 12.24 Contour of M_{xx} for dead and live load pattern 2

12.8.3 Results of Analysis for Dead plus Live Load Pattern 2

Fig. 12.24 shows the contours for the moment about the x-axis. Figs. 12.25a to k show the distribution of moments at eleven sections. Fig. 12.26 shows the contours for the moment about the y-axis.

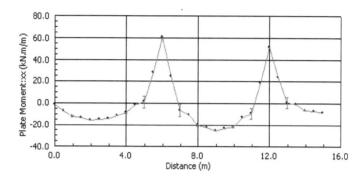

Fig. 12.25a M_{xx} at section 1-1 for dead and live load pattern 2

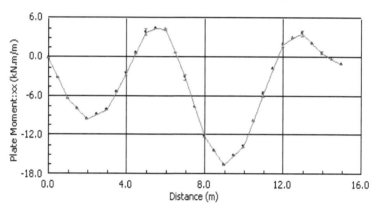

Fig. 12.25b M_{xx} at section 2-2 for dead and live load pattern 2

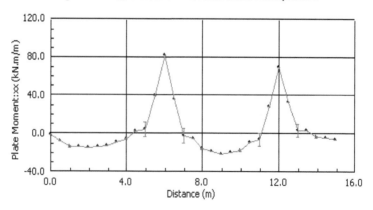

Fig. 12.25c M_{xx} at section 3-3 for dead and live load pattern 2

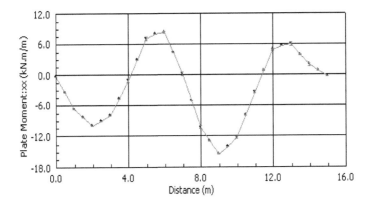

Fig. 12.25d M_{xx} at section 4-4 for dead and live load pattern 2

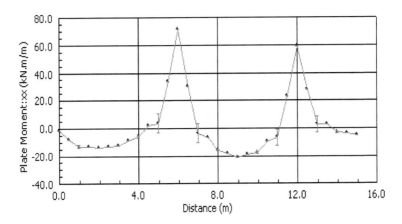

Fig. 12.25e M_{xx} at section 5-5 for dead and live load pattern 2

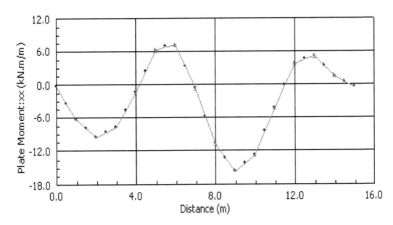

Fig. 12.25f M_{xx} at section 6-6 for dead and live load pattern 2

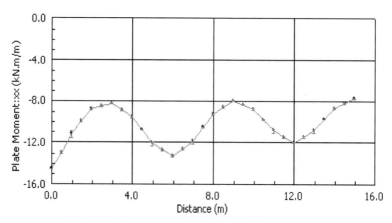

Fig. 12.25g M_{xx} at section 7-7 for dead and live load pattern 2

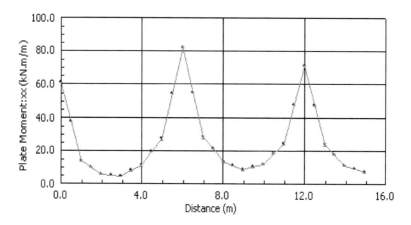

Fig. 12.25h M_{xx} at section 8-8 for dead and live load pattern 2

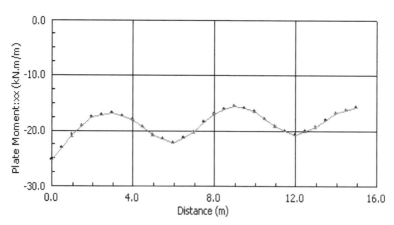

Fig. 12.25i M_{xx} at section 9-9 for dead and live load pattern 2

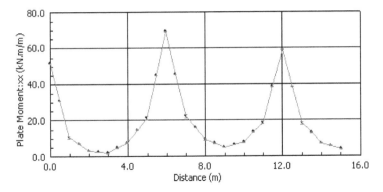

Fig. 12.25j M_{xx} at section 10-10 for dead and live load pattern 2

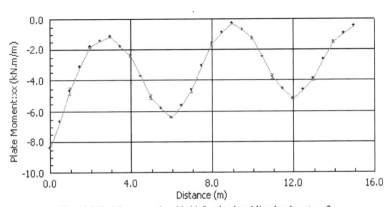

Fig. 12.25k M_{xx} at section 11-11 for dead and live load pattern 2

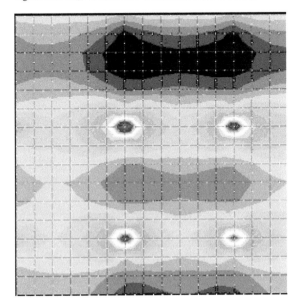

Fig. 12.26 Contour of M_{yy} for dead and live load pattern 2

12.8.4 Results of Analysis for Dead plus Live Load Pattern 3

Fig. 12.27 shows the contours for the moment about the x-axis. Fig. 12.28a to k show the distribution of moments at eleven sections. Fig. 12.29 shows the contours for the moment about the y-axis.

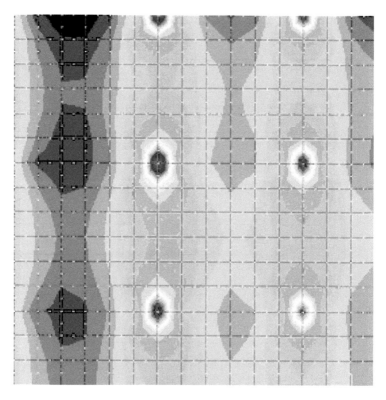

Fig. 12.27 Contour of M_{xx} for dead and live load pattern 3

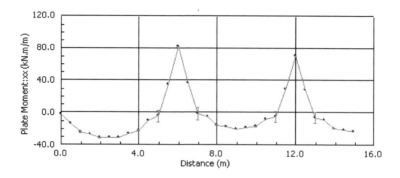

Fig. 12.28a M_{xx} at section 1-1 for dead and live load pattern 3

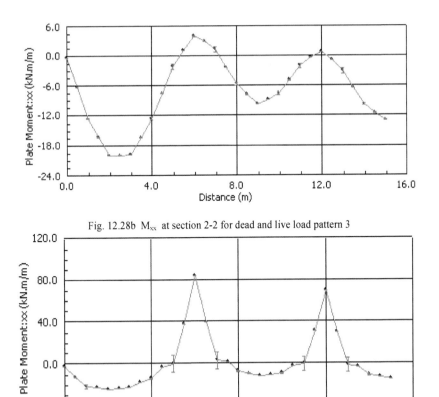

Fig. 12.28b M_{xx} at section 2-2 for dead and live load pattern 3

Fig. 12.28c M_{xx} at section 3-3 for dead and live load pattern 3

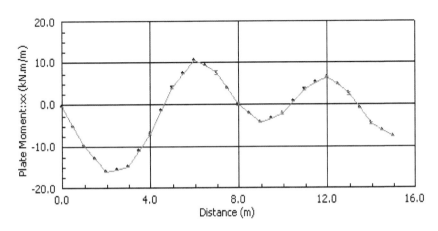

Fig. 12.28d M_{xx} at section 4-4 for dead and live load pattern 3

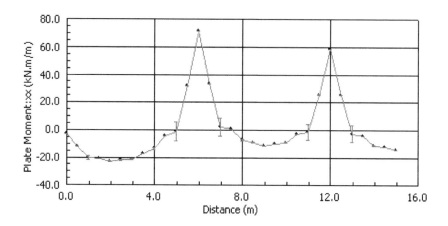

Fig. 12.28e M_{xx} at section 5-5 for dead and live load pattern 3

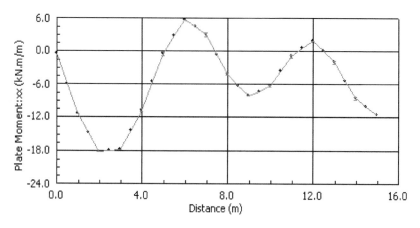

Fig. 12.28f M_{xx} at section 6-6 for dead and live load pattern 3

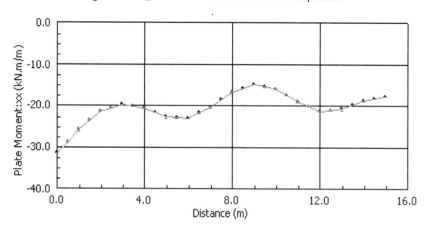

Fig. 12.28g M_{xx} at section 7-7 for dead and live load pattern 3

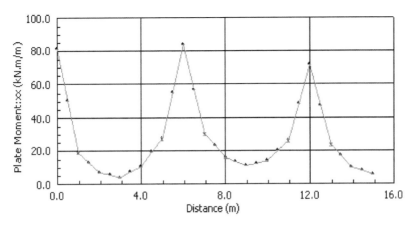

Fig. 12.28h M_{xx} at section 8-8 for dead and live load pattern 3

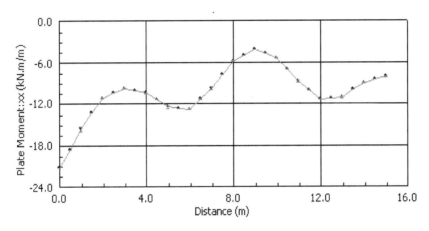

Fig. 12.28i M_{xx} at section 9-9 for dead and live load pattern 3

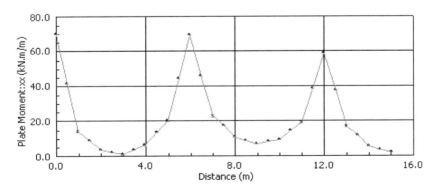

Fig. 12.28j M_{xx} at section 10-10 for dead and live load pattern 3

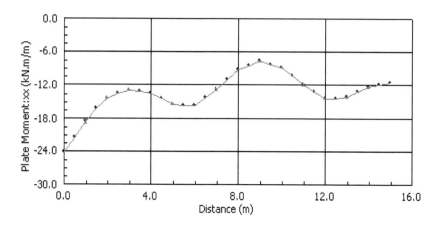

Fig. 12.28k M_{xx} at section 11-11 for dead and live load pattern 3

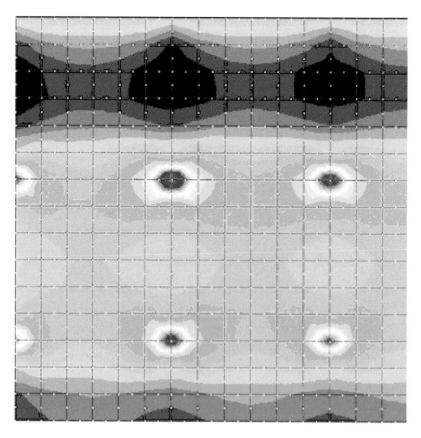

Fig. 12.29 Contour of M_{yy} for dead and live load pattern 3

12.8.5 Results of Analysis for Dead plus Live Load Pattern 4

Fig. 12.30 shows the contours for the moment about the x-axis. Figs. 12.31a-k show the distribution of moments at eleven sections. Fig. 12.32 shows the contours for the moment about the y-axis.

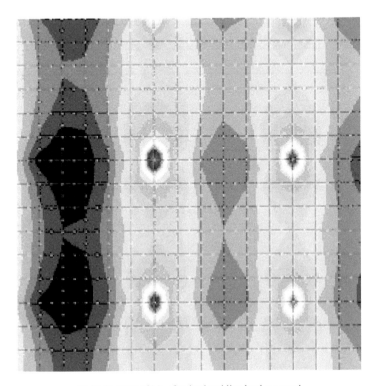

12.30 Contour of M_{xx} for dead and live load pattern 4

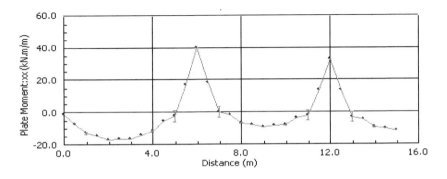

Fig. 12.31a M_{xx} at section 1-1 for dead and live load pattern 4

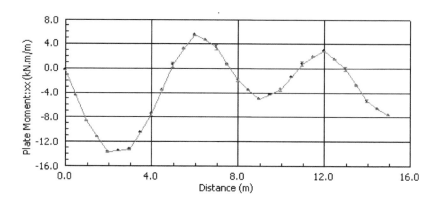

Fig. 12.31b M_{xx} at section 2-2 for dead and live load pattern 4

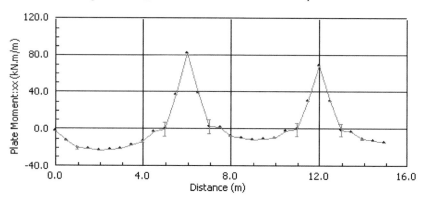

Fig. 12.31c M_{xx} at section 3-3 for dead and live load pattern 4

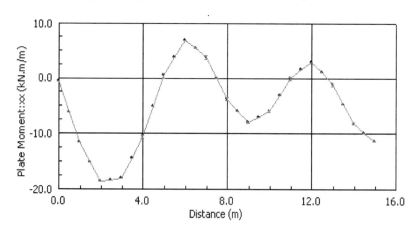

Fig. 12.31d M_{xx} at section 4-4 for dead and live load pattern 4

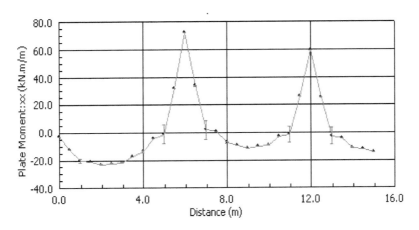

Fig. 12.31e M_{xx} at section 5-5 for dead and live load pattern 4

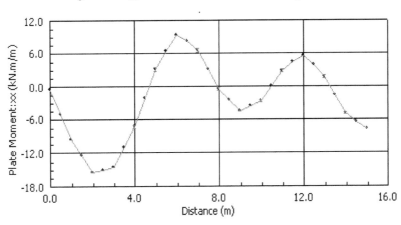

Fig. 12.31f M_{xx} at section 6-6 for dead and live load pattern 4

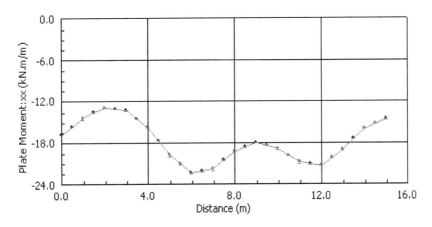

Fig. 12.31g M_{xx} at section 7-7 for dead and live load pattern 4

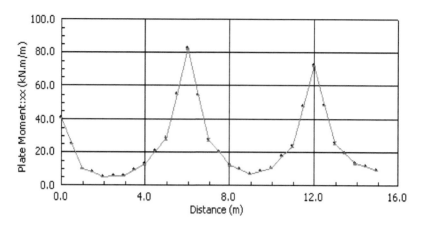

Fig. 12.31h M_{xx} at section 8-8 for dead and live load pattern 4

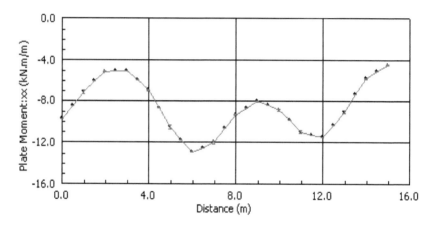

Fig. 12.31i M_{xx} at section 9-9 for dead and live load pattern 4

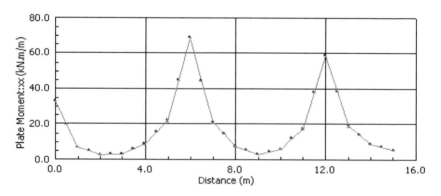

Fig. 12.31j M_{xx} at section 10-10 for dead and live load pattern 4

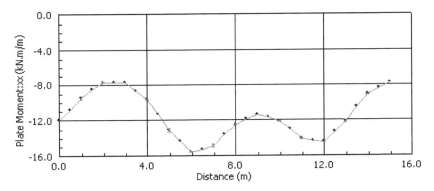

Fig. 12.31k M_{xx} at section 11-11 for dead and live load pattern 4

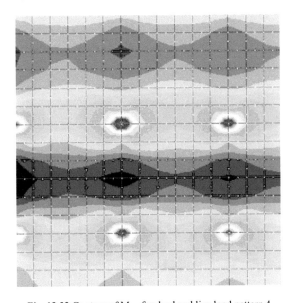

Fig. 12.32 Contour of M_{yy} for dead and live load pattern 4

12.9 FINITE ELEMENT ANALYSIS OF A STRIP OF FLAT SLAB

An approximate analysis of the flat slab can be made by analysing only a part of the slab bounded by the so-called 'zero lines of shear'. Fig. 12.33 shows the symmetrical half of the slab strip which was analysed. The slab was modelled as 250 eight-node isotropic plate elements. Because of symmetry, it was assumed that there was no rotation about the two longer axes and the short right-hand axes of symmetry. The rest of the details were as detailed in section 12.8.

Fig. 12.34a to e show the five load cases analysed. In the figures, the hatched areas are where both dead and live load act and the clear areas are where only dead load acts.

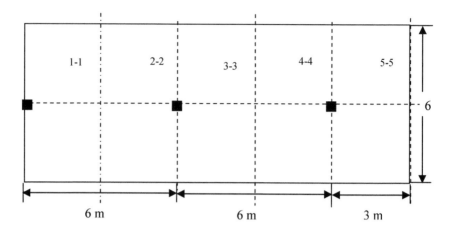

Fig. 12.33 Symmetrical half of slab strip

Fig. 12.34a Dead load only

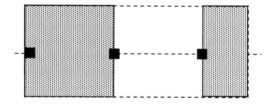

Fig. 12.34b Dead + live load pattern 1

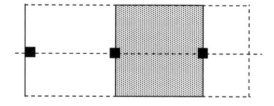

Fig. 12.34c Dead + live load pattern 2

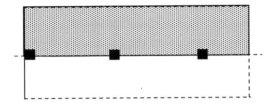

Fig. 12.34d Dead + live load pattern 3

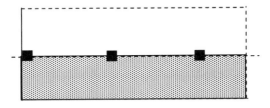

Fig. 12.34e Dead + live load pattern 4

12.9.1 Results of Analysis for Dead Load

Fig. 12.35 shows the contours for the moment about the x-axis. Fig. 12.36a to f show the distribution of moments at six sections. Fig. 12.37 shows the contours for the moment about the y-axis.

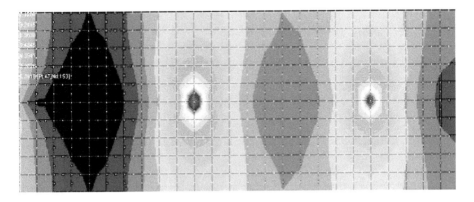

Fig.12.35 Contour of moment M_{xx} for dead load

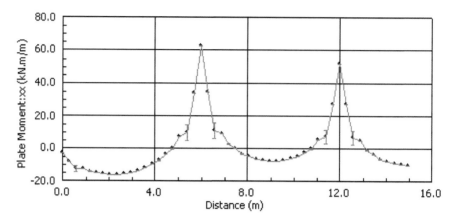

Fig. 12.36a M$_{xx}$ along column line for dead load

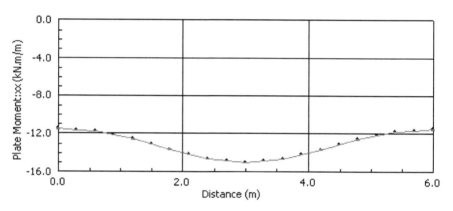

Fig. 12.36b M$_{xx}$ along section 1-1 for dead load

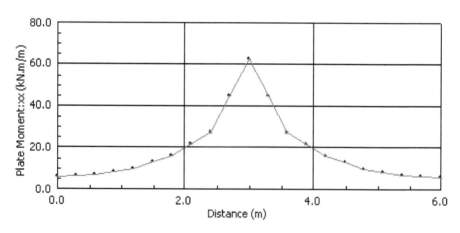

Fig. 12.36c M$_{xx}$ along section 2-2 for dead load

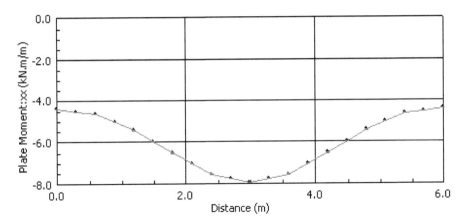

Fig. 12.36d M_{xx} along section 3-3 for dead load

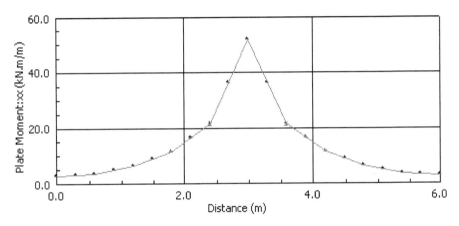

Fig. 12.36e M_{xx} along section 4-4 for dead load

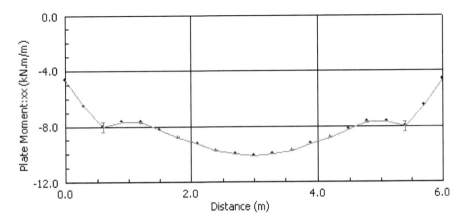

Fig. 12.35f M_{xx} along section 5-5 for dead load

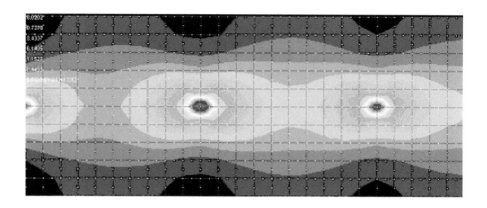

Fig.12.37 Contour of moment M_{yy} for dead load

12.9.2 Results of Analysis for Dead Load plus Live Load Pattern 1

Fig. 12.38 shows the contours for the moment about the x-axis. Fig. 12.39a to f show the distribution of moments at six sections. Fig. 12.40 shows the contours for the moment about the y-axis.

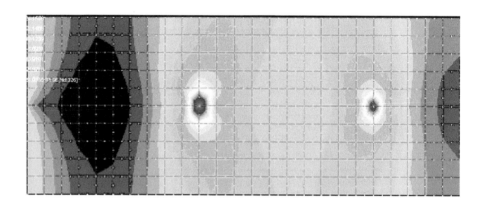

Fig.12.38 Contour of moment M_{xx} for dead plus live load pattern 1

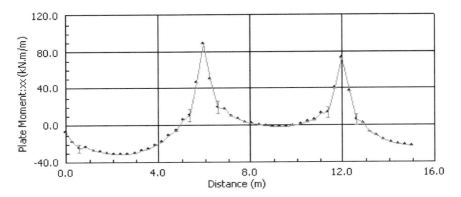

Fig. 12.39a M_{xx} along column line for dead + live load pattern 1

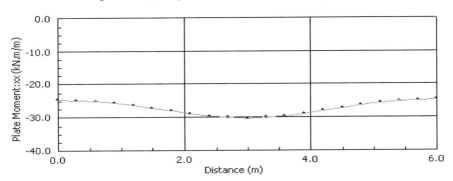

Fig. 12.39b M_{xx} along section 1-1 for dead + live load pattern 1

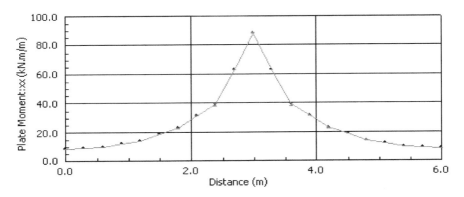

Fig. 12.39c M_{xx} along section 2-2 for dead + live load pattern 1

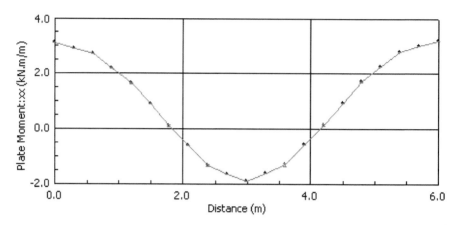

Fig. 12.39d M$_{xx}$ along section 3-3 for dead + live load pattern 1

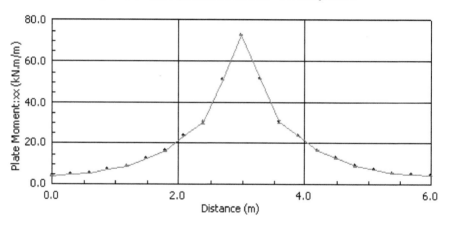

Fig. 12.3 9e M$_{xx}$ along section 4-4 for dead + live load pattern 1

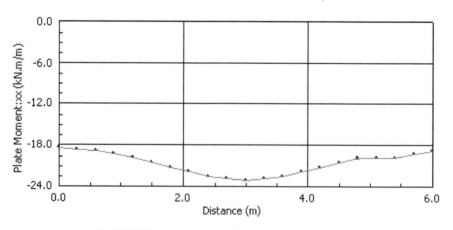

Fig. 12.28f M$_{xx}$ along section 5-5 for dead + live load pattern 1

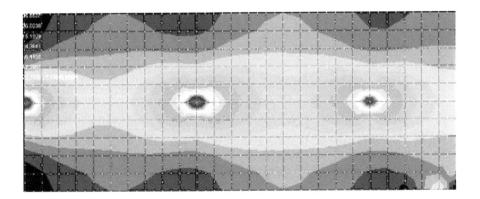

Fig.12.40 Contour of moment M_{yy} for dead plus live load pattern 1

12.9.3 Results of Analysis for Dead Load plus Live Load Pattern 2

Fig. 12.41 shows the contours for the moment about the x-axis. Fig. 12.42a to f show the distribution of moments at six sections. Fig. 12.43 shows the contours for the moment about the y-axis.

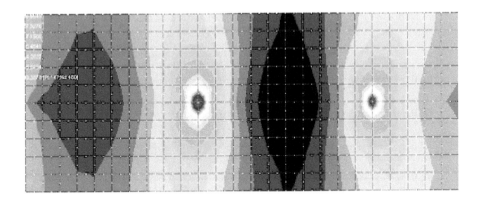

Fig.12.41 Contour of moment M_{xx} for dead + live load pattern 2

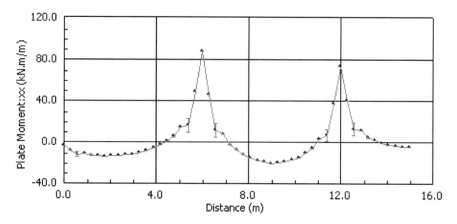

Fig. 12.42a M_{xx} along column line for dead + live load pattern 2

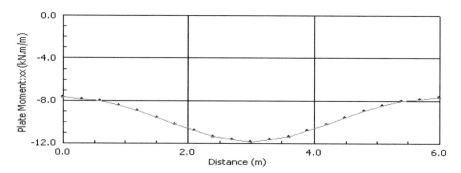

Fig. 12.42b M_{xx} along section 1-1 for dead + live load pattern 2

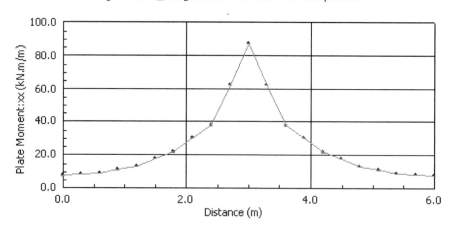

Fig. 12.42c M_{xx} along section 2-2 for dead + live load pattern 2

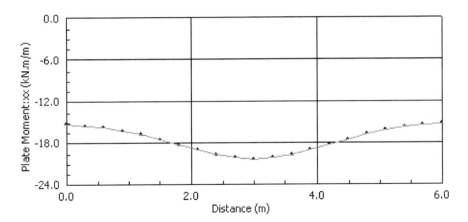

Fig. 12.42d M_{xx} along section 3-3 for dead + live load pattern 2

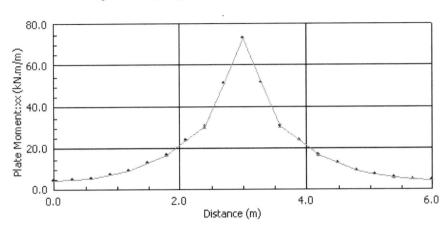

Fig. 12.42e M_{xx} along section 4-4 for dead + live load pattern 2

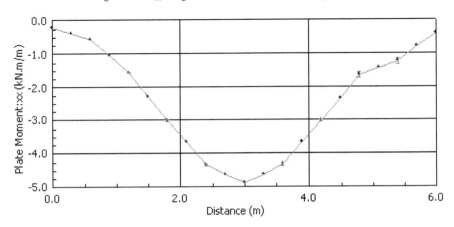

Fig. 12.42f M_{xx} along section 5-5 for dead + live load pattern 2

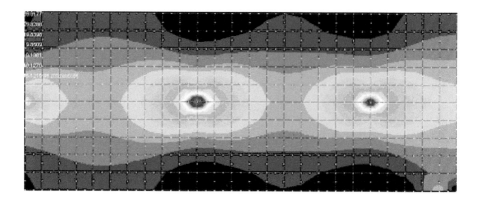

Fig.12.43 Contour of moment M_{yy} for dead + live load pattern 2

12.9.4 Results of Analysis for Dead Load plus Live Load Pattern 3

Fig. 12.44 shows the contours for the moment about the x-axis. Fig. 12.45a to f show the distribution of moments at six sections. Fig. 12.46 shows the contours for the moment about the y-axis.

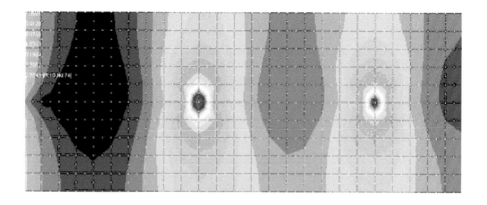

Fig.12.44 Contour of moment M_{xx} for dead + live load pattern 3

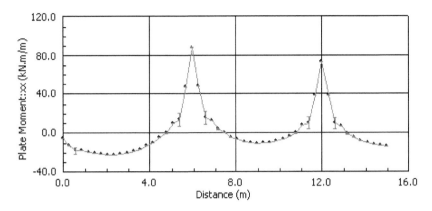

Fig. 12.45a M_{xx} along column line for dead + live load pattern 3

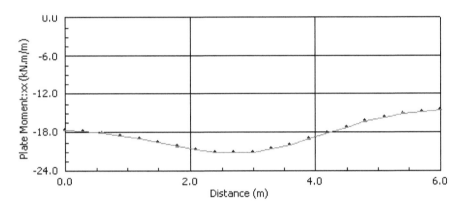

Fig. 12.45b M_{xx} along section 1-1 for dead + live load pattern 3

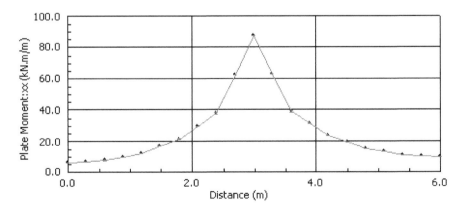

Fig. 12.45c M_{xx} along section 2-2 for dead + live load pattern 3

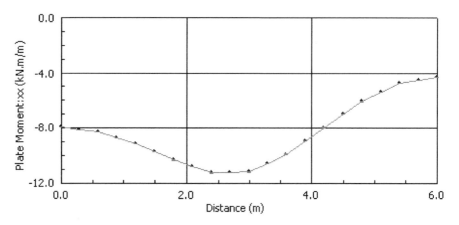

Fig. 12.45d M_{xx} along section 3-3 for dead + live load pattern 3

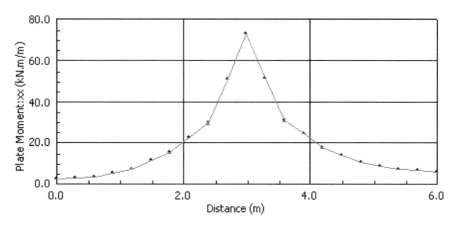

Fig. 12.45e M_{xx} along section 4-4 for dead + live load pattern 3

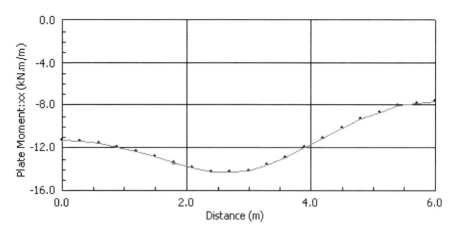

Fig. 12.45f M_{xx} along section 5-5 for dead + live load pattern 3

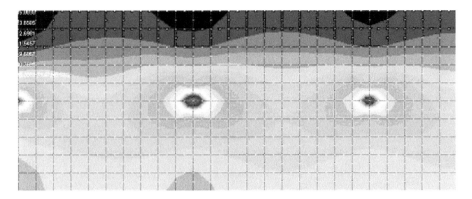

Fig.12.46 Contour of moment M_{yy} for dead + live load pattern 3

12.9.5 Results of Analysis for Dead Load plus Live Load Pattern 4

Fig. 12.47 shows the contours for the moment about the x-axis. Fig. 12.48a to f show the distribution of moments at six sections. Fig. 12.49 shows the contours for the moment about the y-axis.

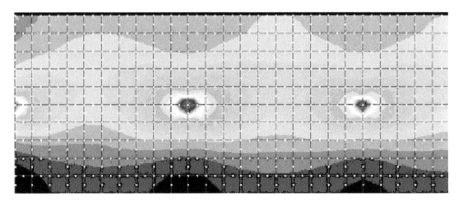

Fig.12.47 Contour of moment M_{xx} for dead + live load pattern 4

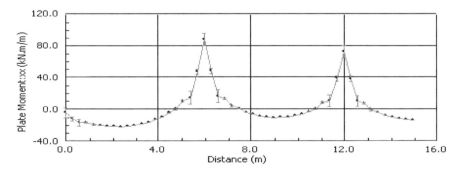

Fig. 12.48a M_{xx} along column line for dead + live load pattern 4

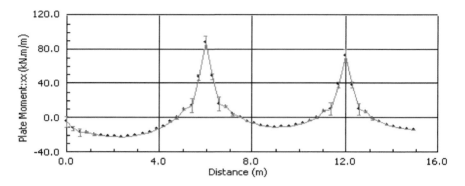

Fig. 12.48 a M_{xx} along column line for dead + live load pattern 4

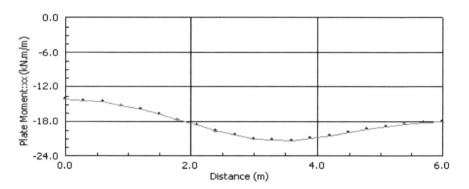

Fig. 12.48b M_{xx} along section 1-1 for dead + live load pattern 4

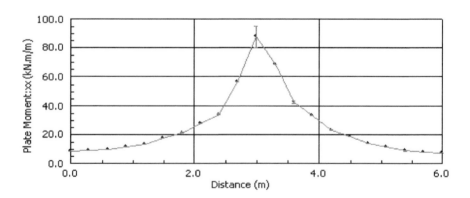

Fig. 12.48c M_{xx} along section 2-2 for dead + live load pattern 4

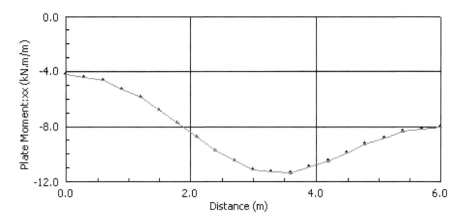

Fig. 12.48d M_{xx} along section 3-3 for dead + live load pattern 4

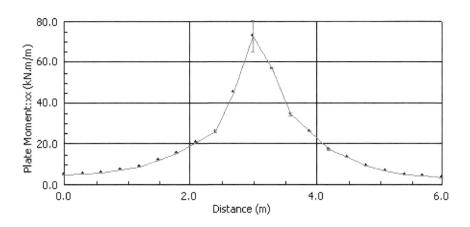

Fig. 12.48e M_{xx} along section 4-4 for dead + live load pattern 4

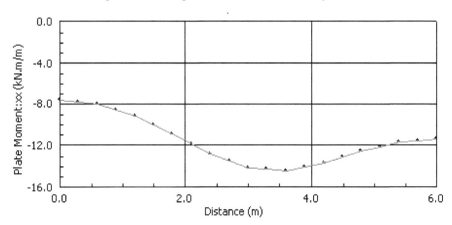

Fig. 12.48f M_{xx} along section 5-5 for dead + live load pattern 4

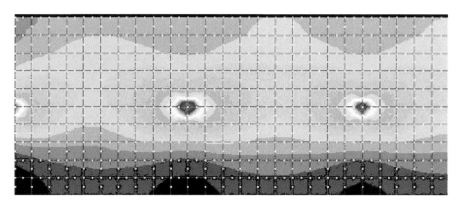

Fig.12.49 Contour of moment M_{yy} for dead + live load pattern 4

12.10 COMPARISON BETWEEN THE RESULTS OF ANALYSIS OF FULL SLAB AND STRIP OF SLAB

Table 12.4 shows a comparison between the results for the M_{xx} moment along six sections as shown in Fig. 12.7 and the section along the column line in the strip analysis as shown in Fig. 12.33. The figures in the table have been obtained by reading off from the graphs and therefore are not very accurate. However, from a comparison of the figures in bold for each load case shown in the table, it can be concluded that the analysis of the strip alone provides a good estimation of the moments obtained when the whole slab is analysed.

12.11 EUROCODE 2 RECOMMENDATIONS FOR EQUIVALENT FRAME ANALYSIS

While it is accepted that the finite element analysis provides a fairly accurate representation of the distribution of moments at the elastic stage, it does not provide a convenient set of moments and shear forces for conventional design procedure. Therefore designers often use a simplified representation of the flat slab system as a rigid-jointed frame. It is recognized that the distribution of moment across the width is not constant and tends to be higher along the column line than that between the columns.

The recommendations from Eurocode 2 are as follows:
- The structure should be divided longitudinally and transversely into frames consisting of columns and sections of slab contained between the centre lines of adjacent panels.
- The stiffness of members may be calculated from their gross cross sections.
- For vertical loading the stiffness may be based on the full width of the panels.

Table 12.4 Comparison between whole slab and strip of slab analysis.

Section	Column 1	Between columns	Column 2	Between columns	Column 3	Between columns
Dead load only						
Section 1-1	0	−15	45	−10	38	−12
Section 2-2	0	−12	3.5	−5	2	−7
Section 3-3	0	−16	60	−10	50	−12
Section 4-4	0	−12	6	−4	3	−7
Section 5-5	0	−15	52	−8	40	−10
Section 6-6	0	−12	6	−4	3	−7
Strip Only	−2	−15	62	−7	55	−8
Dead load + live load pattern 1						
Section 1-1	0	−35	60	−5	52	−25
Section 2-2	0	−25	8	3	3	−20
Section 3-3	0	−30	85	−3	70	−25
Section 4-4	0	−25	10	5	5	−18
Section 5-5	−3	−30	70	0	60	−20
Section 6-6	0	−25	10	6	6	−18
Strip Only	−5	−30	90	0	75	−20
Dead load + live load pattern 2						
Section 1-1	0	−15	60	−25	52	−8
Section 2-2	0	−10	5	−16	5	−2
Section 3-3	0	−15	80	−20	70	−3
Section 4-4	0	−10	9	−15	6	0
Section 5-5	0	−12	75	−20	60	−5
Section 6-6	0	−9	8	−15	5	0
Strip Only	−3	−12	90	−20	75	−3
Dead load + live load pattern 3						
Section 1-1	0	−30	82	−15	75	−20
Section 2-2	0	−20	5	−9	2	−13
Section 3-3	0	−25	85	−10	70	−15
Section 4-4	0	−17	11	−4	8	−8
Section 5-5	0	−22	72	−10	60	−15
Section 6-6	0	−18	6	−8	2	−12
Strip Only	−3	−20	90	−8	75	−12
Dead load + live load pattern 4						
Section 1-1	0	−16	40	−10	35	−10
Section 2-2	0	−16	40	−5	3	−8
Section 3-3	0	−22	82	−10	70	−15
Section 4-4	0	−18	7	−8	4	−12
Section 5-5	0	−22	72	−10	60	−15
Section 6-6	0	−16	10	−5	6	−8
Strip Only	−3	−20	90	−10	72	−7

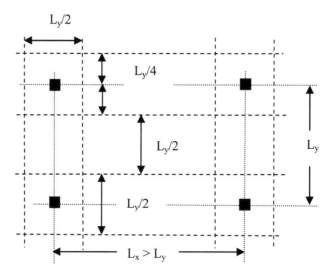

Fig. 12.50 Division of panels in flat slab analysis

- For horizontal loading 40% of this value should be used to reflect the increased flexibility of the column-slab joints in flat slab structures.
- Total load on the panel should be used for the analysis in each direction
- The panel is divided into column and middle strips as shown in Fig. 12.50. The middle strip is $L_y/2$ wide. The rest of the $L_y/2$ width of the slab is divided into two column strips each $L_y/4$ wide.
- If there are column drops $> L_y/3$ wide, the column strips should be taken as the width of drops and the width of the middle strip adjusted accordingly.
- The total bending moment so calculated should be apportioned between column and middle strips as shown in Table 12.5. Note that in the elastic analysis, negative bending moments (causing tension at the top) tend to concentrate towards the centre lines of columns.

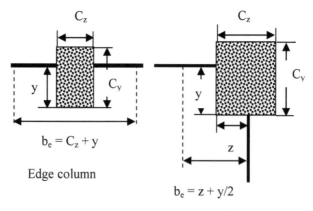

Fig. 12.51 Effective width of flat slab at edge and corner columns

Table 12.5 Simplified apportioning of bending moments for a flat slab

	Negative moments	Positive moments
Column strip	60-80%	50-70%
Middle strip	40-20%	50-30%

Note: The sum of the moments in the column and middle strips, whether negative or positive, should always add up to 100%

Unless there are perimeter beams which are adequately designed for torsion, moments transferred to edge or corner columns should be limited to the moment of resistance of a rectangular section equal to $0.17 \, b_e d^2 \, f_{ck}$ (see Fig. 12.51 for the definition of b_e). The positive moment in the end span should be adjusted accordingly.

12.12 GRILLAGE ANALYSIS FOR IRREGULAR COLUMN LAYOUT

In the case of irregular column layout where one cannot sensibly use equivalent frame analysis, grillage or finite element analysis can be used. In such cases the following simplified approach can be used.
- Analyse the slab with all bays loaded with $\gamma_G \, G_k + \gamma_Q \, Q_k$
- Load a critical bay or bays with $\gamma_G \, G_k + \gamma_Q \, Q_k$ and the rest of the bays with $\gamma_G \, G_k$. Where there may be significant variation in the permanent load between bays, γ_G should be taken as 1 for the unloaded bays.
- Use the information from the second analysis to modify the mid-span and column moments from first analysis.
- The restriction on the moment transfer to edge and corner columns noted in section 12.11 should be followed.

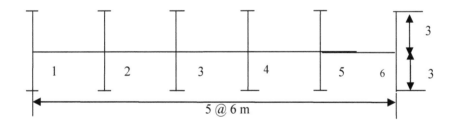

Fig. 12.52 Slab-frame model

12.13 EXAMPLE OF DESIGN OF FLAT SLAB-FRAME

Fig. 12.52 shows a slab with 3 m high columns. It is assumed that the columns are fixed at the far end and no sway is permitted. The slab is considered as a 6 m wide beam and 200 mm thick. The end columns are 300 mm square and the interior columns are 400 mm square.

Take unit weight of concrete = 25 kN/m³:
Self weight = 200 × 10⁻³ × 25 = 5 kN/m²
Partitions = 1.5 kN/m²
Live load = 2.5 kN/m²
The frame is analysed for a bay width of 6 m,
Self weight = 5.0 × 6.0 = 30.0 kN/m
Total load = (5.0 + 1.5 + 2.5) × 6.0 = 54.0 kN/m
Assume material properties as in section 12.3.1.

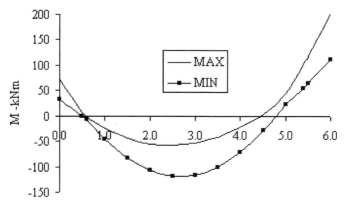

Fig. 12.53 Moment envelop for span 1-2

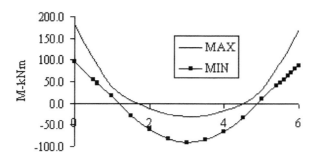

Fig. 12.54 Moment envelop for span 2-3

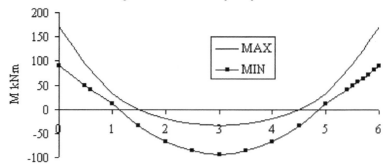

Fig. 12.55 Moment envelop for span 3-4

12.13.1 Results of Analysis of Slab-Frame Model

The rigid-jointed frame is analysed using a computer program. The results are presented in Table 12.6. Figs. 12.53 to 12.55 show the moment envelopes for the symmetrical half of the slab.

Table 12.6 Results for support moments from the analysis of a 2-D frame

Load case	M_{12}	M_{21}/M_{23}	M_{32}/M_{34}	M_{43}/M_{45}	M_{54}/M_{56}	M_6
Case 1: Dead load only	38	111 /96	88/ 90	88/ 90	111/ 96	38
Case 2: Max. 1-2, 3-4, 5-6	72	182/ 111	98/ 150	150/ 98	111/ 182	72
Case 3: Max. 2-3 and 4-5	33	129/ 159	147/ 101	101/ 147	159/129	33
Case 4: Max. Support 2	67	202 /181	138 /98	102/ 149	159 /129	33
Case 5: Max. Support 3	34	126 /150	169 / 172	141 / 95	112 /182	72
Case 6: Max. Support 4	72	182/ 112	95/ 141	172/ 169	150/ 126	34
Case 7: Max. Support 5	33	129/ 159	148/ 102	98/ 138	181/ 202	67

The significant values from the moment envelope calculations are as follows:
Spans 1-2 and 5-6:
Support = 72 kNm at left and 202 kNm at right, Span = 118 kNm
Spans 2-3 and 4-5:
Supports = 181 kNm at left and 169 at right kNm, Span = 90 kNm
Span 3-4:
Supports = 172 kNm at left and 172 kNm at right kNm, Span = 93 kNm

12.13.2 Moment Distribution due to Prestress

In deciding on the prestress in different spans, one need not necessarily attempt to achieve the maximum cable eccentricity in each span as this will lead to varying amounts of prestress, resulting in problems of detailing. It is better to vary the eccentricity so that the prestressing force is constant over the entire structure as this leads to simpler detailing. In many cases this may not be possible. In such cases additional cables can be used to increase the prestressing force in specific spans only.

It is not normal to balance the total load in each span. The reason for this is that if the total load is balanced, then it leads to a large prestressing force. In addition, as the entire live load does not always act on the span, under the action of the large prestressing force and smaller applied load, hogging deflection will occur which might be undesirable under normal working situations.

12.13.3 Cable Profile

Assume cover to steel = 30 mm
Diameter of strand = 16 mm
Centre of strand at say (30 + 16/2) = 38 mm from top or bottom face
Maximum eccentricity (above or below the neutral axis) available
$$\approx 200/2 - 38 \approx 60 \text{ mm}$$
Assume a cable profile as shown in Fig. 12.56. The maximum eccentricity in the span is 60 mm, over the interior supports it is 60 mm and at the end supports the eccentricity is assumed as 20 mm. Keep a constant group of cables over the entire five spans but use extra cables only in the end spans. These extra cables are 'peeled off' and anchored in the first interior spans at say 0.6 m from the support as shown in Fig. 12.56.

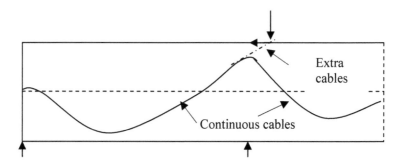

Fig. 12.56 Cable profile

The cable profiles has a convex curvatures at each end over a length of 0.1 L, where L = span. The cable profile is made up of three parabolas as shown in Fig. 6.13 and discussed in Chapter 6, section 6.2.3.1.

The curvatures of the three parabolas are:
Parabola AB: From equation 6.34,

$$\frac{d^2y}{dx^2} = -2\frac{(e_1 + e_3)}{a\,b}$$

Rotation at B = $\dfrac{dy}{dx} = \dfrac{2(e_1 + e_3)}{b}$

Parabola BCD: From equation 6.35,

$$y = -e_3 + \frac{(e_1 + e_3)}{b(b-a)}(b-x)^2$$

$$\frac{dy}{dx} = -2\frac{(e_1 + e_3)}{b(b-a)}(b-x)$$

$$\frac{d^2y}{dx^2} = 2\frac{(e_1 + e_3)}{b(b-a)}$$

At D: $x = (L - c)$, $\dfrac{dy}{dx} = -2\dfrac{(e_1 + e_3)}{b(b-a)}(b-L+c)$

Parabola DE: From equation 6.36,

$$\frac{d^2y}{dx^2} = -2\frac{(e_2 + e_3)}{c(L-b)}$$

Rotation at D $= \dfrac{dy}{dx} = \dfrac{2(e_2 + e_3)}{(L-b)}$

The uniformly distributed acting is given by q = P × curvature. Load q acts downwards where the curvature in convex up and upwards where the curvature is concave up.

12.14 CALCULATION OF LOSS OF PRESTRESS

Before one can determine the number of cables required, it is necessary to calculate the loss of prestress during stressing as well as that due to creep, shrinkage and relaxation.

12.14.1 Calculation of Loss due to Friction and Wobble per Cable

Assume that the cables are stressed from both ends. Take f_{pk} = 1860 MPa, A_{ps} = 140 mm^2, force at stressing = 206.0 kN per strand, μ = 0.19, k = 0.008/m.

(i) Calculate the loss in cables which are continuous over the five spans and stressed with a force of 206 kN.
Span 1-2:
e_1 = 20 mm, e_2 = 60 mm, e_3 = 60 mm, a = 0.1L, c = 0.1L, L = 6 m. Using equations (6.31) to (6.33),

$\alpha = \dfrac{(e_2 + e_3)}{(e_1 + e_3)} \} = 1.5$, $A = 1 - \alpha = -0.5$, $B = (2 - \alpha\dfrac{a}{L} - \dfrac{c}{L}) = 1.75$, $C = (1 - \dfrac{c}{L}) = 0.9$

$\dfrac{b}{L} = \dfrac{B - \sqrt{(B^2 - 4\,A\,C)}}{2A} = 0.455$, b = 0.455 × 6 = 2.73 m, L − b = 3.27 m

Parabola AB: From (6.34),

Rotation at B $= \dfrac{dy}{dx} = \dfrac{2(e_1 + e_3)}{b} = \dfrac{2(20 + 60)\times 10^{-3}}{2.73} = 0.0586\,\text{radians}$

Parabola DE: From (6.35),

Rotation at D $= \dfrac{dy}{dx} = \dfrac{2(e_2 + e_3)}{(L - b)} = \dfrac{2(60 + 60)\times 10^{-3}}{3.27} = 0.0734\,\text{radians}$

P_1 = 206 kN, total angular change = 2(0.0586 + 0.0734) = 0.2640 radians, span = 6 m
$\mu\,(\theta + kx) = 0.19 \times (0.2640 + 0.008 \times 6) = 0.0593$,
$P_2 = P_1\,(1 - 0.0593) = 206 \times 0.94 = 194\,\text{kN}$

Loss of force/metre = (206 – 194)/6 = 2.0 kN/m

Span 2-3: P_2 = 194 kN
e_1 = 60 mm, e_2 = 60 mm, e_3 = 60 mm, a = 0.1L, c = 0.1L, L = 6 m, b= 3 m by symmetry

Parabola AB: Rotation at B = $\dfrac{dy}{dx} = \dfrac{2(e_1 + e_3)}{b} = \dfrac{2(60+60) \times 10^{-3}}{3} = 0.08$ radians

Parabola DE: Rotation at D = $\dfrac{dy}{dx} = \dfrac{2(e_2 + e_3)}{(L-b)} = \dfrac{2(60+60) \times 10^{-3}}{3} = 0.08$ radians

Total angular change = 2(0.08 + 0.08) = 0.32 radians, Span = 6 m
μ (θ + kx) = 0.19 × (0.32 + 0.008 × 6) = 0.07
$P_3 = P_2$ (1 – 0.07) = 194 × 0.93 = 180 kN
Loss of force/metre = (194 – 180)/6 = 2.3 kN/m
Total angular change θ in spans 3-4 and 4-5 as for span 2-3 and in span 5-6 as for span 1-2.
At the middle of span 3-4: Total angular change = 2(0.08) = 0.16 radians, x = 3 m
μ (θ + kx) = 0.19 × (0.16 + 0.008 × 3) = 0.035
$P_{mid-span}/P_3$ = (1 – 0.035), $P_{mid-span}$ = 180 × 0.97 = 174 kN

(ii) Calculate the loss in force in the extra cables in end spans only
The cables do not have a reverse curvature at E. Therefore
Total angular change θ = 2× rotation at B + rotation at D
= 2 × 0.0586 + 0.0734 = 0.1906 radians
Span = 6 m
μ (θ + kx) = 0.19 × (0.1906 + 0.008 × 6) = 0.045,
$P_2 = P_1$ (1 – 0.045) = 206 × 0.96 = 198 kN
Loss of force per metre = (206-198)/6 = 1.3 kN/m

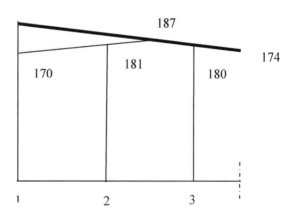

Fig. 12.57 Stress in a single cable before and after transfer

12.14.2 Calculation of Loss due to Wedge Draw-in

Assume $E_s = 195$ kN/mm^2, draw-in, $\Delta = 6$ mm, $A_{ps} = 140$ mm^2.
p = loss per unit length ≈ 2 kN/m . Using (11.20),

$$\ell = \sqrt{\frac{\Delta\, A_{ps}\, E_s}{\text{Loss of prestress per unit length}}} = \sqrt{\frac{(6\times10^{-3})\times140\times195}{2.0}} = 9.05 \text{ m}$$

Loss of force at anchorage = 2 × 9.05 × 2 = 36.0 kN
Force at anchorage = 206 – 36 = 170 kN
Fig. 12.57 shows the force variation in a single cable before and after transfer.
    The average forces in the three spans after transfer are:
Span 1-2 = (170+181)/2 = 176 kN,
Span 2-3 = [(181+187)/2 + (187+180)/2]/2 = 184 kN
Span 3-4 = (180+174)/2 = 177 kN

12.14.3 Calculation of Prestress at Service

Taking relaxation loss aofsay 2.5%, shrinkage strain of say 200×10^{-6}, creep strain of 150×10^{-6}, the loss of force due to creep and shrinkage in the cable with a cross-sectional area of 140 mm^2 as follows:
Creep and shrinkage Loss = $(200 + 150) \times 10^{-6} \times 195\times 10^3 \times 140 \times 10^{-3}$ = 9.6 kN
Relaxation loss ≈ 2.5% of (176 + 184+ 177)/3 = 4.5 kN
Total long term loss = 9.6 + 4.5 = 14 kN
The average forces in the three spans at service are
Span 1-2 = (176 – 14) = 162 kN,
Span 2-3 = (184 – 14) = 170 kN
Span 3-4 = (177 – 14) = 163 kN

12.14.4 Determination of Number of Cables

Try balancing self weight + say 10% of the imposed load.
Load to be balanced = 5.0 + 0.1 × (1.5 + 2.5) = 5.4 kN/m^2
Over a 6 m bay width, load q to be balanced = 5.4 × 6.0 = 32.4 kN/m
End spans:
Drape = 60 + (20 + 60)/2 = 100 mm
P = q L^2/ (8 × drape) = 32.4 × 6.0^2/ (8 × 100 × 10^{-3}) = 1458 kN
Interior span:
Drape = 60 + (60 + 60)/2 = 120 mm
P = q L^2/ (8 × drape) = 32.4 ×6.0^2/ (8×120×10^{-3}) = 1215 kN
No. of cables at transfer: No of cables required over a 6 m wide bay are
End spans = 1458 /176 = 8.3 say 9 cables
Interior spans = 1215/[(170+ 163)/2} = 7.3, say 7 cables
Prestressing forces in the spans at transfer:
Span 1-2: 9 × 176 = 1584 kN

Span 2-3: 7 × 184 = 1288 kN n
Span 3-4: 7 × 177 = 1239 kN
Prestressing forces in the spans at service:
Span 1-2: 9 × 162 = 1458 kN
Span 2-3: 7 × 170 = 1190 kN
Span 3-4: 7 × 163 = 1141 kN

12.15 FIXED END MOMENTS DUE TO PATCH LOADS AND CONCENTRATED FORCE AND COUPLE

(a). Fixed end moments due to patch load as shown in Fig. 6.15 are calculated using the equations (6.43).

(b). Fixed end moments due to concentrated loads and couples as shown in Fig. 12.58 are calculated using the equations

$$M_A = -WL[\alpha_1 \alpha_2^2] , \ M_B = WL[\alpha_1^2 \alpha_2] \tag{12.1}$$

$$M_A = M_c[1 - 4\alpha_1 + 3\alpha_1^2] , \ M_B = M_c[2\alpha_1 - 3\alpha_1^2] \tag{12.2}$$

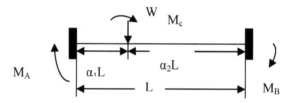

Fig. 12.58 Concentrated load and couple

12.16 EQUIVALENT LOADS AND FIXED END MOMENTS

Equivalent loads are dependent on the curvature of the cable and the force in the cable. If hand calculation method such as the method of moment distribution is used, then it is necessary to calculate the fixed end moments due to equivalent loads. If a frame analysis program is used, then it is not necessary to calculate the fixed end moments as the program will take care of it.

12.16.1 Equivalent Loads and Fixed End Moments at Transfer

Span 1-2:
$e_1 = 20$ mm, $e_2 = 60$ mm, $e_3 = 60$ mm, $L = 6$ m, $a = 0.1L = 0.6$ m, $c = 0.1L = 0.6$ m
Using (6.30) and (6.33),

$$\alpha = \frac{(e_2 + e_3)}{(e_1 + e_3)} = 1.5, \ A = 1 - \alpha = -0.5, \ B = (2 - \alpha\frac{a}{L} - \frac{c}{L}) = 1.75, \ C = (1 - \frac{c}{L}) = 0.9$$

$$\frac{b}{L} = \frac{B - \sqrt{(B^2 - 4AC)}}{2A} = 0.455, b = 0.455 \times 6 = 2.73, (L-b) = 3.27$$

Parabola AB: From equation (6.34),

$$\frac{d^2y}{dx^2} = -2\frac{(e_1 + e_3)}{a\,b} = -2\frac{(20 + 60)\times 10^{-3}}{0.6\times 2.73} = -0.0977\,\text{m}^{-1}$$

Slope at B $= \dfrac{dy}{dx} = \dfrac{2(e_1 + e_3)}{b} = \dfrac{2(20+60)\times 10^{-3}}{2.73} = 0.0586\,\text{radians}$

$P = 1584$ kN, $q = 1584 \times 0.0977 = 154.8$ kNm (downwards)
Substituting in (6.44), $\alpha_1 = 0$, $\alpha_2 = 0.9$, $L = 6.0$,
$M_{12} = -24.30$, $M_{21} = +1.72$
The eccentricity at the left-hand end, e_1, creates a clockwise moment equal to
$Pe_1 = 1584 \times 20 \times 10^{-3} = 31.7$ kNm
It is assumed that the cables are anchored in the slab only.

Parabola BCD: From equation (6.35),

$$\frac{d^2y}{dx^2} = 2\frac{(e_1 + e_3)}{b(b-a)} = 2\frac{(20+60)\times 10^{-3}}{2.73\times(2.73-0.6)} = 0.0275\,\text{m}^{-1}$$

$P = 1584$ kN, $q = 1584 \times 0.0275 = 43.6$ kN/m (upwards),
Substituting in (6.44), $\alpha_1 = 0.1$, $\alpha_2 = 0.1$, $L = 6.0$,
$M_{12} = 123.48$, $M_{21} = -123.48$
Slope at D:

$$\frac{dy}{dx} = -2\,\frac{(e_1 + e_3)}{b(b-a)}(b-L+c)$$

$$= -2\times\frac{(20+60)\times 10^{-3}}{2.73\times(2.73-0.6)}\times(2.73-6.0+0.6) = 0.0735\ \text{radians}$$

As θ is small, $\tan\theta \approx \sin\theta \approx 0.0735$, $\cos\theta \approx 1.0$
The extra cables are peeled off at D and anchored in span 2-3. The force in the peeled-off cables is equal to the difference in the axial forces in span 1-2 and span 2-3.
$P = (1584 - 1288) = 296$ kN, $P\sin\theta = 21.76$ kN, $P\cos\theta = 296$ kN

Parabola DE: From equation (6.36),

$$\frac{d^2y}{dx^2} = -2\frac{(e_2 + e_3)}{c\,(L-b)} = -2\times\frac{(60+60)\times 10^{-3}}{0.6\times 3.27} = -0.1223\,\text{m}^{-1}$$

Slope at D $= \dfrac{dy}{dx} = \dfrac{2(e_2 + e_3)}{(L-b)} = \dfrac{2(60+60)\times 10^{-3}}{(6.0-2.73)} = 0.0734\,\text{radians}$

$P = 1288$ kN, $q = 1288 \times 0.1223 = 157.5$ kN/m (downwards)
Substituting in (6.44), $\alpha_1 = 0.9$, $\alpha_2 = 0$, $L = 6.0$,
$M_{12} = -1.75$, $M_{21} = 24.71$
Total fixed end moments:

$M_{12} = -24.30 + 123.47 - 1.75 = 97.43$ kNm
$M_{21} = 1.72 - 123.47 + 24.71 = -97.05$ kNm
Total angular change θ from A-E is given by
$\theta = 2(0.0586$ slope at B $+ 0.0734$ slope at D$) = 0.2640$ radians

Span 2-3:

$e_1 = 60$ mm, $e_2 = 60$ mm, $e_3 = 60$ mm, $a = 0.1L = 0.6$, $c = 0.1L = 0.6$, $b/L = 0.5$ by symmetry, $b = 3$ m.

Parabola AB: From equation (6.34),

$$\frac{d^2 y}{dx^2} = -2\frac{(e_1 + e_3)}{a\,b} = -2 \times \frac{(60+60) \times 10^{-3}}{0.6 \times 3.0} - 0.1333\,\text{m}^{-1}$$

Slope at B $= \dfrac{dy}{dx} = \dfrac{2(e_1 + e_3)}{b} = \dfrac{2(60+60) \times 10^{-3}}{3.0} = 0.08$ radians

$P = 1288$ kN, $q = 1288 \times 0.1333 = 171.7$ kNm (downwards)
Substituting in (6.44), $\alpha_1 = 0$, $\alpha_2 = 0.9$, $L = 6.0$,
$M_{23} = -26.94$, $M_{32} = +1.91$

Parabola BCD: From equation (6.35),

$$\frac{d^2 y}{dx^2} = 2\frac{(e_1 + e_3)}{b(b-a)} = 2 \times \frac{(60+60) \times 10^{-3}}{3.0 \times (3.0-0.6)} 0.0333\,\text{m}^{-1}$$

$P = 1288$ kN, $q = 1288 \times 0.0333 = 42.89$ kN/m (upwards),
Substituting in (6.44), $\alpha_1 = 0.1$, $\alpha_2 = 0.1$, $L = 6.0$,
$M_{23} = 121.47$, $M_{32} = -121.47$

Parabola DE: From equation (6.36),

$$\frac{d^2 y}{dx^2} = -2\frac{(e_2 + e_3)}{c(L-b)} = -2 \times \frac{(60+60) \times 10^{-3}}{0.6 \times (6.0-3.0)} = -0.1333\,\text{m}^{-1}$$

Slope at D $= \dfrac{dy}{dx} = \dfrac{2(e_2 + e_3)}{(L-b)} = \dfrac{2(60+60) \times 10^{-3}}{(6.0-3.0)} = 0.08$ radians

$P = 1288$ kN, $q = 1288 \times 0.1333 = 171.7$ kNm (downwards)
Substituting in (6.44), $\alpha_1 = 0.9$, $\alpha_2 = 0$, $L = 6.0$,
$M_{23} = -1.91$, $M_{32} = +26.94$
In addition to the above forces, from the three cables peeled off and anchored at 0.6 m from the left-hand support in span 2-3, there is a concentrated downward force of 21.76 kN and an anticlockwise couple equal to $296 \times 100 \times 10^{-3} = 29.60$ kNm. The corresponding fixed end forces are
From (12.1), $W = 21.76$ kN: $\alpha_1 = 0.1$, $\alpha_2 = 0.9$, $L = 6$ m,
$M_{23} = -21.76 \times 6 \times 0.1 \times 0.9^2 = -10.57$,
$M_{32} = 21.76 \times 6 \times 0.1^2 \times 0.9 = 1.18$
From (12.2), $M = 29.6$ kNm anticlockwise couple: $\alpha_1 = 0.1$,
$M_{23} = -29.6 \times (1 - 4 \times 0.1 + 3 \times 0.1^2) = -18.65$
$M_{32} = -29.6 \times (2 \times 0.1 - 3 \times 0.1^2) = -5.03$
The total fixed end moments at the ends are:

$M_{23} = -26.94 + 121.47 - 1.91 - 10.57 - 18.65 = 63.40$
$M_{32} = 1.91 - 121.47 + 26.94 + 1.18 - 5.03 = -96.47$ kNm
Total angular change θ from A-E is given by
$\theta = 2(0.08$ rotation at B $+ 0.08$ rotation at D$) = 0.32$ radians

Span 3-4: $e_1 = 60$ mm, $e_2 = 60$ mm, $e_3 = 60$ mm, $a = 0.1L$, $c = 0.1L$, $b/L = 0.5$ by symmetry and the rest of the data as for span 2-3 except for the contribution from the concentrated force and moment and that the axial force is 1239 kN as opposed to 1288 kN. The corresponding equivalent loads in span 3-4 are 165.2 and -41.26 instead of 171.7 and -42.89 for span 2-3.
The total fixed end moments at the ends are:
$M_{34} = -25.91 + 116.85 - 1.83 = 89.10$
$M_{43} = 1.83 - 116.85 + 25.91 = -89.10$
As for span 2-3, the total angular change θ from A-E $= 0.32$ radians

12.16.2 Equivalent Loads and Fixed End Moments at Service

The equivalent loads and corresponding fixed end moments are calculated as in section 12.11.2. The final results are as follows.

Span 1-2:

Parabola AB:
P = 1458 kN, q = 1458 × 0.0977 = 142.5 kNm (downwards)
$M_{12} = -22.35$, $M_{21} = +1.58$
The eccentricity at the left-hand end, e_1, creates a moment equal to
$Pe_1 = 1458 \times 20 \times 10^{-3} = 29.2$ kNm

Parabola BCD:
P = 1458 kN, q = 1458 × 0.0275 = 40.1 kN/m (upwards)
$M_{12} = 113.66$, $M_{21} = -113.66$

P = (1458 − 1190) = 268 kN, P sin θ = 19.7 kN, P cos θ = 268 kN
Parabola DE:
P = 1190 kN, q = 1190 ×0.1223 = 145.54 kN/m (downwards)
$M_{12} = -1.62$, $M_{21} = 22.84$
Total fixed end moments:
$M_{12} = 89.54$ kNm, $M_{21} = -89.14$ kNm

Span 2-3:

Parabola AB:
P = 1190 kN, q = 1190 × 0.1333 = 158.63 kNm (downwards)
$M_{23} = -24.89$, $M_{32} = +1.76$

Parabola BCD: $\dfrac{d^2y}{dx^2} = 2\dfrac{(e_1 + e_3)}{b(b-a)} = 0.0333\,\text{m}^{-1}$

$P = 1119$ kN, $q = 1119 \times 0.0333 = 37.26$ kN/m (upwards)
$M_{23} = 105.52$, $M_{32} = -105.52$

Parabola DE:
$P = 1190$ kN, $q = 1190 \times 0.1333 = 158.63$ kNm (downwards)
$M_{23} = -1.76$, $M_{32} = +24.89$
In addition to the above forces, from the three cables anchored at 0.6 m from the left-hand support, there is a concentrated downward force of 19.70 kN and an anticlockwise couple equal to $268 \times 100 \times 10^{-3} = 26.80$ kNm. The corresponding fixed end forces are
From (12.1), due to W = 19.70 kN: $M_{23} = -9.56$, $M_{32} = 1.07$
From (12.2), due to 26.8 kNm anticlockwise couple: $\alpha_1 = 0.1$
$M_{23} = -16.89$, $M_{32} = -4.55$
The total fixed end moments at the ends are
$M_{23} = 52.42$, $M_{32} = -82.35$ kNm

Span 3-4: $e_1 = 60$ mm, $e_2 = 60$ mm, $e_3 = 60$ mm, $a = 0.1L$, $c = 0.1L$, $b/L = 0.5$ by symmetry and the rest of data are for span 2-3 except for the contribution from the concentrated force and moment and that the axial force is 1141 kN as opposed to 1190 kN. The corresponding equivalent loads in span 3-4 are 152.1 and -35.73 instead of 158.6 and -37.26 for span 2-3.
The total fixed end moments at the ends are
$M_{34} = 75.63$, $M_{43} = -75.63$

12.16.3 Moment Distribution due to Equivalent Loads at Transfer

The frame is analysed under equivalent loads. The resulting moments are the moments in the slab due to prestress forces. The support moments causing tension at the *bottom* face are
M_1 and $M_6 = 58.94$ kNm, M_2 and $M_5 = 113.4/105.3$ kNm and
M_3 and $M_4 = 93.04/89.85$ kNm
In the span, the maximum moments causing tension at the *top* face are
Span 1-2 and 5-6: 75.01 kNm, Span 2-3 and 4-5: 63.49 kNm and
Span: 3-4: 58.66 kNm
Axial forces are:
Span 1-2 and 5-6: 1584 kN
Span 2-3 and 4-5: 1584 kN for 0.6 m and then 1288 kN
Span: 3-4: P = 1239 kN
Figs. 12.59 to 12.61 show respectively the moment due to equivalent loads resulting from prestress in the slab at transfer in span 1-2 to span 3-4. It should be noted that in these diagrams the load factor of 1.1 on the prestress force has not been included.

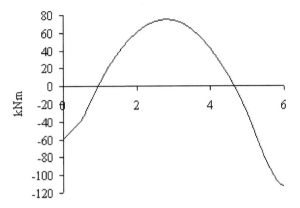

Fig. 12.59 Moment due to equivalent loads at transfer for span 1-2

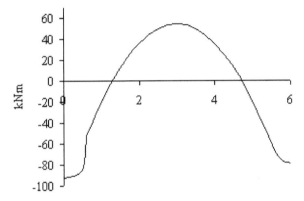

Fig. 12.60 Moment due to equivalent loads at transfer for span 2-3

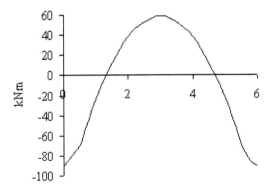

Fig. 12.61 Moment due to equivalent loads at transfer for span 3-4

12.16.4 Moment Distribution due to Equivalent Loads at Service

The frame is analysed under equivalent loads. The resulting moments are the
moments in the slab due to prestress forces. The support moments causing tension
at the *bottom* face are:
M_1 and M_6 = 54.54 kNm, M_2 and M_5 = 102.9/91.89 kNm
M_3 and M_4 = 78.7/76.2 kNm
In the span, maximum moments causing tension at the *top* face are
Span 1-2 and 5-6: 69.4 kNm, Span 2-3 and 4-5: 54.8 kNm
Span: 3-4: 50.8 kNm
Axial forces are
Span 1-2 and 5-6: 1458 kN
Span 2-3 and 4-5: 1458 kN for 0.6 m and then 1190 kN
Span: 3-4: P = 1141 kN
Figs. 12.62 to 12.64 show respectively the moment due to equivalent loads due to a
prestress in the slab at service in span 1-2 to span 3-4. It should be noted that in
these diagrams the load factor on prestress force of 0.9 has not been included.

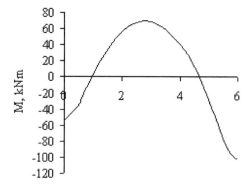

Fig. 12.62 Moment due to equivalent loads at service for span 1-2

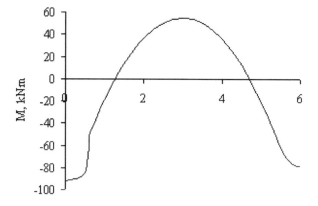

Fig. 12.63 Moment due to equivalent loads at service for span 2-3

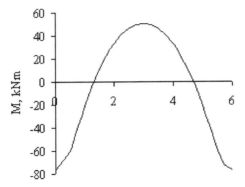

Fig. 12.64 Moment due to equivalent loads at service for span 3-4

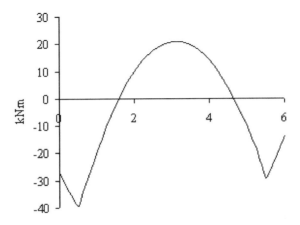

Fig. 12.65 Total moment at transfer for span 1-2

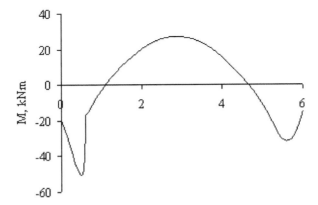

Fig. 12.66 Total moment at transfer for span 2-3

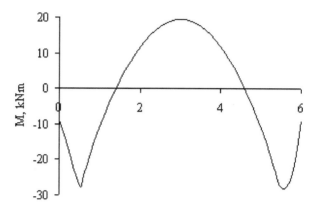

Fig. 12.67 Total moment at transfer for span 3-4

12.16.5 Moment Distribution due to External Load and Equivalent Loads at Transfer

The moment distribution at transfer is obtained by adding to the self-weight moment the moment due to equivalent loads with a load factor of 1.1. Figs. 12.65 to 12.67 show respectively the total moment in the slab at transfer in span 1-2 to span 3-4.

12.16.6 Moment Distribution due to External Load and Equivalent Loads at Service

The moment distribution at service is obtained by adding the moment due to external loads at service to the moment due to equivalent loads at service multiplied by a load factor of 0.9. Figs. 12.68 to 12.70 show respectively the moment in the slab at service in span 1-2 to span 3-4.

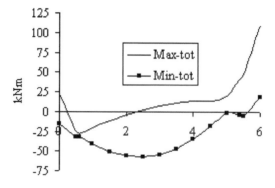

Fig. 12.68 Total moment at service for span 1-2

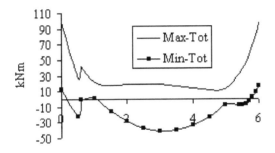

Fig. 12.69 Total moment at service for span 2-3

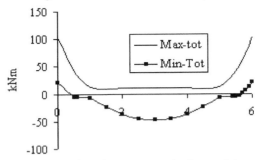

Fig. 12.70 Total moment at service for span 3-4

12.16.7 Stress Distribution in the Slab at Transfer and Service Stages

Before the stress distribution in the slab can be carried out, it is necessary to apportion the total moment to column and middle strips. The following values are used.

The column strip takes 70% and the middle strip takes 30% of the total negative (causing tension at the top face) moment. Similarly the column strip takes 60% and the middle strip takes 40% of the total positive (causing tension at the bottom face) moment.

The stresses at top and bottom faces are calculated from the formula

$\sigma_{top} = -P/A + \{$maximum or minimum moments due to (loads + prestress)$\}/Z$

$\sigma_{bottom} = -P/A - \{$maximum or minimum moments due to (loads + prestress)$\}Z$

$A = 6000 \times 200 = 12 \times 10^4$ mm^2, $Z = 3000 \times 200^2/6 = 20 \times 10^6$ mm^3

Note that axial stress due to prestress will be assumed to be constant over the entire bay but the moment to be used will depend on whether the stress calculation is for the column strip or the middle strip for the transfer or service situations. The total width of the middle or column strip is taken as 3000 mm and the section modulus is calculated for this width.

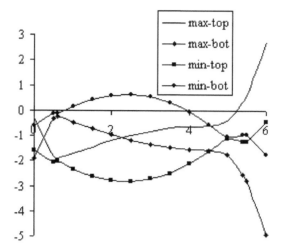

Fig. 12.71 Stress distribution at service in column strip for span 1-2

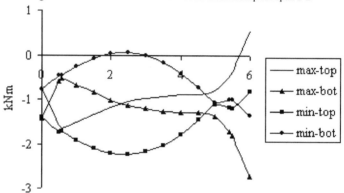

Fig. 12.72 Stress distribution at service in middle strip for span 1-2

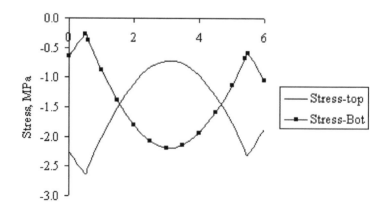

Fig. 12.73 Stress distribution at transfer in column strip for span 1-2

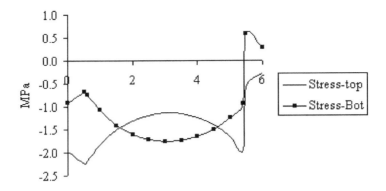

Fig. 12.74 Stress distribution at transfer in middle strip for span 1-2

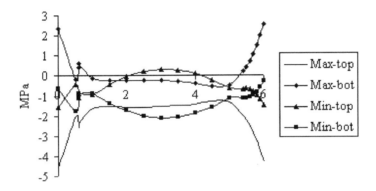

Fig. 12.75 Stress distribution at service in column strip for span 2-3

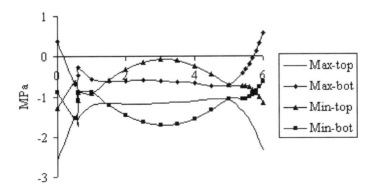

Fig. 12.76 Stress distribution at service in middle strip for span 2-3

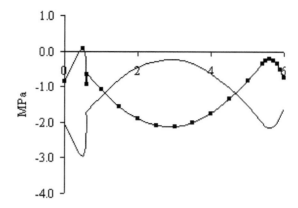

Fig. 12.77 Stress distribution at transfer in column strip for span 2-3

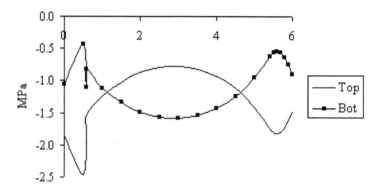

Fig. 12.78 Stress distribution at transfer in middle strip for span 2-3

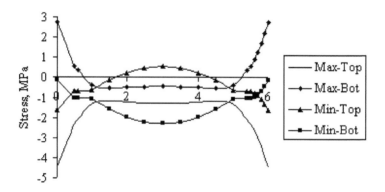

Fig. 12.79 Stress distribution at service in column strip for span 3-4

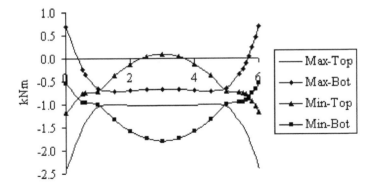

Fig. 12.80 Stress distribution at service in middle strip for span 3-4

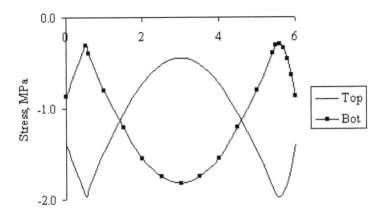

Fig. 12.81 Stress distribution at transfer in column strip for span 3-4

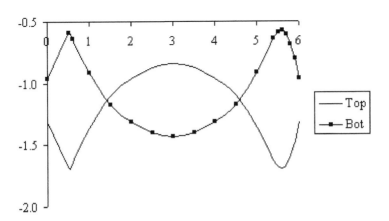

Fig. 12.82 Stress distribution at transfer in middle strip for span 3-4

Fig. 12.71 to Fig. 12.82 show the stress distribution in the middle and column strips at transfer and service stages for all three spans. From section 12.3.1, the permissible stresses are
Transfer: 2.6 MPa and 15.0 MPa in tension and compression respectively
Service: 2.9 MPa and 18.0 MPa in tension and compression respectively
From Figs 12.71 to 12.82, it can be seen that the maximum tensile stress is less than 2.9 MPa in all cases.

12.16.8 Moments in End Columns

Eurocode 2 puts a limit on the maximum moment that can be transmitted to end columns. The maximum moments in the external column are 13.65 kNm and 12.81 at transfer and service stages respectively. The equivalent width b_e from Fig.12.50 is
$C_z = 400$ mm, $y = C_y/2 = 400/2 = 200$mm, $b_e = 400 + 200 = 600$ mm.
Take the effective depth as say 350 mm, $f_{ck} = 30$ MPa, the maximum moment allowed in the column is $0.17\ b_e\ d^2\ f_{ck} = 374.85$ kNm, which is much larger than the applied moment and the design is therefore acceptable.

12.17 ULTIMATE LIMIT STATE MOMENT CALCULATIONS

Having ensured safe design at the serviceability limit state, the next stage is to check that there is sufficient capacity at the ultimate limit state for flexure as well as for shear.

Table 12.7 Results for support moments from the analysis of 2-D frame: ultimate limit state

Load case	M_{12}	M_{21}/M_{23}	M_{32}/M_{34}	M_{43}/M_{45}	M_{54}/M_{56}	M_6
Case 1: Max. 1-2, 3-4, 5-6	109	270/155	136/223	223/136	155/270	109
Case 2: Max. 2-3 and 4-5	44	179/235	219/140	140/320	235/179	44
Case 3: Max. Support 2	101	303/272	205/136	142/220	235/179	44
Case 4: Max. Support 3	45	175/221	255/259	209/131	157/270	109
Case 5: Max. Support 4	109	270/156	131/209	259/255	221/175	45
Case 6: Max. Support 5	44	179/235	220/142	136/204	272/303	101

12.17.1 Moment Envelopes

Moment envelopes are calculated using a load factor of 1.35 for a permanent (dead) load if it is unfavourable and 1.0 if it is favourable. Similarly for a variable (imposed) load a load factor of 1.50 is used if it is unfavourable and 0.0 if it is favourable.

Table 12.7 summarises the results of analysis and Figs. 12.84 to 12.86 show the moment envelopes.

The significant values from the moment envelope calculations are
Spans 1-2 and 5-6:
Support = 109 kNm at left and 303 kNm at right, Span = 179 kNm
Spans 2-3 and 4-5: Supports = 272 kNm at left and 255 at right kNm, Span = 137 kNm
Span 3-4:
Supports = 259 kNm at left and at right kNm, Span = 142 kNm

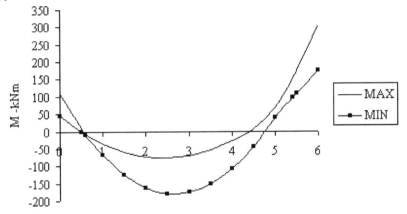

Fig. 12.83 Moment envelope at ultimate for span 1-2

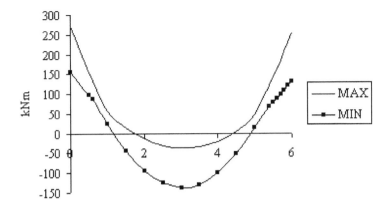

Fig. 12.84 Moment envelope at ultimate for span 2-3

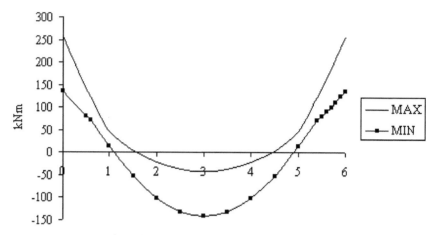

Fig. 12.85 Moment envelope at ultimate for span 3-4

12.17.2 Parasitic Moments

As explained in section 6.7.3.6, Chapter 6, when a continuous structure is prestressed, reactions (moments and shear forces) at supports alter, thus inducing at a section moments additional to that due to prestress and eccentricity at the section. However, these parasitic moments are dependent on the stiffness of the members of the structure. As the stiffness of a member decreases with increase in cracking, there is a corresponding reduction in the parasitic moments. If the structure is sufficiently ductile, then sufficient plastic hinges form at collapse to convert a statically indeterminate structure into a statically determinate structure. As a consequence, the entire parasitic moments disappear.

Generally at the service stage, a prestressed concrete structure is much less cracked compared to a similar reinforced concrete structure and therefore when checking stresses at a section, it is necessary to take the parasitic moments into account. However, at the ultimate stage, the importance of parasitic moments is highly dependent on the ductility of the structure. If the structure is ductile, then the parasitic moments are of less importance. On the other hand, if the structural behaviour is less ductile, then the parasitic moments can be of importance. As a conservative step, when checking the moment capacity of a section at the ultimate stage, it is prudent to include the parasitic moments if they are of the same sign as the moments due to an external load.

The value of the parasitic moment at a section is the difference between the moment due to prestress and eccentricity alone at the section and that due to the equivalent loads acting on the continuous structure.

12.17.3 Parasitic Moments: Example

In the present example, the prestress at service and the corresponding eccentricity at different sections along with the moment due to equivalent loads acting on the frame are shown in Table 12.8. Note that at end 2, two cables in span 1-2 are 'peeled' off and anchored in span 2-3. That is the reason why there are two pairs of values of prestress and corresponding eccentricity.

Fig. 12.86 shows the distribution of the parasitic moment. Except in span 1-2, the parasitic moments are very small. Even in span 1-2, compared with a moment at mid-span of 270 kNm due to external loads at the ultimate stage, the corresponding parasitic moment is only $(23 + 13)/2 = 18$ kNm,   which is a negligible amount.

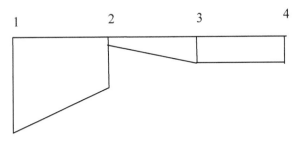

Fig. 12.86 Parasitic moments

Table 12.8 Parasitic moment calculations

Span	Section	P, kN	e, mm	ΣPe, kNm	Equivalent load analysis moment, kNm	Parasitic moment, kNm
1-2	1	1458	20	29.2	54.5	25.3
	2	1190	60	89.9	102.9	13.0
		268	69			
2-3	2	1190	60	89.9	91.9	2.0
		268	69			
	3	1190	60	71.4	78.7	7.3
3-4	3	1141	60	68.5	76.1	7.6
	4	1141	60	68.5	76.2	7.6

12.17.4 Ultimate Moment Capacity

The ultimate moment capacity at a section is computed as explained in Chapter 7. The first step is to determine the neutral axis depth x by ensuring that the total tensile and compressive forces balance.

Concrete: The rectangular stress block in bending is taken as having a depth of 0.8x. Taking $f_{ck} = 30$ MPa, $\gamma_m = 1.5$, the constant stress in the stress block is $f_{cd} = f_{ck}/\gamma_m = 20$ MPa.

The maximum compressive strain in concrete is $\varepsilon_{cu3} = 3.5 \times 10^{-3}$.

Steel: The stress-strain relationship for prestressing steel is taken as elastic-perfectly plastic. Taking $f_{pk} = 1860$ MPa, $f_{p0.1k} = 0.88\ f_{pk}$ and $\gamma_m = 1.15$, the maximum stress $f_{pd} = f_{p0.1k}/\gamma_m = 1423$ MPa. Young's modulus $E_s = 195$ GPa. The strain ε_b due to bending in steel at a depth d from the compression face is

$$\varepsilon_b = \varepsilon_{cu3} \times (d - x)/x$$

The prestress σ_{pe} in steel is given by $\sigma_{pe} = P_s/A_{ps}$.

The total stress σ_s in steel is given by

$$\sigma_s = (\sigma_{pe} + \varepsilon_b \times E_s) \leq 1423 \text{ MPa}$$

(1). Right-hand support section in span 1-2:

There are a total of nine cables. Seven cables are at 160 mm and two cables are at 169 mm from the compression face. The total prestress is 1458 kN. The area of cross section of each cable is 140 mm^2.

$\sigma_{pe} = 1458 \times 10^3/ (9 \times 140) = 1157$ MPa

If x = 19 mm, bay width = 6 m

Total compressive force C = $20 \times 0.8 \times 19 \times 6000 \times 10^{-3} = 1824$ kN

The bending strains in steel are

At 160 mm, $\varepsilon_b = 3.5 \times 10^{-3} \times (160 - 19)/19 = 25.97 \times 10^{-3}$

$\sigma_{s1} = (1157 + \varepsilon_b \times E_s) = 6221 \leq 1423$ MPa

$T_1 = 7 \times 1423 \times 140 \times 10^{-3} = 1395$ kN

At 169 mm, $\varepsilon_b = 3.5 \times 10^{-3} \times (169 - 19)/19 = 27.63 \times 10^{-3}$

$\sigma_{s2} = (1157 + \varepsilon_b \times E_s) = 6545 \leq 1423$ MPa

$T_2 = 2 \times 1423 \times 140 \times 10^{-3} = 398$ kN

Total tensile force T = 1395 + 398 = 1793 kN

$C \approx T$

$M_u = (-1824 \times 0.5 \times (0.8 \times 19) + 1395 \times 160 + 398 \times 169) \times 10^{-3} = 277$ kNm

From the moment envelope, the required moment at that section is 303 kNm.

(2). Right-hand support section in span 3-4:

There are a total of seven cables at 160 mm from the compression face. The total prestress is 1119 kN. The area of cross section of each cable is 140 mm^2.

$\sigma_{pe} = 1119 \times 10^3/ (7 \times 140) = 1142$ MPa

If x = 15 mm, bay width = 6 m

Total compressive force C = $20 \times 0.8 \times 15 \times 6000 \times 10^{-3} = 1440$ kN

The bending strains in steel are

At 160 mm, $\varepsilon_b = 3.5 \times 10^{-3} \times (160 - 15)/15 = 35.0 \times 10^{-3}$

$\sigma_{s1} = (1142 + \varepsilon_b \times E_s) = 7967 \leq 1423$ MPa

$T = 7 \times 1423 \times 140 \times 10^{-3} = 1395$ kN

$C \approx T$

$M_u = (-1440 \times 0.5 \times (0.8 \times 15) + 1395 \times 160) \times 10^{-3} = 215$ kNm

From the moment envelope, the required moment at that section is 259 kNm. Quite clearly, the section does not have sufficient moment capacity. The following are a few of the steps that can be taken to enhance the ultimate moment capacity:

- Redesign the section using a thicker slab.
- Enhance the prestressing force ensuring that stress limits are not violated.
- Use additional unstressed steel. This generally improves the ductility of the section.

Adopt the use of unstressed steel with three 20 mm bars provided at 160 mm from the compression face, with a maximum yield stress of 500 MPa and a material safety factor of 1.25.

i. Right-hand support section in span 1-2:

If x = 23 mm, bay width = 6 m

Total compressive force C = $20 \times 0.8 \times 23 \times 6000 \times 10^{-3} = 2208$ kN

The bending strains in steel are

At 160 mm, $\varepsilon_b = 3.5 \times 10^{-3} \times (160 - 23)/23 = 20.85 \times 10^{-3}$

$\sigma_{s1} = (1157 + \varepsilon_b \times E_s) = 6221 \leq 1423$ MPa

$T_1 = 7 \times 1423 \times 140 \times 10^{-3} = 1395$ kN

At 169 mm, $\varepsilon_b = 3.5 \times 10^{-3} \times (169 - 23)/23 = 22.22 \times 10^{-3}$

$\sigma_{s2} = (1157 + \varepsilon_b \times E_s) = 6545 \leq 1423$ MPa

$T_2 = 2 \times 1423 \times 140 \times 10^{-3} = 398$ kN

Unstressed steel at 160 mm: Area $= 3 \times 314 = 943$ mm^2

Allowable stress $= 500/1.15 = 435$

$\varepsilon_b = 3.5 \times 10^{-3} \times (160 - 23)/23 = 20.85 \times 10^{-3}$

$\sigma_{s3} = (0 + \varepsilon_b \times E_s) = 4066 \leq 435$ MPa

$T_3 = 435 \times 943 \times 10^{-3} = 410$ kN

Total tensile force $T = 1395 + 398 + 410 = 2203$ kN

$C \approx T$

$M_u = (-2208 \times 0.5 \times (0.8 \times 23) + 1395 \times 160 + 398 \times 169 + 410 \times 160) \times 10^{-3}$
$= 336$ kNm

From the moment envelope, the required moment at that section is 303 kNm.

ii. Right hand support section in span 3-4:

There are a total of seven cables at 160 mm from the compression face. The total prestress is 1119 kN. The area of cross section of each cable is 140 mm^2.

$\sigma_{pe} = 1119 \times 10^3 / (7 \times 140) = 1142$ MPa

If $x = 19$ mm, bay width $= 6$ m

Total compressive force $C = 20 \times 0.8 \times 19 \times 6000 \times 10^{-3} = 1824$ kN

The bending strains in steel are:

At 160 mm, $\varepsilon_b = 3.5 \times 10^{-3} \times (160 - 19)/19 = 25.97 \times 10^{-3}$

$\sigma_{s1} = (1142 + \varepsilon_b \times E_s) = 6207 \leq 1423$ MPa

$T_1 = 7 \times 1423 \times 140 \times 10^{-3} = 1395$ kN

Unstressed steel at 160 mm: Area $= 3 \times 314 = 943$ mm^2

Allowable stress $= 500/1.15 = 435$

$\varepsilon_b = 3.5 \times 10^{-3} \times (160 - 19)/19 = 25.97 \times 10^{-3}$

$\sigma_{s2} = (0 + \varepsilon_b \times E_s) = 5065 \leq 435$ MPa

$T_2 = 435 \times 943 \times 10^{-3} = 410$ kN

$T = 1395 + 410 = 1805$ kN

$C \approx T$

$M_u = (-1824 \times 0.5 \times (0.8 \times 19) + 1805 \times 160) \times 10^{-3} = 275$ kNm

From the moment envelope, the required moment at that section is 259 kNm.
Design is satisfactory.

12.18 DETAILING OF STEEL

There are various possibilities for detailing tendons. Fig. 12.87 shows some possible arrangements. From a construction point of view, it is normal to lay the tendons as fully banded in one direction and evenly distributed in the other direction. This generally minimizes the amount of weaving of tendons and hence simplifies the laying procedure.

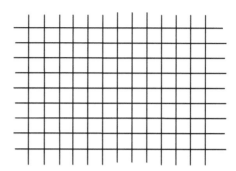

Fig. 12.87a Uniform distribution of tendons

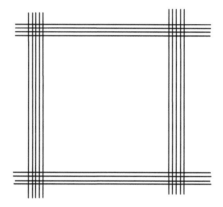

Fig. 12.87b Fully banded distribution of tendons

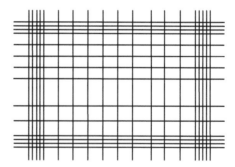

Fig. 12.87c A mix of banded and uniform distribution of tendons

It is recommended that, as shown in Fig. 12.88, where there is no shear reinforcement at the slab-column junction,

- At internal columns, at least two tendons should pass within 0.5 × width of column.
- At external columns, at least one tendon should pass within 0.5 × width of column.

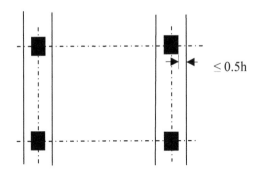

$\leq 0.5h$

Fig. 12.88 Placing of minimum number of tendons at an internal column

Fig. 12.89 shows a photograph of a banded layout. Fig. 12.90 shows the detail at the anchoring end and Fig. 12.91 shows the cables being stressed.

Fig. 12.89 Fully banded layout of tendons
Photo Courtesy VSL

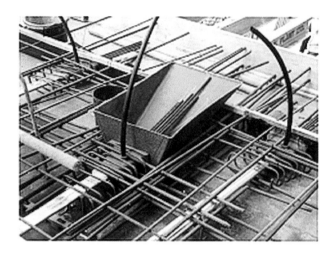

Fig. 12.90 Details at the anchoring end
Photo Courtesy VSL

Fig. 12.91 Stressing the cables
Photo Courtesy VSL

12.19 EUROCODE 2 RECOMMENDATIONS FOR DETAILING OF STEEL

The Eurocode recommendations are as follows.

(a). In solid slabs:

(i). Minimum steel: $A_{s,min} = 0.26\dfrac{f_{ctm}}{f_{yk}}b_t\,d \geq 0.0013b_t\,d$

b_t = mean width of tension zone

(ii). Maximum steel: A_s, max = 0.04 A_c
(iii). In one-way slabs secondary transverse reinforcement of not less than 20% of principal reinforcement.

(b). In flat slabs:

(i). At internal columns, top reinforcement of 50% of the total required area of reinforcement needed for the negative moment (moment causing tension on the top face) should be placed with in a width of one eight of the panel width on either side of the support.
(ii). At internal columns, bottom reinforcement of at least two bars in each orthogonal direction should be provided and these bars should pass through the column.

12.20 REFERENCES TO EUROCODE 2 CLAUSES

The following clauses in Eurocode 2 have been referred to.
Effective span: 5.3.2.2
Flexural reinforcement, solid slabs: 9.3.1.1
Flat slabs: 9.4.1

CHAPTER 13

DESIGN FOR PUNCHING SHEAR

13.1 PUNCHING SHEAR FAILURE

In a flat slab where the slab is supported by individual columns as described in Chapter 12, shear failure takes place in the form of a cone surrounding the column. Fig. 13.1a shows an idealized typical punching shear failure and Fig. 13.1b shows an actual punching shear failure. As both shear force and moment about two axes act at the column, this is a highly stressed area and particular attention should be paid to designing this critical section. Design is done only at the ultimate limit state.

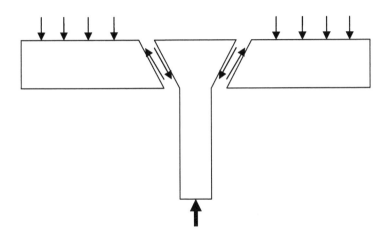

Fig. 13.1a Idealized punching shear failure

13.2 PUNCHING SHEAR STRESS CALCULATION

Consider a portion of the slab along with an isolated column as shown in Fig. 13.2a. If V_{ED} is the force in the column due to loads acting on the slab at the ultimate limit state, the column force is resisted by shear stresses v_{ED} distributed around the edges of the slab. In general the distribution of the shear stress is not constant, but for design purposes, it is taken as constant and is given by

$$v_{ED} = \frac{V_{ED}}{u\,d} \qquad (13.1)$$

where d = effective depth of the slab
u = perimeter of the portion of the slab considered
Similarly, consider a portion of the slab along with an isolated column as shown in Fig. 13.2b. If M_{ED} is the moment in the column due to loads acting on the slab at

Fig. 13.1b Punching shear failure in a slab at a support

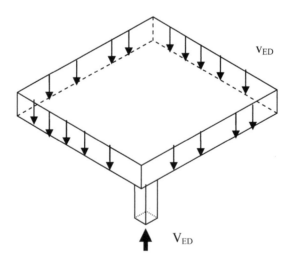

Fig. 13.2a Shear stress distribution due to shear force

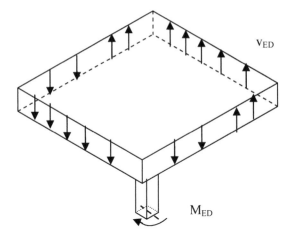

Fig. 13.2b Shear stress distribution due to moment

the ultimate limit state, the column moment is resisted by shear stresses v_{ED} distributed around the edges of the slab. Similar to bending stresses the shear stress v_{ED}
changes sign on the two symmetrical halves of the slab. In general the magnitude of the shear stress is not constant, but for design purposes, it is taken as constant.

The shape of the slab considered for calculation of the shear stress varies depending on the shape of the column and the presence of local thickening of the slab near columns known as column capitals. The perimeter of the portion of the slab considered is known as the critical perimeter.

13.3 CRITICAL SHEAR PERIMETER

The basic control perimeter is at a distance of two times the effective depth of the slab from the face of the column as shown in Fig. 13.3. The effective depth d can be taken as the **average** of the effective depths in two orthogonal directions.

Depending on the shape of the column cross section, the shape of the critical perimeter in plan varies as shown in Fig. 13.4.
i. In the case of a circular columns, the critical perimeter U_1 is of the same shape as the column and is equal to
$$u_1 = \pi (D + 4d) \tag{13.2}$$
where D is diameter of the column
ii. In the case of rectangular columns, the critical perimeter is similar to the shape of the column, except with rounded corners and is equal to
$$u_1 = 2(C_1 + C_2) + 4\pi d \tag{13.3}$$
where C_1 and C_2 are side dimensions of the column

For other than circular and rectangular columns, the basic idea of the critical perimeter being at a distance of twice the effective depth of the slab from the sides of the column and rounded at the corners as shown in Fig. 13.5 is adopted.

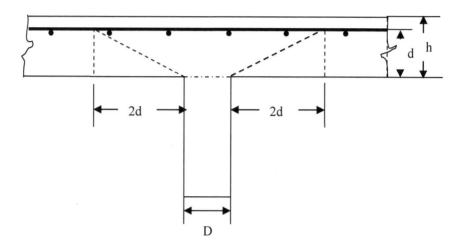

Fig. 13.3 Cross section of a punching shear failure

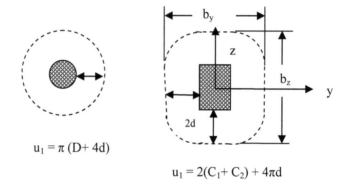

$$u_1 = \pi (D + 4d)$$

$$u_1 = 2(C_1 + C_2) + 4\pi d$$

Fig. 13.4 Critical perimeter for circular and rectangular columns

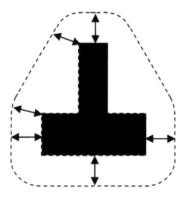

Fig. 13.5 Critical perimeter for an I-column

In the case of columns near a corner or near an edge, free edges are excluded in calculating the length of the perimeter as shown in Fig. 13.6

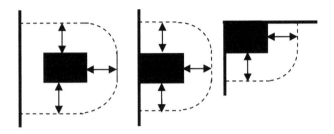

Fig. 13.6 Control perimeter for edge and corner columns

13.4 EFFECT OF HOLES NEAR THE COLUMN

If holes exist near a column at a distance from the perimeter of the column not exceeding six times the effective depth of the slab, then an allowance has to be made for the reduction in the length of the critical perimeter. The reduction is the length of the perimeter contained between two tangents drawn from the centre of the column to the outline of the opening as shown in Fig. 13.7a and b.
Two possible situations occur:
- The longer side of the hole (ℓ_2) is parallel to the longer side (C_2) of the column as shown in Fig. 13.7. In this case the critical perimeter length is *reduced* by

$$u = \ell_2 \times [(0.5C_1 + 2d)/ (a + 0.5C_1) \qquad (13.4)$$

- The shorter side (ℓ_1) of the hole is parallel to the longer side (C_2) of the column as shown in Fig. 13.8. In this case the critical perimeter length is *reduced* by

$$u = \sqrt{(\ell_1\ell_2)} \times [(0.5C_1 + 2d)/ (a + 0.5C_1) \qquad (13.5)$$

where a $\leq$ 6d is the distance from the surface of the column to the edge of the hole. C_1 = smaller dimension of the column.

13.4.1 Example

Calculate the effective basic perimeter length for the following data:
Column 300 × 400 mm, effective depth of slab d = 200 mm, distance from the surface of column to the hole a = 3d = 600 mm, hole 400 × 600 mm.
Length of basic perimeter without hole:
$u_1 = 2(C_1 + C_2) + 4 \pi d = 2(300 + 400) + 4 \pi \times 200 = 3913$ mm
Reduction in length u of the perimeter due the effect of hole:
Case 1: Long side of the hole (ℓ_2) parallel to the long side (C_2) of column
$u = \ell_2 \times [(0.5C_1 + 2d)/ (a + 0.5C_1)$
$= 600 \times (0.5 \times 300 + 2 \times 200)/ \{600 + 0.5 \times 300) = 440$ mm

Net length = 3913 − 440 = 3473 mm

Case 2: Short side of the hole (ℓ_1) parallel to the long side (C_2) of the column

$u = \sqrt{(\ell_1\ell_2)} \times [(0.5C_1 + 2d)/(a + 0.5C_1)$

$= \sqrt{(400 \times 600)} \times (0.5\times300 + 2\times 200)/\{600 + 0.5\times 300) = 359$ mm

Net length = 3913 − 359 = 3554 mm

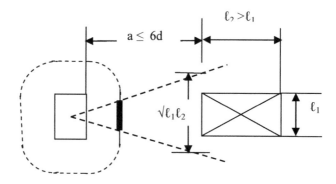

Fig. 13.7 The shorter side of the hole is parallel to the longer side of the column

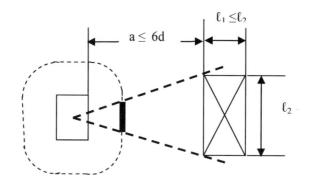

Fig. 13.8 The longer side of the hole is parallel to the longer side of the column

13.5 COLUMNS WITH CAPITALS

The presence of capitals will affect the section where the shear stresses are checked.

i. Circular column heads: If $\ell_H < 2h_H$ (See Fig. 13.9 for notation), then the shear stress needs to be checked at a section **outside** column head at a radius of

$r_{cont} = 0.5c + 2d + \ell_H$ (13.6)

In effect the presence of a head the shifts the control radius by the 'length' ℓ_H of the capital.

Example: Calculate the radius of the control perimeter r_{cont} using the following data.

Rectangular column: 300×400 mm, effective depth of slab $d = 200$ mm
Depth of column head $h_H = 300$ mm, $2h_H = 600$ mm.
Dimensions of the head: $\ell_{H1} = 400$mm, $\ell_{H2} = 530$ mm **both less than** 600 mm.
$\ell_1 = c_1 + 2\ \ell_{H1} = 300 + 2 \times 400 = 1100$ mm
$\ell_2 = c_2 + 2\ \ell_{H2} = 400 + 2 \times 530 = 1460$ mm
$\ell_2/\ell_1 = 1.33 < 1.5$
$r_{cont} = 2d + 0.56\ell_1\sqrt{(\ell_2/\ell_1)} = 2 \times 200 + 0.56 \times 1100 \times \sqrt{(1460/1100)} = 1110$ mm
Total length $U = 2\pi\ r_{cont} = 6972$ mm

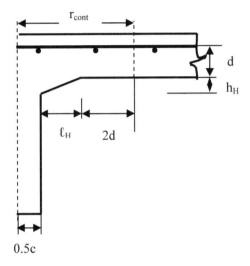

Fig. 13.9 Column with circular capital

ii. Rectangular column with rectangular column heads: If $\ell_H < 2h_H$ (see Fig. 13.10 for notation) of overall dimension $\ell_1 \times \ell_2$, then the shear stress needs to be checked at a section **outside** column head at a radius of
$$r_{cont} = 2d + 0.56\ell_1\sqrt{(\ell_2/\ell_1)}, \qquad (\ell_2/\ell_1) \le 1.5$$
$$r_{cont} = 2d + 0.69\ell_1, \qquad\qquad (\ell_2/\ell_1) > 1.5 \qquad (13.7)$$

Note that for a rectangular column without drops, the control perimeter is as shown in Fig. 13.4. However when considering a rectangular column with a capital, the control perimeter is taken as circular.

iii. Circular column with enlarged heads: If $\ell_H > 2(d + h_H)$ then the shear stress needs to be checked at sections **both inside as well as outside** the column head at a radius of (see Fig. 13.11)
$$r_{cont,\ ext} = 0.5c + 2d + \ell_H, \text{ effective depth} = d$$
$$r_{cont,\ int} = 0.5c + 2(d + h_H), \text{ effective depth} = d + h_H - h_H \times 2(d + h_H)/\ell_H \qquad (13.8)$$

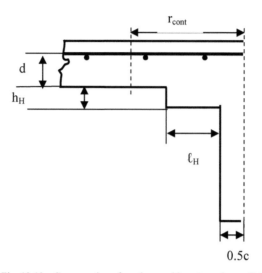

Fig. 13.10 a Cross section of a column with rectangular capital

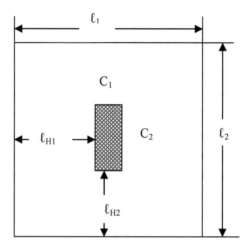

Fig. 13.10 b Plan of the rectangular capital

iv. Rectangular column with enlarged heads: If ℓ_{H1} and $\ell_{H2} > 2(d + h_H)$ and $\ell_{H1} < \ell_{H2}$ then the shear stress needs to be checked at sections **both inside as well as outside** the column head.

Eurocode 2 does not clearly specify sections at which the shear stresses need to be checked. As a conservative approach, inside the drop, the shear stress is checked on circular perimeter with a radius r_{int} given by
$$r_{int} = 0.5C_1 + 2(d + h_H)$$
For r_{ext}, follow the same rules as when ℓ_{H1} and ℓ_{H2} were both less than $2(d+h_H)$.

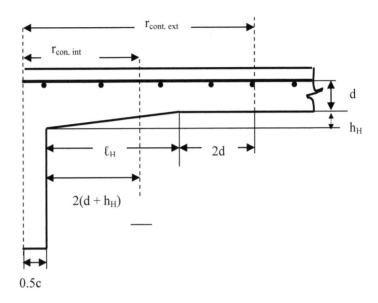

Fig. 13.11 Column with circular capital

Example: Calculate the radius of the control perimeter r_{cont} using the following data:

Rectangular column: 300×400 mm, effective depth of slab d = 200 mm

Depth of column head $h_H = 300$ mm, $2(d + h_H) = 1000$ mm.

Dimensions of the head: $\ell_{H1} = 1200$mm, $\ell_{H2} = 1600$mm both greater than 1000 mm.

$\ell_1 = C_1 + 2 \ \ell_{H1} = 300 + 2 \times 1200 = 2700$ mm

$\ell_2 = C_2 + 2 \ \ell_{H2} = 400 + 2 \times 1600 = 3600$ mm

$\ell_2/\ell_1 = 1.33 < 1.5$

$r_{cont, \ ext} = 2d + 0.56 \ \ell_1 \sqrt{} \ (\ell_2/\ell_1) = 2 \times 200 + 0.56 \times 2700 \times \sqrt{} \ (3600/2700)$
$= 3518$ mm

$r_{cont, \ ext}$ is greater than $0.5\ell_2$ and lies outside the capital.

Total length $U = 2\pi \ r_{cont, \ ext} = 22104$ mm

13.6 CALCULATION OF PUNCHING SHEAR STRESS v_{ED} UNDER THE ACTION OF A MOMENT

The shear stress v_{ED} due to moment can be calculated as follows.

(a). Circular column: As shown in Fig. 13.4, the radius r of the control perimeter is given by

$r = 0.5D + 2d$

where D = diameter of the column.

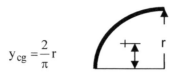

$$y_{cg} = \frac{2}{\pi} r$$

Fig. 13.12 Centroid of a circular arc

As shown in Fig. 13.12, the centroid of a quarter-circular arc is at a distance y_{cg} from the centre, where $y_{cg} = (2/\pi)\, r$. Taking moment about the horizontal centroidal axis of the column,

$M_{ED} = v_{ED} \times d \times$ Circumference of semicircular arc $\times (2\, y_{cg})$

$\quad = v_{ED} \times d \times \pi\, r \times \{2 \times (2/\pi)\, r\} = v_{ED} \times d \times 4r^2$

Substituting $r = 0.5\, D + 2d$,

$\quad M_{ED} = v_{ED} \times d \times W_1$

where $W_1 = (D + 4d)^2$

$$v_{ED} = \frac{M_{ED}}{d \times 4r^2} = \frac{M_{ED}}{d \times W_1} \tag{13.9}$$

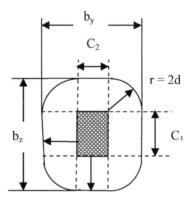

Fig. 13.13 Calculation of W_1 for a rectangular column

(b). Rectangular column: Assuming that the moment acts about the axis parallel to column side C_2, taking moments about the horizontal centroidal axis, then from Fig. 13.13

$M_{ED} = v_{ED} \times d \times 2 \times [C_2 \times (0.5C_1 + r) + 2\{0.5C_1 \times 0.25C_1\}$
$\quad\quad\quad + \pi\, r \times \{0.5C_1 + (2/\pi)\, r\}]$

where the first term is due to v_{ED} on column side C_2, the second term is due to v_{ED} on column sides C_1 and the last term is due to v_{ED} on the two quarter-circular arcs. Simplifying,

$$M_{ED} = v_{ED}\, d\, [\frac{C_1^2}{2} + C_1 C_2 + 2 C_2\, r + \pi r C_1 + 4r^2]$$

$$M_{ED} = v_{ED} \times d \times W_1$$

$$W_1 = [\frac{C_1^2}{2} + C_1 C_2 + 2 C_2\, r + \pi r C_1 + 4r^2]$$

Substituting $r = 2d$,

$$W_1 = [\frac{C_1^2}{2} + C_1 C_2 + 4 C_2 d + 2\pi C_1 d + 16 d^2]$$ (13.10)

13.7 PUNCHING SHEAR STRESS UNDER SHEAR FORCE AND MOMENT ACTING SIMULTANEOUSLY

The shear stress v_{ED} under the action of the shear force V_{ED} alone is given by

$$v_{ED} = \frac{V_{ED}}{u_i d}$$

The shear stress v_{ED} under the action of the moment M_{ED} is given by

$$v_{ED} = \frac{M_{ED}}{W_1 d}$$

When both shear force and moment act simultaneously,

$$v_{ED} = \frac{V_{ED}}{u_i d} + \frac{M_{ED}}{W_1 d} = \frac{V_{ED}}{u_i d}[1 + \frac{M_{ED}}{V_{ED}} \frac{u_i}{W_1}]$$

Eurocode 2 approximates the above equation to

$$v_{ED} = \frac{V_{ED}}{u_i d}[1 + k\frac{M_{ED}}{V_{ED}} \frac{u_i}{W_1}] = \beta\frac{V_{ED}}{u_i d}$$

$$\beta = [1 + k\frac{u_i e}{W_1}], \; e = \frac{M_{ED}}{V_{ED}}$$ (13.11)

(a). In the case of circular columns, $k = 0.6$,
$u_i = \pi(D + 4d)$, $W_1 = (D + 4d)^2$; therefore

$$\beta = [1 + 0.6\pi\frac{e}{(D + 4d)}]$$ (13.12)

(b). In the case of rectangular columns,
$u_i = 2(C_1 + C_2) + 4\pi d$
If the moment acts about the y-axis parallel to the smaller side C_2 of the column,

$$e = e_z = M_{EDy}/V_{ED}, \; W_1 = W_{1y} = [\frac{C_1^2}{2} + C_1 C_2 + 4 C_2 d + 2\pi C_1 d + 16 d^2]$$

If the moment acts about the z-axis parallel to the longer side C_1 of the column,

$$e = e_y = M_{EDz}/V_{ED}, \; W_1 = W_{1z} = [\frac{C_2^2}{2} + C_1 C_2 + 4 C_1 d + 2\pi C_2 d + 16 d^2]$$ (13.13)

The values of k are shown in Table 13.1.

Table 13.1 Values of k for rectangular columns

C_1/C_2	≤ 0.5	1.0	2.0	≥ 3.0
k	0.45	0.60	0.70	0.80

13.7.1 Special Cases of Shear Force and Moment Acting Together

Eurocode 2 gives information on several special cases of loading.
1. If an internal rectangular column subjected to an axial load V_{ED} and moments M_{EDy} and M_{EDz} about the y-axis (parallel to the smaller side C_2) and the z-axis (parallel to the larger side C_1) respectively, then
$e_y = M_{EDz}/V_{ED}$, $e_z = M_{EDy}/V_{ED}$, $b_y = C_2 + 4d$, $b_z = C_1 + 4d$
An approximate expression for β is given by

$$\beta = 1 + 1.8 \sqrt{[(\frac{e_y}{b_z})^2 + (\frac{e_z}{b_y})^2]} \tag{13.14}$$

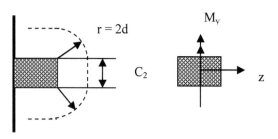

Fig. 13.14 Edge column subjected moment and shear force

2. For an edge slab as shown in Fig. 13.14 with the moment M_{EDy} applied about an axis parallel to the slab edge
   - Eccentricity $e_z = M_{yED}/V_{ED}$ is towards the interior of the slab
   - $M_{EDz} = 0$
   - Use a smaller value for the contour perimeter u_1^* where
     $$u_1^* = C_2 + 2\pi d + \text{lesser of } (3d, C_1) \tag{13.15}$$
   - Assume that the shear stress v_{ED} is uniformly distributed

$$v_{ED} = \frac{V_{ED}}{u_1^* d}$$

3. In addition to the conditions in (2) above, if the moment M_{EDz} is not zero, then
$$\beta = \frac{u_1}{u_1^*} + k \frac{u_1}{W_1} e_y$$
$$u_1 = 2C_1 + C_2 + 2\pi d$$
$$W_1 = [\frac{C_2^2}{4} + C_1 C_2 + 4 C_1 d + \pi C_2 d + 8d^2]$$
$$e_y = M_{EDz}/V_{ED} \tag{13.16}$$

k = From Table 13.1, using $0.5 C_1/C_2$ instead of C_1/C_2.

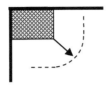

Fig. 13.15 Corner column

4. Corner columns: If the eccentricity is towards the interior of slab, then, as shown in Fig. 13.15

$$v_{ED} = \frac{V_{ED}}{u_1^* d}$$

$$u_1^* = \text{lesser of } (0.5C_1, 1.5d) + \text{lesser of } (0.5C_2, 1.5d) + \pi d \qquad (13.17)$$

5. If in a corner column the eccentricity is towards the exterior, then

$$v_{ED} = \beta \frac{V_{ED}}{u_i d}$$

$$\beta = [1 + k \frac{u_i e}{W_1}]$$

$$e = \frac{\sqrt{[M_{EDy}^2 + M_{EDz}^2]}}{V_{ED}}$$

$$u_i = C_1 + C_2 + \pi d \qquad (13.18)$$

6. If the lateral stability is not dependent on the frame action between slab and columns, then the following approximate values of β can be used:
- Interior column, $\beta = 1.15$
- Edge column, $\beta = 1.4$
- Corner column, $\beta = 1.5$ $\qquad (13.19)$

13.8 PUNCHING SHEAR STRESS CHECKS

The punching shear stress is checked at the face of the column and also at the basic control perimeter.

(a). At the column perimeter, the maximum shear stress v_{ED} should not exceed $v_{RD, max}$ so that the compression struts are not crushed:

$$v_{ED} = \beta \frac{V_{ED}}{u_0 d} \leq v_{RD, max}$$

The 'column perimeter' u_0 is given by
Interior column: $u_0 = 2(C_1 + C_2)$
Edge column: $u_0 = C_2 + 3d \leq (2C_1 + C_2)$
$\qquad\qquad C_2 = $ column dimension parallel to the free edge
Corner column: $u_0 = 3d \leq (C_1 + C_2)$

$$v_{RD,max} = 0.3[1 - \frac{f_{ck}}{250}] \times f_{cd} \qquad (13.20)$$

(b). Along the control section, if v_{ED} is less than $v_{RD,c}$, then no shear reinforcement is required. $v_{RD,c}$ is given by (9.6) in Chapter 9. It is repeated here for convenience:

$$v_{Rd,c} = [C_{Rd,c} \, k \, (100\rho_1 \, f_{ck})^{\frac{1}{3}} + k_1 \, \sigma_{cp}] \geq (v_{min} + k_1 \, \sigma_{cp})$$

where

$\gamma_c = 1.5$, $C_{Rd,c} = 0.18/\gamma_c$, $k = 1 + \sqrt{\dfrac{200}{d}} \leq 2.0$

d = effective depth to steel in tensile zone

$\sigma_{cp} = 0.5(\sigma_{cy} + \sigma_{cz})$, $\sigma_{cy} = N_{Edy}/A_{cy}$, $\sigma_{cz} = N_{Edz}/A_{cz}$, $\sigma_{cp} \leq 0.2 \, f_{cd}$

N_{Edy}, N_{Edz} = longitudinal forces across the full bay for internal columns and the longitudinal forces across the cross section for edge columns.

Average stress in stress block, $f_{cd} = \alpha_{cc} \, f_{ck}/\gamma_c$, $\alpha_{cc} = 0.85$, $\gamma_c = 1.5$

$v_{min} = 0.035 \, k^{\frac{3}{2}} \, \sqrt{f_{ck}}$, $\quad k_1 = 0.1$

$\rho_1 = \sqrt{(\rho_{1y} \, \rho_{1z})} \leq 0.02$

ρ_{1y} and ρ_{1z} are calculated using the total area of bonded tensile steel which extends a distance at least equal to (anchorage length + d) beyond the section considered and using a width equal to column width plus 3d on either side.

The perimeter u_{out} beyond which no shear reinforcement is required is given by

$$u_{out} = \beta \frac{V_{ED}}{v_{RD,c} \times d} \qquad (13.21)$$

(c). Where shear reinforcement is required, it is calculated using the following equation:

$$v_{Rd,cs} = 0.75 v_{Rd,c} + 1.5 \frac{d}{s_r} A_{sw} \, f_{fwd,ef} \frac{1}{u_1 d} \sin\alpha \qquad (13.22)$$

A_{sw} = area of one perimeter of shear reinforcement around the column in mm^2

s_r = radial spacing of perimeters of shear reinforcement in mm

$f_{ywd,ef} = 250 + 0.25 \, d \leq f_{ywd}$ in MPa

d = mean effective depth in mm

α = angle of inclination between the shear reinforcement and main steel (For vertical links, $\alpha = 90°$, $\sin\alpha = 1$)

The shear reinforcement is detailed as shown in Fig. 13.16.

- The outermost perimeter of shear reinforcement should be placed at a distance not greater than 1.5d *within* u_{out}.
- The shear reinforcement should be provided in at least two perimeters of link legs and the spacing of the perimeters should not exceed 0.75d.
- The spacing link legs around a perimeter within the first control perimeter at 2d from the loaded area should not exceed 1.5d.

- For perimeters outside the first control perimeter the spacing link legs around a perimeter should not exceed 2d.
- If s_r and s_t are respectively the spacing of vertical links in the radial and tangential directions, the area of a link leg link $A_{sw, min}$ is given by

$$1.5 \frac{A_{sw, min}}{s_r \, s_t} \geq 0.08 \frac{\sqrt{f_{ck}}}{f_{yk}}$$

- The vertical component of those prestressing tendons passing within a distance of 0.5d of the column may be included in the shear calculations.
- The distance between the face of a support and the nearest shear reinforcement taken into account in the design should not exceed 0.5d. This distance is measured at the level of the tensile reinforcement.
- If a single line of bent down bars is provided then the ratio $d/s_r = 0.67$

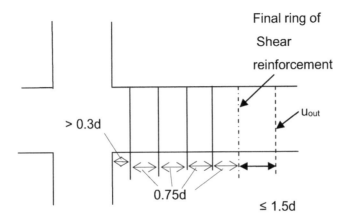

Fig. 13.16 Detailing of shear reinforcement

13.9 EXAMPLE OF PUNCHING SHEAR CAPACITY DESIGN

Calculate the punching shear capacity of a 200 mm thick slab at an internal column. The slab is supported by 400 mm square internal column at a spacing of 6 m both ways. It is prestressed both ways by seven cables each of 140 mm² in cross sectional area providing a force of 1141 kN. It is subjected to a working live load of 4.0 kN/m². Take concrete grade as C30/37.
Taking unit weight of concrete as 25 kN/m³,
Self weight = 25 × 200 × 10^{-3} = 5 kN/m²
At the ultimate limit state, using load factors of 1.35 for permanent load and 1.5 for variable load, the total load being supported is
q = (5 × 1.35 + 4 ×1.5) = 12.75 kN/m²
If the centre-to-centre distance of the columns is 6 m, the total load on a column is
V = 12.75 × 6 × 6 = 451 kN
Taking cover of 38 mm and an overall depth of 200 mm for the slab, the effective depth d is

$d = 200 - 38 = 162$ mm

(a). Check the shear stress around the column perimeter to ensure that the depth of slab is sufficient

Assume that frame stability is not dependent on frame action. Also a certain value of moment will act due to any imbalance of loads. Treat it as case of combined shear and moment.

$V_{ED} = 451$ kN, $\beta = 1.15$, from (13.19) for internal column

$u_0 = 4 \times 400 = 1600$ mm

$v_{ED} = 1.15 \times 451 \times 10^3 / (1600 \times 162) = 2.0$ MPa

Calculate $v_{RD, max}$ from (13.20):

$$v_{RD,max} = 0.3[1 - \frac{f_{ck}}{250}] \times 17 = 0.3[1 - \frac{30}{250}] \times 17.0 = 4.5 \text{ MPa}$$

$v_{RD, max} > v_{ED}$. Therefore the slab depth is sufficient.

(b). Calculate the shear stress at the basic control perimeter

From (13.3), the basic control perimeter $u_1 = 2(C_1 + C_2) + 4\pi d$

$$= 2(400 + 400) + 4 \times \pi \times 162$$
$$= 3636 \text{ mm}$$

shear stress $v_{ED} = 1.15 \times (451 \times 10^3)/ (3636 \times 162) = 0.88$ MPa

(c). Calculate the value of $v_{RD, c}$

In the case of sections where the flexural tensile stress is greater than the permissible concrete tensile stress f_{ctd}, the design value of punching shear resistance of a slab without punching shear reinforcement along the control section considered is given by $v_{Rd,c}$. Since over the column the section would have cracked, $v_{RD, c}$ is given by equation (9.6):

$$C_{Rd, c} = 0.12, \quad k = 1 + \sqrt{\frac{200}{d}} = 1 + \sqrt{\frac{200}{162}} = 2.11 \leq 2.0$$

Take k = 2.0, $k_1 = 0.10$, $f_{cd} = 0.85 \times 30/1.5 = 17.0$ MPa

$$v_{min} = 0.035 k^{\frac{3}{2}} \sqrt{f_{ck}} = 0.035 \times 2^{\frac{3}{2}} \times \sqrt{30} = 0.54 \text{ MPa}$$

A_{sl} = area of tensile steel which extends a distance at least equal to $(\ell_{bd} + d)$ beyond the section considered

A_{sl} is the area of 7-cables each of 140 mm^2 in cross section:

$A_{sl} = 7 \times 140 = 980$ mm^2

$\rho_{1y} = \rho_{1z} = 980/ (6 \times 10^3 \times 162) = 0.10 \times 10^{-2} \leq 2.0 \times 10^{-2}$

$$v_{Rd,c} = [C_{Rd,c} k (100 \rho_t f_{ck})^{\frac{1}{3}} + k_1 \sigma_{cp}] \geq (v_{min} + k_1 \sigma_{cp}]$$

$$= [0.12 \times 2.0 \times (0.10 \times 30)^{\frac{1}{3}} + 0.1 \times 0.95] \geq (0.54 + 0.1 \times 0.95]$$

$$= [0.35 + 0.1\} \geq [0.54 + 0.1]$$

$$= 0.45 \geq 0.65$$

$v_{Rd,c} = 0.65$ MPa

$(v_{ED} = 0.88) > (v_{Rd, c} = 0.65)$. Therefore shear reinforcement is required.

(d). Calculate u_{out} beyond which shear reinforcement is *not* required.
From equation (13.21),

$$u_{out} = \beta \frac{V_{ED}}{v_{RD,c} \times d} = 1.15 \times \frac{451 \times 10^3}{0.65 \times 162} = 4926 \text{ mm}$$

Calculate the distance u_{out} from the column face:
$u_{out} = 4926 = 2(C_1+C_2) + 2\pi a = 2(400 + 400) + 2\pi a$, a = 529 mm.
(a – 1.5 d) = 286 mm. The outermost perimeter of shear reinforcement should be placed at a distance not greater than 286 mm from the column face.

(e). Calculate the positions of shear reinforcement
(i). The first ring must be at a distance greater than 0.3d = 49 mm from the column face (see Fig. 13.16).
(ii). At least two rings of shear reinforcement with a spacing between rings not exceeding 0.75d = 122 mm must be placed between a distance greater than 49 mm and 286 mm from the column face.
(iii). Place the last ring at 286 mm and the inner ring at (286 – 0.75d) = 164 mm from the column face. However, as the distance of 164 mm is greater than 0.5d = 81 mm from the column face, it will be necessary to provide an additional ring of reinforcement at a distance less than 0.5d = 81 mm from the column face. Therefore place a third ring at 80 mm from the second ring. Adjust the spacing between the rings to be equal at (286 – 80)/2 = 103 mm < 0.75d. Therefore arrange the three rings as follows:
First ring at 80 mm from column face
Second ring at (80 + 103) = 183 mm from column face
Third ring at (183 + 103) = 286 mm from column face
Fig. 13.17 shows the arrangement of the three rings.

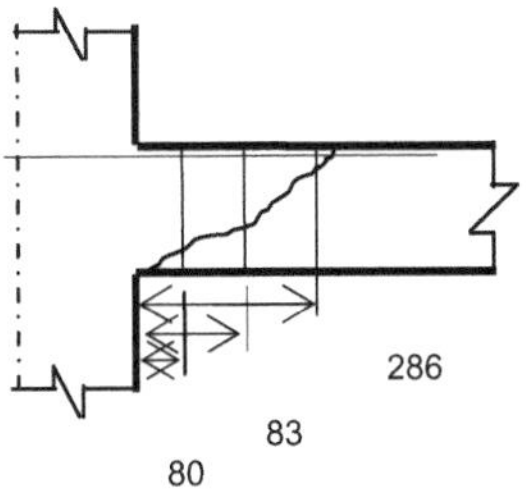

Fig. 13.17 Arrangement of shear reinforcement

(f). Calculate the total area of shear reinforcement required

$$v_{Rd,cs} = 0.75 v_{Rd,c} + 1.5 \frac{d}{s_r} A_{sw} f_{fwd,ef} \frac{1}{u_1 d} \sin \alpha$$

$v_{RD, cs}$ = applied shear stress = v_{RD} = 0.88 MPa
$v_{RD, c}$ = 0.65 MPa
u_1 = 3636 mm

f_{yw} = yield stress of shear link steel = 500 MPa, γ_s = 1.15, f_{ywd} = 500/1.15 = 435
$f_{fwd, ef}$ = 250+ 0.25d = 250 + 0.25 × 162 = 291 MPa < (f_{ywd} = 435)
s_r = radial spacing of links = 103 mm
sin α = 1.0
A_{sw} = 337 mm^2 to be provided in each of three rings. Assuming 8 mm single-leg link, cross sectional area of one link = 50 mm^2
(i).Radius of the first ring, r = (80 + C_1/2) = 280 mm
Maximum circumferential or tangential spacing s_t of links = 1.5d = 243 mm
No. of links = 2πr/ Spacing of links = 2 × π × 280/ 243 = 7.2, say 8 links
Total area provided = 50 × 8 = 400 mm^2 > (A_{sw} = 337 mm^2)
(ii).Radius of the second ring, r = (183 + C_1/2) = 383 mm
Maximum circumferential or tangential spacing s_t of links = 1.5d = 243 mm
No. of links = 2πr/ Spacing of links = 2 × π × 383/ 243 = 9.9 say 10 links
Total area provided = 50 × 10 = 500 mm^2 > (A_{sw} = 337 mm^2)
(iii). Radius of the third ring, r = (286 + C_1/2) = 486 mm
Maximum circumferential or tangential spacing s_t of links = 1.5d = 243 mm
No. of links = 2πr/ Spacing of links = 2 × π × 486/ 243 = 12.5, say 13 links
Total area provided = 50 × 13 = 650 mm^2 > (A_{sw} = 337 mm^2)
Note: One can reduce the total area of links by using 6 mm links with a cross sectional area of 28 mm^2 giving a total area of 28 × 13 = 364 mm^2 which is greater than A_{sw} = 337 mm^2. Howeve, this might not be suitable for, simplicity of construction it is undesirable to use too many different diameters of links, with the possibility of the wrong size of links being used.

(g). Check if the area of a link satisfies minimum requirements

$$\frac{A_{sw, min}}{s_r s_t}(1.5\sin \alpha +\cos \alpha)\geq 0.08\frac{\sqrt{f_{ck}}}{f_{yk}}$$

f_{ck} = 30 MPa, f_{yk} = 500 MPa, s_r = 103 mm, s_t = 243 mm α = 90°, sin α = 1, cos α. = 0.
$A_{sw, min} \geq 17$ mm^2
$A_{sw, min}$ = 50 mm^2 for an 8 mm diameter link is greater than the minimum area required

13.10 REFERENCES TO EUROCODE 2 CLAUSES

In this chapter, the following clauses in Eurocode 2 have been referred to:
Punching: 6.4
Load distribution and basic control perimeter: 6.4.2
Control perimeter near an opening: 6.4.2. (3)
Control perimeter for loaded areas close to or at edge or corner: 6.4.2. (4)
Slabs with enlarged column head: 6.4.2. (8)
Punching shear calculation: 6.4.3
Punching shear resistance of slabs without shear reinforcement: 6.4.4
Punching shear resistance of slabs with shear reinforcement: 6.4.5

CHAPTER 14

LOADING ON BUILDINGS

14.1 INTRODUCTION

Structures are subjected to various types of loading caused by gravity, wind, earthquake, blasts, explosion and so on. The nature of their variation in time and space, the frequency of their occurrence and the possibility of several types of loads acting at the same time at their maximum values are all highly variable and almost impossible to predict with any certainty. Similar uncertainty attaches to strength of materials as well. The only sensible approach that one can take is to work in terms of statistical variations of loads and strengths of materials.

Fig. 14.1 shows qualitatively the statistical variation of load and strength of material. The variation is usually represented by what are known as normal distribution curves.

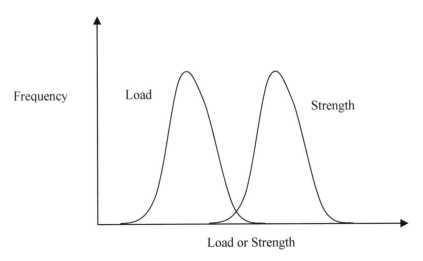

Fig. 14.1 Normal distribution of load and strength

The two properties which govern the shape of the curve are the mean value and standard deviation as shown in Fig. 14.2. The mean value is the value of the variable, which occurs with the greatest frequency. The standard deviation is defined as the value of the variable (mean ± standard deviation) beyond which not more than 16% of the data lie. Note that towards the right-hand tail of the load distribution curve, the load increases but its frequency of occurrence decreases. In a similar manner towards the left-hand tail of the strength distribution curve, the strength decreases but its frequency of occurrence decreases. It is clear from the figure that the statistical possibility of a high value of load coinciding with the low value of strength is an event of fairly low frequency. In practice very often there is not sufficient reliable data to know with any great certainty the statistical distribution of loads and strength. A certain amount of common sense and experience are used in coping with the variation. Generally one deals with what are known as characteristic values of loads and strength and makes sure that there is only a 5% probability of ***loads exceeding*** their characteristic value and similarly there is only a 5% probability of the ***strength falling below*** their characteristic value. Unfortunately the characteristic loads and strengths are themselves subject to a great deal of uncertainty. This is taken care of in design by ***multiplying*** the characteristic loads by factors called load factors, which are greater than unity. Similarly the characteristic strengths are ***divided*** by factors called material safety factors, which are greater than unity. For example it is possible to predict with reasonable confidence the self-weight of a structural element as opposed to predicting the forces due to wind load acting on an element. Therefore one can use a smaller load factor for self-weight loads than for wind loads. Similarly the strength variation of a material like steel is likely to take place in a narrower range compared to the strength variation of a material like concrete. It therefore is logical to use a material safety factor for steel which is smaller than that used for concrete.

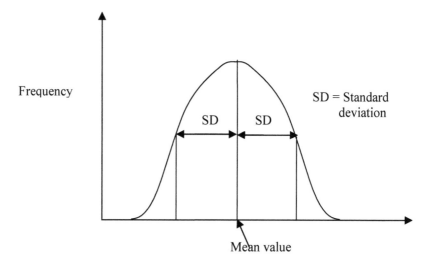

Fig. 14.2 Mean and standard deviation of normal distribution of load and strength.

14.2 LIMIT STATES

Amongst many requirements that a structure has to satisfy during its life such as that it is easy and economical to construct and maintain, is readily modified to suit a change in intended use, has a reasonably long life, does not deteriorate rapidly, is aesthetically pleasing, is robust such that failure in a small part does not cause excessive damage to the entire structure and so on, two of the most important and basic are

- During its normal use, it performs satisfactorily. For example it does not deform or crack excessively or cause alarming vibration and so on. In design this is achieved by prescribing the limits to these factors beyond which the structure becomes unserviceable. The serviceability limit state (SLS) is the condition beyond which the structure fails to satisfy its intended purpose.

- The structure must also posses an acceptable margin of safety against collapse caused by loads that are much greater than that acting during normal use. The ultimate limit state (ULS) is the state beyond which the structure or some part of it is unsafe for its intended purpose.

14.3 CLASSIFICATION OF ACTIONS

Actions (forces) are classified according to their variation in time as permanent, variable or accidental. Table 14.1 shows the classification of typical loads.

Table 14.1 Classification of actions

Permanent action	Variable action	Accidental action
a. Self weight of structures	a. Imposed loads	a. Explosions
b. Prestressing force	b. Snow loads	b. Fire
	c. Wind loads	c. Impact from vehicles
c. Water and earth pressure	d. Indirect actions e.g. temperature effects	
d. Indirect actions e.g. settlement of supports		

14.4 CHARACTERISTIC VALUES OF ACTIONS

The characteristic value of an action can be specified as a mean value, an upper or a lower value or a nominal value that does not refer to any known statistical distribution.

The characteristic value of a permanent action can be specified by a single value G_k, if the variability of G is small. If the variability is not small then two values should be used: an upper value $G_{k, \text{ superior}}$ which is the 95% fractile and a lower value $G_{k, \text{ inferior}}$ which is the 5% fractile of the statistical distribution of G.

In the case of variable actions, the following representative values can be specified depending upon their use in specific design situations.

- Characteristic value Q_k
- A combination value represented as a product $\psi_0 Q_k$. This is used for the verification of ultimate limit states and irreversible serviceability limit

states. For example in the case of imposed loads in buildings, the value of ψ_0 can vary from 1.0 for storage areas to 0.70 in most other areas such as residential, office and shopping areas. In the case of wind loads $\psi_0 = 0.60$.

- A frequent value represented as a product $\psi_1\, Q_k$. This is used for the verification of ultimate limit states involving accidental actions and for the verification of reversible serviceability limit states. For example in the case of imposed loads in buildings, the value of ψ_1 can vary depending on the area under consideration. For example, $\psi_1 = 0.90$ for storage areas, $\psi_1 = 0.70$ for shopping areas and $\psi_1 = 0.50$ for residential areas. In the case of wind loads $\psi_1 = 0.20$.
- A quasi-permanent value represented as a product $\psi_2\, Q_k$. This is used for the verification of ultimate limit states involving accidental actions and for the verification of reversible serviceability limit states. Quasi-permanent values are also used for the calculation of long-term effects such as creep deformations. For example in the case of imposed loads in buildings, the value of ψ_2 can vary depending on the area under consideration. For example, $\psi_2 = 0.80$ for storage areas, $\psi_2 = 0.60$ for shopping areas and $\psi_2 = 0.30$ for residential and office areas. In the case of wind loads $\psi_2 = 0$.

Table 14.1 Recommended values of combination ψ factors for variable action in buildings

Imposed load in buildings:	Combination ψ_0	Frequent ψ_1	Quasi-permanent ψ_2
Domestic, residential and office areas and traffic areas with vehicle weight W $30 < W \leq 160$ kN	0.7	0.5	0.3
Congregation, shopping and traffic areas with vehicle weight $\leq$ 30 kN	0.7	0.5	0.6
Storage areas	1.0	0.9	0.8
Wind loads on buildings	0.6	0.2	0

14.5 DESIGN VALUES OF ACTIONS

The design value of action is the characteristic value multiplied by a load factor γ appropriate to the action considered. Thus the design value F_d of an action F_k is given by

$$F_d = \gamma_f\, \psi\, F_k \qquad\qquad (14.1)$$

where F_k = characteristic value of the action

γ_f = load factor which accounts for the possibility of unfavourable

deviation of the action from the representative value
$\psi = 1.0$ or ψ_0, ψ_1, ψ_2 depending on whether it is characteristic value, combination value, frequent value or a quasi-permanent value which is relevant

14.6 COMBINATION OF ACTIONS

In the case of buildings, combination of actions may be based on **not more than two variable** actions. For example permanent actions such as gravity loading and prestressing force can be combined with say imposed loading and wind. Note also that each combination of actions should include a leading variable action or an accidental action. When a number of variable actions are possible each action should be considered in turn as a leading action and combined with the other actions considered as accompanying variables.

14.6.1 Combination of Actions for ULS

(i). When considering persistent and transient design situations Eurocode 1 gives two options as follows:

Option 1:

$$F_d = \sum_{j\geq1} \gamma_{G,j} G_{k,j} \; "+" \gamma_P P \; "+" \gamma_{Q,1} Q_{k,1} \; "+" \sum_{i>1} \gamma_{Q,i} \, \psi_{0,i} Q_{k,i} \tag{14.2}$$

Option 2: Less favourable of

$$(a). \; F_d = \sum_{j\geq1} \gamma_{Gj} G_{kj} \; "+" \gamma_P P \; "+" \gamma_{Q,1} \psi_{0,1} Q_{k,1} \; "+" \sum_{i>1} \gamma_{Q,i} \, \psi_{0,i} Q_{k,i} \tag{14.3}$$

$$(b). \; F_d = \sum_{j\geq1} \xi_j \gamma_{Gj} G_{kj} \; "+" \gamma_P P \; "+" \gamma_{Q,1} Q_{k,1} \; "+" \sum_{i>1} \gamma_{Q,i} \, \psi_{0,i} Q_{k,i} \tag{14.4}$$

(ii). When considering accidental design situations

$$F_d = \sum_{j\geq1} G_{kj} \; "+" P \; "+" A_d \; "+" (\psi_{1,1} \text{ or } \psi_{2,1}) Q_{k,1} \; "+" \sum_{i>1} \psi_{2,i} Q_{k,i} \tag{14.5}$$

Note: Partial load factors γ are not included as this event occurs when a structure is in use.

(iii). When considering seismic design situations

$$F_d = \sum_{j\geq1} G_{kj} \; "+" P \; "+" A_{Ed} \; "+" \sum_{i\geq1} \psi_{2,i} Q_{k,i} \tag{14.6}$$

Note: Partial load factors γ are not included as this event occurs when a structure is in use.

Note in the above:
"+" implies 'to be combined with'
Σ implies 'the combined effect of'
P is the representative value of the prestress action
A_d is the accidental action
G_{kj} is characteristic value of a permanent action
Q_{k1} is characteristic value of the leading variable action
Q_{kj} is characteristic values of accompanying variable action

ξ_j is a reduction factor for unfavourable permanent action G

14.6.2 Values of γ Factors

In the case of persistent and transient design situations, the values of partial safety factors to be used depend on whether the design situation involves verifying equilibrium of the system or if it involves design of sections without considering geotechnical actions.

(i). The values of partial safety factors γ to be used when *verifying equilibrium*

Permanent actions: Unfavourable: $\gamma_{Gj, \, sup} \, G_{kj, \, sup}, \, \gamma_{Gj, \, sup} = 1.10$

Favourable: $\gamma_{Gj, \, inf} \, G_{kj, \, inf}, \, \gamma_{Gj, \, inf} = 0.90$

Leading variable action: $\gamma_{Q1} \, Q_{k, \, 1}, \, \gamma_{Q1} = 1.50$ where unfavourable and 0 when favourable

Other variable actions: $\gamma_{Qi} \, \psi_{0, \, i} \, Q_{k, \, i}, \gamma_{Qi} = 1.50$ where unfavourable and 0 when Favourable   (14.7)

(ii). The values of partial safety factors γ to be used when *designing sections*

Use Choice 1 *or* Choice 2A *and* Choice 2B.

Choice 1:

Permanent actions: Unfavourable: $\gamma_{Gj, \, sup} \, G_{kj, \, sup}, \, \gamma_{Gj, \, sup} = 1.35$

Leading variable action: $\gamma_{Q1} \, Q_{k,1}, \, \gamma_{Q1} = 1.50$ where unfavourable and 0 when favourable

Other variable actions: $\gamma_{Qi} \, \psi_{0, \, i} \, Q_{k, \, i}, \, \gamma_{Qi} = 1.50$ where unfavourable and 0 when Favourable   (14.8)

Choice2A:

Permanent actions: Unfavourable: $\gamma_{Gj, \, sup} \, G_{kj, \, sup}, \, \gamma_{Gj, \, sup} = 1.35$

Main variable action: $\gamma_{Q1} \, \psi_{0, \, 1} \, Q_{k, \, 1}, \, \gamma_{Q1} = 1.50$ where unfavourable and 0 when favourable

Other variable actions: $\gamma_{Qi} \, \psi_{0, \, i} \, Q_{k, \, i}, \, \gamma_{Qi} = 1.50$ where unfavourable and 0 when favourable   (14.9)

Choice2B:

Permanent actions: Unfavourable: $\xi \, \gamma_{Gj, \, sup} \, G_{kj, \, sup}, \, \xi = 0.85. \, \gamma_{Gj, \, sup} = 1.35$

Leading variable action: $\gamma_{Q1} \, Q_{k, \, 1}, \, \gamma_{Q1} = 1.50$ where unfavourable and 0 when favourable

Other variable actions: $\gamma_{Qi} \, \psi_{0, \, i} \, Q_{k, \, i}, \, \gamma_{Qi} = 1.50$ where unfavourable and 0 when favourable.   (14.10)

(iii). The values of partial safety factors γ to be used when *verifying Equilibrium and designing of sections*

Permanent actions: Unfavourable: $\gamma_{Gj, \, sup} \, G_{kj, \, sup}. \, \gamma_{Gj, \, sup} = 1.0$

Leading variable action: $\gamma_{Q1} \, Q_{k, \, 1}, \, \gamma_{Q1} = 1.30$ where unfavourable and 0 when favourable

Other variable actions: $\gamma_{Qi} \, \psi_{0, \, i} \, Q_{k, \, i}. \, \gamma_{Qi} = 1.30$ where unfavourable and 0 when favourable   (14.11)

Note: The characteristic values of all **permanent actions** from one source such as the self weight of structures are multiplied by $\gamma_{G,\,Sup}$ if the effect of the action is unfavourable and by $\gamma_{G,\,Inf}$ if the effect of the action is favourable.

14.6.3 Examples of the Use of γ Factors

Example 1: Fig. 14.3 shows a simply supported beam with an overhang. The characteristic loads are
Permanent characteristic uniformly distributed dead load $g_k = 11.0$ kN/m,
Variable characteristic uniformly distributed live load $q_k = 20.0$ kN/m.
Variable characteristic concentrated live load $Q_k = 30.0$ kN

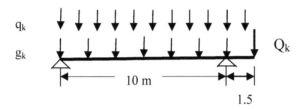

Fig. 14.3 Simply supported beam with an overhang

(i). Checking stability
Instability will occur when the left-hand reaction goes into 'tension'. Therefore check for minimum reaction. The minimum reaction will occur when the cantilever carries the maximum load and there is minimum load in the simply supported length. Equilibrium is checked using equation (14.2).
Permanent actions:
The dead load is favourable when acting on the simply supported part but unfavourable when acting on the cantilever part. Therefore $\gamma_{Gj,\,inf} = 0.90$ for the load between supports and $\gamma_{Gj,\,sup} = 1.10$ for the load on the cantilever.
Let the leading variable be the concentrated load and it is unfavourable. Therefore $\gamma_{Q1} = 1.50$
The other leading variable is the uniformly distributed load, which is also unfavourable when acting on the cantilever part. Therefore $\gamma_{Qi} = 1.50$. It is favourable when acting on the simply supported part. Therefore $\gamma_{Qi} = 0$.
  The loads on the beam are:
Permanent action due to dead load:
$\gamma_{Gj,\,inf}\, G_{kj} = 0.90 \times 11.0 = 9.9$ kN/m
$\gamma_{Gj,\,sup}\, G_{kj} = 1.10 \times 11.0 = 12.1$ kN/m
In terms of the variable load, there are two possible choices.
Option 1a: Treat the concentrated load as the leading variable:
Leading variable $= \gamma_{Q1}\, Q_{k,\,1} = 1.5 \times 30 = 45$ kN
Accompanying variable $= \gamma_{Qi}\, \psi_{0,\,i}\, Q_{k,\,i} = 1.5 \times 0.7 \times 20 = 21.0$ kN/m
Note that $\psi_{0,\,i}$ is assumed to be 0.7. These loads act on the cantilever part only.
The loading on the beam is shown in Fig. 14.4.

Total load $= 9.9 \times 10.0 + 45 + (21.0 + 12.1) \times 1.5 = 193.65$kN
$V_2 = [9.9 \times 10.0^2/2 + 45 \times 11.5 + (21.0 + 12.1) \times 1.5 \times (11.5 - 1.5/2)]/10$
   $= 154.62$ kN
$V_1 = 193.65 - V_2 = 39.03$ kN
Option 1b: Treat the distributed load as the leading variable:
Leading variable $= \gamma_{Q1} Q_{k,1} = 1.5 \times 20 = 30$ kN/m. This acts on the cantilever part only.
Accompanying variable $= \gamma_{Qi} \psi_{0,i} Q_{k,i} = 1.5 \times 0.7 \times 30 = 31.5$ kN.
Note that $\psi_{0,i}$ is assumed to be 0.7. It has a zero value on the simply supported part.
Total load $= 9.9 \times 10.0 + 31.5 + (30.0 + 12.1) \times 1.5 = 193.65$kN
$V_2 = [9.9 \times 10.0^2/2 + 31.5 \times 11.5 + (30.0 + 12.1) \times 1.5 \times (11.5 - 1.5/2)]/10$
   $= 153.61$ kN
$V_1 = 193.65 - V_2 = 40.04$ kN

Table 14.1 Support reaction

Load cases	Leading variable	V_1, kN
Option 1a	Concentrated	39.03
Option 1b	Distributed	40.04

Clearly Option 1a leads to the smaller value of $V_1 = 39.27$ and is therefore the critical combination.

Fig. 14.4 Loading on the beam for minimum value of V_1

ii. Design hogging bending moment over the right-hand support
The loads on cantilever are all unfavourable.
Option 1:   Use equations (14.2) together with (14.8):
$$F_d = \sum_{j \geq 1} \gamma_{G,j} G_{k,j} \; "+" \gamma_P P \; "+" \gamma_{Q,1} Q_{k,1} \; "+" \sum_{i>1} \gamma_{Q,i} \; \psi_{0,i} Q_{k,i}$$
The permanent dead load is unfavourable; therefore $\gamma_{Gj, sup} = 1.35$.
$\gamma_{Gj, sup} G_{kj, sup} = 1.35 \times 11.0 = 14.85$ kN/m
Choice 1: Let the leading variable be the concentrated load and it is unfavourable. Therefore $\gamma_{Q1} = 1.50$.
$\gamma_{Q1} Q_{k1} = 1.5 \times 30 = 45$ kN
The ther leading variable is the uniformly distributed load which is also unfavourable when acting on the cantilever part. Therefore $\gamma_{Qi} = 1.50$.
$\gamma_{Qi} \psi_{0,i} Q_{kl} = 1.5 \times 0.7 \times 20 = 21.0$ kN/m
Moment over support $= 14.85 \times 1.5^2/2 + 21.0 \times 1.5^2/2 + 45.0 \times 1.5 = \mathbf{107.83\ kNm}$

Choice 2: Let the leading variable be the uniformly distributed load and it is unfavourable. Therefore $\gamma_{Q1} = 1.50$.
$\gamma_{Q1}\, Q_{k1} = 1.5 \times 20 = 30$ kN/m kN
The other leading variable is the concentrated load which is also unfavourable when acting on the cantilever part. Therefore $\gamma_{Qi} = 1.50$.
$\gamma_{Qi}\, \psi_{0,i}\, Q_{kl} = 1.5 \times 0.7 \times 30 = 31.5$ kN
Moment over support $= 14.85 \times 1.5^2/2 + 30.0 \times 1.5^2/2 + 31.5 \times 1.5 = 97.71$ kNm

Option 2a: Use equations (14.3) together with (14.9):
$$F_d = \sum_{j \geq 1} \gamma_{Gj}\, G_{kj} \; "+" \gamma_P\, P \; "+" \gamma_{Q,1}\, \psi_{0,1}\, Q_{k,1} \; "+" \sum_{i>1} \gamma_{Q,i}\, \psi_{0,i}\, Q_{k,i}$$
The permanent dead load is unfavourable, therefore $\gamma_{Gj,\,sup} = 1.35$.
$\gamma_{Gj,\,sup}\, G_{kj,\,sup} = 1.35 \times 11.0 = 14.85$ kN/m
In this case if ψ_0 is taken as 0.7 for both variable actions, then it does not matter which is considered as leading and which as accompanying.
Therefore $\gamma_{Q1} = 1.50$.
$\gamma_{Q1}\, \psi_{0,1}\, Q_{k1} = 1.5 \times 0.7 \times 30 = 31.5$ kN
The other leading variable is the uniformly distributed load which is also unfavourable when acting on the cantilever part. Therefore $\gamma_{Qi} = 1.50$.
$\gamma_{Qi}\, \psi_{0,i}\, Q_{kl} = 1.5 \times 0.7 \times 20 = 21.0$ kN/m
Moment over support $= 14.85 \times 1.5^2/2 + 21.0 \times 1.5^2/2 + 31.5 \times 1.5 = 87.58$ kNm

Option 2b: Use equations (14.4) together with (14.10).
$$F_d = \sum_{j \geq 1} \xi_j\, \gamma_{Gj}\, G_{kj} \; "+" \; \gamma_{Q,1}\, Q_{k,1} \; "+" \sum_{i>1} \gamma_{Q,i}\, \psi_{0,i}\, Q_{k,i}$$
Permanent actions: Unfavourable: $\xi\, \gamma_{Gj,\,sup}\, G_{kj,\,sup}$. $\xi = 0.85$, $\gamma_{Gj,\,sup} = 1.35$,
$\xi\, \gamma_{Gj,\,sup}\, G_{kj,\,sup} = 0.85 \times 1.35 \times 11.0 = 12.62$ kN/m

Choice 1: Let the leading variable be the concentrated load and it is unfavourable. Therefore $\gamma_{Q1} = 1.50$. $\gamma_{Q1}\, Q_{k1} = 1.5 \times 30 = 45$ kN
The other leading variable is the uniformly distributed load which is also unfavourable when acting on the cantilever part. Therefore $\gamma_{Qi} = 1.50$.
$\gamma_{Qi}\, \psi_{0,i}\, Q_{kl} = 1.5 \times 0.7 \times 20 = 21.0$ kN/m
Compared to Choice 1a, this is not a critical loading.
Moment over support $= 12.62 \times 1.5^2/2 + 21.0 \times 1.5^2/2 + 45.0 \times 1.5 = 105.32$ kNm
Choice 2: Let the leading variable be the uniformly distributed load and it is unfavourable. Therefore $\gamma_{Q1} = 1.50$.
$\gamma_{Q1}\, Q_{k1} = 1.5 \times 20 = 30$ kN/m kN
The other leading variable is the concentrated load which is also unfavourable when acting on the cantilever part. Therefore $\gamma_{Qi} = 1.50$.
$\gamma_{Qi}\, \psi_{0,i}\, Q_{kl} = 1.5 \times 0.7 \times 30 = 31.5$ kN
Moment over support $= 12.62 \times 1.5^2/2 + 30.0 \times 1.5^2/2 + 31.5 \times 1.5 = 95.20$ kNm
The results are shown in Table 14.2 from which the maximum bending moment over the support is 107.83 kNm.

Table 14.2 Moment over support

Load cases	Leading variable	M, kNm
Option 1, Choice 1	Concentrated	107.83
Option 1, Choice 2	Distributed	97.71
Option 2A	Concentrated or distributed	87.58
Option 2B, Choice 1	Concentrated	105.32
Option 2B, Choice 2	Distributed	95.20

iii. Design sagging bending moment at mid-span
Proceeding as in (ii) for the hogging moment over support,
Option 1:   Use equations (14.2) together with (14.8):
The permanent dead load is unfavourable; therefore $\gamma_{Gj,\,sup} = 1.35$.
$\gamma_{Gj,\,sup}\, G_{kj,\,sup} = 1.35 \times 11.0 = 14.85$ kN/m

Choice 1: Let the leading variable be the concentrated load and it is favourable.
Therefore $\gamma_{Q1} = 0$.
$\gamma_{Q1}\, Q_{k1} = 0$kN
The other leading variable is the uniformly distributed load which is also
unfavourable when acting on the simply supported part. Therefore $\gamma_{Qi} = 1.50$.
$\gamma_{Qi}\, \psi_{0,\,i}\, Q_{kI} = 1.5 \times 0.7 \times 20 = 21.0$ kN/m
There is no variable load on the cantilever part as this is a favourable load.
$V_2 = [14.85 \times 11.5^2/2 + 21.0 \times 10^2/2]/10 = 203.20$ kN
$V_1 = [14.85 \times 11.5 + 21.0 \times 10] - 203.20 = 177.58$ kN
$M = 177.58 \times 10/2 - (14.85 + 21.0) \times 5^2/2 = 439.77$ kNm
Choice 2: Let the leading variable be the uniformly distributed load and it is
unfavourable. Therefore $\gamma_{Q1} = 1.50$.
$\gamma_{Q1}\, Q_{k1} = 1.5 \times 20 = 30$ kN/m kN
The other leading variable is the concentrated load which is favourable. Therefore
$\gamma_{Qi} = 0$.
$\gamma_{Qi}\, \psi_{0,\,i}\, Q_{kI} = 0$ kN
There is no variable load on the cantilever part as this is a favourable load.
$V_2 = [14.85 \times 11.5^2/2 + 30.0 \times 10^2/2]/10 = 248.20$ kN
$V_1 = [14.85 \times 11.5 + 30.0 \times 10] - 248.20 = 222.58$ kN
$M = 222.58 \times 10/2 - (14.85 + 30.0) \times 5^2/2 = \mathbf{552.27\ kNm}$
Option 2A:
The permanent dead load is unfavourable; therefore $\gamma_{Gj,\,sup} = 1.35$.
$\gamma_{Gj,\,sup}\, G_{kj,\,sup} = 1.35 \times 11.0 = 14.85$ kN/m
In this case if ψ_0 is taken as 0.7, the leading variable is the uniformly distributed
load which is also unfavourable. Therefore $\gamma_{Qi} = 1.50$.
$\gamma_{Qi}\, \psi_{0,\,i}\, Q_{kI} = 1.5 \times 0.7 \times 20 = 21.0$ kN/m
This is identical to Option 1, Choice 1.
Option 2B:
Permanent actions: Unfavourable: $\xi\, \gamma_{Gj,\,sup}\, G_{kj,\,sup}$. $\xi = 0.85$, $\gamma_{Gj,\,sup} = 1.35$,
$\xi\, \gamma_{Gj,\,sup}\, G_{kj,\,sup} = 0.85 \times 1.35 \times 11.0 = 12.62$ kN/m
Let the leading variable be the uniformly distributed load and it is unfavourable.
Therefore $\gamma_{Q1} = 1.50$.
$\gamma_{Q1}\, Q_{k1} = 1.5 \times 20 = 30$ kN/m

This is certainly less critical than Option 1, Choice 2. Therefore the maximum bending moment at mid-span is 552.27 kNm

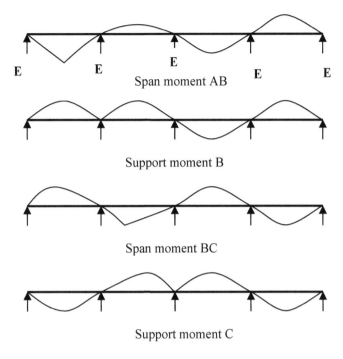

Fig. 14.5 Qualitative influence lines

Example 2: A four span continuous beam is loaded with permanent actions G_k and variable action Q_k. Indicate the various load combinations for maximum sagging moment in spans and hogging moment at supports.

It is useful to sketch qualitatively the influence lines for span and support moments to assist in suitable placing of the loads. Fig. 14.5 shows four influence lines.

Fig. 14.6 shows the various imposed load positions for obtaining the maximum sagging moments in span and hogging moments at support. In all cases the permanent load $G_{k, sup}$ is multiplied by γ_{sup}. Since the load factor for variable action is zero if it is favourable, no variable load is applied in the favourable parts of the influence line.

Fig 14.7 shows the various imposed load positions for obtaining the maximum sagging moments in span and hogging moments at support. In all cases the permanent load $G_{k, inf}$ is multiplied by γ_{inf} and no variable action load is applied in the favourable parts of the influence line.

Example 3: Fig. 14.8 shows a building frame subjected to the following characteristic loads:
Permanent load G_k
Variable action imposed load Q_k Wind load W_k.

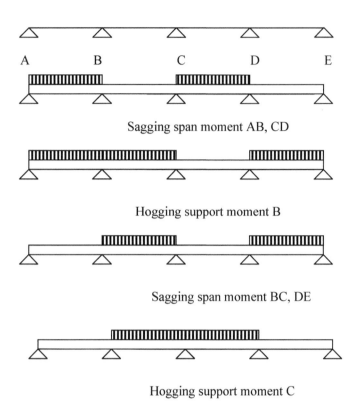

Fig. 14.6 Imposed loading for maximum moments sagging in span and hogging at support

Assume load factor for imposed load is $\gamma_Q = 1.5$
ψ_0 for imposed load $= 0.7$
ψ_0 for wind load $= 0.5$
$\gamma_{G, Inf} = 0.90$, $\gamma_{G, sup} = 1.1$
Indicate the various load cases to be considered for checking equilibrium.
Solution: It is clear overturning will occur under wind load combined with maximum load on the overhanging part of the frame combined with minimum load on the rest of the frame. The load combinations are used according to equation (14.2).
Case 1: Treat wind load as the leading variable action:
Leading variable action, wind load: $\gamma_{Q, 1} Q_{k, 1} = 1.5\ W_k$
Accompanying variable action, imposed load: $\gamma_{Q, i}\ \psi_{k, I}\ Q_{k, I} = (1.5\ \text{or}\ 0) \times 0.7 \times Q_k$
$= (1.05\ \text{or}\ 0)\ Q_k$
γ factor of 1.5 applies to the load in the overhanging part and zero in the rest of the frame.
Permanent load: $\gamma_{i,}\ G_{k, i} = (0.90\ \text{or}\ 1.1) \times G_k$
γ factor of 1.1 applies to the load in the overhanging part and 0.90 in the rest of the frame.
Fig. 14.9 shows the loading on the frame.

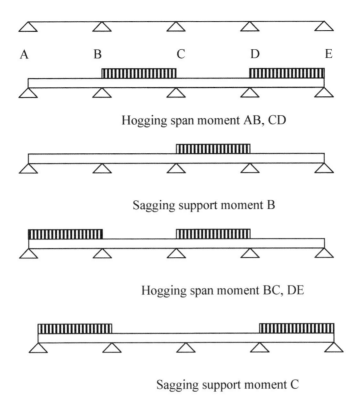

Hogging span moment AB, CD

Sagging support moment B

Hogging span moment BC, DE

Sagging support moment C

Fig. 14.7 Imposed loading for maximum moments hogging in span and sagging at support

Case 2: Treat imposed load as the leading variable action.

Leading variable action, imposed load: $\gamma_{Q,1} Q_{k,1} = (1.5 \text{ or } 0) Q_k$

γ factor of 1.5 applies to the load in the overhanging part and zero in the rest of the frame.

Accompanying variable action, wind load: $\gamma_{Q,i} \psi_{k,1} Q_{k,1} = 1.5 \times 0.5 \times W_k$
$$= 0.75 W_k$$

Permanent load: $\gamma_{i,} G_k, i = (0.90 \text{ or } 1.1) \times G_k$

γ factor of 1.1 applies to the load in the overhanging part and 0.90 in the rest of the frame. Fig. 14.10 shows the loading on the frame.

14.7 COMBINATION OF ACTIONS FOR SLS

(i).*The characteristic combination*: When considering the function of a structure and damage to structural/non-structural elements and irreversible limit states:

$$F_d = \sum_{j\geq1} G_{kj} \; "+" \; P \; "+" \; Q_{k,1} \; "+" \; \sum_{i>1} \psi_{0,i} Q_{k,i} \tag{14.11}$$

(ii). *The frequent combination*: When considering the comfort of the occupants, the function of the machinery, avoiding ponding of water, i.e. reversible limit states:

$$F_d = \sum_{j\geq 1} G_{kj} \; "+" \; P \; "+" \; \psi_{1,1} Q_{k,1} \; "+" \; \sum_{i>1} \psi_{2,i} Q_{k,i} \tag{14.12}$$

(iii). *The quasi-permanent combination*: When considering the long-term effects, e.g. creep effects, and the appearance of a structure:

$$F_d = \sum_{j\geq 1} G_{kj} \; "+" \; P \; "+" \; \sum_{i\geq 1} \psi_{2,i} Q_{k,i} \tag{14.13}$$

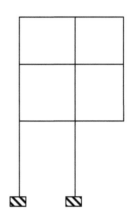

Fig. 14.8 Building frame with an overhang

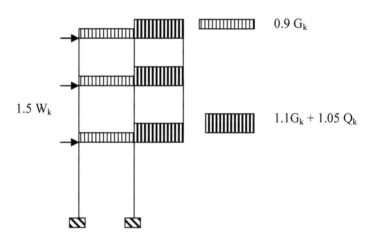

Fig. 14.9 Case 1 loading

14.8 REFERENCES TO EUROCODE 1 CLAUSES

In this chapter the following clauses of Eurocode 1 have been referred to.
Eurocode 1: Actions on structures
Part 1.1: General actions: Densities, self-weight, imposed loads for buildings.
Section 6: Imposed loads on buildings.

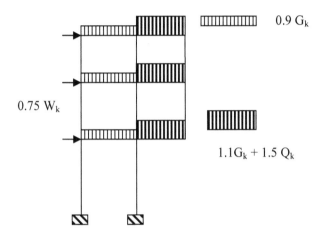

Fig. 14.10 Case 2 loading

CHAPTER 15

LOADING ON BRIDGES

15.1 INTRODUCTION

The loads acting on road bridges come mainly from
- Vehicular traffic imposing vertical loads.
- Horizontal loads from vehicular traffic due to centrifugal forces and breaking forces
- Crowd loading

It has to be remembered that loading models used in codes do not represent actual loads; rather they reflect the effect of actual loads on bridges.

15.2 NOTIONAL LANES

The carriageway width w is in general measured between the kerbs. The width of one notional lane is taken as 3 m. The number of notional lanes and their width are related to the carriageway width w, as shown in Table 15.1.

Table 15.1 Number and width of notional lanes

Carriageway width, w	Number of notional lanes	Width of notional lane w_ℓ	Width of remaining area
w < 5.4 m	$n_1 = 1$	3 m	w – 3 m
5.4 m ≤ w < 6.0 m	$n_1 = 2$	w/2	0
w ≥ 6 m	$n_1 = $ integer(w/3)	3 m	$w - 3 \times n_1$
Example: If w = 11 m, $n_1 = $ integer(11/3) = 3, remaining area = 11 – 3 × 3 = 2 m			

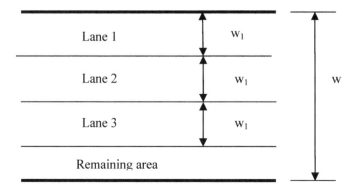

Fig. 15.1 Example of lane numbering

If there is more than one lane, the lane giving the most unfavourable effect is numbered Lane number 1; the lane giving the second most unfavourable effect is numbered Lane number 2, and so on. Fig. 15.1 shows an example of lane numbering.

15.3 LOAD MODELS

There are four models given in Eurocode 1. They are:
Load Model 1: Concentrated and uniformly distributed loads which cover most of the effects of the traffic of lorries and cars. This model is used for general and local verifications.
Load Model 2: A single-axle load applied on specific tyre contact areas which covers the dynamic effects of normal traffic on short (in the 3 to 7m range) structural members.
Load Model 3: A set of assemblies of axle loads representing special vehicles transporting abnormal loads, for example loads due to industrial traffic.
Load Model 4: This represents crowd loading and is intended for general verification. It is used only for transient design situations.

15.3.1 Load Model 1

Load model 1 consists of two partial systems. They are
- A double axle concentrated load called tandem system TS, each axle having a weight equal to $\alpha_Q Q_k$, where α_Q is an adjustment factor. Only one complete set of the concentrated load is applied per notional lane. Fig. 15.2 shows the details of the tandem load. The footprint of a wheel is 0.4 m square. For spans greater than 10 m, each tandem system in each lane can be replaced by a single-axle concentrated load equal to the total weight of two axles.
- A uniformly distributed load equal to $\alpha_Q q_k$, where α_Q is an adjustment factor and q_k is the weight per square meter of the notional lane. For a notional lane width of 3 m, the load per meter run in the lane is $3 \times \alpha_Q q_k$.

Table 15.2 Load model 1: Characteristic values of loads

Location	Tandem system (TS)	UDL system	
	Axle load, Q_{ik} (kN)	q_{ik} (kN/m^2)	Load in kN/m for a 3 m wide lane
Lane no. 1	300	9	27
Lane no. 2	200	2.5	7.5
Lane no. 3	100	2.5	7.5
Other lanes	0	2.5	7.5
Remaining area	0	2.5	7.5

Table 15.2 shows the characteristic values of loads including an allowance for dynamic effects.

The minimum values of α_Q suggested are:

Lane 1: $\alpha_Q \geq 0.8$,

Rest of lanes: $\alpha_Q \geq 1.0$

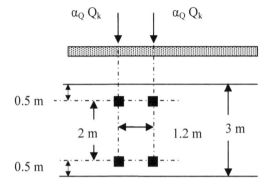

Fig. 15.2 Details of a tandem load

15.3.2 Load Model 2

This load model consists of a single **axle load** 400 β kN, where $\beta \geq 0.8$ which should be applied at any location on the carriage way. When relevant it is permissible to use only one **wheel load** equal to one half of the axle load. Fig. 15.3 shows the positioning of the load. The contact surface of **each twin wheel** is 0.6 m $\times$ 0.35 m, with 0.35 m parallel to the longitudinal axis.

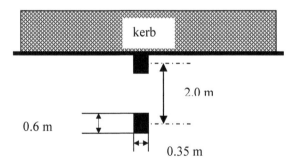

Fig. 15.3 Load model 2

15.3.3 Load Model 3

This model applies to special vehicles. The total weight of vehicles can vary from 600 kN to 3600 kN. The number of axles can vary from four to 15. The axles are generally spaced at 1.5 m. The width of the vehicle varies from 3.0 m for 150 kN and 200 kN axle lines and 4.5 m for 240 kN axle lines.

15.3.4 Load Model 4

This represents crowd loading generally taken at 5 kN/m². The load is applied to all relevant parts of the length and width of the bridge deck, including central reservations. This loading is considered only when checking for transient design situations.

15.4 DISPERSAL OF CONCENTRATED LOAD

For local verification associated with Load models 1 and 2, wheel loads are taken as uniformly distributed over the contact area and dispersed through the pavement and concrete slab at 45° as shown in Fig. 15.4.

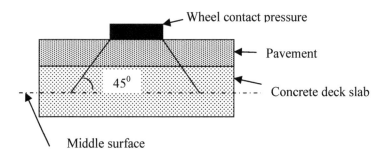

Fig. 15.4 Dispersal of wheel loads

15.5 HORIZONTAL FORCES

In addition to the vertical forces discussed in section 15.3, due to acceleration and deceleration (breaking) of the vertical forces, and centrifugal forces due to vehicles going round a bend, horizontal forces are applied to the bridge. These forces should be taken into account in the design.

15.5.1   Breaking Forces

A breaking force of $Q_{\ell k}$ should be taken as a longitudinal force acting at the surface of the carriage way. It is calculated as a fraction of the maximum vertical loads to be applied to lane number 1 in Load model 1. It is limited to a maximum of 900 kN for the entire width of the bridge:

$$180 \; \alpha_{Q1} \leq \{Q_{\ell k} = 1.2 \times [\alpha_{Q1} \times Q_{1k}] + 0.10 \times [\alpha_{q1} \times q_{1k}] \times (w_1 \times L)\} \leq 900 \; kN$$

where L = length of the deck or part of the bridge under consideration

w_1 = width of the lane (normally 3 m)

As an example, if α_{Q1} and α_{q1} are both unity, loads for Lane 1are Q_{1k} = 300 kN, q_{1k} = 9 kN/m², w_1 = 3 m, and L = 30 m, then

$180 \leq \{Q_{\ell k} = 1.2 \times [1 \times 300] + 0.10 \times [1 \times 9] \times (3 \times 30) = 441\} \leq 900$ kN

$Q_{\ell k}$ = 441 kN

The force due to acceleration has the same value as the breaking force except that it is of the opposite sign.

15.5.2 Centrifugal forces

A centrifugal force Q_{tk} should be taken as a concentrated transverse force acting at the finished carriage level and radially to the axis of the carriageway and of a value as shown in Table 15.3 and acting at any deck cross section.

Table 15.3 Characteristic values of centrifugal forces

Centrifugal force, kN	Horizontal radius r of carriageway centreline, m
$Q_{tk} = 0.4 \sum\limits_{i=1}^{3} \alpha_{Qi} \, Q_{ik}$	r < 200 m
$Q_{tk} = 80 \dfrac{\sum\limits_{i=1}^{3} \alpha_{Qi} \, Q_{ik}}{r}$	$200 \leq r \leq 1500$
$Q_{tk} = 0$	r > 1500 m

Example: Taking α_{Qi} = 1.0, Q_{1k} = 300 kN, Q_{2k} = 200 kN, Q_{3k} = 100 kN,
Q_{ik} = 0, i > 3

If r = 180 m, Q_{tk} = 0.4× (300 + 200 + 100+ 0) = 240 kN

If r = 400 m, Q_{tk} = 80× (300 + 200 + 100+ 0)/400 = 120 kN

15.6 LOADS ON FOOTWAYS, CYCLE TRACKS AND FOOT BRIDGES

Three *mutually exclusive* classes of vertical loading with their characteristic values are as follows:

- Uniformly distributed load q_{fk} = 5 kN/m²
- A concentrated load Q_{fwk} = 10 kN acting on a square of side 0.10 m
- Load of a service vehicle to be defined separately on any individual project

The corresponding horizontal force in the longitudinal direction is:

Larger of 10% of the total uniformly distributed load and 60% of the total weight of the service vehicle.

15.7 GROUPS OF TRAFFIC LOADS

Eurocode 1 specifies various groups of loads which need to be considered as acting simultaneously. Some typical ones are:

- Characteristic values from Load model 1 are combined with $3kN/m^2$ vertical load on footways and cycle tracks.
- A uniformly distributed vertical load of $3 kN/m^2$ to represent frequent vehicular traffic load is combined with characteristic longitudinal horizontal forces due to braking/acceleration forces and transverse centrifugal force.

Table 15.4 gives the complete details of groups of traffic loads to be considered for multi-component action.

Table 15.4 Assessment of groups of traffic loads

	Carriageway						Footway and cycle tracks
Load type	Vertical forces				Horizontal forces		Vertical forces only
Load system	LM1 TS and UDL	LM2 Single axle	LM3 Special vehicles	LM4 Crowd loading	Braking and acceleration forces	Centrifugal and transverse forces	UDL load
gr1a	c.v.				Undefined	Undefined	$3 kN/m^2$
gr1b		c.v.					
gr2	$3 kN/m^2$				c.v.	c.v.	
gr3							c.v.
gr4				c.v.			$3 kN/m^2$
gr5	Annex A		c.v.				

Note: **c.v** = characteristic value. In each group the c.v. forces shown are to be treated as the dominant component action.

15.8 COMBINATION OF ACTIONS FOR ULS

When considering persistent and transient design situations two options are given as follows.

Option 1:

$$F_d = \sum_{j \geq 1} \gamma_{G,j} G_{k,j} \ "+" \gamma_P \, P \ "+" \gamma_{Q,1} Q_{k,1} \ "+" \sum_{i>1} \gamma_{Q,i} \, \psi_{0,i} Q_{k,i} \qquad (15.1)$$

Option 2: Less favourable of

$$(a). \ F_d = \sum_{j \geq 1} \gamma_{Gj} G_{kj} \ "+" \gamma_P \, P \ "+" \gamma_{Q,1} \psi_{0,1} \, Q_{k,1} \ "+" \sum_{i>1} \gamma_{Q,i} \, \psi_{0,i} Q_{k,i} \qquad (15.2)$$

$$(b). \ F_d = \sum_{j \geq 1} \xi_j \, \gamma_{Gj} G_{kj} \ "+" \gamma_P \, P \ "+" \gamma_{Q,1} Q_{k,1} \ "+" \sum_{i>1} \gamma_{Q,i} \, \psi_{0,i} Q_{k,i} \qquad (15.3)$$

is in use.

Note in the above:
"+" implies 'to be combined with'
Σ implies 'the combined effect of'
P is the representative value of the prestress action
G_{kj} is the characteristic value of a permanent action
Q_{k1} is the characteristic value of the leading variable action
Q_{kj} is the characteristic values of accompanying variable actions
ξ_j is a reduction factor for unfavourable permanent action G

15.9 VALUES OF γ FACTORS

In the case of persistent and transient design situations, the values of partial safety factors to be used areas followa
Permanent actions:
Unfavourable: $\gamma_{Gj, sup} G_{kj, sup}, \gamma_{Gj, sup} = 1.35$
Favourable: $\gamma_{Gj, inf} G_{kj, inf}, \gamma_{Gj, inf} = 1.0$
Leading variable action: $\gamma_{Q1} Q_{k, 1}$
$\gamma_{Q1} = 1.35$ where unfavourable and 0 when favourable due to loads from road or pedestrian traffic.
Other variable actions: $\gamma_{Qi} \psi_{0, i} Q_{k, i}$
$\gamma_{Qi} = 1.50$ when unfavourable and 0 when favourable
$\xi = 0.85$

15.10 VALUES OF ψ FACTORS FOR ROAD BRIDGES

Table 15.5 shows the values of ψ factors to be used for road bridges.

Table 15.5 Values of ψ factors for road bridges

Load combination		ψ_0	ψ_1	ψ_2
gr1a	Tandem System (TS)	0.75	0.75	0
	UDL	0.40	0.40	0
	Pedestrian and cycle track	0.40	0.40	0
gr1b	Single axle	0	0.75	0
gr2	Horizontal forces	0	0	0
gr3	Pedestrian loads	0	0	0
gr4	Crowd loading	0	0.75	0
gr5	Special vehicles	0	0	0

Note: Factors $\alpha_{Qi}, \alpha_{qi}, \alpha_{qr}, \beta_Q$ are all taken as unity.

15.11 COMBINATION OF ACTIONS FOR SLS

(i). *The characteristic combination*: When considering the function of a structure and damage to structural/non-structural elements and irreversible limit states:

$$F_d = \sum_{j \geq 1} G_{kj} \quad "+" \quad P \quad "+" \quad Q_{k,1} \quad "+" \quad \sum_{i>1} \psi_{0,i} Q_{k,i} \qquad (15.4)$$

(ii). *The frequent combination*: When considering the comfort of the occupants, the function of the machinery, avoiding ponding of water, i.e. reversible limit states:

$$F_d = \sum_{j \geq 1} G_{kj} \quad "+" \quad P \quad "+" \quad \psi_{1,1} Q_{k,1} \quad "+" \quad \sum_{i>1} \psi_{2,i} Q_{k,i} \qquad (15.5)$$

(iii). *The quasi-permanent combination*: When considering the long-term effects, e.g. creep effects, and the appearance of a structure:

$$F_d = \sum_{j \geq 1} G_{kj} \quad "+" \quad P \quad "+" \quad \sum_{i \geq 1} \psi_{2,i} Q_{k,i} \qquad (15.6)$$

15.12 REFERENCES TO EUROCODE 1 CLAUSES

In this chapter the following clauses to Eurocode 1 have been referred to.
Eurocode 1: Actions on structures
Part 2: Traffic loads on bridges
Section 2: Classification of actions
Section 3: Design situations
Section 4: Road traffic actions and other actions specifically for road bridges.

CHAPTER 16

ANALYSIS AND DESIGN OF BRIDGE DECKS

16.1 INTRODUCTION

Many types of bridges are constructed in reinforced and prestressed concrete, but the two most common forms are the contiguous bridge and the box beam bridge.

In a contiguous beam bridge, pre-tensioned precast prestressed concrete inverted T-beams are placed close together and connected by an in-situ reinforced concrete deck slab as shown in Fig. 16.1. The main advantage of this form of bridge is that minimum false work is needed, as the precast beams are supported directly on the bearings placed on the abutments and the formwork needed for casting the deck slab is supported directly on the precast beams without needing any additional support. This form is particularly attractive for simply supported spans up to about 30 m.

In-situ concrete slab

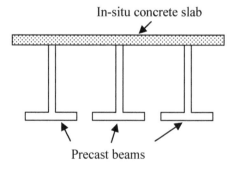

Precast beams

Fig. 16.1 Contiguous beam-slab bridge

In a box beam bridge, the deck consists of one or more boxes connected by a slab as shown in Fig. 16.2 to Fig. 16.4. Box sections are good at resisting torsion and they are generally used for spans greater than about 35 m. In order to give access for inspection and maintenance, the depth of the box needs to be at least about 1.8 m. For spans up to about 50 m, the deck is usually built on false work but for longer spans, 'cantilever form of construction' is used as explained in section 16.1.1. It is generally preferable to minimize the number of cells in the cross section as each cell requires additional formwork, which demands at least about 6 m spacing between the webs. It is also more economical to provide a small number of thick webs rather than a large number of thin webs because of the reduction in effective thickness doe to the presence of strands.

The spacing between the webs is governed by the bending resistance of the top flange to live loading. It is also important to make sure that the cantilever projections are not too long as otherwise they might induce a large bending moment in the webs. Single boxes are preferred when the width of the deck does not exceed about 13m. If the width is between 13m and 18 m then a two-cell single box as shown in Fig. 16.3 is used, although this are not popular because of the complexities during construction. However, if the width is between 18 and 25 m then two separate boxes with a connecting deck slab as shown in Fig. 16.4 are used. These dimensions need not be strictly adhered to as there are examples of single box bridges with a deck width of nearly 30 m with the side cantilevers nearly 7 m long. Trapezoidal boxes are used for aesthetic reasons, with the inclination of the web to the vertical in the range of 10° to 15°. They have the additional advantage of reducing the width of the bottom flange, which is generally fairly thick as it is designed to resist compressive stresses induced during cantilever construction.

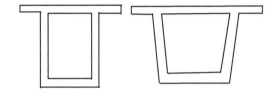

Fig. 16.2 Single-box: cross section of the deck

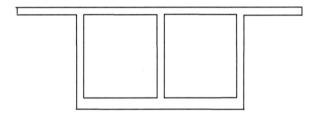

Fig. 16.3 Twin-box: cross section of the deck

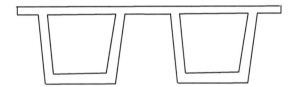

Fig. 16.4 Two separate boxes: cross section of the deck

16.1.1 Balanced Cantilever Construction

When the spans exceed about 35 m and the area under the bridge is inaccessible, a balanced cantilever method is adopted. Fig. 16.5 illustrates the basic steps in this form of construction.

Step 1: As shown in Fig. 16.5a, an intermediate pier is cast and a small part of the deck is cast. Temporary props are used to prevent the deck from rotating.

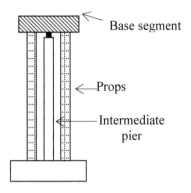

Fig. 16.5a Casting of base segment

Step 2: As shown in Fig. 16.5b, additional segments are added to the base segment. When attaching additional segments, as far as possible, symmetry is maintained so that large unbalanced moments are not created. The additional segments can be precast segments or cast in-situ using travelling formwork.

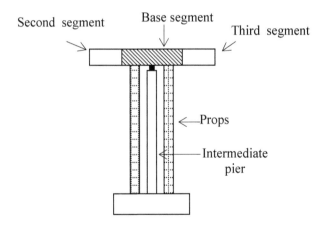

Fig. 16.5b Casting of additional segments

Step 3: As shown in Fig. 16.5c, additional segments are attached to the base segment using post-tensioning. Ducts are provided in the segments when they are cast so that post-tensioning cables can be threaded through. Most often shear keys are provided to improve shear resistance and also for proper location purposes. In the casting yard, segments are normally match cast against the previous segment so that a perfect fit is obtained. Epoxy resin is used to join the segments although it is not relied upon to provide any structural strength.   The design is made such that the joints are always in compression.

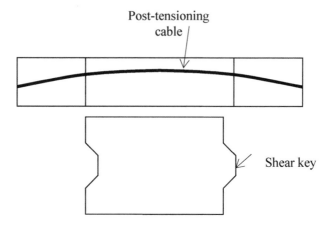

Fig. 16.5c Attaching segments through post-tensioning

Step 4: As shown in Fig. 16.5d, the cantilevers from opposite ends are joined by a stiching segment. Post-tensioning cables are provided in the bottom flange to resist the sagging moment due to traffic loads.

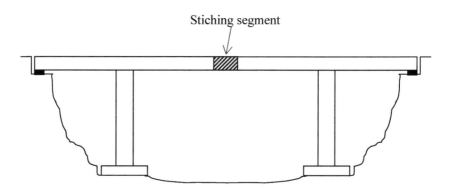

Fig. 16.5d Stiching segment cast

Fig. 16.6 shows a photograph of an additional segment being erected. Fig. 16.7 shows the reinforcement cage for an additional unit.

Fig. 16.6 Erection of an additional unit
(Photo: Courtesy of CCL)

16.2 METHODS OF ANALYSIS

Over the years many approximate and reasonably accurate methods of analysis of bridge decks have been developed. Most of them are based on the assumption that the material behaviour is elastic. However, with the advent of efficient computer

Fig. 16.7 Reinforcement for a segment
Photo: Courtesy of CCL Ltd

programs and computers with reasonably cheap and large memory, today only two methods are commonly used. They are
- Grillage analysis
- Finite element analysis

It has to be remembered that all methods of analysis in a way approximate the actual complex behaviour to a manageable simplified behaviour. The error used in such approximations can be large or small depending on the problem being analysed and the approximations involved in the model. Before choosing a method of analysis, the engineer has to have a clear understanding of
- The important aspects of the behaviour of the bridge which need to be included in any analysis
- The limitations of the particular method of analysis chosen
- Economy in terms of the cost of analysis
- Suitability of the output, particularly the 'stresses' for design purposes.

For example a finite element analysis using 3-D brick elements will certainly capture most of the complexities of behaviour in all forms of bridge deck but not only will the analysis be expensive in terms of data preparation and cost of analysis but also the output of stresses will so large that a great deal of time will be wasted simply in 'converting' the output into a form suitable for the design of reinforcement. As another example, in the case of box girders, if the out-of-plane bending moments are not important, then a simplified analysis by including only the in-plane behaviour of the elements of the box girder could lead to an economical solution. In some instances, it might be better to use a 'crude' model to simulate the overall behaviour, and then use another refined model to model only a small part of the structure to capture the 'local' effects. For example to simulate the bending moments in a deck slab subjecting to concentrated wheel loads, a finite element analysis for only a small part of the deck slab with the boundary conditions obtained from a 'crude' analysis will provide the necessary information needed for design.

16.3 GRILLAGE ANALYSIS

One of the most popular methods of analysis, particularly for a beam-slab type of bridge, is Grillage analysis. A grillage consists of a set of intersecting beams (generally in the horizontal plane) with the load applied normal to the plane of the grillage. Loads are resisted by a combination of bending and twisting moments and shear force. This is relatively inexpensive to use and the output is easy to interpret. It has been used in the analysis of a very wide variety of bridge types. The general recommendations for modelling a beam-slab bridge are as follows:

- Place longitudinal members to coincide with main beams. This will facilitate easy interpretation of the member forces for design.
- Locate the beams such that it is easy to place the nodes at the actual location of the supports.
- Transverse beams which model the deck slab should have spacing approximately similar to the spacing of longitudinal beams, although up to three times the spacing is acceptable. The second moment of area is $bt^3/12$ and the torsion constant is $bt^3/6$, where b = width and t = thickness of the deck slab representing the transverse beam. The reason for this is given in section 16.4.
- The second moment of area of the longitudinal beams is calculated in the normal way. For calculation of torsion constant see details in section 16.4.
- If there are transverse diaphragms (which in present-day practice is unusual), then an effective width of the deck slab should be assumed to act with the diaphragm acting as either a T- or an L-beam.

16.3.1 Aspects of Behaviour Ignored in Grillage Analysis

It is always important when using any method of analysis to be aware of the aspects of behaviour ignored and their effects on design. In the case of Grillage analysis, one of the important aspects ignored is the membrane action of the deck slab.

Consider a set of beams connected by a continuous deck slab. During the modelling of the longitudinal beams for grillage analysis, the continuity is disrupted and the beams are treated as individual T-beams. The beams are subjected to varying moments and the slab contracts in the longitudinal direction by varying amounts. This differential contraction will be resisted by in-plane shear forces as shown in Fig. 16.8. The resultant shear forces thus generated in the beam must be balanced by axial force in the beam as shown in Fig. 16.9.

The effect of the actions described is to

- Increase in-plane shear force in the deck slab over and above that obtained from the moment distribution along the span by grillage analysis.
- The axial force in the beam will shift the neutral axis away from the centroidal axis.
- The axial force together with the in-plane shear will cause a moment over and above that calculated from the grillage analysis.

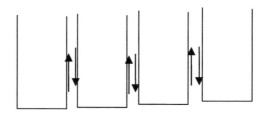

Fig. 16.8 In-plane discontinuity of deck slab

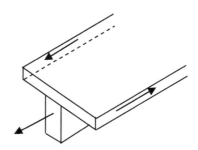

Fig. 16.9 Axial force in the beam due to in-plane shear forces in the slab

Construction joint

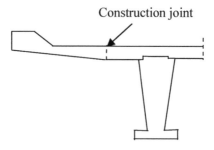

Fig. 16.10 Outer beam modelling

16.3.2 Edge Stiffening

In most bridge decks, edge stiffening is required for the parapets. However if the parapets are made to serve the structural function as edge beams, they can create many problems. For example, it is essential to ensure that they are not damaged by the impact of a vehicle and they are laterally stable against possible buckling. It is for these reasons that upstand parapets are made discontinuous along their length by expansion joints. This reduces cracking due to thermal and differential shrinkage.

One important point to remember is to make the projected part of the slab beyond the outer beam part of the effective width of the flange of the outer beam as shown in Fig. 16.10. It should not be treated as a separate beam because the grillage model assumes that the displacements all refer to a common centroid. Sometimes in modelling, in order to position the loads properly, edge beams of very small sectional properties are used being aware of the fact that they are not structural.

16.4 TORSIONAL CONSTANT

Fig. 16.11 shows the cross section of a narrow rectangle subjected to torsion. The distribution of shear stresses in the cross section shows that most of the stress is in the width direction but acting at a small lever arm. The stresses in the direction of the thickness are mainly confined to the ends but act at a very large lever arm almost equal to the width of the section. Because of this stress distribution, one half of the torsion resistance comes from the shear stresses in the longitudinal direction and the other half from the shear stresses in the direction of the thickness. The torsion constant J of a narrow rectangle of width b and thickness t is given by

$$J = bt^3/3$$

The resultant of the vertical shear stress at each edge acts at a distance of approximately 0.3t from the edge of the slab.

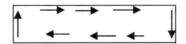

Fig. 16.11 Torsional shear stress distribution

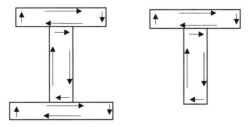

Fig. 16.12 Torsion shear stress distribution in open sections

In an open section such as a T- or I-beam made up of an assembly of thin narrow rectangles as shown in Fig. 16.12, the shear stress distribution in the individual rectangles is largely as in an independent thin rectangle. There is of course a slight change at the junction, but its effect is very marginal.

Fig. 16.13 shows the torsion shear stress distribution in an inverted T-beam with a long in-situ slab. The torsion constant contribution by the slab to the total torsion constant is $bt^3/3$, where b and t are respectively the width and thickness of the slab. If in modelling the beam, only a width b_1 of the slab is included as forming the flange of the composite beam, then because the width b_1 does not include the torsion shear stress parallel to the thickness of the slab, the contribution of the width b_1 of the slab to the total torsion constant of the beam is $b_1 t^3/6$.

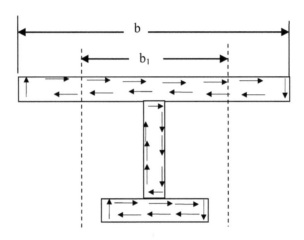

Fig. 16.13 Torsion shear stress distribution in a composite beam

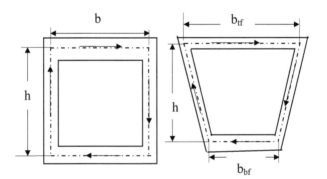

Fig. 16.14 Single- cell closed sections

16.4.1 Torsional Constant of Solid Sections

The torsion constant J of a solid section without re-entrant corners is given approximately by

$$J \approx \frac{A^4}{40\,I_p}$$

where A = area of cross section, I_p = polar moment of inertia
$I_p = I_{xx} + I_{yy}$, x and y are an orthogonal set of axes in the cross section
In the case of a rectangular cross section of size b × d,
$A = bd$, $I_p = (bd^3 + db^3)/12$, $J \approx 0.3\ b^3\ d^3 / (b^2 + d^2)$

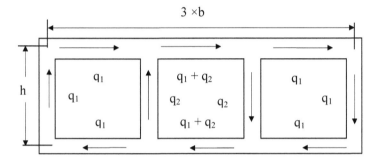

Fig. 16.15 Multi-cell closed sections

16.4.2 Torsional Constant of Thin-walled Closed Hollow Sections

In the case of thin-walled single-cell closed sections, the stress distribution in the walls is almost constant across the thickness. The stress distribution, ignoring the complex state of stress at re-entrant corner, is as shown in Fig. 16.14. The torsion constant J for a single trapezoidal cell section is given by the general expression

$$J = 4\frac{A^2}{\oint \dfrac{ds}{t}}, \quad A = 0.5\,(b_{tf} + b_{bf})\,h, \quad \oint\frac{ds}{t} = \frac{2s}{t_w} + \frac{b_{tf}}{t_{tf}} + \frac{b_{bf}}{t_{bf}}$$

where s = inclined web length
In the case of multiple hollow sections, the stress distribution is as shown in Fig. 16.15. Reasonable answers can be obtained by ignoring the effect of intermediate webs and treating it as a single-cell section. As an example, if the height and width of all the three cells are 1.5b × b and the walls are of constant thickness t, the accurate and approximate values of J are $9.22\ b^3t$ and $9.0\ b^3t$ respectively giving an error of about 2%. The stress in the intermediate webs is only 23% of the stress in the end webs while the stress in the flanges of the intermediate cell is 1.23 times the stress in the flanges of the end cells. As can be seen, shear flow (product of constant shear stress and thickness) of q_1 flowing round the whole tube provides a torsion resistance T_1 equal to $2 \times q_1 \times 3b \times h$ while the inner cell provides a torsion of resistance T_2 equal to $2 \times q_2 \times b \times h$. Since the shear flow q_2 is very small (only 23%) compared to q_1, the error involved in omitting the intermediate webs is small.

16.5 EXAMPLE OF ANALYSIS OF A BEAM AND SLAB BRIDGE DECK

Fig. 16.16 shows the cross section of a bridge deck.   The details are as follows:
Span: 24 m simply supported
Carriageway: 9 m wide
Footpaths: 1.65 m wide both sides
Parapet upstands: 0.35 m wide × 0.2m high
The deck consists of 13 beams at 1 m c/c with an in-situ reinforced concrete slab
160 mm thick.
The footpath is 120 mm thick and the black top on the carriageway is 50 mm thick.
The precast beam and the slab are made from C40/50 and C25/30 concrete
respectively.

16.5.1 Bending Properties of Precast Beam

Fig. 16.17 shows a simplified cross section of the precast beam. In actual practice,
the corners will be chamfered and re-entrant corners rounded. These details are
ignored in the following calculations.

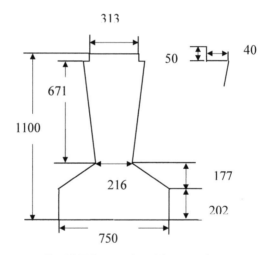

Fig. 16.17 Cross section of the precast beam

i. Area of cross section, A:
A = 750 ×202 + {(750 + 216)/2} × 177 + {(393 + 216)/2} × 671 + {313×50}
  =   {151.50 + 85.49 + 204.32 + 15.65} × 10^3 = 456.96 × 10^3 mm^2
Taking unit weight of concrete as 25/kN/m^3, self weight q per unit length is
  q = 456.96 × 10^3 × 10^{-6} × 25 = 11.42 kN/m
ii. First moment of area about the soffit:
A × y_b = 750 × 202 × (202/2 = 101) + 216 ×177× (202 + 177/2 = 290.5)
+ 0.5 × (750 – 216 = 534) × 177 × (202 + 177/3 = 261)
+ 216 ×671 × {1100 –50 – 671/2 = 714.5)
+ 0.5 × (393 – 216 = 177) × 671 × {1100 –50 – 671/3 = 826.3)

+ 313 ×50 × (1100 – 50/2 = 1075)

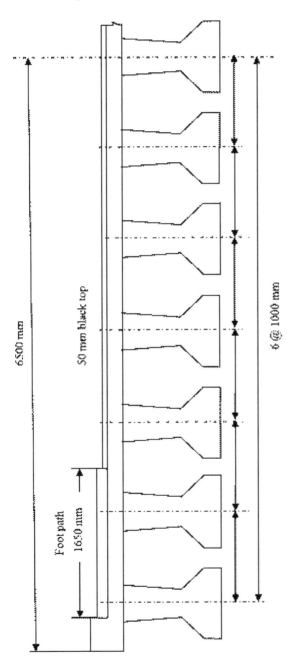

Fig 16.16 Cross section of a symmetric half of the bridge deck

$A \times y_b = \{15.30 + 11.11 + 12.33 + 103.08 + 49.07 + 16.83\} \times 10^6$
$= 207.71 \times 10^6 \, mm^3$

iii. Position of the centroid:

$y_b = A \times y_b / A = 455 \, mm, \; y_t = 1100 - y_b = 645 \, mm$

iv. Second moment of area I about the centroidal axis:

$I = 750 \times 202^3/12 + 750 \times 202 \times (y_b - 101 = 354)^2$
$+ 216 \times 177^3/12 + 216 \times 177 \times (y_b - 290.5 = 163.5)^2$
$+ 534 \times 177^3/36 + \{0.5 \times 534 \times 177\} \times (y_b - 261 = 194)^2$
$+ 216 \times 671^3/12 + 216 \times 671 \times (y_b - 714.5 = -260.5)^2$
$+ 177 \times 671^3/36 + \{0.5 \times 177 \times 671\} \times (y_b - 826.3 = -372.3)^2$
$+ 313 \times 50^3/12 + 313 \times 50 \times (y_b - 1075 = -620)^2$
$= (0.05 + 1.90 + 0.01 + 0.10 + 0.01 + 0.18 + 0.54 + 0.98 + 0.15 + 0.82 + 0 + 0.60)$
$= 5.35 \times 10^{10} \, mm^4$

v. Section moduli:

$Z_b = I/y_b = 117.58 \times 10^6 \, mm^3, \; Z_t = I/y_t = 82.95 \times 10^6 \, mm^3$

16.5.2 Section Properties of Interior Composite Beam

Since the young's moduli for slab and precast beam concrete are different, it is necessary to use an 'effective' width for the slab which is smaller than the actual width.   From equation (3.4) from Chapter 3,
f_{cm} for slab = 25 + 8 = 33 MPa, f_{cm} for beam = 40 + 8 = 48 MPa

$E_{slab} = E_{cm} = 22\,(0.1 \times f_{cm})^{0.3} = 22 \times (3.3)^{0.3} = 31.5 \, GPa$

$E_{Beam} = E_{cm} = 22\,(0.1 \times f_{cm})^{0.3} = 22 \times (4.8)^{0.3} = 35.2 \, GPa$

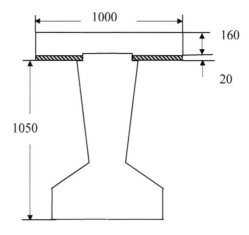

Fig. 16.18 Cross section of interior composite beams

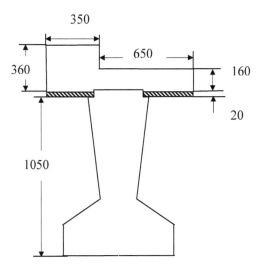

Fig. 16.19 Cross section of end composite beams

Modular ratio $\alpha = E_{slab}/E_{beam} = 31.5/35.2 = 0.895$
Assume 20 mm thick permanent form work and a 160 mm thick slab. The slab width associated with a beam is the same as the spacing of beams which is 1000 mm. The key of 50 mm projects $(50 - 20) = 30$ mm into the slab. Fig. 16.18 shows the cross section of an interior composite beam.

i. Area of cross section, A_{comp}:
Effective area of slab, $A_{slab} = \alpha \times \{1000 \times 130 + (1000 - 313 = 687) \times 30\}$
$$= \alpha \times 150.61 \times 10^3 = 137.80 \times 10^3 \text{ mm}^2$$
Area of composite beam, $A_{comp \ interior} = A_{beam} + A_{slab} = (456.96 + 137.80) \times 10^3$
$$= 591.76 \times 10^3 \text{ mm}^2$$
Taking unit weight of concrete as 25 kN/m³, self weight q per metre is
$q = (456.96 + 150.61) \times 10^3 \times 10^{-6} \times 25 = 15.19$ kN/m

ii. First moment of area about the soffit:
$A_{comp \ interior} \times y_{b \ comp \ interior} = A_{beam} \times y_b + \alpha \times \{1000 \times 130\} \times (1100 + 130/2 = 1165)$
$$+ \alpha \times 687 \times 30 \times (1100\_30/2 = 1085)$$
$$= (207.71 + 135.55 + 20.01) \times 10^6 = 363.27 \times 10^6 \text{ mm}^3$$

iii. Position of the centroid:
$y_{b \ comp \ interior} = 614$ mm, $y_{t \ comp \ interior} = (1100 + 130 - 614) = 616$ mm

iv. Second moment of area I about the centroidal axis:
$I_{comp \ interior} = I_{beam} + A_{beam} \times (y_{b \ comp} - y_b = 159)^2 + \alpha \times 1000 \times 130^3/12$
$+ \alpha \times 1000 \times 130 \times (y_{b \ comp} - 1165 = -551)^2 + \alpha \times 687 \times 30^3/12$
$+ \alpha \times 687 \times 30 \times (y_{b \ comp} - 1085 = -471)^2$
$= (5.35 + 1.16 + 0.02 + 3.53 + 0 + 0.41) \times 10^{10} = 10.46 \times 10^{10} \text{ mm}^4$

v. Section moduli:

$Z_{b\text{ comp interior}} = I_{comp}/y_{b\text{ comp interior}} = 170.36 \times 10^6 \text{ mm}^3$

$Z_{t\text{ comp interior}}$ to top of precast $= I_{comp\text{ interior}} / (1100 - y_{b\text{ comp interior}}) = 215.23 \times 10^6 \text{ mm}^3$

$Z_{t\text{ comp interior}}$ to top of slab $= I_{comp\text{ interior}} / (1100 - y_{b\text{ comp interior}} + 130)$
$$= 169.81 \times 10^6 \text{ mm}^3$$

$Z_{t\text{ comp interior}}$ to bottom of slab $= I_{comp\text{ interior}} / (1050 + 20 - y_{b\text{ comp interior}})$
$$= 229.39 \times 10^6 \text{ mm}^3$$

16.5.3 Section Properties of End Composite Beam

Fig. 16.19 shows the cross section of an end composite beam. The only difference from the interior composite beam is the 350 wide × 200 high parapets upstand. The section properties can therefore be computed using the properties of the interior beam as the basis.

i. Area of cross section, A_{comp}:

$A_{comp\text{ end}} = A_{comp\text{ interior}} + \alpha \times (350 \times 200 = 70 \times 10^3) = 654.48 \times 10^3 \text{ mm}^2$
Taking unit weight of concrete as 25 kN/m^3, self weight q is
$q = (456.96 + 150.61 + 70) \times 10^3 \times 10^{-6} \times 25 = 16.94 \text{ KN/m}$

ii. First moment of area about the soffit:

$A_{\text{comp end}} \times y_{b\text{ comp end}} = A_{comp\text{ interior}} \times y_{b\text{ comp interior}}$
$$+ \alpha \times 350 \times 200 \times (1100 + 130 + 200/2 = 1330)$$
$$= (363.27 + 83.42) \times 10^6 = 446.69 \times 10^6 \text{ mm}^3$$

iii. Position of the centroid:

$y_{\text{b comp end}} = 683$ mm, $y_{t\text{ comp end}} = (1100 + 130 + 200 - 683) = 747$ mm

iv. Second moment of area I about the centroidal axis:

$I_{comp\text{ end}} = I_{comp\text{ interior}} + A_{comp\text{ interior}} \times (y_{b\text{ comp interior}} - y_{b\text{ comp end}} = 69)^2$
$+ \alpha \times 350 \times 200^3/12 + \alpha \times 300 \times 200 \times (y_{b\text{ comp end}} - 1330 = -647)^2$
$= (10.46 + 0.28 + 0.02 + 2.25) \times 10^{10} = 13.01 \times 10^{10} \text{ mm}^4$

v. Section moduli:

Section moduli of end composite beam:
$Z_{b\text{ comp end}} = I_{comp\text{ end}}/y_{b\text{ comp end}} = 190.48 \times 10^6 \text{ mm}^3$,
$Z_{t\text{ comp end}}$ to top of precast $= I_{comp\text{ end}} / (1100 - y_{b\text{ comp end}}) = 312.0 \times 10^6 \text{ mm}^3$

16.5.4 Torsion Constant for Composite Beam

The shear stress distribution in the cross section of the composite beam will be approximately as shown in Fig. 16.20. This can be simplified to the distribution shown in Fig. 16.21. Note that only the horizontal shear stresses are shown in the slab because the slab associated with the beam is only a part of the total slab width.

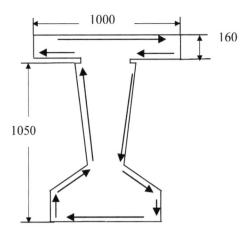

Fig. 16.20 Shear stress distribution in the cross section

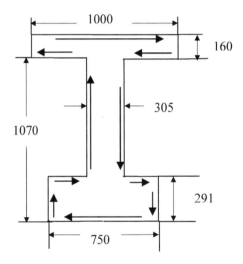

Fig. 16.21 Simplified shear stress distribution in the cross section

The dimensions of the three rectangles are:
Slab: 1000 × 160
Equivalent rectangle for bottom trapezium:
Width ≈ 750 mm, depth ≈ 202 + 177/2 = 291 mm
Equivalent rectangle for the web:
width ≈ (393 + 216)/2 = 305 mm, height ≈ 1070 – 291 = 779 mm
The contributions to the torsion constant J from the three rectangles are:
(i). Slab, $J_1 = 1000 \times 160^3/6 = 6.83 \times 10^8$ mm^4
(ii). Bottom trapezium, $J_2 = 0.3 \times 750^3 \times 291^3 / (750^2 + 291^2) = 48.19 \times 10^8$ mm^4
(iii). Middle rectangle: $J_3 = 0.3 \times 779^3 \times 305^3 / (779^2 + 305^2) = 57.49 \times 10^8$ mm^4

$J = J_1 + J_2 + J_3 = 1.125 \times 10^{10}$ mm^4

This value can be accepted for internal as well as for end composite beams as the contribution to the torsion constant from the parapet upstand will be insignificant. Note that J/ $I_{comp.\ Interior}$ = 0.11. This shows that as J is an order of magnitude smaller than I, torsion plays only a small part in load distribution in the deck.

16.5.5 Alternative Expressions for Approximate Value of J for Thin Rectangular Cross Section

There are a large number of alternative expressions given for calculating an approximate value for J for rectangular cross sections of width b and thickness d, b > d. Two of them are:

i. $\qquad J \approx b d^3 \dfrac{0.3}{[1+(\dfrac{d}{b})^2]}$

ii. $\qquad J \approx bd^3 [\dfrac{1}{3} - 0.21 \dfrac{d}{b} \{1 - 0.083 (\dfrac{d}{b})^4\}]$

The first expression is due to St. Venant and the second one is due to Ghali and Neville. Fig. 16.22 shows a comparison between the J computed from the two expressions. It shows that there is very little difference between the values of J as given by the two expressions. Considering the approximations made in dividing the actual cross sections into a series of individual rectangles, the differences become irrelevant.

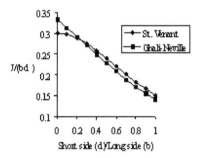

Fig. 16.22 J for thin rectangular sections from two expressions

16.5.6 Section Properties of Transverse Beams

It is assumed that there are no end diaphragms. Therefore the 160 mm thick deck slab acts as the transverse beam. Assuming that the transverse beams are spaced at 2 m centres, the section properties of the beams can be calculated. However, the second moment of area of the beams has to be multiplied by the modular ratio $\alpha = E_{slab}/ E_{beam} = 0.895$ so that the property refers to E_{beam}.

$I = \alpha \times 2000 \times 160^3/12 = 6.11 \times 10^8$ mm^4, $J = 2000 \times 160^3/6 = 13.65 \times 10^8$ mm^4
At supports, the width of the slab is in 1 m. Therefore
$I = \alpha \times 1000 \times 160^3/12 = 3.06 \times 10^8$ mm^4, $J = 1000 \times 160^3/6 = 6.83 \times 10^8$ mm^4
Fig. 16.23 shows the grillage mesh used.

Fig. 16.23 Grillage mesh

16.5.7 Material Properties

The elastic moduli used in the analysis are:
Young's modulus $E_{Beam} = 22\,(0.1 \times f_{cm})^{0.3} = 22 \times (0.1 \times 48)^{0.3} = 35.2$ GPa
Poisson's ratio $\nu = 0.2$
Shear modulus $G = E_{beam}/(1 + \nu) = 14.67$ GPa

16.5.8 Calculation of Live Loads and Bending Moment Distribution in Beam Elements at SLS

The beams are numbered from 1 to 13 as shown in Fig. 16.24. The tributary areas are also shown in enclosing rectangles. In all six analyses are carried out. In cases 1 to 3, lanes A, B and C are treated as lanes 1, 2 and 3 respectively. Similarly in cases 4 to 6, lanes A, B and C are treated as lanes 2, 1 and 3 respectively. The idea is to check which of the two systems will lead to larger moments. In order to assist in debonding, in each of the two syatems of lane numbering used, three cases of wheel load positions have been investigated by placing the wheel load at mid-span, third span and sixth span respectively. Note that if lanes A, B and C are treated as lanes 3, 2 and 1 respectively, the results will be just a mirror image of results for the first three cases.

(i). Case 1: SLS Analysis with lanes A, B and C treated as lanes 1, 2 and 3 respectively. Wheel loads at mid-span.

All γ factors are taken as unity. The ψ_0 values for UDL due to vehicular traffic and pedestrian load is 0.40. For Tandem load $\psi_0 = 0.75$.

For tandem load, $\alpha_Q = 0.9$ for lane 1 and $\alpha_Q = 1.0$ for lanes 2 and 3

(a). Super dead load on foot path due to 120 mm black top weighing 24 kN/m^3
$$= 120 \times 24 \times 10^{-3} = 2.88 \text{ kN/m}^2$$

(b). Super dead load on lanes due to 50 mm black top weighing 24 kN/m^3
$$= 50 \times 24 \times 10^{-3} = 1.2 \text{ kN/m}^2$$

(c). Foot path load $= 3$ kN/m$^2 \times (\psi_0 = 0.40) = 1.2$ kN/m^2

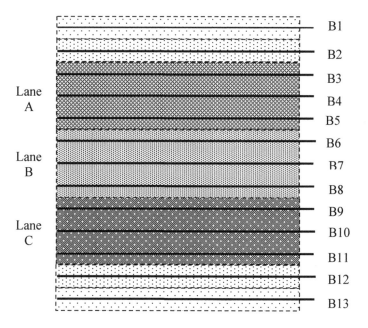

Fig. 16.24 Longitudinal beams and their tributary areas

(i). Case 1: SLS Analysis with lanes A, B and C treated as lanes 1, 2 and 3 respectively. Wheel loads at mid-span.

All γ factors are taken as unity. The ψ_0 values for UDL due to vehicular traffic and pedestrian load is 0.40. For Tandem load $\psi_0 = 0.75$.

For tandem load, $\alpha_Q = 0.9$ for lane 1 and $\alpha_Q = 1.0$ for lanes 2 and 3,

(a). Super dead load on foot path due to 120 mm black top weighing 24 kN/m^3
$$= 120 \times 24 \times 10^{-3} = 2.88 \text{ kN/m}^2$$

(b). Super dead load on lanes due to 50 mm black top weighing 24 kN/m^3
$$= 50 \times 24 \times 10^{-3} = 1.2 \text{ kN/m}^2$$

(c). Foot path load $= 3$ kN/m$^2 \times (\psi_0 = 0.40) = 1.2$ kN/m^2

(i). Uniformly distributed loads on beams 1 and 13:

Total load due to super dead load and foot path load $= (2.88 + 1.20) = 4.08$ kN/m^2

Excluding 350 mm wide parapet upstands,
Tributary width = 1000 – 350 = 650 mm
Load per m = $4.08 \times 650 \times 10^{-3} = 2.652$ kN/m

(ii). Uniformly distributed loads on beams 2 and 12:
Tributary width = spacing of beams = 1000 mm
Load per m = $4.08 \times 1000 \times 10^{-3} = 4.08$ kN/m

(iii). Beams 3-5 occupy lane A = lane 1:
(a). Uniformly distributed load = 9 kN/m², $\psi_0 = 0.40$
Tributary width = spacing of beams = 1000 mm.
Super dead load on lanes due to 50 mm black top = 1.2 kN/m²
Load per m = $(9 \times 0.40 + 1.2) \times 1000 \times 10^{-3} = 4.8$ kN/m
(b). The four wheels each have a load of
$$(300/2) \times (\alpha_Q = 0.9) \times (\psi_0 = 0.75) = 101.25 \text{ kN.}$$
As shown in Fig.16.25a, these loads, because of their spacing, act only on beams 3
and 5. If the load is placed symmetrical with respect to the mid-span of the bridge,
the concentrated loads do not coincide with the node positions of the grillage mesh
as shown in Fig. 16.25b. For analysis, the wheel loads have to be transferred on to
the nodes:
Loads on the nodes away from centre line are: $101.25 \times 0.6/2.0 = 30.38$ kN
Load on the node on the centre line is $2 \times (101.25 – 30.38) = 141.75$ kN

(iv). Beams 6 to 8 occupy lane B = lane 2:
(a). Uniformly distributed load = 2.5 kN/m², $\psi_0 = 0.40$
Tributary width = spacing of beams = 1000 mm.
Load per m = $(2.5 \times (\psi_0 = 0.40) + 1.2) \times 1000 \times 10^{-3} = 2.2$ kN/m
(b). The four wheels each have a load of
$(200/2) \times (\alpha_Q = 1.0) \times (\psi_0 = 0.75) = 75$ kN.
These loads, because of their spacing, act only on beams 6 and 8. As in the case of
beams 3 and 5,
Loads on the nodes away from centre line are: $75 \times 0.6/2.0 = 22.5$ kN
Load on the node on the centre line is $2 \times (75 – 22.5) = 105.0$ kN

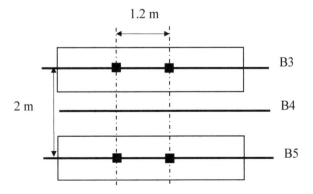

Fig. 16.25a Position of wheel loads

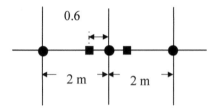

0.6

2 m   2 m

Fig. 16.25b Wheel loads and position of nodes

(v). Beams 9 to11 occupy lane C = lane 3:
(a). Uniformly distributed load = 2.5 kN/m², $\psi_0 = 0.40$
Tributary width = spacing of beams = 1000 mm
Load per m = $(2.5 \times (\psi_0 = 0.40) + 1.2) \times 1000 \times 10^{-3} = 2.2$ kN/m
(b). The four wheels each have a load of
$(100/2) \times (\alpha_Q = 1.0) \times (\psi_0 = 0.75) = 37.5$ kN
These loads, because of their spacing, act only on beams 6 and 8. As in the case of
beams 3 and 5,
Loads on the nodes away from centre line are $37.5 \times 0.6/2.0 = 11.25$ kN
Load on the node on the centre line is $2 \times (37.5 - 11.25) = 52.5$ kN
Fig. 16.26 shows the bending moment distribution along the span in beams 1 to 7.
From beams 1 to 5, the bending moments keep increasing but after beam 5, the
bending moment starts to decrease. The maximum moment of 856 kNm occurs at
mid-span in beam 5.

(ii). Case 2: SLS analysis with lanes A, B and C treated as lanes 1, 2 and 3 respectively. Wheel loads at third span
Loads as in case 1 except that the nodal loads act at the third span as opposed to
being at mid-span. Fig. 16.27 shows the bending moment distribution along the
span in beams 1-7. From beams 1-5, the bending moments keep increasing but
after beam 5, the bending moment starts to decrease. The maximum moment of
777 kNm occurs at the third span in beam 5.

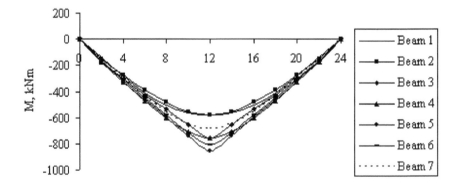

Fig. 16.26 Bending moment distribution in longitudinal beams 1-7, case 1, SLS.

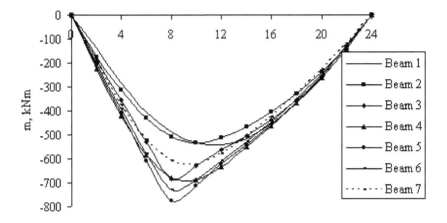

Fig. 16.27 Bending moment distribution in longitudinal beams 1-7, case 2, SLS.

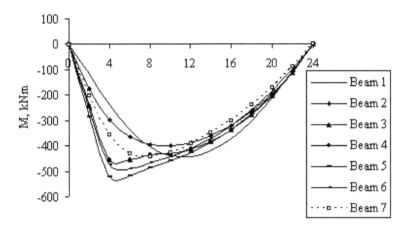

Fig. 16.28 Bending moment distribution in longitudinal beams 1-7, case 3, SLS.

(iii). Case 3: SLS analysis with lanes A, B and C treated as lanes 1, 2 and 3 respectively. Wheel loads at sixth span

Loads as in case 2 except that the nodal loads act at the sixth span as opposed to being at third span. Fig. 16.28 shows the bending moment distribution along the span in beams 1 to 7. From beams 1 to 5, the bending moments keep increasing but after beam 5, the bending moment starts to decrease. The maximum moment of 521 kNm occurs at the sixth span in beam 5.

Extracting the maximum bending moments at sixth (521), third (777) and mid-span (856) sections from cases 1 to 3 loading and fitting a symmetric parabola, the maximum bending moment distribution is as shown in Fig. 16.29.

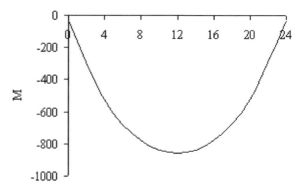

Fig. 16.29 Maximum bending moment distribution due to case 1, 2 and 3 loadings.

(iv). Case 4: SLS analysis with lanes A, B and C treated as lanes 2, 1 and 3 respectively. Wheel loads at mid-span.
Load calculation is as for case 1 except that loads on beams 3 to 5 and beams 6 to 8 are interchanged. Fig. 16.30 shows the bending moment distribution along the span in beams 1 to 7. From beams 1 to 5, the bending moments keep increasing but after beam 6, the bending moment starts to decrease. The maximum moment of 861 kNm occurs at mid-span in beam 6.

(v). Case 5: SLS analysis with lanes A, B and C treated as lanes 2, 1 and 3 respectively. Wheel loads at third span.
Load calculation as for case 4 except wheel loads are at third span rather than at mid-span. Fig. 16.31 shows the bending moment distribution along the span in beams 1 to 7. From beams 1-5, the bending moments keep increasing but after beam 6, the bending moment starts to decrease. The maximum moment of 781kNm occurs at the third span in beam 6.

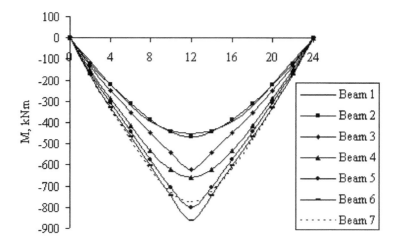

Fig. 16.30 Bending moment distribution in longitudinal beams 1-7, case 4, SLS

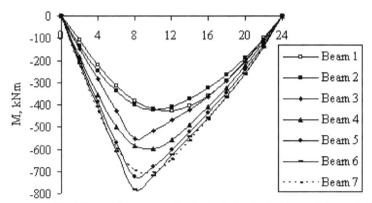

Fig. 16.31 Bending moment distribution in longitudinal beams 1-7, case 5, SLS.

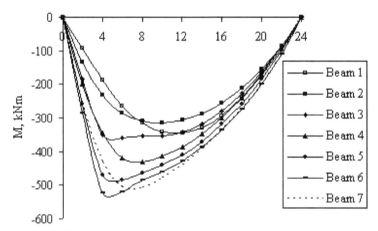

Fig. 16.32 Bending moment distribution in longitudinal beams 1-7, case 6, SLS.

(vi). Case 6: SLS analysis with lanes A, B and C treated as lanes 2, 1 and 3 respectively. Wheel loads at sixth span.
Load calculation is as for case 5 except that the wheel loads are at the sixth span rather than at mid-span. Fig. 16.32 shows the bending moment distribution along the span in beams 1 to 7. From beams 1 to 5, the bending moments keep increasing but after beam 6, the bending moment starts to decrease. The maximum moment of 522 kNm occurs at the sixth span in beam 6.

Extracting the maximum bending moments at the sixth (522), third (781) and mid-span (861) sections from cases 4 to 6 loading and fitting a symmetric parabola, the maximum bending moment distribution is as shown in Fig. 16.33. Clearly bending moments due to cases 4-6 are marginally higher than that due to cases 1 to 3. This bending moment distribution can be used to decide on debonding of prestressing tendons.

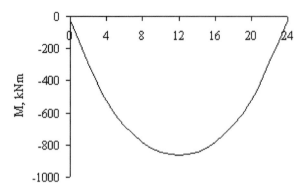

Fig. 16.33 Maximum bending moment distribution due to cases 4, 5 and 6.

16.6 STRESSES DUE TO SHRINKAGE OF SLAB

Using the calculations from section 4.5.3, Chapter 4, shrinkage stresses are calculated only for the interior beams as they are the most heavily loaded.
Modular ratio = 0.895
Force F in the slab to restrain free shrinkage of slab = 1072
Distance to centroid of slab from the centroid of the composite beam is a, where
$a = (y_{t\,comp} - 160/2) \times 10^{-3} = (616 - 80) \times 10^{-3} = 0.536$ m
$F \times a = 574.59$ kNm
Substituting in equations (4.20) to (4.23),
Stresses in the slab:

$$\sigma_{top} = \frac{1072 \times 10^3}{160 \times 10^3} - \left[\frac{1072 \times 10^3}{591.76 \times 10^3} + \frac{574.59 \times 10^6}{169.81 \times 10^6}\right] \times 0.895 = 2.1\,\text{MPa}$$

$$\sigma_{bottom} = \frac{1072 \times 10^3}{160 \times 10^3} - \left[\frac{1072 \times 10^3}{591.76 \times 10^3} + \frac{574.89 \times 10^6}{229.30 \times 10^6}\right] \times 0.895 = 2.8\,\text{MPa}$$

Stresses in the precast beam:

$$\sigma_{top} = -\frac{1072 \times 10^3}{591.76 \times 10^3} - \frac{574.89 \times 10^6}{215.20 \times 10^6} = -4.5\,\text{MPa}$$

$$\sigma_{top} = -\frac{1072 \times 10^3}{591.76 \times 10^3} + \frac{574.89 \times 10^6}{215.23 \times 10^6} = 0.9\,\text{MPa}$$

16.7 THERMAL STRESSES IN THE COMPOSITE BEAM

For calculating the stresses at the serviceability limit state and also to determine the prestress and eccentricity, it is necessary to calculate the thermal stresses due to heating as well as cooling. Since the internal beams are the ones most heavily stressed, thermal stresses are calculated only for internal beams.

16.7.1 Thermal Stresses: Heating

Using the data in section 4.7.1, Chapter 4,
h = 1050 +20 +160 = 1230 mm
0.3h = 369 mm, h_1 = 0.3h ≤ 150 mm, h_1 = 150 mm
h_2 =0.3h but ≥100 mm and ≤ 250 mm, h_2 = 250 mm
h_3 = 0.3h but ≤(100 mm + surfacing depth in mm), h_3 = 200 mm
From Table 4.2, interpolating for h = 1.23 m,
ΔT_1 = 13.0°C, ΔT_2 = 3.0°C, ΔT_3 = 2.5°C
$E_{slab} = E_{cm} = 22\,(0.1 \times f_{cm})^{0.3} = 22 \times (3.3)^{0.3} = 31.5\,GPa$

$E_{Beam} = E_{cm} = 22\,(0.1 \times f_{cm})^{0.3} = 22 \times (4.8)^{0.3} = 35.2\,GPa$
Coefficient of thermal expansion $\alpha_T = 10 \times 10^{-6}/°C$,
Corresponding restraining stresses are $\sigma = -\alpha_T \times E_c \times \Delta T$
Table 16.1 shows the restraining stresses at various levels in the cross section.

Table 16.1 Restraining stresses in the cross section: Heating

Level from top, mm	E_c, GPa	ΔT, °C	σ, MPa
0(slab)	31.5	13	−4.10
130 (slab)	31.5	4.3	−1.36
150 (slab)	31.5	3.0	−0.95
160 (slab)	31.5	2.9	−0.91
130 (key)	35.2	4.3	−1.51
150 (key)	35.2	3.0	−1.06
180 (key)	35.2	2.6	−0.93
180 (beam web)	35.2	2.6	−0.93
400 (beam web)	35.2	0	0
1030	35.2	0	0
1230	35.2	2.5	−0.88

Having obtained the restraining stresses, the next step is to calculate the forces F in different segments of the cross section.

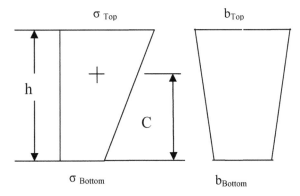

Fig. 16.34 Position of the resultant

If as shown in Fig. 16.34, both stresses and widths vary linearly over a depth h such that the stresses at top and bottom are σ_{top} and σ_{bottom} and the corresponding widths are b_{top} and b_{bottom} respectively, the total force F is equal to

$$F = \frac{h}{2}[\sigma_{top} \times b_{top} + \sigma_{bottom} \times b_{bottom}]$$

The centroid lies from the base at a distance of C, where

$$C = \frac{h}{6} \times [\frac{\sigma_{top} \times (3\,b_{top} + b_{bottom}) + \sigma_{bottom} \times (b_{top} + b_{bottom})}{\{\sigma_{top} \times b_{top} + \sigma_{bottom} \times b_{bottom})}]$$

It is not necessary that $\sigma_{top} > \sigma_{bottom}$ or $b_{top} > b_{bottom}$.
The values of the forces are shown in Table 16.2.

Table 16.2 Restraining forces in the cross section: Heating

Region	σ_{top} MPa	σ_{Bottom} MPa	Area	F, kN
slab	−4.10	−1.36	1000 × 130	−354.90
	−1.36	−0.95	687 × 20	−15.87
	−0.95	−0.91	687 × 10	−6.39
Key	−1.51	−1.06	313× 20	−8.04
	−1.06	−0.93	313 × 30	−9.34
Beam web	−0.93	0	b_t= 393, b_b = 335*, h =220	−40.20
Beam bottom flange	0	−2.5	750 × 200	−187.5

*The width of 335 mm in web = 216 + (393−216) × (671−220)/ 671 = 335 mm

Having calculated the forces, the next step is to calculate the moment due to the restraining forces about the neutral axis of the beam. Table 16.3 shows the details of the calculation of the moment due to force F. Note that if the stress distribution is as shown in Fig. 16.41, the resultant is at a distance C from the base of the element.

Table 16.3 Calculation of moment: Heating

F, kN	h mm	C mm	Lever arm mm	Moment kNm
−354.90	130	75.9	y_t − h + C= 562	−199.42
−15.87	20	10.6	y_t −130 − h + C= 477	−7.56
−6.39	10	5.0	y_t − 150 − h + C= 461	−2.95
−8.04	20	10.6	y_t − 160 − h + C= 447	−3.59
−9.34	30	15.3	y_t − 180 − h + C= 421	−3.94
−40.20	220	141.3	y_t − 210 − h + C= 327	−13.15
−187.5	200	66.7	−(y_b−C) = −547	+102.62
Σ−622.24				Σ−128.0

h = 1230 mm, y_b = 614 mm, y_t = 616 mm.

The last step is the calculation of the final stress in the cross section as the sum of the initial restraining stress plus the stress due to axial force and moment in order to produce a self-equilibrating stress system. Table 16.4 shows the details of stress calculation. Fig. 16.35 shows the final stresses. The maximum tensile stress in the web is equal to 4.3 MPa.

Table 16.4 Thermal stress calculation in precast beam: Heating

Position from top	σ MPa	$-\Sigma F/A$ MPa	y mm	$-(\Sigma M/I)y$	Final stress MPa
130 (key)	-1.51	1.05	486	0.60	0.14
150 (key)	-1.06	1.05	466	0.57	0.56
180 (Key)	-0.93	1.05	436	0.53	0.65
400 (web)	0	1.05	216	0.26	1.31
616 (N.Axis)	0	1.05	0	0	1.05
1010(Neck)	0	1.05	-394	-0.48	0.57
1230(Soffit)	-0.85	1.05	-614	-0.75	-0.55

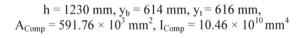

$$h = 1230 \text{ mm}, y_b = 614 \text{ mm}, y_t = 616 \text{ mm},$$
$$A_{Comp} = 591.76 \times 10^3 \text{ mm}^2, I_{Comp} = 10.46 \times 10^{10} \text{ mm}^4$$

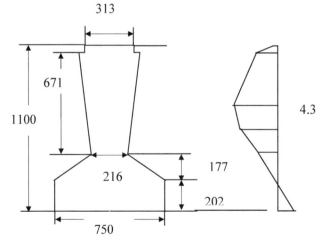

Fig.16.35 Final stresses: Heating

16.7.2 Thermal Stresses: Cooling

Using the data in section 4.7.2, Chapter 4,
h = 1230 mm, 0.2h = 246 mm, $h_1 = h_4 = 0.2h \leq 250$ mm
0.25h = 308 mm, $h_2 = h_3 = 0.25h$ but ≥ 200 mm
Therefore: $h_1 = h_4 = 250$ mm, $h_2 = h_3 = 308$ mm
From Table 4.3, interpolating for h = 1.23 m,
$\Delta T_1 = -8 - (8.4 - 8) \times (1.23 - 1.0/ (1.5 - 1.0) = 8.2 \ ^0C,$

$\Delta T_2 = -1.5 + (1.5 - 0.5) \times (1.23 - 1.0) / (1.5 - 1.0) = 1.0 \,^{\circ}\text{C}$,
$\Delta T_3 = -1.5 + (1.5 - 1.0) \times (1.23 - 1.0) / (1.5 - 1.0) = 1.3 \,^{\circ}\text{C}$,
$\Delta T_4 = -6.3 - (6.5 - 6.3) \times (1.23 - 1.0) / (1.5 - 1.0) = 6.4 \,^{\circ}\text{C}$,
Table 16.5 shows the restraining stresses at various levels in the cross section.

Table 16.5 Restraining stresses in the cross section: Cooling

Level from top, mm	E_c, GPa	ΔT, $^{\circ}C$	σ, MPa
0(slab)	31.5	8.2	2.58
130 (slab)	31.5	4.5	1.42
160 (slab)	31.5	3.6	1.13
130 (key)	35.2	4.5	1.58
180 (key)	35.2	3.0	1.06
180 (beam web)	35.2	3.0	1.06
250 (ΔT_2)	35.2	1.0	0.35
558(beam web)	35.2	0	0
672 (beam web)	35.2	0	0
851 (neck)	35.2	0.8	0.28
980 (ΔT_3)	35.2	1.3	0.46
1028 (start of trapezium)	35.2	2.3	0.81
1230 (soffit)	35.2	6.4	2.25

Having calculated the restraining stresses, the next step is to calculate the forces F in different segments of the cross section as shown in Table 16.6. Note that the widths in web come from interpolation.

Table 16.6 Restraining forces in the cross section: Cooling

Region	σ_{top}, MPa	σ_{Bottom}, MPa	Area	F, kN
Slab	2.58	1.42	1000 × 130	260.00
	1.42	1.13	687 × 30	26.28
Key	1.58	1.06	313× 50	20.66
Beam web (top part)	1.06	0.35	b_t= 393, b_b = 375*, h =70	19.17
Beam web (from top part to zero ΔT)	0.35	0	b_t= 375, b_b = 293*, h =308	20.21
Beam web (from neck to lower zero ΔT)	0	0.28	b_t= 263, b_b = 216*, h =180	5.44
From ΔT_3 up to neck	0.28	0.46	b_t= 216, b_b = 605*, h =128	21.68
Beam web (bottom trapezium)	0.46	0.81	b_t= 605, b_b = 750*, h =48	21.26
Beam bottom flange	0.81	2.25	750 × 202	231.80

Having calculated the forces, the next step is to calculate the moment due to the restraining forces about the neutral axis of the beam. Table 16.7, shows the details of the calculation of the moment due to force F.

Table 16.7 Calculation of moment: Cooling

F, kN	h	C	Lever arm	Moment
	mm	mm	mm	kNm
260.00	130	71.3	$y_t - h + C = 557$	144.90
26.28	30	15.6	$y_t - 130 - h + C = 472$	12.39
20.66	50	26.6	$y_t - 130 - h + C = 463$	9.56
19.17	70	40.8	$y_t - 180 - h + C = 407$	7.80
20.21	308	194.1	$y_t - 250 - h + C = 252$	5.10
5.44	180	66.5	$-(y_b - 378 - C) = -170$	−0.92
21.68	128	45.9	$-(y_b - 250 - C) = -318$	−6.90
21.26	48	20.6	$-(y_b - 202 - C) = -391$	−8.32
231.80	202	85.2	$-(y_b - C) = -528.8$	−122.58
$\Sigma626.50$				$\Sigma41.03$

$h = 1230$ mm, $y_b = 614$ mm, $y_t = 616$ mm

The last step is the calculation of the final stress in the cross section as the sum of the initial restraining stress plus the stress due to axial force and moment in order to produce a self equilibrating stress system. Table 16.8 shows the details of stress calculation. Fig. 16.36 shows the final stresses.

Table 16.8 Thermal stress calculation in precast beam: Cooling

Position from top mm	Σ, MPa	$-\Sigma F/A$, MPa	Y, mm	$-(\Sigma M/I)y$, MPa	Final stress, MPa
130 (key top)	1.58	−1.06	486	−0.19	0.33
180 (key bottom)	1.06	−1.06	436	−0.17	−0.17
250 (web, top part 1)	0.35	−1.06	366	−0.14	−0.85
558 (web, top part 2)	0	−1.06	58	−0.02	−1.06
616 (Neutral axis)	0	−1.06	0	0	−1.06
672 (web above neck)	0	−1.06	−56	0.02	−1.04
852 (neck)	0.28	−1.06	−236	0.09	−0.69
980 (inside trapezium)	0.46	−1.06	−364	0.14	−0.46
1028 (end bottom flange)	0.81	−1.06	−412	0.16	−0.09
1230 (soffit)	2.25	−1.06	−614	0.24	1.43

$h = 1230$ mm, $y_b = 614$ mm, $y_t = 616$ mm

$A_{Comp} = 591.76 \times 10^3$ mm^2, $I_{Comp} = 10.46 \times 10^{10}$ mm^4

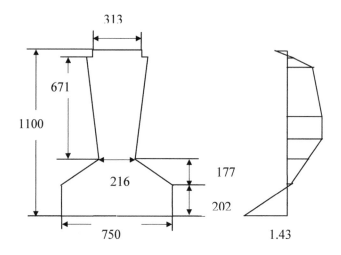

Fig.16.36 Final stresses: Cooling

16.8 STRESS DISTRIBUTION AT SLS DUE TO EXTERNAL LOADS

Tables 16.9 summarises the moments due to 'live loads' and dead loads at SLS as well as at ULS conditions.

Table 16.9 Summary of moments at SLS and ULS

Section	SLS moments, kNm			ULS moments, kNm	
	Live loads	Self weight only	Selfweight + slab	Live loads	Self weight + slab
Mid-span	861	822.24	1093.68	1440.4	1756.0
Third span	781	730.88	972.16	1308.5	1560.9
Sixth span	522	450.80	607.60	877.2	975.6

Using the following section properties, stresses at top and bottom at SLS conditions due to moments shown in Table 16.9 are calculated. Precast section only: $Z_b = 117.58 \times 10^6$ mm^3, $Z_t = 82.95 \times 10^6$ mm^3
Composite section (Interior):
$Z_b = 170.36 \times 10^6$ mm^3, $Z_{t\,top\,of\,precast} = 215.23 \times 10^6$ mm^3
The results are shown in Table 16.10.

Table 16.10 Summary of stresses at SLS due to loads

Section	Stresses at SLS, MPa					
	Live loads		Self weight only		Self weight + slab	
	σ_{top}	σ_{bottom}	σ_{top}	σ_{bottom}	σ_{top}	σ_{bottom}
Mid-span	−4.00	5.05	−9.91	6.99	−13.19	9.30
Third span	−3.63	4.59	−8.81	6.22	−11.72	8.27
Sixth span	−2.43	3.06	−5.44	3.83	−7.33	5.17

The stresses due to shrinkage of the cast in-situ slab and thermal gradients in the cross section are shown in Table 16.11.

Table 16.11 Stresses due to self equilibrating forces

Cause	σ_{top}	σ_{bottom}
Shrinkage of slab	−4.5	0.9
Thermal, Heating	0.14	−0.55
Thermal, cooling	0.33	1.43

16.9 MAGNEL DIAGRAMS

Precast section: $A = 456.96 \times 10^3$ mm^2, $y_b = 455$ mm, $y_t = 655$ mm,
   $Z_b = 117.58 \times 10^6$ mm^3, $Z_t = 82.95 \times 10^6$ mm^3
Permitted position of tendons: Table 16.12 shows the permitted position of cables in the cross section.

Table 16.12 Fixed cable positions

Level	Height above soffit, mm	No. of strands
A	60	10
B	110	14
C	160	12
D	210	10
E	260	8
H	1000	2

Permissible stresses: $f_{ck} = 40$ MPa. At transfer $f_{cki} = 30$ MPa

At transfer: $f_{tc} = -0.6\, f_{cki} = -18$ MPa, $f_{tt} = 0.30\, f_{cki}^{(2/3)} = 2.9$ MPa

At service: $f_{sc} = -0.6\, f_{ck} = -24$ MPa, $f_{st} = 0.30\, f_{ck}^{(2/3)} = 3.5$ MPa

Load factors for prestress: $\gamma_{Superior} = 1.05$, $\gamma_{Inferior} = 0.95$

Loss of prestress: Assume 10% at transfer and 25% at service.

$\eta = (1 - 0.25)/ (1 - 0.10) = 0.83$

From Chapter 4, the necessary Magnel equations are (4.5) and (4.6) at transfer and (4.18) and (4.19) at service.

(i). Mid-span: Substituting the numerical values and simplifying the Magnel equations are:

$$-218.84 + 1.206 \, e \leq 10.17 \times (10^8/P_s)$$
$$-218.84 - 0.851 \, e \geq -19.84 \times (10^8/P_s)$$
$$-218.84 + 1.206 \, e \geq -2.78 \times (10^8/P_s)$$
$$-218.84 - 0.851 \, e \leq -13.87 \times (10^8/P_s)$$

Fig. 16.37 shows the Magnel diagram for mid-span.

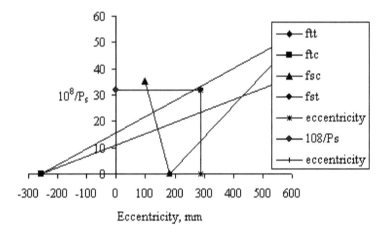

Fig. 16.37 Magnel diagram for mid-span

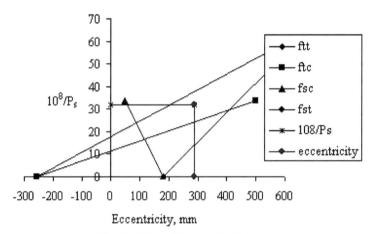

Fig. 16.38 Magnel diagram for third span

(ii). Third span: Substituting the numerical values and simplifying the Magnel equations are

$$-218.84 + 1.206 \text{ e} \le 9.29 \times (10^8/P_s)$$
$$-218.84 - 0.851 \text{ e} \ge -19.22 \times (10^8/P_s)$$
$$-218.84 + 1.206 \text{ e} \ge -4.72 \times (10^8/P_s)$$
$$-218.84 - 0.851 \text{ e} \le -12.29 \times (10^8/P_s)$$

Fig. 16.38 shows the Magnel diagram for the third span.

(iii). Sixth span: Substituting the numerical values and simplifying the Magnel equations are

$$-218.84 + 1.206 \text{ e} \le 6.61 \times (10^8/P_s)$$
$$-218.84 - 0.851 \text{ e} \ge -17.33 \times (10^8/P_s)$$
$$-218.84 + 1.206 \text{ e} \ge -10.61 \times (10^8/P_s)$$
$$-218.84 - 0.851 \text{ e} \le -7.42 \times (10^8/P_s)$$

Fig. 16.39 shows the Magnel diagram for the sixth span.

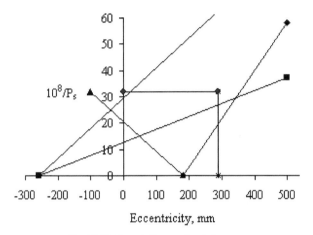

Fig. 16.39 Magnel diagram for sixth span

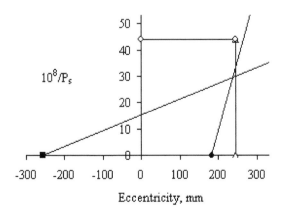

Fig. 16.40 Magnel diagram for support section

(iv). Support: Substituting the numerical values and simplifying the Magnel equations are

$$-218.84 + 1.206 \text{ e} \leq 2.30 \times (10^8/P_s)$$
$$-218.84 - 0.851 \text{ e} \geq -14.29 \times (10^8/P_s)$$

Fig. 16.40 shows the Magnel diagram for the support section.

Choosing 15 mm diameter 7-wire tendons with a cross sectional area of 140 mm^2 and if f_{pk} = 1860 MPa, stress at jacking is approximately 0.77 f_{pk} as explained in Chapter 3.

Force at jacking per tendon is 0.77 × 1860 × 140 × 10^{-3} = 200.5 kN

Maximum eccentricity allowable is y_b – 60 = 455 – 60 = 395 mm
For this values of eccentricity, from Magnel diagram at mid-span section,
$10^8/P_s$ = 36. P_s = 2778 kN.
Force at jacking ≈ P_s/0.75 = 3703 kN
No. of tendons required = 3703/200.5 = 18.5, say 20 to maintain symmetry.
These can be arranged as follows: Level A = 10, Level B = 8 and level H = 2. Two tendons are placed at the top to assist in tying shear links. They also act as a safety feature, in case during tensioning the beam fails explosively.
Resultant position of the tendons from the soffit is
Resultant = (2 × 1000 + 10 × 60 + 8 × 110)/20 = 174 mm
Eccentricity = y_b – 174 = 281 mm
From Magnel diagram, $10^8/P_s$ = 32, P_s = 3125 kN,
Force at jacking ≈ P_s/0.75 = 4167 kN
No. of tendons required = 4167/200.5 = 20.8, say 22 to maintain symmetry.
These can be arranged as follows: Level A = 10, Level B = 10 and level H = 2.
Resultant position of the tendons from the soffit is
Resultant = (2 × 1000 + 10 × 60 + 10 × 110)/22 = 168 mm
Eccentricity = y_b – 168 = 287 mm
From Magnel diagram, $10^8/P_s$ = 32.0, e= 287, P_s = 3125 kN.
This arrangement is suitable at mid span, third span and sixth span as shown in Figs. 16.37 to 16.39. However this arrangement is unsuitable at support section as the point corresponding to eccentricity = 287 mm and $10^8/P_s$ = 32 is outside the feasible region. Therefore debonding is required.   As it is desirable to keep as much prestress as possible in order to assist shear capacity, try removing the minimum number of cables from row A to push the point inside the feasible region. As a trial if six cables are removed from row A, then
Resultant = (2 × 1000 + 4 × 60 + 10 × 110)/16 = 209 mm
Eccentricity = y_b – 209 = 246 mm
$10^8/P_s$ = 32 for 22 cables. Therefore for 16 cables, $10^8/P_s$ = 32 × (22/16) = 44. As can be seen from Fig. 16.40, the point (e = 246 mm, $10^8/P_s$ = 44) is inside the Magnel diagram at the support section. P_s at the support section = 2273 kN.

It is worth pointing out that prestress can be applied in stages, although this does complicate the operations on site. For example when concrete is sufficiently strong it is possible to apply say only 50% of the prestress so as to prevent the formation of shrinkage and early thermally induced cracks. The rest of the prestress can be applied later.

16.9.1 Stress Checks

Having determined the prestress and eccentricity at different sections, the next stage is to check the stresses at different sections. From section 16.10,
Service:
$P_s = 3125$ kN, e = 287 mm at mid-span, third span and sixth span sections
$P_s = 2273$ kN, e = 246 mm at support section
$\gamma_{inferior} = 0.95$, $f_{sc} = -24$ MPa, $f_{st} = 3.5$ MPa
Transfer:
Using $P_t = P_s/\eta$, $\eta = 0.833$,
$P_t = 3752$ kN, e = 287 mm at mid-span, third span and sixth span sections.
$P_t = 2729$ kN, e = 246 mm at support section.
$\gamma_{superior} = 1.05$, $f_{tc} = -18$ MPa, $f_{tt} = 2.9$ MPa
Stresses at transfer and SLS conditions are shown in Tables 16.13 and 16.14. Comparing the permitted stresses to actual stresses, it can be concluded that the design is satisfactory.

Table 16.13 Summary of stresses at transfer

Section	P_t, kN	e, mm	Prestress*		Self weight		Total	
			σ_{top}	σ_{bottom}	σ_{top}	σ_{bottom}	σ_{top}	σ_{bottom}
Mid-span	3752	287	4.53	−16.50	−9.91	6.99	−5.38	−9.51
3rd Span	3752	287	4.53	−16.50	−8.81	6.22	−4.28	−10.28
6th span	3752	287	4.53	−16.50	−5.44	3.83	−0.91	−12.67
Support	2729	246	2.01	−11.10	0	0	2.01	−11.10

*Uses $\gamma_{inferior} = 0.95$.

Table 16.14 Summary of stresses at SLS

Section	Prestress#		Live loads		Self weight + slab		Total*	
	σ_{top}	σ_{bottom}	σ_{top}	σ_{bottom}	σ_{top}	σ_{bottom}	σ_{top}	σ_{bottom}
Mid-span	4.17	−15.19	−4.95	6.25	−13.19	9.30	−18.14	2.69
3rd Span	4.17	−15.19	−4.49	5.68	−11.72	8.27	−16.21	1.09
6th span	4.17	−15.19	−3.10	3.93	−7.33	5.17	−10.43	−3.76
Support	1.86	−10.22	0	0	0	0	−2.31	−7.89

*Stresses due to shrinkage of slab and thermal cooling from Table 16.11 equal to $\sigma_{top} = -4.5 + 0.33 = -4.17$ MPa and $\sigma_{bottom} = 0.90 + 1.43 = 2.33$ MPa have been added.
#uses $\gamma_{superior} = 1.05$.

At the mid-span section, with stresses at top and bottom respectively -18.14 and 2.69, the zero-stress axis is at 159 mm from the soffit of a composite beam 1230 mm high. This means that the bottom 159 mm of the beam is in tension.

16.10 CALCULATION OF LOADS AND BENDING MOMENT DISTRIBUTION IN BEAM ELEMENTS AT ULS

In the case of SLS, all γ factors are taken as unity. However, at ULS appropriate γ factors have to be included in the calculations of loads.
Dead load; γ_G: 1.35 unfavourable, 1.0 favourable
Leading variable: γ_{Q1}: 1.35 unfavourable, 0 favourable
Other variable actions: γ_{Qi}: 1.50 unfavourable, 0 favourable
The ψ_0 values for UDL due to vehicular traffic and pedestrian load is 0.40. For a Tandem load $\psi_0 = 0.75$.
For a Tandem load, $\alpha_Q = 0.9$ for lane 1 and $\alpha_Q = 1.0$ for Lanes 2 and 3
(a). Super dead load on footpath due to 120 mm black top weighing 24 kN/m^3
$$= 120 \times 24 \times 10^{-3} = 2.88 \text{ kN/m}^2$$
(b). Super dead load on lanes due to 50 mm black top weighing 24 kN/m^3
$$= 50 \times 24 \times 10^{-3} = 1.2 \text{ kN/m}^2$$
Footpath load = 3 kN/m^2
As the system is simply supported, all loads are unfavourable. Therefore for all load cases, $\gamma_G = 1.35$, $\gamma_{Q1} = 1.35$ and $\gamma_{Qi} = 1.50$

(i). Case 1: ULS analysis with lanes A, B and C treated as lanes 1, 2 and 3 respectively. Wheel loads at mid-span
(i). Uniformly distributed loads on Beams 1 and 13:
Total load due to super dead load and foot path load =
$$= (\gamma_G \times 2.88 + \gamma_{Qi} \times (\psi_0 = 0.40) \times 3.0) = 6.14 \text{ kN/m}^2$$
Excluding 350 mm wide parapet upstands,
Tributary width = 1000 − 350 = 650 mm
Load per m = $6.14 \times 650 \times 10^{-3} = 4.0$ kN/m

(ii). Uniformly distributed loads on Beams 2 and 12:
Tributary width = spacing of beams = 1000 mm
Load per m = $6.14 \times 1000 \times 10^{-3} = 6.14$ kN/m

(iii). Beams 3 to 5 occupy lane A = lane 1:
Uniformly distributed load = 9 kN/m^2, $\psi_0 = 0.40$,
Tributary width = Spacing of beams = 1000 mm
Super dead load on lanes due to 50 mm black top = 1.2 kN/m^2
Load per m = $[\gamma_{Q1} \times (\psi_0 = 0.40) \times 9 + \gamma_G \times 1.2] \times 1000 \times 10^{-3} = 6.48$ kN/m
The four wheels each have a load of

$\gamma_{Q1} \times (300/2) \times (\alpha_Q = 0.9) \times (\psi_0 = 0.75) = 136.69$ kN
Loads on the nodes away from the centre line $= 136.69 \times 0.6/2.0 = 41.01$ kN
Load on the node on the centre line $= 2 \times (136.60 - 41.01) = 191.36$ kN
(iv). Beams 6 to 8 occupy lane B = lane 2:
Uniformly distributed load $= 2.5$ kN/m^2, $\psi_0 = 0.40$
Tributary width = Spacing of beams = 1000 mm
Load per m $= [\gamma_{Q1} \times (\psi_0 = 0.40) \times 2.5 + \gamma_G \times 1.2] \times 1000 \times 10^{-3} = 2.97$ kN/m
The four wheels each have a load of
$\gamma_{Q1} \times (200/2) \times (\alpha_Q = 1.0) \times (\psi_0 = 0.75) = 101.25$ kN
Loads on the nodes away from centre line $= 101.25 \times 0.6/2.0 = 30.38$ kN
Load on the node on the centre line $= 2 \times (101.25 - 30.38) = 141.75$ kN
(v). Beams 9 to 11 occupy lane C = lane 3:
Uniformly distributed load $= 2.5$ kN/m^2, $\psi_0 = 0.40$
Tributary width = Spacing of beams = 1000 mm
Load per m $= [\gamma_{Q1} \times (\psi_0 = 0.40) \times 2.5 + \gamma_G \times 1.2] \times 1000 \times 10^{-3} = 2.97$ kN/m
The four wheels each have a load of
$\gamma_{Q1} \times (100/2) \times (\alpha_Q = 1.0) \times (\psi_0 = 0.75) = 50.63$ kN
Loads on the nodes away from centre line $= 50.63 \times 0.6/2.0 = 15.19$ kN
Load on the node on the centre line $= 2 \times (50.63 - 15.19) = 70.88$ kN

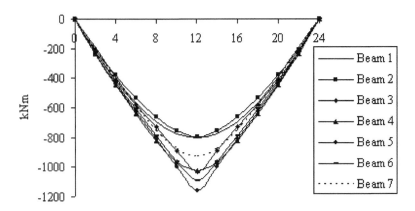

Fig. 16.41 Bending moment distribution in longitudinal beams 1 to 7, case 1, ULS

Fig. 16.41 shows the bending moment distribution along the span in beams 1 to 7. From beams 1 to 5, the bending moments keep increasing but after beam 5, the bending moment starts to decrease. The maximum moment of 1158 kNm occurs at mid-span in beam 5.

(ii). Case 2: ULS analysis with lanes A, B and C treated as lanes 1, 2 and 3 respectively. Wheel loads at the third span
Loads as in case 1 except that the nodal loads act at the third span as opposed to being at mid-span. Fig. 16.42 shows the bending moment distribution along the span in beams 1-7. From beams 1 to 5, the bending moments keep increasing but

after beam 5, the bending moment starts to decrease. The maximum moment of 1050 kNm occurs at the third span in beam 5.

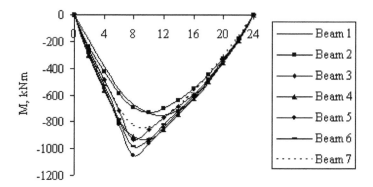

Fig. 16.42 Bending moment distribution in longitudinal beams 1 to 7, case 2, ULS

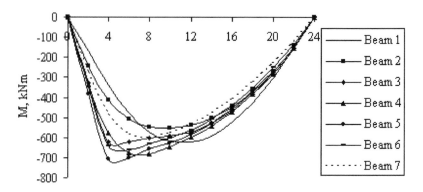

Fig. 16.43 Bending moment distribution in longitudinal beams 1 to 7, case 3, ULS

(iii). Case 3: ULS analysis with lanes A, B and C treated as lanes 1, 2 and 3 respectively. Wheel loads at the sixth span
Loads as in Case 2 except that the nodal loads act at the sixth span as opposed to being at mid-span. Fig. 16.43 shows the bending moment distribution along the span in beams 1 to 7. From beams 1 to 5, the bending moments keep increasing but after beam 5, the bending moment starts to decrease. The maximum moment of 705 kNm occurs at the sixth span in beam 5.

Extracting the maximum bending moments at sixth (705), third (1050) and mid-span (1158) sections from cases 1-3 loading and fitting a symmetric parabola, the maximum bending moment distribution is as shown in Fig. 16.44.

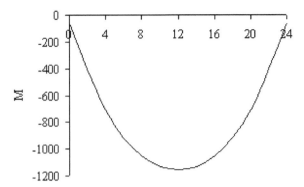

Fig. 16.44 Maximum bending moment distribution due to case 1, 2 and 3 loadings.

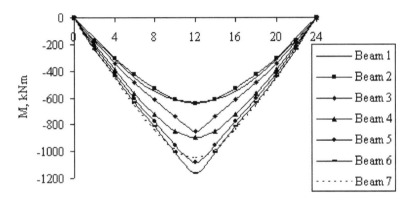

Fig. 16.45 Bending moment distribution in longitudinal beams 1 to 7, case 4, ULS

(iv). Case 4: ULS analysis with lanes A, B and C treated as lanes 2, 1 and 3 respectively. Wheel loads at mid-span.

Load calculation as for Case 1 except loads on beams 3 to 5 and beams 6 to 8 are interchanged. Fig. 16.45 shows the bending moment distribution along the span in beams 1 to 7. From beams 1 to 6, the bending moments keep increasing but after beam 6, the bending moment starts to decrease. The maximum moment of 1164 kNm occurs at mid-span in beam 6.

(v). Case 5: ULS analysis with lanes A, B and C treated as lanes 2, 1 and 3 respectively. Wheel loads at the third span

Loads as in Case 4 except that the nodal loads act at the third span as opposed to being at mid-span. Fig. 16.46shows the bending moment distribution along the span in beams 1 to 7. From beams 1-5, the bending moments keep increasing but after beam 5, the bending moment starts to decrease. The maximum moment of 1056 kNm occurs at third span in beam 5.

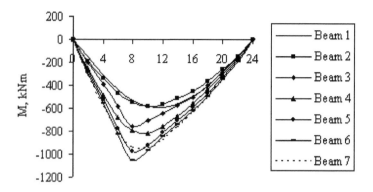

Fig. 16.46 Bending moment distribution in longitudinal beams 1 to 7, case 5, ULS

(vi). Case 6: ULS analysis with lanes A, B and C treated as lanes 2, 1 and 3 respectively. Wheel loads at the sixth span
Loads as in Case 5 except that the nodal loads act at the sixth span as opposed to being at third span. Fig. 16.47 shows the bending moment distribution along the span in beams 1 to 7. From beams 1 to 6, the bending moments keep increasing but after beam 6, the bending moment starts to decrease. The maximum moment of 705 kNm occurs at sixth span in beam 6.

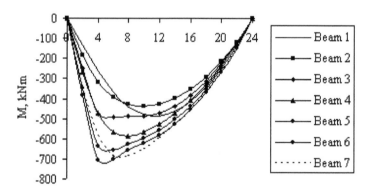

Fig. 16.47 Bending moment distribution in longitudinal beams 1 to 7, case 6, ULS

Extracting the maximum bending moments at the sixth (705), third (1056) and mid-span (1164) sections from cases 1 to 3 loading and fitting a symmetric parabola, the maximum bending moment distribution is as shown in Fig. 16.48. Clearly bending moments due to cases 4 to 6 are marginally higher than that due to cases 1 to 3. This bending moment distribution can be used to decide for checking the ultimate bending capacity required.

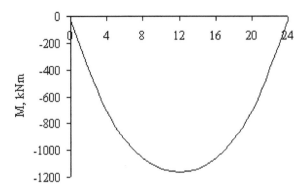

Fig. 16.48 Maximum bending moment distribution due to case 4, 5 and 6.

16.11 SELF-WEIGHT MOMENTS

At the stage when the beams are being prestressed, the self weight acting is simply the self weight of the precast beam at 11.42 kN/m. This gives in a simply supported span of 24 m bending moments of (450.80, 730.88 and 822.24) at the sixthspan, third span and mid-span respectively. Similarly at the serviceability limit state, the total dead loads due to the precast beam plus the weight of the slab are 15.19 and 16.94 kN/m for interior and end composite beams respectively. This is borne by simply supported beams over a span of 24 m. This leads to bending moments in kNm of (607.6, 972.16, and 1093.68) in interior beams and (677.6, 1084.16, 1219.68) at the sixth, third and mid-span sections in exterior beams respectively. All these act on individual beams without any interaction between the beams. However, at ultimate limit state it is assumed that one has to take into account the interaction between the beams. Using a load factor $\gamma_G = 1.35$, the self weights of composite beam are 20.51 and 22.87 kN/m for interior and end composite beams respectively. As the torsional constant is quite small compared to the second moment of area, there is likely to be minimal interaction. Fig. 16.49 shows the bending moment distribution. The maximum moments are 1756 kN/m and 1476 kN/m in end and interior beams respectively. If the interactions had been ignored, then the corresponding moments would have been 1647 kNm and 1477 kNm respectively.

16.12 ULTIMATE MOMENT CAPACITY: MID-SPAN SECTION

Combining the moments due to live load and self weight, the moment envelope at ULS is as shown in Fig. 16.50.

Basic data for the internal composite beam are as follows.

(i). Concrete beam: $f_{ck} = 40$ MPa, $\gamma_m = 1.5$, $\eta = 1$, $f_{cd} = f_{ck}/\gamma_m = 26.67$ MPa, $\lambda = 0.8$, $\eta\, f_{cd} = 26.7$ MPa, $\varepsilon_{cu3} = 3.5 \times 10^{-3}$

(ii). Concrete slab: $f_{ck} = 25$ MPa, $\gamma_m = 1.5$, $\eta = 1$, $f_{cd} = f_{ck}/\gamma_m = 16.68$ MPa, $\lambda = 0.8$, $\eta\, f_{cd} = 16.7$ MPa, $\varepsilon_{cu3} = 3.5 \times 10^{-3}$

(iii.) Steel: f_{pk} = 1860 MPa, f_{pd} = 1424 MPa, E_s = 195 GPa

Seven-wire strand with an effective cross sectional area of 140 mm^2.

Prestressing details: 24 strands are located at the following distances from the soffit: 10 at 60 mm, 12 at 110 mm and 2 at 1000 mm. The cables are spaced at 50 mm c/c.

Total prestress at service P_s = 3125 kN

Stress σ_{pe} due to prestress force in the strand = P_s/Area of 22 stressed strands
$$= 3125 \times 10^3 / (22 \times 140) = 1015 \text{ MPa}$$

Prestrain $\varepsilon_{pe} = \sigma_{pe} / E_s = 1015 / (195 \times 10^3) = 5.20 \times 10^{-3}$

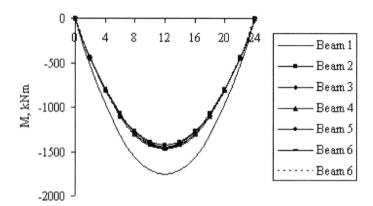

Fig. 16.49 Self weight moments at ULS

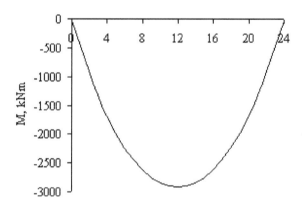

Fig. 16.50 Total moment envelope at ULS

Step 1: Estimate a value for the neutral axis depth:

i. Total tensile force: Assume all 20 strands in the bottom three rows 'yield', i.e. stress is the maximum of f_{pd}, equal to 1424 MPa. Area of each strand = 140 mm^2. Therefore total tensile force T = $20 \times 140 \times 1424 \times 10^{-3}$ = 3987 kN

ii. Total compressive force: Assuming a rectangular stress block, the uniform compressive stress = ηf_{cd} = 26.7 MPa.

(a). Total compressive force in the top 130 mm of slab = $16.7 \times 1000 \times 130 \times 10^{-3}$
$$= 2167$$

(b). Total compressive force in the bottom 30 mm of slab
$$= 16.7 \times (1000 - 313) \times 30 \times 10^{-3} = 343.6 \text{ kN}$$

(c). Total compressive force in the key of the beam
$$= 26.67 \times 313 \times 50 \times 10^{-3} = 417.3 \text{ kN}$$

Total compressive force from (a) + (b) + (c) = 2928 kN

The above is smaller than the tensile force T from (i). The stress block will extend into the web of the beam.

Taking an average width of web = (393 + 212)/2 = say 303 mm, the depth of the web in compression will be

a = $(3987 - 2928) \times 10^3 / (26.67 \times 303)$ = say 131 mm

Taking λ = 0.8, $\lambda x > (160 + 20 + 131)$ or x > 389 mm

Step 2: Iteration 1: Assume x = 350 mm.

i. Compressive force in web:

Stress block depth s = λx = 0.8 × 350 = 280 mm

Depth d_x of web in compression = 280 − 160 − 20 = 100 mm

Width B_x of web at the bottom of the stress block:

B_x = 212 + (393 − 212) × (1 − 100/671) = 366 mm

Compressive force in web = $26.67 \times (393 + 366)/2 \times 100 \times 10^{-3}$ = 1012 kN

Total compressive force C from Slab + Key + Web = 2928 + 1012 = 3940 kN

ii. Total tensile force:

Bending strain ε_b in strand at d from the compression face = $\varepsilon_{cu3} \times (d - x)/x$

Depth of composite beam = 1230 mm

The tensile force calculation is shown in Table 16.15. T = 4206 kN

The difference between the total tensile force and compressive force
$$= 4206 - 3940 = 266 \text{ kN}$$

The compressive force is too small, indicating that the neutral axis depth is larger than 350 mm.

Table 16.15 Calculation of total tensile force

Distance from soffit, mm	d mm	ε_b ×10^{-3}	ε_s × 10^{-3}	σ_s MPa	No. of strands	T kN
60	1170	8.20	13.40	1424	10	1994
110	1120	7.70	12.90	1424	10	1994
1000	230	-1.20	4.00	780	2	218
SUM					22	4206

$\varepsilon_{cu3} = 3.5 \times 10^{-3}$, $\varepsilon_{pe} = 5.20 \times 10^{-3}$, x = 350 mm

Step 3: Iteration 2: Assume x = 400 mm.

i. Compressive force in web:

Stress block depth s = λx = 0.8 × 400 = 320 mm

Depth d_x of web in compression = 320 − 160 − 20 = 140 mm

Width B_x of web at the bottom of the stress block:

$B_x = 212 + (393 - 212) \times (1 - 140/671) = 355$ mm
Compressive force in web $= 26.67 \times (393 + 355)/2 \times 140 \times 10^{-3} = 1397$ kN
Total compressive force C from Slab + Key + Web $= 2928 + 1397 = 4325$ kN
ii. Total tensile force:
Detailed calculations are shown in Table 16.16. T $= 4191$ kN
T $= 4571$ kN
The difference between the total tensile force and compressive force
$$T - C = 4191 - 4325 = -134 \text{ kN}$$
    The compressive force is too large, indicating that the neutral axis depth is smaller than 450 mm.

Table 16.16 Calculation of total tensile force

Distance from soffit, mm	d mm	ε_b $\times 10^{-3}$	ε_s $\times 10^{-3}$	σ_s MPa	No. of strands	T kN
60	1170	6.74	11.94	1424	10	1994
110	1120	6.30	11.50	1424	10	1994
1000	230	-1.49	3.71	724	2	203
SUM					22	4191

$\varepsilon_{cu3} = 3.5 \times 10^{-3}$, $\varepsilon_{pe} = 5.20 \times 10^{-3}$, x = 400 mm

Step 4: Interpolation: From the two values of and the corresponding difference between the total tensile and compressive forces, the value of x for which the difference is zero can be determined by interpolation.

x	T- C
350	266
400	-134

$x = 350 + (400 - 350) \times 266/ (266 + 134) = 383$ mm

Step 5: Final calculation:
Assume x $= 383$ mm
i. Compressive force in web:
Stress block depth s $= \lambda x = 0.8 \times 383 = 307$ mm
Depth d_x of web in compression $= 307 - 160 - 20 = 127$ mm
Width B_x of web at the bottom of the stress block:
$B_x = 212 + (393 - 212) \times (1 - 127/671) = 359$ mm
Compressive force in web $= 26.67 \times (393 + 359)/2 \times 127 \times 10^{-3} = 1273$ kN
Total compressive force C from Slab + Key + Web $= 2928 + 1273 = 4201$ kN
ii. Total tensile force:
Depth of composite beam $= 1230$ mm
Detailed calculations are shown in Table 16.17. T $= 4196$ kN

Table 16.17 Calculation of total tensile force

Distance from soffit, mm	d mm	ε_b $\times 10^{-3}$	ε_s $\times 10^{-3}$	σ_s, MPa	No. of strands	T, kN
60	1170	7.19	12.39	1424	10	1994
110	1120	6.74	11.94	1424	10	1994
1000	230	−1.40	3.80	741	2	208
SUM					22	4196

The difference between the total tensile force and compressive force
= 4196 – 4201 = -5 kN
This is small enough to be ignored. In fact the correct value of x = 433 mm. The 'error' is because of linear interpolation.
Step 6: Ultimate moment: The ultimate moment is obtained by calculating the moment of all forces, tensile as well as compressive about the soffit.
(a). Total compressive force in the top 130 mm of slab = 2167 kN
Lever arm from soffit = 1230 – 130/2 = 1165 mm
(b). Total compressive force in the bottom 30 mm of slab = 343.6 kN
Lever arm from soffit = 1230 – 130 – 30/2 = 1085 mm
(c). Total compressive force in the key of the beam = 417.3 kN
Lever arm from the soffit = 1100 – 50/2 = 1075 mm
(d). Compressive force in the web = 1273 kN
Depth of stress block in the web d_x = 127 mm
Width of the web at the bottom edge of stress block B_x = 359 mm
Lever arm = $1100 - 50 - (d_x/3) \times [1 + B_x / (393 + B_x)]$ = 988 mm
iii. Tensile force forces at various level act at their position from the soffit as shown in Table 16.17.
The ultimate bending moment M_u is given by
M_u = [2167 × 1165 + 344 ×1085 + 417 × 1075 + 1273 × 988 – 1994 × 60
 – 1994 × 110 – 208 × 1100] × 10^{-3} = 4036 kNm
Applied moments at ULS: Self weight + Slab = 1476 kNm and Live loads = 1164 kNm. The total applied moment is therefore 2640 kNm, which is much less than the value of 4036 kNm for M_u.

16.13 ULTIMATE SHEAR FORCE

Loads used for determining the maximum shear force will be same as the corresponding cases for determining the maximum bending moment as used in section 16.10. The only difference will be the positioning of the loads as dictated by the influence line.
Fig. 16.51 shows the influence for shear force in a simply supported beam. In order to obtain the maximum shear force, the uniformly distributed load which is unfavourable needs to cover the part of the span to the right of the section and the , the uniformly distributed load which is favourable needs to cover the part of the span to the left of the section. Any wheel load needs to be at the section and to the

right of the section. The bridge structure is not exactly a simply supported beam. However, the amount of interaction is not large and hence for influence line purposes, it can be treated as a simply supported beam.

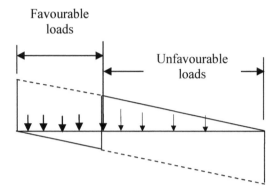

Fig. 16.51 Influence line for shear force

16.13.1 Analysis to Determine the Maximum Shear Force along the Span: Cases 1 to 4

The structure will be analysed for eight load cases. Cases 1 to 4 cover the load cases to determine maximum shear force at four sections viz. mid-span, third span, sixth span and support. In these cases lanes A, B and C are treated respectively as lanes 1, 2 and 3. Similarly cases 5 to 8 are similar to cases 1 to 4 except that lanes A, B and C are treated respectively as lanes 2, 1 and 3. Results are given for beams 1 to 7 only. Load calculations will very similar to that for bending moment calculations at ULS. However, in bending moment calculations, all loads were unfavourable. But in the case of shear force calculations, loads to the left of the section are favourable and those to the right of the section are unfavourable. The values of load factors used are as follows.

Dead load: γ_G: 1.35 unfavourable, 1.0 favourable
Leading variable: γ_{Q1}: 1.35 unfavourable, 0 favourable
Other variable actions: γ_{Qi}: 1.50 unfavourable, 0 favourable
ψ_0 values for UDL due to vehicular traffic and pedestrian load is 0.40. For Tandem load $\psi_0 = 0.75$.
For tandem load, $\alpha_Q = 0.9$ for lane 1 and $\alpha_Q = 1.0$ for lanes 2 and 3
(a). Super dead load on foot path due to 120 mm black top weighing 24 kN/m^3
$$= 120 \times 24 \times 10^{-3} = 2.88 \text{ kN/m}^2$$
(b). Super dead load on lanes due to 50 mm black top weighing 24 kN/m^3
$$= 50 \times 24 \times 10^{-3} = 1.2 \text{ kN/m}^2$$

Footpath load $= 3$ kN/m^2
Case 1: ULS analysis with lanes A, B and C treated as lanes 1, 2 and 3 respectively. Uniformly distributed load covers the right half of the span and self weight covers the left half of the span. Wheel loads to right of mid-span.
Note: Uniformly distributed load includes all loads including self weight.

(i). Uniformly distributed loads on Beams 1 and 13:
Excluding 350 mm wide parapet upstands, the tributary width is equal to
$1000 - 350 = 650$ mm
Total load due to self weight only $= \gamma_G \times 15.71 = 21.21$ kN/m
Total load due to super dead load and footpath load :
$\gamma_G \times 2.88 + \gamma_{Qi} \times (\psi_0 = 0.40) \times 3.0 = 5.51$ kN/m^2
Load per metre including self weight $= (5.51 \times 650 \times 10^{-3}) + 21.21 = 24.79$ kN/m

(ii). Uniformly distributed loads on Beams 2 and 12:
Total load due to self weight only =
$$= \gamma_G \times 14.20 = 19.17 \text{ kN/m}$$
Total load due to self weight, super dead load and foot path load
$$= [\gamma_G \times (2.88 + 14.20) + \gamma_{Qi} \times (\psi_0 = 0.40) \times 3.0] = 24.68 \text{ kN/m}^2$$
Tributary width = Spacing of beams = 1000 mm
Load per m $= 24.68 \times 1000 \times 10^{-3} = 24.68$ kN/m

(iii). Beams 3 to 5 occupy lane A = lane 1:
Uniformly distributed load $= 9$ kN/m^2, $\psi_0 = 0.40$
Tributary width = spacing of beams = 1000 mm.
Super dead load on lanes due to 50 mm black top $= 1.2$ kN/m^2
Load per m $= [\gamma_{Q1} \times (\psi_0 = 0.40) \times 9 + \gamma_G \times (1.2 + 14.20)] \times 1000 \times 10^{-3}$
$\qquad = 25.65$ kN/m
Total load due to self weight only $= 19.17$ kN/m
The four wheels each have a load of $\gamma_{Q1} \times (300/2) \times (\alpha_Q = 0.9) \times (\psi_0 = 0.75)$
$\qquad = 136.69$ kN.
Loads on the nodes to the right of centre line $= 136.69 \times 1.2/2.0 = 82.01$ kN
Load on the node on the centre line $= 2 \times 136.69 - 82.01 = 191.36$ kN

(iv). Beams 6 to 8 occupy lane B = lane2:
Uniformly distributed load $= 2.5$ kN/m^2, $\psi_0 = 0.40$,
Tributary width = spacing of beams = 1000 mm.
Load per m $= [\gamma_{Q1} \times (\psi_0 = 0.40) \times 2.5 + \gamma_G \times (1.2 + 14.20)] \times 1000 \times 10^{-3}$
$\qquad = 22.14$ kN/m
Total load due to self weight only $= 19.17$ kN/m
The four wheels each have a load of $\gamma_{Q1} \times (200/2) \times (\alpha_Q = 1.0) \times (\psi_0 = 0.75)$
$\qquad = 101.25$ kN.
Loads on the nodes to the left of centre line $= 101.25 \times 1.2/2.0 = 60.75$ kN
Load on the node on the centre line $= 2 \times (101.25) - 60.75 = 141.75$ kN

(v). Beams 9 to 11 occupy lane C = lane 3:
Uniformly distributed load $= 2.5$ kN/m^2, $\psi_0 = 0.40$
Tributary width = Spacing of beams = 1000 mm.
Load per m $= [\gamma_{Q1} \times (\psi_0 = 0.40) \times 2.5 + \gamma_G \times (1.2 + 14.20)] \times 1000 \times 10^{-3}$
$\qquad = 22.14$ kN/m
Total load due to self weight only $= 19.17$ kN/m
The four wheels each have a load of
$\gamma_{Q1} \times (100/2) \times (\alpha_Q = 1.0) \times (\Psi_0 = 0.75) = 50.63$ kN.
Loads on the nodes to the left of centre line $= 50.63 \times 1.2/2.0 = 30.38$ kN

Load on the node on the centre line = 2 × (50.63) – 30.38 = 70.88 kN
  Fig. 16.52 shows the shear force distribution along the span in beams 1 to 7.
The maximum shear force of 117 kN occurs at mid-span in beam 5.

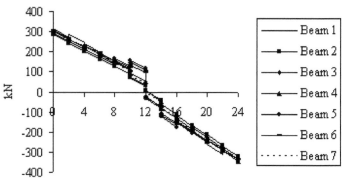

Fig. 16.52 Shear force diagram, case 1

***Case 2: ULS analysis with lanes A, B and C treated as lanes 1, 2 and 3
respectively. Uniformly distributed load covers the right two thirds of the span
and self weight covers the left one third of the span. Wheel loads to the right of
third span.***
Note: Uniformly distributed load includes all loads including self weight.
Loads are as in case 1 except that the nodal loads act at the third span and the
uniformly distributed load covers two thirds of the span to the right of the section.
Fig. 16.53 shows the shear force distribution along the span in beams 1 to 7. The
maximum shear force of 222 kN occurs at the third span in beam 5.

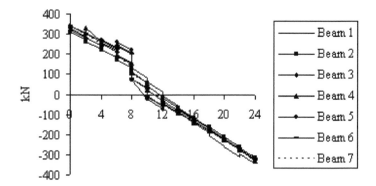

Fig. 16.53 Shear force diagram, case 2

***Case 3: ULS analysis with lanes A, B and C treated as lanes 1, 2 and 3
respectively. Uniformly distributed load covers the right six seventh of the span
and self weight covers the left one sixth of the span. Wheel loads to the right of
the sixth span.***
Note: Uniformly distributed load includes all loads including self weight.

Loads are as in case 1 except that the nodal loads act at the sixth span and the uniformly distributed load cover six sevenths of the span to the right of the section. Fig. 16.54 shows the shear force distribution along the span in beams 1 to 7. The maximum shear force of 338 kN occurs at the sixth span in beam 5.

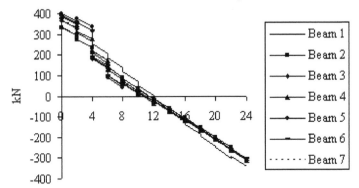

Fig. 16.54 Shear force diagram, case 3

Case 4: ULS analysis with lanes A, B and C treated as lanes 1, 2 and 3 respectively. Uniformly distributed load covers the entire span and wheel loads to the right of the support.

Note: Uniformly distributed load includes all loads including self weight.
Loads are as in case 1 except that the nodal loads act at the support and the uniformly distributed load cover the entire span. Fig. 16.55 shows the shear force distribution along the span in beams 1 to 7. The maximum shear force of 615 kN occurs at support in beam 5.

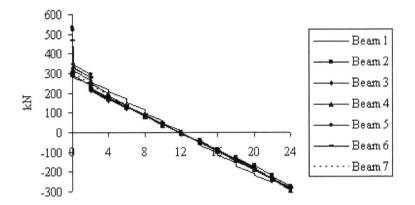

Fig. 16.55 Shear force diagram, case 4

Thus for design, the maximum shear forces in kN are as follows: support = 615, sixth span = 338, third span = 222 and mid-span =117.

16.13.2 Analysis to Determine the Maximum Shear Force along the Span: Cases 5 to 8

As already mentioned, cases 5 to 8 are respectively similar to cases 1 to 4 except that lanes A, B and Care treated as lanes 2, 1 and 3 respectively.

Case 5. Fig. 16.56 shows the shear force distribution along the span in beams 1 to 7. The maximum shear force at mid-span is 117 kN in beam 6.

Case 6: Fig. 16.57 shows the shear force distribution along the span in beams 1 to 7. The maximum shear force at the third span is 222 kN in beam 6.

Case 7: Fig. 16.58 shows the shear force distribution along the span in beams 1 to 7. The maximum shear force at the sixth span is 337 kN in beam 6.

Case 8: Fig. 16.59 shows the shear force distribution along the span in beams 1 to 7. The maximum shear force at the support is 537 kN in beam 6.

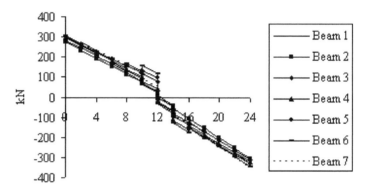

Fig. 16.56 Shear force diagram, case 5

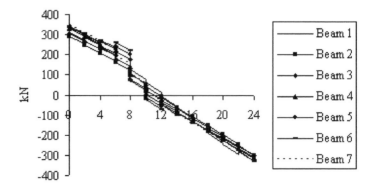

Fig. 16.57 Shear force diagram, case 6

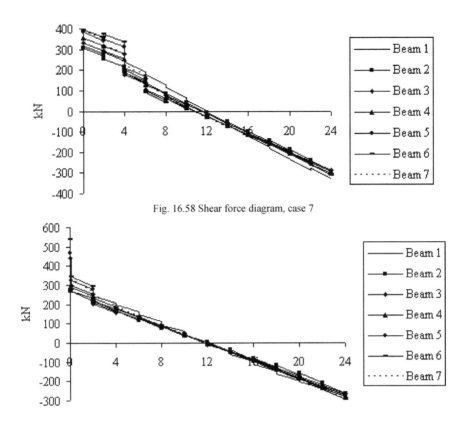

Fig. 16.58 Shear force diagram, case 7

Fig. 16.59 Shear force diagram, case 8

Thus for design, the maximum shear forces in kN are as follows:
support = 537 (536), sixth span = 337 (338), third span = 222 (222) and
mid-span = 117 (117).
Clearly there is very little difference in the maximum shear force calculated from
cases 1 to 4, given in parenthesis or cases 5 to 8.

Table 16.18 Design Shear Force

Section	All loads, kN	Self weight + slab only, kN
Support	537	297.5
Sixth Span	338	198.3
Third Span	222	99.2
Mid-span	117	0

16.13.3 Summary of Results

Table 16.18 summarises the results of the analysis. The combined self-weight and
live load shear forces are computed using a grillage analysis. The self-weight only

shear forces are computed by ignoring any interaction between the beams and taking the self-weight of the outer composite beam. The error in ignoring any interaction will be insignificant. Fig. 16.60 shows the shear envelope for live loading.

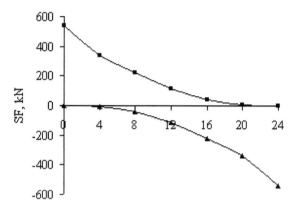

Fig. 16.60 Design shear force envelope

16.13.4 Design of Shear Reinforcement

The detailed calculations below are similar to section 9.6, Chapter 9.
(a). Calculation of $V_{Rd, c}$ at sections without shear reinforcement and uncracked in bending
The value of $V_{Rd, c}$ will be calculated by limiting the principle elastic tensile stress at the neutral axis of the composite section to permissible tensile stress f_{ctd}:

$$f_{ctm} = 0.3 \times 40^{0.667} = 3.51 \text{ MPa}, \quad f_{ctk,0.05} = 0.7 f_{ctm} = 2.46$$

$$f_{ctd} = \alpha_{ct} \frac{f_{ctk,0.05}}{\gamma_m} = 1.0 \times \frac{2.46}{1.5} = 1.6 \text{ MPa}$$

Calculate σ_{cp} due to prestress at supports at the neutral axis of the composite section:

$$\sigma_{cp} = \frac{P_s}{A} - \frac{P_s e}{I} \times (y_{comp} - y_{beam})$$

$$= \frac{2273 \times 10^3}{457 \times 10^3} - \frac{2273 \times 10^3 \times 246}{5.35 \times 10^{10}} \times (614 - 455) = 4.97 - 1.66 = 3.31 \text{ MPa}$$

Calculate the shear stress τ that can be applied in combination with σ_{cp} without exceeding the permissible tensile stress equal to f_{ctd}. Assume $\alpha_1 = 0.5$.

$$\tau = \sqrt{(f_{ctd}^2 + \alpha_1 \sigma_{cp} f_{ctd})} = \sqrt{(1.6^2 + 0.5 \times 3.31 \times 1.6)} = 2.28 \text{ MPa}$$

In calculating the shear stress at the neutral axis of the composite section, the contribution to shear stresses from V_{dead} and V_{live} are calculated from the formulae

$$\tau_{dead} = V_{dead} \times \frac{S_{beam}}{I_{beam} \times b_w}$$

$$\tau_{live} = V_{live} \times \frac{S_{comp\ beam}}{I_{comp\ beam} \times b_w}$$

The value of S_{beam} is the first moment of area about the *neutral axis of the precast beam* of the area of the precast beam above the neutral axis of the composite beam. Using the data in Fig. 16.61,
Calculation of S_{beam}:
Width of web at the neutral axis of the composite beam:
$= 216 + (393 - 216) \times (671 - 436)/671 = 278$ mm
$S_{beam} = [313 \times 50 \times (159 + 436 + 50/2) = 9.70 \times 10^6$ mm³ for the key
$+ 278 \times 436 \times (159 + 436/2) = 45.70 \times 10^6$ mm³ for the web rectangle
$+ 0.5 \times (393 - 278) \times 436 \times (159 + 0.67 \times 436) = 11.27 \times 10^6$ mm³ for the web triangle]

$S_{beam} = 66.67 \times 10^6$ mm³
$V_{dead} = 298$ kN, $I_{beam} = 5.35 \times 10^{10}$ mm⁴, $b_w = 278$ mm

$$\tau_{dead} = V_{dead} \times \frac{S_{beam}}{I_{beam} \times b_w} = 298 \times 10^3 \times \frac{66.67 \times 10^6}{5.35 \times 10^{10} \times 278} = 1.34 \ \text{MPa}$$

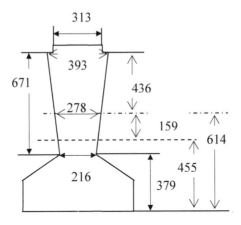

Fig. 16.61 Computation of S_{beam}

$S_{comp.\ beam}$ is the first moment of area about the *neutral axis of the composite beam* of the composite beam area above the neutral axis of the composite beam. In order to get the effective area, a modular ratio $\alpha = 0.895$ is used for the slab area. The first moment of area of the beam portion will be as for S_{beam}, except that the moments are taken about the neutral axis of the composite beam rather than about the neutral axis of the beam:
$S_{comp.\ beam} = [313 \times 50 \times (436 + 50/2) = 7.22 \times 10^6$ mm³ for the key
$+ 278 \times 436 \times (436/2) = 26.42 \times 10^6$ mm³ for the web rectangle
$+ 0.5 \times (393 - 278) \times 436 \times (0.67 \times 436) = 7.29 \times 10^6$ mm³ for the web triangle

$+ \alpha \{1000 \times 130 \times (436 + 50 + 130/2)\} = 64.11 \times 10^6$ slab area above the key

$+ \alpha \{(1000 - 313) \times 50 \times (436 + 50/2)\} = 14.17 \times 10^6$ slab area excluding the key]

$S_{comp.\ beam} = 119.21 \times 10^6\ mm^3$

$V_{live} = (537 - 298) = 239\ kN,\ I_{beam\ comp} = 10.46 \times 10^{10}\ mm^4,\ b_w = 278\ mm$

$$\tau_{live} = V_{live} \times \frac{S_{comp\,beam}}{I_{comp\ beam} \times b_w} = 239 \times 10^3 \times \frac{119.21 \times 10^6}{10.46 \times 10^{10} \times 278} = 0.98\ \ MPa$$

Calculate the portion ξ of τ_{live} together with τ_{dead} that are equal to the maximum permissible shear stress τ:

$\tau = 2.28 = 1.34 + \xi \times 0.98,\ \xi = 0.96$

$V_{Rd,c} = V_{dead} + \xi \times V_{live} = 298.0 + 0.96 \times 239.0 = 527.3\ \ kN$

$V_{RD,\,c}$ is almost equal to the applied value of 537 kN. Therefore only nominal shear links are required.

An approximate value can be calculated as

$$\tau = 2.28 = V_{RD,c} \times \frac{S_{comp.\ beam}}{I_{comp.\ beam} \times b_w} = V_{RD,c} \times 10^3 \times \frac{119.21 \times 10^6}{10.46 \times 10^{10} \times 278}$$

$V_{RD,c} = 556.2\ kN$

The approximate value of $V_{RD,\,c}$ is 5.5% higher than the accurate value.

(b). Start of cracked section

$$f_{ctd} = \gamma_{inf\ erior}\{-\frac{P_s}{A} - \frac{P_s\ e}{Z_b}\} + [\frac{M_{dead\,mid-span}}{Z_b} + \frac{M_{live\ mid-span}}{Z_{b\ comp}}] \times 4\frac{x}{L}(1 - \frac{x}{L})$$

From the moment envelope, at ULS

$M_{dead} = 1756\ (1647)$ kNm. The figure in parenthesis refers to the value when interaction between beams is ignored at ULS.

$M_{live} = 1164$ kNm

$P_s = 3125\ (2273)$ kN, $e = 281(246)$ mm. Figures in parenthesis refer to values at the support.

$\gamma_{Inferior} = 0.95$

$A = 456.96 \times 10^3\ mm^2,\ Z_b = 117.58 \times 10^6\ mm^3,\ Z_{b\ comp} = 170.36 \times 10^6\ mm^3$

Using prestress values at the sixth span onwards and $M_{dead} = 1647$ kNm,

$1.6 = -6.50 - 7.10 + [59.74 + 26.82] \times (x/L) \times (1 - x/L)$

$(x/L) \times (1 - x/L) = 0.1756,\ x/L = 0.23$ and 0.7728

If using prestress values near the support,

$1.6 = -4.73 - 4.52 + [59.74 + 26.82] \times (x/L) \times (1 - x/L)$

$(x/L) \times (1 - x/L) = 0.1254,\ x/L = 0.16$ and 0.84

It is reasonable to accept that the sections beyond $x/L = 0.23$ will be cracked.

(c). Shear capacity of sections without shear reinforcement and cracked in bending

In the case of sections where the flexural tensile stress is greater than f_{ctd}, the shear capacity of the section is given by equation (9.6), Chapter 9:

$$V_{Rd,c} = [C_{Rd,c}\, k(100\rho_t\, f_{ck})^{\frac{1}{3}} + k_1\sigma_{cp}]b_w d \ge (v_{min} + k_1\sigma_{cp})b_w d$$

$C_{Rd,c} = 0.12$, $k_1 = 0.15$

Ignoring the two strands at level H, there are 22 strands; d, the effective depth to steel in the tensile zone, is given by

d = Total depth − (10 ×60 + 12× 110)/22 = 1230 − 87 = 1143 mm

$$k = 1 + \sqrt{\frac{200}{d}} \le 2.0, \quad k = 1.42 < 2.0$$

Average stress in the stress block, $f_{cd} = \alpha_{cc}\, f_{ck}/\gamma_c$, $\alpha_{cc} = 0.85$, $\gamma_c = 1.5$.
Using $f_{ck} = 40$ MPa, $f_{cd} = 22.7$ MPa
Using $\gamma_{inferior} = 0.95$, $\sigma_{cp} = N_{Ed}/A = P_s/A = 6.50^* \le$ (0.2 $f_{cd} = 4.53$)
Take $\sigma_{cp} = 4.53$ MPa

$$v_{min} = 0.035\, k^{\frac{3}{2}}\sqrt{f_{ck}} = 0.035 \times 1.42^{\frac{3}{2}} \times \sqrt{40} = 0.38$$

A_{sl} =Area of tensile steel which extends a distance at least equal to the anchorage length + d
Ignoring the two strands at level H, there are 22 strands each of 140 mm^2 cross-sectional area of steel in the tensile zone $A_{sl} = 3080$ mm^2
b_w = Width of web = Average width = (393 + 216)/2 = 305 mm
$\rho_t = A_{sl}/ (b_w \times d) = 3080/ (305 \times 1143) = 0.009 \le 0.02$

$$V_{Rd,c} = [C_{Rd,c}\, k(100\rho_t\, f_{ck})^{\frac{1}{3}} + k_1\sigma_{cp}]b_w d \ge [v_{min} + k_1\sigma_{cp}]b_w d$$

$$= \{[0.12 \times 1.42 \times (100 \times 0.009 \times 40)^{\frac{1}{3}} + 0.15 \times 4.53] \ge [0.38 + 0.15 \times 4.53]\} \times 305 \times 1143 \times 10^{-3}$$
$$= \{[0.56 + 0.68] \ge 1.06\} \times 348.6 = 432.3 \text{ kN}$$

Comparing the permissible values of shear resistance of 527 kN at sections uncracked in flexure and 432 kN at sections cracked in flexure with the design values of shear force as shown in the shear envelope in Fig. 16.60, it is clear that only nominal shear reinforcement is required at all sections.

(d). Interfacial shear between cast in-situ slab and precast beam
The design shear resistance at the interface $v_{Rd, i}$ is given by the equation (9.23), Chapter 9:
$v_{Rdi} = c\, f_{ctd} + \mu\, \sigma_n + \rho\, f_{yd}\, (\mu \sin \alpha + \cos \alpha) \le 0.5\, v\, f_{cd}$
where c = 0.45 and μ = 0.7 for surfaces with exposed aggregate. A = 90° for vertical stirrups, $\rho = A_s/A_i$
A_s = Area of reinforcement crossing the joint including ordinary shear reinforcement with adequate anchorage on both sides of the interface.
A_i = Area of the joint.
$f_{yd} = (f_{yk}/\gamma m, \gamma_m = 1.15)$ = Design tensile strength of reinforcement crossing the joint
f_{ck} for slab concrete = 25 MPa

$f_{cd} = 25/1.5 = 16.7$ MPa, $f_{ctk, 0.05} = 0.21 \times 25^{2/3} = 1.8$, $f_{ctd} = 1.8/1.5 = 1.2$ MPa

$v = 0.6 \times (1 - 25/250) = 0.54$

Taking $f_{yk} = 500$ MPa, $f_{yd} = 500/1.15 = 435$ MPa

q_{dead}= Taking dead load moment at mid-span as 1647 kNm for a span of 24 m,

$q_{dead} = 1647 \times 8/24^2 = 22.48$ kN/m

Similarly for a live load moment of 1164 kNm, $q_{live} = 1164 \times 8/24^2 = 16.17$ kN/m

$\sigma_n = (q_{dead} + q_{live}) \times 10^3 / [1000 \times$ (Contact width = Web width = 393 mm)]

$= 0.10$ MPa

Assuming minimum shear reinforcement,

$$\rho_w = [\frac{A_{sw}}{s \times b_w \times \sin \alpha} \geq 0.08 \frac{\sqrt{f_{ck}}}{f_{yk}} = 0.08 \frac{\sqrt{25}}{500} = 0.08\%]$$

Taking $\alpha = 90°$ for vertical stirrups,

$v_{Rdi} = 0.45 \times 1.2 + 0.6 \times 0.10 + 0.08 \times 10^{-2} \times 435 \times (0.6 \times 1.0 + 0)$

$\leq 0.5 \times 0.54 \times 16.7$

$v_{Rdi} = 0.81 \leq 4.5$ MPa

Note that in the expression for V_{EDi}, although the shear force V_{Ed} and the lever arm z refer to the section where the interface shear stress v_{Edi} is being calculated, it is thought reasonable to use the lever arm calculated at mid-span for sections other than that at mid-span.

As shown in section 16.11, at ULS the neutral axis depth is 383 mm. The total compressive force is 4196 kN and the force in the slab is 2928 kN. The ultimate moment capacity M_u is 4036 kNm,

$\beta = C_{slab}/C_{concrete} = 2928/4196 = 0.70$, b_i = contact width = 393 mm

Lever arm $z = M_u$/Total compressive force = $4036 \times 10^6 / (4196 \times 10^3) = 962$ mm

V_{Ed} = Shear force at support which is the maximum value = 537 kN

$$v_{Edi} = \beta \frac{V_{Ed}}{z\,b_i} = 0.70 \times \frac{537 \times 10^3}{962 \times 393} = 0.99 \text{ MPa} > (v_{Rdi} = 0.81 \text{ MPa})$$

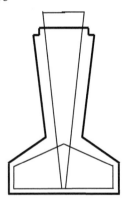

Fig. 16.62 Shear link

Provide more than minimum shear reinforcement to improve shear resistance:
$V_{Rdi} = 0.45 \times 1.2 + 0.6 \times 0.10 + \rho \times 435 \times (0.6 \times 1 + 0) \geq 0.99$
$\rho \geq 0.15\%$
Two-legged 12 mm diameter links giving $A_{sw} = 226$ mm^2 at a spacing not
exceeding approximately 0.75 d, with d = 1143 mm (if the cables at level H are
ignored) gives the maximum spacing $s_{max} = 857$mm, b_i = Contact width= 393 mm.
$\rho = A_{sw}/(s \times b_i) = 226/(s \times 393) \geq 0.15\%$
$s \leq 383$ mm, say s = 400 mm
Provide two-legged 12 mm diameter single-piece shear links as shown in Fig.
16.62 at a spacing of 400 mm.

16.14 DESIGN OF A POST-TENSIONED BOX GIRDER BRIDGE

Fig. 16.63 shows the cross section of the box girder. In order to keep the
calculations simple, it is assumed that there is no variation in the cross section
along the span, although in practice often the cross sections at mid-span and at the
support tend to be different. It is used in a two span continuous beam of 40 m per
span.

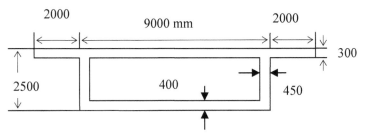

Fig. 16.63 Cross section of the box girder

The cross-sectional properties of the bridge are as follows:
Area of cross section A:
$A = 2 \times 2000 \times 300 + 9000 \times 2500$
$- (9000 - 2 \times 450 = 8100) \times (2500 - 300 - 400 = 1800)$
$= (1.20 + 22.50 - 14.58) \times 10^6 = 9.12 \times 10^6$ mm^2
Taking the unit weight of concrete as 24 kN/m^3,
Self weight $= 9.12 \times 10^{-6} \times 24 = 218.88$ kN/m
Taking moments about the soffit, the first moment of area is
$A \times y_b = [1.2 \times (2500 - 300/2 = 2350) + 22.50 \times (2500/2 = 1250)$
$- 14.58 \times (1800/2 + 400 = 1300)] \times 10^6$
$= 11.991 \times 10^9$ mm^3
$y_b = 1315$ mm, $y_t = 2500 - y_b = 1185$ mm
The second moment of area I is
$I = 2 \times 2000 \times 300^3/12 + 1.2 \times 10^6 \times (2350 - y_b)^2$
$+ 9000 \times 2500^3/12 + 22.50 \times 10^6 \times (1250 - y_b)^2$
$- 8100 \times 1800^3/12 - 14.58 \times 10^6 \times (1300 - y_b)^2$
$= (0.009 + 1.286 + 11.719 + 0.095 - 3.937 - 0.003) \times 10^{12}$
$I = 9.168 \times 10^{12}$ mm^4

$Z_t = I/y_t = 7737 \times 10^6 \text{ mm}^3$, $Z_b = I/y_b = 6972 \times 10^6 \text{ mm}^3$

Note: Because of the very wide flange, in order to account for the shear lag phenomenon, it is sensible to check the effective width of the flange both at top as well as bottom. Using Fig. 16.55 and the rules for the effective width of T-beams,

$b_1 = 2 \text{ m}$, $b_2 = 9/2 - 0.450 = 4.05 \text{ m}$, $b_w = 0.45 \text{ m}$, $\ell_1 = \text{span} = 40 \text{ m}$,

$\ell_0 = \text{end span} = 0.85 \, \ell_1 = 34 \text{ m}$

$b_{\text{eff, 1}} = 0.2 \, b_1 + 0.1 \, \ell_0 \leq 0.2\ell_0$ and $\leq b_1$

$\qquad = 0.2 \times 2 + 0.1 \times 34 = 3.8\text{m} > 2\text{m}. \quad b_{\text{eff, 1}} = 2 \text{ m}$

$b_{\text{eff, 2}} = 0.2 \, b_2 + 0.1 \, \ell_0 \leq 0.2\ell_0$ and $\leq b_2$

$\qquad = 0.2 \times 4.05 + 0.1 \times 34 = 4.21 \text{ m} > 4.05, \quad b_{\text{eff, 2}} = 4.05 \text{ m}$

$b_{\text{eff}} = b_{\text{eff, 1}} + b_{\text{eff, 1}} + b_w = 2.0 + 4.05 + 0.45 = 6.5 \text{ m}$

This is exactly equal to half the width of the top flange. Taking two beams together, one can use the entire width of the top flange as effective.

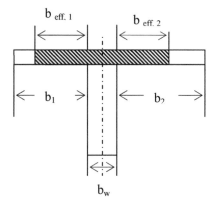

Fig. 16.64 Effective width of T-beam

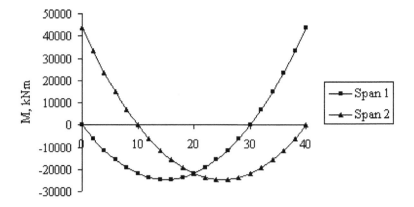

Fig. 16.65 Moment distribution due to self weight

6.14.1 Calculation of Moments at SLS

(i). Self weight moments: Self weight = 218.88 kN/m
Support moment = 218.88 × 40²/8 = 43776 kNm
Fig. 16.65 shows the moment distribution due to self weight.

(ii). Live Loading: The live loadings to be used are as detailed in section 16.5.8.
All γ factors are taken as unity. The ψ_0 value for UDL due to vehicular traffic and
pedestrian load is 0.40. For Tandem load $\psi_0 = 0.75$.
For tandem load, $\alpha_Q = 0.9$ for lane 1 and $\alpha_Q = 1.0$ for lanes 2 and 3.
(a). The superdead load on footpath due to 120 mm black top weighing 24 kN/m³
is equal to $120 \times 24 \times 10^{-3} = 2.88$ kN/m²
(b). The superdead load on lanes due to 50 mm black top weighing 24 kN/m³ is
equal to $50 \times 24 \times 10^{-3} = 1.2$ kN/m²
(c). Footpath load = 3 kN/m² × ($\psi_0 = 0.40$) = 1.2 kN/m²
(d). Lane A: UDL = = 9 kN/m² × ($\psi_0 = 0.40$) = 3.6 kN/m²
   Tandem load: four wheel loads each W = (300/2) × ($\alpha_Q = 0.9$) × ($\psi_0 = 0.75$)
$$= 101.25 \text{ kN}$$
(e). Lane B: UDL = = 2.5 kN/m² × ($\psi_0 = 0.40$) = 1.0 kN/m²
   Tandem load: four wheel loads each W = (200/2) × ($\alpha_Q = 1.0$) × ($\psi_0 = 0.75$)
$$= 75.0 \text{ kN}$$
(f). Lane C: UDL = = 2.5 kN/m² × ($\psi_0 = 0.40$) = 1.0 kN/m²
   Tandem load: four wheel loads each W = (100/2) × ($\alpha_Q = 1.0$) × ($\psi_0 = 0.75$)
$$= 37.5 \text{ kN}$$

Load placement for maximum bending moment: Fig. 16.66 shows the influence
lines for bending moment at mid-span and Fig. 16.67 shows the influence line for
the support moment for a continuous beam with two equal spans.

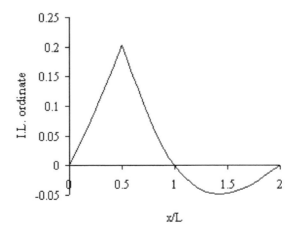

Fig. 16.66 Influence line for mid-span bending moment

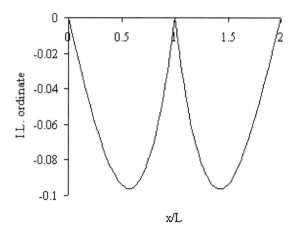

Fig. 16.67 Influence line for support bending moment

Maximum mid-span bending moment:
Uniformly distributed load: Load span 1 with full load and span 2 with only super dead load due to black top. Wheel loads at mid-span. Fig. 16.68 shows the live load arrangement.

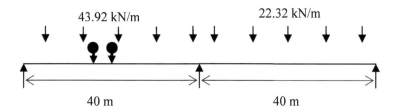

Fig. 16.68 Loading for maximum mid-span moment.

(a). Span 1, UDL:
(i). 120 mm black top over the two cantilevers 2 m wide:
    UDL = 2× 2.88 × 2.0 = 11.52 kN/m
(ii). 50 mm black top over the 9 m wide box:
    UDL= 1.20 × 9.0 = 10.80 kN/m
(iii). Footpath: 1.2 kN/m² acts over two cantilevers each 2 m wide:
    UDL = 2× 1.2 × 2.0 = 4.80 kN/m
(iv). UDL over lanes 1, 2 and 3: (3.6 + 1.0 + 1.0) × 3 = 16.8 kN/m
Total = 11.52 + 10.80 + 4.80 + 16.81 = 43.92 kN/m

(b). Span 1, wheel loads:
Axle loads over lanes 1, 2 and 3:
Total = 2 × (101.25 + 75 + 37.5) = 427.5 kN
The two axles are separated by 1.2 m. The maximum effect is obtained by placing one axle at mid span and another at 1.2 m to the left of mid-span.

(c). Span 2, UDL:

(i). 120 mm black top over the two cantilevers 2 m wide:

UDL = 2× 2.88 × 2.0 = 11.52 kN/m

(ii). 50 mm black top over the 9 m wide box:

UDL = 1.20 × 9.0 = 10.80 kN/m

Total = 11.52 + 10.80 = 22.32 kN/m

Fixed end moments:

Fig. 16.69 shows the fixed end moments due to a uniformly distributed load q and a concentrated load W:

$$M_{12} = q \frac{L^2}{12} + W \frac{a\, b^2}{L^2}, \quad M_{21} = q \frac{L^2}{12} + W \frac{a^2\, b}{L^2}$$

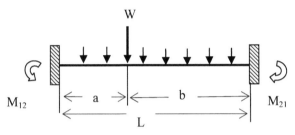

Fig. 16.69 A fixed end beam

From the above equations, the fixed end moments in spans 1 and 2 can be determined:

$M_{12} = - [43.92 \times 40^2/12 + 427.5/40^2 \times (18.8 \times 21.2^2 + 18.8^2 \times 21.2)]$

$\quad = - [5856 + 2258 + 2002] = -10116$ kNm,

$M_{21} = [43.92 \times 40^2/12 + 427.5/40^2 \times (18.8^2 \times 21.2 + 18.8 \times 21.2^2)]$

$\quad = [5856 + 2002 + 2258] = 10116$ kNm

$M_{23} = -22.32 \times 40^2/12 = -2976$ kNm

$M_{32} = 22.32 \times 40^2/12 = 2976$ kNm

The self weight fixed end moments are: $M_{12} = 0$, $M_{21} = -M_{23} = 43776$, $M_{32} = 0$.

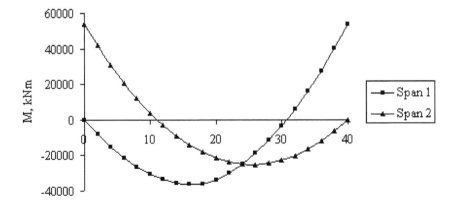

Fig. 16.70 Bending moment distribution at SLS for maximum mid-span moment

Carrying out the moment distribution process, the support moments are
$M_{12} = 0$, $M_{21} = 53595$, $M_{23} = -53595$, $M_{32} = 0$.
Fig. 16.70 shows the bending moment distribution.

Maximum support bending moment:
Uniformly distributed load: Both spans will carry a full load of 43.92 kN/m.
Axle loads: It appears from the influence line shown in Fig. 16.67, that one set of
axles needs to be at mid-span and another set at 1.2 m to the left of mid-span. This
is the same arrangement as for maximum bending moment at mid-span shown in
Fig. 16.68.
Fixed end moments: From mid-span moment calculations,
$M_{12} = -10116$ kNm, $M_{21} = 10116$ kNm
$M_{23} = -43.92 \times 40^2/12 = -5856$ kNm
$M_{32} = 43.92 \times 40^2/12 = 5856$ kNm
The fixed end moments due to self weight are:
$M_{12} = 0$, $M_{21} = 43776$ kNm, $M_{23} = -43776$, $M_{32} = 0$.
Carrying out the moment distribution process, the support moments are:
$M_{12} = 0$, $M_{21} = 55755$, $M_{23} = -55755$, $M_{32} = 0$.
Fig. 16.71 shows the bending moment distribution.

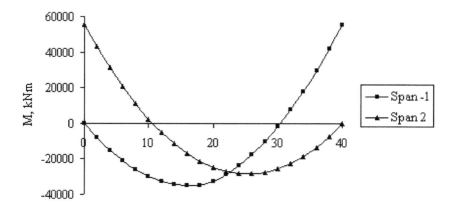

Fig. 16.71 Maximum support SLS bending moment distribution

16.14.2 Thermal Stresses: Heating

Using the data in section 4.7.1, Chapter 4,
h = 2500 mm
$0.3h = 750$ mm, $h_1 = 0.3h \le 150$ mm, $h_1 = 150$ mm
$h_2 = 0.3h$ but ≥ 100 mm and ≤ 250 mm, $h_2 = 250$ mm
$h_3 = 0.3h$ but $\le (100$ mm + surfacing depth in mm$)$, $h_3 = 200$ mm
From Table 4.2, interpolating for $h \ge 0.8$ m,
$\Delta T_1 = 13.0°C$, $\Delta T_2 = 3.0°C$, $\Delta T_3 = 2.5°C$
$E_{Beam} = E_{cm} = 22(0.1 \times f_{cm})^{0.3} = 22 \times (4.8)^{0.3} = 35.2$ GPa

Coefficient of thermal expansion $\alpha_T = 10 \times 10^{-6}/^\circ C$. Corresponding restraining stresses are $\sigma = -\alpha_T \times E_c \times \Delta T$, $\alpha_T \times E_c = 0.352$.
Table 16.19 shows the restraining stresses at various levels in the cross section.

Table 16.19 Restraining stresses in the cross section: Heating

Level from top, mm	$\Delta T, ^\circ C$	σ, MPa
0 (Top flange)	13	-4.58
150 (Top flange)	3.0	-1.06
300 (Top flange)	1.2	-0.42
300 (Web)	1.2	-0.42
400 (Web)	0	0
2300 (Bot. flange)	0	0
2500 (Bot. flange)	2.5	-0.88

Table 16.20 Restraining forces in the cross section: Heating

Region	σ_{top}, MPa	σ_{Bottom}, MPa	Area	F, kN
Top flange	-4.38	-1.06	13000 × 150	-5304
	-1.06	-0.42	13000 × 150	-820
Web	-0.42	0	100× 2 × 450	-19
Bottom flange	0	-0.88	9000 × 200	-792
				Σ-6935

Having obtained the restraining stresses, the next step is to calculate the forces F in different segments of the cross section. The values of the forces are shown in Table 16.20.

Table 16.21 Calculation of moment: Heating

F, kN	h mm	C mm	Lever arm mm	Moment kNm
-5304	150	90	$y_t - h + C = 1125$	-5968
-820	150	86	$y_t - 150 - h + C = 971$	-796
-19	100	33	$y_t - 300 - h + C = 818$	-16
-792	200	67	$-(y_b - C) = -1248$	988
Σ-6935				Σ-5792

Having calculated the forces, the next step is to calculate the moment due to the restraining forces about the neutral axis of the beam. Table 16.21, shows the details of the calculation of the moment due to force F. Note that if the stress distribution is as shown in Fig. 16.41, the resultant is at a distance C from the base of the element.

The last step is the calculation of the final stress in the cross section as the sum of the initial restraining stress plus the stress due to axial force and moment in order to produce a self-equilibrating stress system. Table 16.22 shows the details of the stress calculation. Fig. 16.35 shows the final stresses. The maximum tensile stress is in the web and is equal to 1.26 MPa.

Table 16.22 Thermal stress calculation in precast beam: Heating

Position from top	Σ, MPa	$-\Sigma$ F/A, MPa	Y, mm	$-(\Sigma$ M/I)y	Final stress, MPa
0 (top flange)	−4.58	0.76	1185	0.75	−3.07
150 (top flange)	−1.06	0.76	1035	0.65	0.35
300 (top flange)	−0.42	0.76	885	0.56	0.90
400 (web)	0	0.76	785	0.50	1.26
1185 (neutral axis)	0	0.76	0	0	0.76
2300 (bot. flange)	0	0.76	−1115	−0.70	0.06
2500 (bot. flange)	−0.88	0.76	−1315	−0.83	−0.95

16.14.3 Thermal Stresses: Cooling

Using the data in section 4.7.2, Chapter 4,
h = 2500 mm, 0.2h = 500 mm, $h_1 = h_4 = 0.2h \leq 250$ mm
0.25h = 625 mm, $h_2 = h_3 = 0.25h$ but $\geq$ 200 mm
Therefore: $h_1 = h_4 = 250$ mm, $h_2 = h_3 = 625$ mm
$\Delta T_1 = -8.4^\circ C$, $\Delta T_2 = -0.5^\circ C$, $\Delta T_3 = -1.0^\circ C$, $\Delta T_4 = -6.5^\circ C$
Table 16.23 shows the restraining stresses at various levels in the cross section.

Table 16.23 Restraining stresses in the cross section: Cooling

Level from top, mm	ΔT, $^\circ C$	σ, MPa
0 (top flange)	8.4	2.96
250 (top flange)	0.5	0.18
300 (top flange)	0.46	0.16
875 (web)	0	0
1615 (web)	0	0
2100 (bottom flange)	0.76	0.27
2250 (botttom flange)	1.0	0.35
2500 (soffit)	6.5	2.29

After calculating the restraining stresses, the next step is to calculate the forces F in different segments of the cross section as shown in Table 16.24

The next step is to calculate the moment due to the restraining forces about the neutral axis of the beam. Table 16.25, shows the details of the calculation of the moment due to force F.

The last step is the calculation of the final stress in the cross section as the sum of the initial restraining stress plus the stress due to axial force and moment in order to produce a self equilibrating stress system. Table 16.26 shows the details of stress calculation.

Table 16.24 Restraining forces in the cross section: Cooling

Region	σ_{top}, MPa	σ_{Bottom}, MPa	Area	F, kN
Top flange	2.96	0.18	13000 × 250	5103
	0.18	0.16	13000 × 50	26
Web (top part)	0.16	0	2 × 450× 575	41
Web (bottom part)	0	0.27	2 × 450× 475	58
Bottom flange	0.27	0.35	9000 × 150	42
	0.35	2.29	9000 × 250	297
				Σ5567

Table 16.25 Calculation of moment: Cooling

F, kN	H, mm	C, mm	Lever arm, mm	Moment, kNm
5103	250	184	$y_t - h + C = 1119$	5710
26	50	26	$y_t - 250 - h + C = 911$	24
41	575	383	$y_t - 300 - h + C = 693$	28
58	475	158	$-(y_b - 400 - C) = -757$	−44
42	150	72	$-(y_b - 250 - C) = -993$	−42
297	250	94	$-(y_b - C) = -1221$	−364
Σ5567				Σ5312

Table 16.26 Thermal stress calculation in precast beam: Cooling

Position from top, mm	Σ, MPa	$-\Sigma F/A$, MPa	Y, mm	$-(\Sigma M/I)y$, MPa	Final stress, MPa
0 (top flange)	2.96	−0.61	1185	−0.69	1.66
250 (top flange)	0.18	−0.61	935	−0.54	−0.97
300 (top flange)	0.16	−0.61	885	−0.51	−0.96
875 (web)	0	−0.61	310	−0.18	−0.79
1185 (neutral axis)	0	−0.61	0	0	−0.61
1615 (web)	0	−0.61	−440	0.26	−0.35
2100 (bot. flange)	0.27	−0.61	−915	0.53	0.19
2250 (bot. flange)	0.35	−0.61	−1065	0.62	0.28
2500 (soffit)	2.29	−0.61	−1315	0.76	2.44

16.14.4 Determination of Prestress and Eccentricity

Preliminary choice of prestressing force:

Initially design the support section in order to get an approximate value of the prestress and eccentricity needed.

At the support section: $M_{self\,weight} = 43776$ kNm, SLS moment = 55755 kNm

Assuming $f_{ck} = 40$ MPa, $f_{cki} = 25$ MPa, the permissible stresses at transfer and service are: $f_{tt} = 2.6$ MPa, $f_{tc} = -15.0$ MPa, $f_{st} = 3.5$ MPa, $f_{sc} = -24$ MPa

Load factors are: $\gamma_{inferior} = 0.95$, $\gamma_{Superior} = 1.05$

Thermal stresses due to cooling are 1.66 MPa and 2.44 MPa at the top and bottom faces respectively.

Loss of prestress: 10% at transfer, 25% long term.

Substituting in equations (4.5), (4.6), (4.9) and (4.10), the Magnel equations are

$-1/A + e/Z_t = 6.53/P_s,$
$-1/A - e/Z_b = -16.89/P_s,$
$-1/A + e/Z_t = -19.43/P_s,$
$-1/A - e/Z_b = -7.29/P_s$

Fig. 16.72 shows the Magnel diagram.

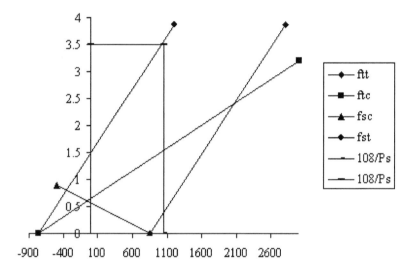

Fig. 16.72 Magnel diagram

Maximum eccentricity at support: $y_t = 1185$ mm. Assuming a cover of say 60 mm, a duct of 120 mm external diameter and a shift in the centroid of the cables of 19 mm say, maximum eccentricity $1185 - 60 - 120/2 - 19 \approx 1046$ mm

From the Magnel diagram for this value of eccentricity, $10^8/P_s = 3.5$, $P_s = 28571$ kN

Force at jacking $P_j = P_s/0.75 = 38095$ kN

Taking for the 0.6 inch (15 mm) nominal diameter 7-wire strands $f_{pk} = 1860$ MPa and an area of cross section of 150 mm² and stress at jacking as 0.77 f_{pk}, the force per jack at stressing is 215 kN. The number of strands needed is $38095/215 \approx 178$. Using anchorages suitable for say 30 strands, the number of cables needed is six with 30 strands in each.

The maximum eccentricity available at the mid-span is approximately

$(y_b = 1315) - 60 - 120/2 - 19 \approx 1176$ mm.

Ignoring the reverse curvature over the support region,

Drape $\approx 1176 + 1046/2 = 1700$ mm

Equivalent load $= 8 \times P \times$ Drape$/L^2 = 8 \times 28571 \times 1700 \times 10^{-3}/40^2 = 243$ kN/m

The prestressing cables apply an upward equivalent load of approximately 243 kN/m. The moment at the support and at mid-span are approximately:

$M_{support} = 243 \times 40^2/8 = 48600$ kNm, $M_{Mid-span} = 243 \times 40^2/16 = 24300$ kNm

Effective eccentricities are

$e_{mid-span} = 24300 \times 10^3/P_s = 851$ mm, $e_{support} = 48600 \times 10^3/P_s = 1701$ mm

Stresses due to prestress and external loading at SLS, using equations (4.7) and (4.8) are

$$\sigma_{top} = \{-\frac{P_s}{A} + \frac{P_s\, e_{effective}}{Z_t}) \times \gamma_{Inferior} - \frac{M_{SLS}}{Z_t} + 1.66$$

$$\sigma_{bottom} = \{-\frac{P_s}{A} - \frac{P_s\, e_{effective}}{Z_b}) \times \gamma_{Inferior} + \frac{M_{SLS}}{Z_b} + 2.44$$

$P_s = 28571$ kN, $\gamma_{inferior} = 0.95$, the effective eccentricities are:

At mid-span e $= 851$ mm and at the support: e $= -1701$ mm

Case 1: Loading for maximum mid-span moment in span 1

At mid-span: M $= 34052$ kNm in span 1 and 21443 kNm in span 2

At support|: M $= -53603$ kNm

Mid-span in span 1: $\sigma_{top} = -2.98 + 2.99 - 4.40 + 1.66 = -2.7$ MPa
$\sigma_{bottom} = -2.98 - 3.31 + 4.88 + 2.44 = 1.0$ MPa

Mid-span in span 2: $\sigma_{top} = -2.98 + 2.99 - 1.30 + 1.66 = 0.4$ MPa
$\sigma_{bottom} = = -2.98 - 3.31 + 3.07 + 2.44 = -0.8$ MPa

At support: $\sigma_{top} = -2.98 - 5.97 + 6.93 + 1.66 = -0.4$ MPa
$\sigma_{bottom} = -2.98 + 6.62 - 7.68 + 2.44 = -1.6$ MPa

Case 2: Loading for maximum support moment

At mid-span: M $= 32976$ kNm in span 1 and 24682 kNm in span 2

At support|: M $= -55755$ kNm

Mid-span in span 1: $\sigma_{top} = -2.98 + 2.99 - 4.26 + 1.66 = -2.6$ MPa
$\sigma_{bottom} = -2.98 - 3.31 + 4.73 + 2.44 = 0.9$ MPa

Mid-span in span 2: $\sigma_{top} = -2.98 + 2.99 - 1.50 + 1.66 = 0.2$ MPa
$\sigma_{bottom} = -2.98 - 3.31 + 3.53 + 2.44 = -0.3$ MPa

At support: $\sigma_{top} = -2.98 - 5.97 + 7.21 + 1.66 = -0.1$ MPa
$\sigma_{bottom} = -2.98 + 6.62 - 8.00 + 2.44 = -1.9$ MPa

All stresses appear satisfactory. One can now do an accurate stress calculation taking into account the reverse curvature over the support:

End support $e_1 = 0$, mid-span $e_3 = 1176$ mm, middle support $e_2 = 1046$ mm

$P_s = 28571$ kN and reverse curvature starts at span/10 from the support.
  Substituting $\lambda = 0.5$, $\beta = 0.1$ in equations (6.8), (6.9) and (6.10) from section 6.2.2, the equivalent loads in the three segments are

$$\text{Segment 1: } q_1 = -\frac{2}{L^2}\frac{(e_1+e_3)}{(1-\lambda)^2}P_s = -168.0\,\text{kN}/\text{m}$$

$$\text{Segment 2: } q_2 = -\frac{2}{L^2}\frac{(e_3+e_2)}{\lambda(\lambda-\beta)}P_s = -396.8\,\text{kN}/\text{m}$$

$$\text{Segment 3: } q_3 = \frac{2}{L^2}\frac{(e_3+e_2)}{\lambda\beta}P_s = 1587.2\,\text{kN}/\text{m}$$

Fig. 16.73 shows the equivalent loads acting on the beam.

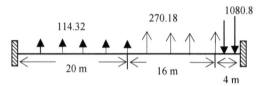

Fig. 16.73 Equivalent loads due to prestress

The fixed end moments at the supports can be computed from equation (6.44) in section 6.3 as shown in Table 16.27.

Table 16.27 Fixed end moments in span 1

$\alpha_1 L$	$\alpha_2 L$	α_1	α_1	q	M_{Left}	M_{Right}
0	20	0	0.5	−168.0	15400.0	−7000.0
20	4	0.5	0.1	−396.8	16337.6	−33606.3
36	0	0.9	0	1587.2	−783.0	11068.1
					Σ30954.6	Σ−29538.2

If the left support is simply supported, $M_{\text{Right}} = -29538.2 - 0.5 \times 30954.6$.
                                                                                     $= -45015.5$ kNm
Using $\gamma_{\text{inferior}} = 0.95$ for prestress, the moment at the middle support is equal to
$-45015.5 \times 0.95 = 42765$ kNm
   Fig 16.74 shows the distribution of the prestress moment in span 1. Because of the fact that the cable profile is symmetrical about the middle support, there is no rotation at the support. Therefore, in span 2 the moment due to prestress will be the mirror image of that in span 1. Fig. 16.75 shows the distribution of shear force due to equivalent loads caused by prestress.

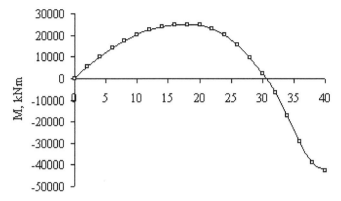

Fig. 16.74 Prestress moment in span 1

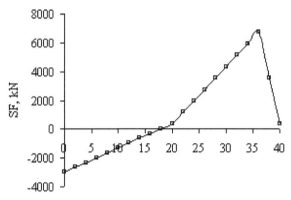

Fig. 16.75 Prestress shear force in span 1

16.14.5 Stress Calculation at SLS

Figs. 16.76 to 16.83 show the bending moment distribution and stress distribution in spans 1 and 2 for the two main loading cases considered. The calculations have not included the thermal stresses due to cooling. The state of stress at any section is calculated using the equations

$$\sigma_{top} = \{-\frac{P_s}{A} \pm \frac{P_s\, e_{effective}}{Z_t}) \times \gamma_{Inferior} \mp \frac{M_{SLS}}{Z_t} + 1.64$$

$$\sigma_{bottom} = \{-\frac{P_s}{A} \mp \frac{P_s\, e_{effective}}{Z_t}) \times \gamma_{Inferior} \pm \frac{M_{SLS}}{Z_t} + 2.44$$

Note: Use the upper sign at sections in span positions and the lower sign at support sections.

For the two cases of loading considered, the states of stress are satisfactory at SLS. In the above calculations, the loss of prestress has been allowed for by using a 25% reduction from the jacking load. More accurate loss calculations as detailed in Chapter 11 should be carried out before accepting the prestress and eccentricity provided. It is worth pointing out that because of the varying eccentricity along the span, it is necessary to investigate other loading patterns which can cause maximum moment at other sections not considered.

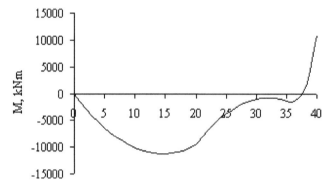

Fig. 16.76 Total moment distribution in span 1, case 1

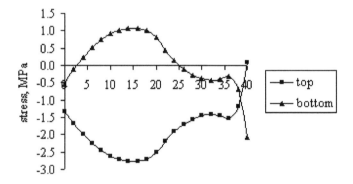

Fig. 16.77 Stress distribution in span 1, case 1

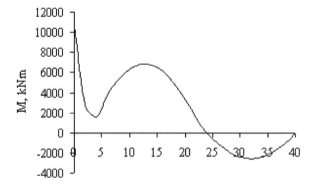

Fig. 16.78 Total moment distribution in span 2, case 1

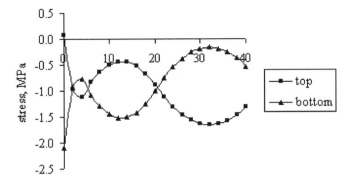

Fig. 16.79 Stress distribution in span 2, case 1

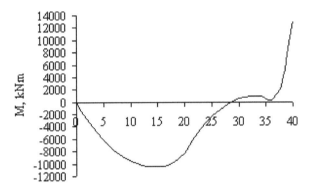

Fig. 16.80 Total moment distribution in span 1, case 2

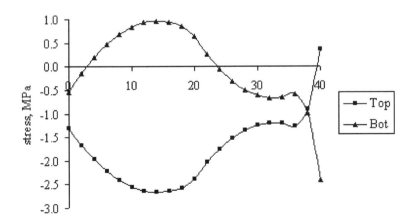

Fig. 16.81 Stress distribution in span 1, case 2

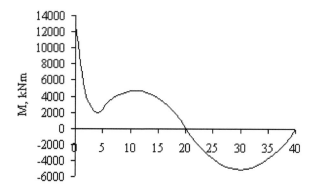

Fig. 16.82 Total moment distribution in span 2, case 2

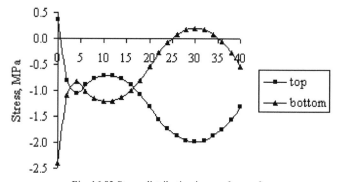

Fig. 16.83 Stress distribution in span 2, case 2

16.14.6 Calculation of Moment at ULS

The moment distribution at ULS is calculated in a manner similar to that at SLS.
The main difference is the inclusion of load factors as detailed in section 16.5.9.
In the following SLS loads from section 16.13.1 are multiplied by appropriate γ
factors.

(i). Self-weight loading:
Self weight = (γ_G = 1.35) × 218.88 = 295.5 kN/m
Self weight = (γ_G = 1.0) × 218.88 = 218.88 kN/m

(ii). Live loading: Using the loads calculated in section 16.13.1 and applying load
factors of γ_Q = 1.35 if unfavourable and 0 if favourable, the maximum and
minimum uniformly distributed loads are respectively 43.92 × 1.35 = 59.29 kN/m
and 22.32 kN/m
Tandem loads: Using a load factor of 1.35, fthe our concentrated wheel loads W
are
Lane A: W = (γ_Q = 1.35) × 101.25 =136.69 kN
Lane B: W = (γ_Q = 1.35) × 75.0 = 101.25 kN

Lane C: W = (γ_Q = 1.35) × 37.5 = 50.63 kN

(a). Maximum mid-span bending moment:
Uniformly distributed live load: Span 1 is loaded with full load and span 2 is loaded with only super dead load due to black top. These loads act in addition to self weight.
Span 1:
UDL = 295.5 + 59.29 = 354.68 kN/m
Axle loads over lanes 1, 2 and 3:
Total = 2 × (136.89 + 101.25 + 50.63) = 577.54 kN
The two axles are separated by 1.2 m. The maximum effect is obtained by placing one axle at mid span and another at 1.2 m to the left of mid-span.
Span 2: UDL = 218.88 + 22.32 = 241.2 kN/m
Fig. 16.84 shows the live load arrangement.

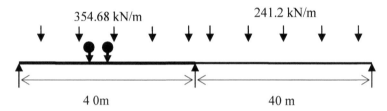

354.68 kN/m   241.2 kN/m

4 0m   40 m

Fig. 16.84 Loading for maximum mid-span moment at ULS

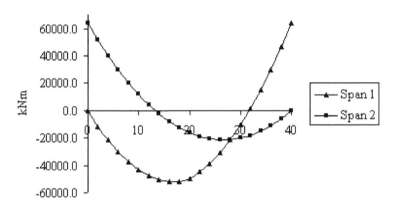

Fig. 16.85 Bending moment distribution at ULS for case 1

Fixed end moments:
M_{12} = − [354.68 × 40^2/12 + 577.54/40^2 × (18.8 ×21.2^2+ 18.8^2 ×21.2)]
  = − [47291 + 5755] = −53046 kNm,
M_{21} = [354.68 × 40^2/12 + 577.54/40^2 × (18.8^2 ×21.2+ 18.8 ×21.2^2)]
  = [47291 + 5755] = 53046 kNm,
M_{23} = −241.2 × 40^2/12 = −32160 kNm
M_{32} = 241.2 × 40^2/12 = 32160 kNm

Carrying out the moment distribution process, the support moments are
$M_{12} = 0$, $M_{21} = 64165$, $M_{23} = -64165$, $M_{32} = 0$.
Fig. 16.85 shows the bending moment distribution.

(b). Maximum support bending moment:
Uniformly distributed load: Both spans will carry a full load of 354.68 kN/m.
Axle loads: It appears from the influence line shown in Fig. 16.67, that one set of
axles needs to be at mid-span and another set at 1.2 m to the left of mid-span. This
is the same arrangement as for maximum bending moment at mid-span. Fig. 16.86
shows the loading arrangement.

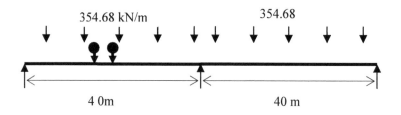

Fig. 16.86 Loading for maximum support moment at ULS

Fixed end moments:
From mid-span moment calculations,
$M_{12} = -53046$ kNm, $M_{21} = 53046$ kNm
$M_{23} = -354.68 \times 40^2/12 = -47291$ kNm
$M_{32} = 354.68 \times 40^2/12 = 47291$ kNm
Carrying out the moment distribution process, the support moments are:
$M_{12} = 0$, $M_{21} = 75253$, $M_{23} = -75253$, $M_{32} = 0$.
Fig. 16.87 shows the bending moment distribution.

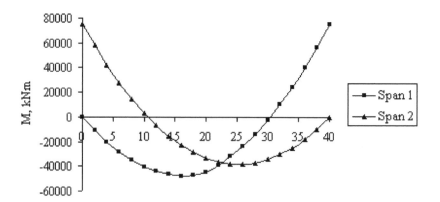

Fig. 16.87 Bending moment distribution at ULS for case 2

16.14.7 Calculation of Moment Capacity at ULS

Fig. 16.88 shows the arrangement of the cables in the span and support sections. The cables are in 120 mm ducts with a cover of 60 mm and the shift in the centroid of the cables is about 19 mm. From the top or bottom surface the centroid of the cables are therefore at a distance of
60 + 120/2 + 19 = 139 mm
In the mid-span and support sections the cables are at a distance of
2500 – 139 = 2361 mm from the compression face.

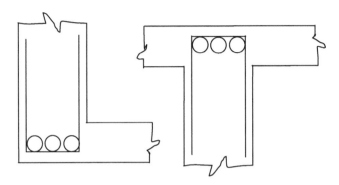

Fig. 16.88 Cable arrangement at mid-span and support sections

(i). Ultimate moment capacity at mid-span:
Choosing neutral axis depth x = 157 mm, the stress block depth = 0.8x = 126 mm. Stress block is inside the top flange.
f_{ck} = 40 MPa, γ_m = 1.5, f_{cd} = 40/1.5 = 26.7 MPa
Stress in stress block = η f_{cd} = 26.67 as η = 1
Total compressive force C = (2 × 2000 + 9000) × 126 × 26.67 × 10^{-3}
$$= 43735 \text{ kN}$$
P_s = 29412 kN, Total area of steel = 180 strands at 150 mm² each = 27000 mm²
Prestress σ_{pe} = 28571× 10^3/ 27000 = 1058 MPa
Young's modulus E_s = 195 GPa
Prestrain ε_{pe} = 1058/ (195 × 10^3) = 5.43 × 10^{-3}
Bending strain = 3.5 × 10^{-3} × (d/x – 1) = 3.5 × 10^{-3} × (2361/157 – 1) = 49.1 × 10^{-3}
Total strain = (5.43 + 49.1) × 10^{-3} = 54.53 × 10^{-3}
Taking f_{pe} = 1860 MPa, γ_m = 1.15, f_{pd} = 1860/1.15 = 1617 MPa
f_{pd}/E_s = 8.29 × 10^{-3}
Steel yields and total tensile force T is given by
T = f_{pd} × 27000 × 10^{-3} = 43659 kN
C – T ≈ 0
Lever arm z = 2361 – 126/2 = 2298 mm
Ultimate moment capacity M_u = C × z × 10^{-3} = 100328 kNm
The applied moment at mid-span is about 52055 kNm. The provided moment capacity is more than adequate. Note that the parasitic moment has been ignored in this calculation. Its value is only 7400 kNm at mid-span and is therefore not significant.

(ii). Ultimate moment capacity at support:

Proceeding as in the case of mid-span, choosing neutral axis depth x = 227 mm

Stress block depth = 0.8x = 182 mm

Total compressive force C = 9000 × 182 × 26.67 × 10^{-3} = 43689 kN

Bending strain = 3.5 × 10^{-3} × (d/x − 1) = 3.5 × 10^{-3} × (2361/ 152 − 1) = 50.87 × 10^{-3}

Total strain = (5.43 + 50.87) × 10^{-3} = 56.30 × 10^{-3}

Steel yields and total tensile force T is given by

T = f_{pd} × 27000 × 10^{-3} = 43659 kN

C − T ≈ 0

Lever arm z = 2361 − 182/2 = 2270 mm

Ultimate moment capacity M_u = C × z × 10^{-3} = 99106 kNm

The applied moment at mid-span is about 75253 kNm. The beam has enough ultimate moment capacity.

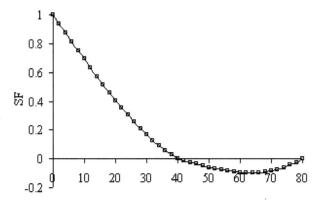

Fig. 16.89 Influence line for shear force at support

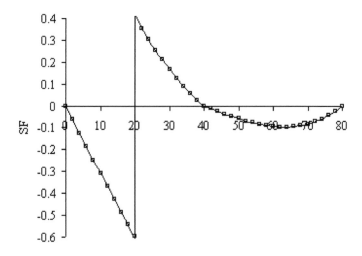

Fig. 16.90 Influence line for shear force at mid-span

16.14.8 Calculation of Shear Force at ULS

Fig. 16.89 and Fig. 16.90 show respectively the influence line for shear force at support and at mid-span.

(i). Shear force at support:
Span 1 is loaded with a uniformly distributed load of 354.68 kN/m and span 2 with 241.2 kNm. In addition, two concentrated loads 577.54 kN each are placed one to the immediate right of the support and the other at 1.2 m from the support. The fixed end moments are

$M_{12} = - (354.68 \times 40^2/12 + 577.54 \times 38.8^2 \times 1.2/40^2) = -47943$ kNm
$M_{21} = (354.68 \times 40^2/12 + 577.54 \times 38.8 \times 1.2^2/40^2) = 47311$ kNm
$M_{23} = -241.2 \times 40^2/12 = -32160$ kNm
$M_{32} = 241.2 \times 40^2/12 = 32160$ kNm

After balancing the moments, the support moments are
$M_{12} = 0$, $M_{21} = 59761$ kNm, $M_{23} = -59761$ kNm, $M_{32} = 0$.
Reaction at the left support:
$V_1 = 354.68 \times 40/2 + 577.54 \times 38.8/40 + 577.54 - 59761/40$
$\qquad\qquad = 6737$ kN (Clockwise shear force)
Reaction at the right support:
$V_2 = V_1 - 354.68 \times 40 - 2 \times 577.54 = -8605$ kN

(ii). Maximum positive shear force at mid-span:
The left half of span 1 and whole of span 2 are loaded with a UDL of 241.2 kN/m and the right half of span 1 is loaded with 354.68 kN/m. In addition, of the two concentrated loads each of 577.54 kN, one is placed just to the right of mid-span and another at 1.2 m to the right of the first load.

$M_{12} = -241.2 \times 40^2/12 \times \{1 - 0.5^3 \times (4 - 3 \times 0.5)\}$
$\qquad -354.7 \times 40^2/12 \times \{0.5^3 \times (1 + 3 \times 0.5)\} - 577.54 \times 40/8$
$\qquad -577.54 \times 21.2 \times 18.8^2/40^2 = - (22110 + 14778 + 2888 + 2705)$
$\qquad = -42481$ kNm
$M_{21} = 241.2 \times 40^2/12 \times \{0.5^3 \times (1 + 3 \times 0.5)\}$
$\qquad + 354.7 \times 40^2/12 \times \{1 - 0.5^3 \times (4 - 3 \times 0.5)\} + 577.54 \times 40/8$
$\qquad + 577.54 \times 21.2^2 \times 18.8/40^2 = (10050 + 32512 + 2888 + 3050)$
$\qquad = 48500$ kNm
$M_{23} = -241.2 \times 40^2/12 = -32160$, $M_{32} = 241.2 \times 40^2/12 = 32160$
Balancing the moments, $M_{12} = 0$, $M_{21} = -58991$, $M_{23} = 58991$, $M_{32} = 0$
Reaction at left $V_1 = 241.2 \times 20 \times 30/40 + 354.68 \times 20 \times 10/40 + 577.54/2$
$\qquad\qquad + 577.54 \times 18.8/40 - 58991/40$
$\qquad\qquad = \{3618 + 1773 + 289 + 271 - 1475\} = 4476$ kN
Reaction at right $V_2 = V_1 - 241.2 \times 20 - 354.68 \times 20 - 2 \times 577.54$
$\qquad\qquad = -8597$ kN
Mid-span positive shear force $= V_1 - 241.2 \times 20 = -348$ kN

(iii). Positive shear force envelope:
Similar calculations to the above can be carried out at different sections and the resulting envelope is shown in Fig. 16.91.

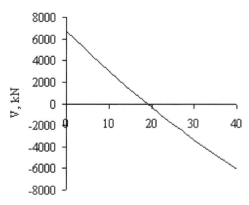

Fig. 16.91 Positive shear force envelope

(iv). Maximum negative shear force at mid-span:
The left half of span 1 and the whole of span 2 are loaded with a UDL of 354.68 kN/m and the right half of span 1 is loaded with 241.2 kN/m. In addition, of the two concentrated loads each of 577.54 kN, one is placed just to the left of mid-span and another at 1.2 m to the left of the first load.

$M_{12} = -354.7 \times 40^2/12 \times \{1 - 0.5^3 \times (4 - 3 \times 0.5)\}$
$\qquad -241.2 \times 40^2/12 \times \{0.5^3 \times (1 + 3 \times 0.5)\} - 577.54 \times 40/8$
$\qquad -577.54 \times 18.8 \times 21.2^2/40^2 = -(32514 + 10050 + 2888 + 3050)$
$\qquad = -48502$ kNm

$M_{21} = 354.7 \times 40^2/12 \times \{0.5^3 \times (1 + 3 \times 0.5)\}$
$\qquad +241.2 \times 40^2/12 \times \{1 - 0.5^3 \times (4 - 3 \times 0.5)\} + 577.54 \times 40/8$
$\qquad + 577.54 \times 21.2 \times 18.8^2 /40^2 = (14779 + 22110 + 2888 + 2705)$
$\qquad = 42482$ kNm

$M_{23} = -354.7 \times 40^2/12 = -47293, M_{32} = 354.7 \times 40^2/12 = 47293$

Balancing the moments, $M_{12} = 0, M_{21} = -68847, M_{23} = 68847, M_{32} = 0$

Reaction at left $V_1 = 354.68 \times 20 \times 30/40 + 241.2 \times 20 \times 10/40 + 577.54/2$
$\qquad\qquad + 577.54 \times 21.2/40 - 68847/40$
$\qquad\qquad = \{5320 + 1206 + 289 + 306 - 1721\} = 5400$ kN

Reaction at right $V_2 = V_1 - 354.68 \times 20 - 241.2 \times 20 - 2 \times 577.54$
$\qquad\qquad = -7673$ kN

Mid-span shear force $= V_1 - 354.7 \times 20 - 577.54 - 577.54 = -2849$ kN

(v). Negative shear force envelope:
Similar calculations to the above can be carried out at different sections and the resulting envelope is shown in Fig. 16.92.

The shear force diagram due to equivalent loads caused by prestress is shown in Fig. 16.75. The final design envelope, which is the algebraic sum of the shear force due to applied loads and that due to equivalent loads, is shown in Fig.16.93. From the shear force envelope, the shear forces at key sections are

Left support: 3668 kN
At mid-span: 2460 kN
At middle support: 9623 kN

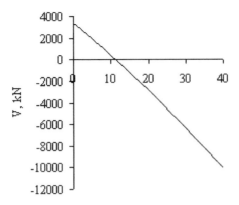

Fig. 16.92 Negative shear force envelope

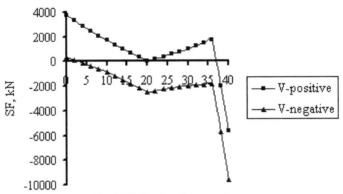

Fig. 16.93 Net shear force envelopes

16.14.9 Calculation of Twisting Moment at ULS

Twisting moment is produced by live loads acting eccentrically with respect to the vertical axis of symmetry.

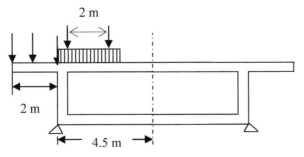

Fig. 16.92 Loads causing twisting moment

As shown in Fig. 16.92, maximum torsion is generated when
(i). Footpath load at ULS = 1.62 kN/m^2. It acts on one footpath only.
Torsion moment = 1.62 × 2 × (4.5 + 2/2) = 17.82 kNm/m
(ii). Lane 1:
(a). UDL = 4.86 kN/m^2.
Torsion moment = 4.86 × (Lane width = 3) × (4.5 – 3/2) = 43.74 kNm/m
(b). Four wheels each of 136.69 kN. The two axles are separated by 2 m.
Torsion moment = 2× 136.69 × (4.5 – 0.5) + 2× 136.69 × (4.5 – 2.5)
                          = 1640 kNm
Over one span total torsion moment = (17.82 + 43.74) × 40 + 1640 = 4102 kNm

The twisting moment is mainly resisted by the reactions at the supports. 50% of the total twisting moment is resisted by each support.
The shear flows in the webs are calculated from equation (9.27) in Chapter 9. The centre-line dimensions of the box are:
Height = 2500 – 300/2 – 400/2 = 2150 mm
Width = 9000 – 450 = 8550 mm
Area enclosed by centre line, A_k = 8550 × 2150 = 18.38 × 10^6 mm^2
Twisting moment T_{Ed} = 0.5 × 4102 = 2051 kNm
Shear flow q = T_{ED}/ (2 × A_k) = 2051 × 10^6 / (2 × 18.38× 10^6) = 56 N/mm
Shear force in the webs due to twisting moment = 56 ×2150 × 10^{-3} = 120.4 kN
Shear force in the flanges due to twisting moment = 56 ×8550 × 10^{-3} = 478.8.4 kN

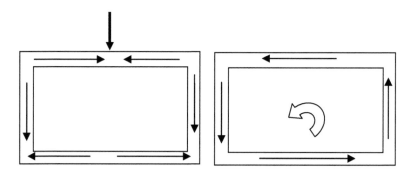

Fig. 16.93 Shear forces due to vertical loads and torsion

As shown in Fig. 16.93, the total shear force in the webs is the algebraic sum of the shear force due to vertical loads plus the shear force due to torsion. The shear force in one web is one half of the total shear force from self weight and live loads. The net shear force in a web from the shear force due to torsion and the shear force due to vertical loads is
At left support: 3756/ 2 +120.4 = 1998 kN
At mid-span: 2470/2 + a small amount from torsion = 1235 kN
At middle support: 9635/ 2 +120.4 = 4938 kN

16.14.10 Design of Shear and Torsional Reinforcement

Design and checking for shear reinforcement were detailed in Chapter 9. The steps to be followed are as follows.

(a). Calculation of $V_{Rd, c}$ at sections without shear reinforcement and uncracked in bending

The value of $V_{Rd, c}$ will be calculated by limiting the principle elastic tensile stress at the neutral axis of the composite section to permissible tensile stress f_{ctd}.
$f_{ck} = 40$ MPa, $f_{ctd} = 1.6$ MPa
(i). From equation (9.1), the compressive stress σ_{cp} due to prestress at supports at the neutral axis of the section is

$$\sigma_{cp} = \frac{P_s}{A} = \frac{28571 \times 10^3}{9.12 \times 10^6} = 3.13 \text{ MPa}$$

(ii). From equation (9.3), the shear stress τ that can be applied in combination with σ_{cp} without exceeding the permissible tensile stress equal to f_{ctd}. Assuming $\alpha_1 = 0.5$,

$$\tau = \sqrt{(f_{ctd}^2 + \alpha_1 \sigma_{cp} f_{ctd})} = \sqrt{(1.6^2 + 0.5 \times 3.13 \times 1.6)} = 2.25 \text{ MPa}$$

(iii). From equation (9.4), $\tau = V \times \dfrac{S}{I \times b_w}$

The value of S is the first moment of one half of the area about the *neutral axis of the beam* of the area of the beam above the neutral axis of the beam. Taking $y_t = 1185$ mm,
$S = (2000 + 4500) \times 300 \times (1185 - 300/2) = 2.02 \times 10^9$ mm^3 for the top flange
$\quad + (1185 - 300)^2 \times 450/2 = 0.176 \times 10^9$ mm^3 for the web
$\quad = 2.20 \times 10^9$ mm^3
$I_{beam} = 9.168 \times 10^{12}$ mm^4, $b_w = 450$ mm

$$\tau = 2.25 = V_{Rd, c} \times \frac{S}{I \times b_w} = V_{RD, c} \times 10^3 \times \frac{2.02 \times 10^9}{9.168 \times 10^{12} \times 450}$$

$$V_{RD, c} = 4595 \text{ kN}$$

From the net shear force envelope shown in Fig. 16.93, the shear force at the left support $V_{ED} = 1998$ kN which is less than $V_{Rd, c}$. Therefore no shear reinforcement is required.

(b). Start of cracked section
The start of the cracked section is calculated by limiting the maximum tensile stress at the soffit to f_{ctd} using the equation

$$f_{ctd} = -\frac{\gamma_{inf\ erior} \times P_s}{A} + [\frac{M - \gamma_{inf\ erior} \times P_s \times e_{effective}}{Z_b}]$$

Fig. 16.94 and Fig. 16.95 show respectively for cases 1 and 2 the net moment due to external loads and prestress acting at an effective eccentricity. The load factor on prestress is taken as $\gamma_{Inferior} = 0.95$.

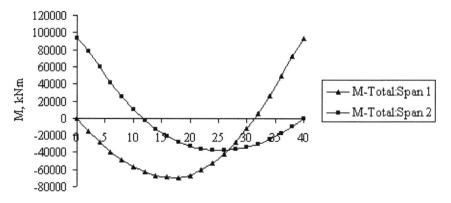

Fig. 16.94 Total moment in case 1

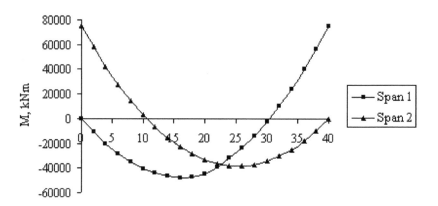

Fig. 16.95 Total moment in case 2

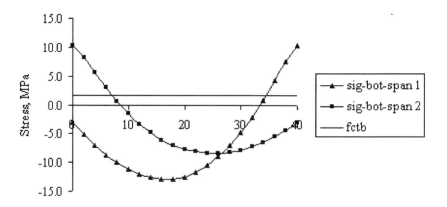

Fig. 16.96 Tensile stress at the soffit in case 1

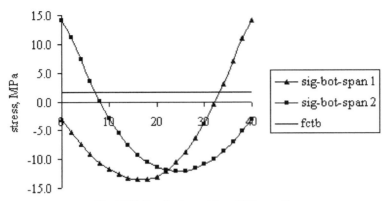

Fig. 16.97 Tensile stress at the soffit in case 2

Similarly Fig. 16.96 and Fig. 16.97 show respectively for cases 1 and 2 the stress distribution at the soffit. Clearly for both cases, the tensile stress limit of f_{ctd} = 1.6 MPa is exceeded only near the middle support for a distance of approximately 8 m. This region should be treated as cracked in flexure.

(c). Shear capacity of sections without shear reinforcement and cracked in bending

In the case of sections where the flexural tensile stress is greater than f_{ctd}, the shear capacity of the section is given by equation (9.6):

$C_{Rd, c} = 0.12$, $k_1 = 0.15$
$d \approx 2361$ mm

$$k = 1 + \sqrt{\frac{200}{d}} \leq 2.0, \quad k = 1.29 < 2.0$$

Taking f_{ck} = 40 MPa, $f_{cd} = f_{ck}/\gamma_m = 40/1.5 = 26.7$ MPa
$\sigma_{cp} = P_s/A = (\gamma_{inferior} = 0.95) \times 28571 \times 10^3/(9.12 \times 10^6) = 3.0 \leq (0.2 f_{cd} = 5.3)$
Taking $\sigma_{cp} = 3.0$ MPa

$$v_{min} = 0.035 k^{\frac{3}{2}} \sqrt{f_{ck}} = 0.035 \times 1.29^{\frac{3}{2}} \times \sqrt{40} = 0.32$$

A_{sl} = Area of tensile steel which extends a distance at least equal to anchorage length + d.
A_{sl} = 180 strands of 150 mm^2 = 27000 mm^2
b_w = Width of two webs = 2 × 450 mm
$\rho_t = A_{sl}/(b_w \times d) = 27000/(2 \times 450 \times 2361) = 0.013 \leq 0.02$
Substituting in equation (9.5), $V_{Rd, c}$ per web is

$$V_{Rd,c} = [C_{Rd,c} k (100 \, \rho_t \, f_{ck})^{\frac{1}{3}} + k_1 \, \sigma_{cp}] b_w d \geq [v_{min} + k_1 \, \sigma_{cp}] b_w d$$

$$V_{Rd,c} = \{[0.12 \times 1.29 \times (100 \times 0.013 \times 40)^{\frac{1}{3}} + 0.15 \times 3.0]$$

$$\geq [0.32 + 0.15 \times 3.0]\} \times 450 \times 2361 \times 10^{-3}$$

$$V_{Rd,c} = \{[0.58 + 0.45] \geq 0.77\} \times 1063 = 1095 \, kN$$

At the mid-span section, applied shear force $V_{ED} = 1235$ kN $> V_{RD,c}$, indicating that shear reinforcement is needed even at mid-span.

(d). Check adequacy of section

Taking $f_{ck} = 40$ MPa, $f_{cd} = f_{ck}/\gamma_m = 40/1.5 = 26.7$ MPa

$$v_1 = 0.6[1 - \frac{f_{ck}}{250}] = 0.50$$

$\sigma_{cp} = P_s/A = (\gamma_{inferior} = 0.95) \times 28571 \times 10^3/ (9.12 \times 10^6) = 3.0$ MPa

$$r = \frac{\sigma_{cp}}{f_{cd}} = \frac{3.0}{26.7} = 0.112$$

$$\alpha_{cw} = 1 + r = 1.112, \quad 0 < r \le 0.25$$

The maximum stress in the compression struts is σ_c,

$\sigma_c = \alpha_{cw} \, v_1 \, f_{cd} = 1.112 \times 0.50 \times 26.67 = 14.83$ MPa

Taking $z = 2270$ mm at the support as calculated in section 16.13.5 and equating the maximum applied shear force of 4932 kN per web to $V_{Rd,\,max}$,

$$V_{Rd,max} = \sigma_c b_w z \, \frac{1}{(\cot\theta + \tan\theta)} = 14.83 \times 450 \times 2270 \times 10^{-3} \times \frac{1}{(\cot\theta + \tan\theta)}$$

$$15149 \frac{1}{(\cot\theta + \tan\theta)} = 4938 \text{ kN}$$

$(\cot\theta + \tan\theta) = 3.07$

$t^2 - 3.07t + 1 = 0, \ t = \cot\theta$

Solving for $\cot\theta = 2.70 > 2.5$ permitted

Restricting $\cot\theta = 2.5, \ \theta = 21.8°$

$$V_{Rd,max} = \sigma_c b_w z \, \frac{1}{(\cot\theta + \tan\theta)} = 14.83 \times 450 \times 2270 \times 10^{-3} \times \frac{1}{(2.5 + 0.4)} = 5224 \text{ kN}$$

From (9.30), using the minimum thickness 300 mm for the top flange,

$T_{Rd,\,max} = 2 v \alpha_{cw} f_{cd} A_k t_{ef,i} \sin\theta \cos\theta$

$\qquad = 2 \times 0.50 \times 1.112 \times 26.7 \times 18.38 \times 10^6 \times 300 \times 0.3714 \times 0.93 \times 10^{-6}$

$\qquad = 56547$ kNm

Check adequacy using equation (9.29):

$$\frac{T_{ED}}{T_{Rd,\,max}} + \frac{V_{ED}}{V_{Rd,\,max}} = \frac{2051}{56547} + \frac{4938}{5224} = 0.98 \le 1.0$$

The section is adequate.

(e). Design shear reinforcement

The shear force resisted by vertical links is given by the equation (9.11)

$$V_{RD,s} = \frac{A_{sw}}{s} z \, f_{ywd} \cot\theta$$

Take $f_{ywd} = 460/1.15 = 400$ MPa, the design yield strength of shear reinforcement. Assuming 16 mm diameter two leg shear links in webs, $A_{sw} = 402$ mm^2.

(i). At the middle support section,

$4938 \times 10^3 = (402/s) \times 2270 \times 400 \times 2.5, \ s = 185$ mm

Provide links at 180 mm c/c.
(ii). At the mid-span section,
$1235 \times 10^3 = (402/s) \times 2298 \times 400 \times 2.5$, s =748 mm
Maximum spacing = 0.75 d = 0.75 × (d ≈ 2361) = 1771 mm.
Provide links at 750 mm c/c.

16.14.11 Longitudinal Reinforcement to Resist Torsion

Longitudinal reinforcement is required, in addition to link reinforcement to resist torsion. From equation (9.28), the centreline dimensions of the box are
Height = 2500 − 300/2 − 400/2 = 2150 mm
Width = 9000 − 450 = 8550 mm
Area enclosed by centre line, A_k = 8550 × 2150 = 18.38 × 10^6 mm^2
Twisting moment, T_{Ed} = 0.5 × 4102 = 2051 kNm
u_i = 2 × (2150 + 8550) = 21.4 × 10^3 mm
f_{yk} = 500 MPa, f_{yd} = 500/1.15 = 435 MPa
cot θ = 2.5

$$\Sigma A_{s\ell} = \frac{T_{ED}}{2A_k} \frac{u_k}{f_{yd}} \cot\theta = \frac{2051 \times 10^6}{2 \times 18.38 \times 10^6} \times \frac{21.4 \times 10^3}{435} \times 2.5 = 6862 \text{ mm}^2$$

Using 25 mm bars, provide 14 bars around the box, ΣA_{sl} = 6874 mm^2.

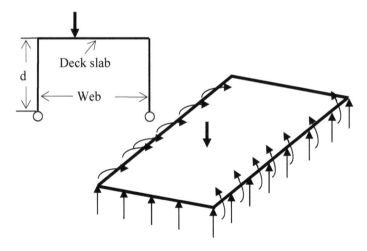

Fig. 16.98 Modelling of deck slab

16.14.12 Stress Analysis of the Deck

While an overall analysis of the box girder carried out using a finite element program can be used to obtain the detailed stress distribution in the deck slab, it is often too expensive. Frequently a simplified procedure such as the one suggested

here will in most cases be more than adequate. The deck can be analysed as a plate
supported on the webs. The restraint imposed on the deck slab by the web can be
simulated by providing rotational restraints as shown in Fig. 16.98. The stiffness
of the rotational springs is equal to 3EI/L, where L = centre-line depth of the web,
I = bt³/12, where t = thickness of the web, b = average width of the finite elements
on either side of a node and E = Young's modulus of the concrete. Fig. 16.99
shows the contour of the moment along the width of the deck and Fig. 16.100 the
distribution of the moment on a section containing the wheel load, when the deck
is subjected to maximum span bending moment as detailed in section 16.13.1.

The deck slab needs to be checked not only for bending stresses but also for
resistance to punching shear failure under wheel loads. The minimum slab
thickness should be at least 200 mm in order to avoid excessive reinforcement.

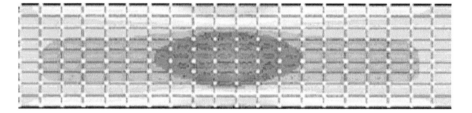

Fig. 16.99a Contour of moment M$_{xx}$ on the loaded part

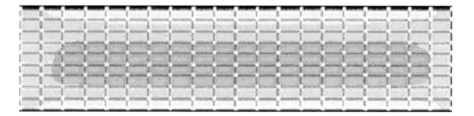

Fig. 16.99b Contour of moment M$_{xx}$ on the unloaded part

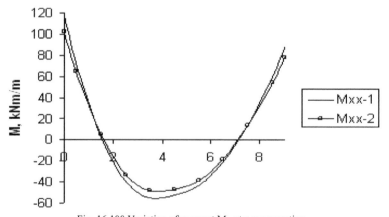

Fig. 16.100 Variation of moment M$_{xx}$ at a cross section

16.15 EUROCODE 2 RULES FOR REINFORCEMENT AT ANCHORAGES

In post-tensioned beams, the cables are anchored at the ends of the beam. Even in flanged beams such as T- or I-beams, the end part of the beam where the cables are anchored are often thickened to create a rectangular section. Fig. 16.101 shows in a simplified form that as the concentrated force in the anchor diffuses into the beam, it creates a tensile force which can cause bursting. In addition elastic analysis shows that tensile stresses normal to the anchor force cause spalling. As shown in Fig. 16.102, bursting force is resisted by steel reinforcement in the form of links or spirals and additional reinforcement at the face of the end block is needed to resist spalling.

Eurocode 2 suggests that a strut-tie model can be used to calculate the bursting force, as will be explained in Chapter 17. The force P can be assumed to disperse at an angle of 2β, where
$\beta = \tan^{-1}(2/3) = 33.7°$.
The force F in the concrete struts will be $F \approx 0.5\,P\,\sec\beta = 0.6\,P$ and the tie force will be $F\sin\beta \approx 0.33\,P$.

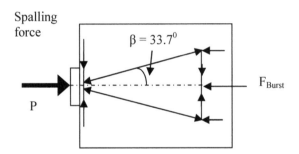

Fig. 16.91 Simplified force system in end blocks

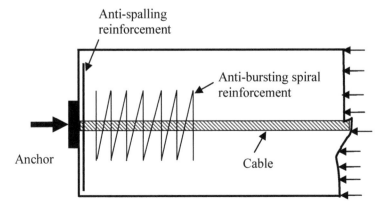

Fig. 16.102 Reinforcement in end blocks

If the dimensions of the anchor plate are a × a′, the dimensions c × c′ of the associate rectangle as shown in Fig. 16.103 are approximately 1.25 times the corresponding dimensions of the anchor plate. It is important that the associated rectangle should remain inside the concrete of the beam. In the case of multiple anchorages, the associate rectangles should not overlap. The associate prism represents approximately where the local stresses change from very high values to a reasonable value for concrete under uni-axial compression.

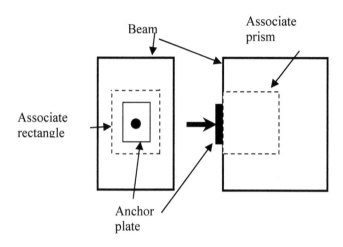

Fig. 16.103 Associate rectangle and anchor plate

The first check is about the bearing stress. It is required that bearing stress P_{max}/ (Area of associate rectangle) ≤ 0.6 f_{ck} (t)
where f_{ck} (t) = concrete cylinder strength at the time of stressing
P_{max} = Cross sectional area of cable × MIN {0.8 f_{pk}, 0.9 $f_{p0.1k}$}
Since $f_{p0.1k}$ ≈ 0.85 f_{pk}, 0.9 $f_{p0.1k}$ ≈ 0.77 f_{pk} will govern the maximum permissible stress in the cable at the time of jacking
As an example, if f_{pk} = 1862 MPa, the cross-sectional area of a cable with nineteen 13 mm seven-wire strands is 1839 mm²:
P_{max} = 1839 × 0.77 f_{pk} = 2637 kN
If the dimensions of the anchor plate are 280 mm square, then the dimensions of the associate rectangle are 1.25 × 280 = 350 mm square.
Bearing stress = 2637 × 10³/ (350 × 350) = 21.5 MPa ≤ 0.6 f_{ck} (t)
f_{ck} (t) = 35.9 MPa
The concrete at the time of stressing should have a minimum cylinder strength of about 36 MPa.
The minimum amounts of reinforcement to prevent bursting is as follows:

A_s ≥ 0.15 P_{max} × ($\gamma_{p, unfav}$ = 1.2) / (f_{yd} = 300 MPa) = 1583 mm²
Note that if f_{yd} = 300 MPa, then there is no need to check crack widths. Assuming 12 mm bars as links, the area of a two-leg link = 226 mm². Approximately seven links will suffice. These are distributed over a length of 1.2 × max {c, c′}. In this

case they are distributed over a length of $1.2 \times 350 = 420$ mm. Seven links at 70 mm c/c will provide the required area of reinforcement.

Reinforcement to prevent spalling should not be less than
$A_s \geq 0.03\ P_{max} \times (\gamma_{p,\ unfav} = 1.2) / (f_{yd} = 300\ \text{MPa}) = 317\ \text{mm}^2$
Four 12 mm bars both horizontally and vertically will provide the required area of reinforcement.

16.16 EXTERNAL AND INTERNAL TENDONS: A COMPARISON

In the example in section 16.13, the design was carried out on the assumption that the tendons used will be grouted internal tendons. Grouting will give good protection against corrosion when it is done well. Otherwise there is the danger of cables failing due to corrosion and it is not possible to replace the tendons. In recent years many bridges have been constructed using external tendons. The advantages of external tendons over internal tendons are several. Some of the important ones are:
- Tendons are replaceable
- Friction losses minimal and therefore the effective force in tendon will be larger leading to economy.
- As no internal ducts are needed, the whole process of fixing reinforcement and casting is considerably simplified. In addition as there is no need to allow for holes in the webs, they can be made thinner.
- Future upgrading of the bridge is simplified.

However, as opposed to the advantages listed above, the following are some of the drawbacks of external tendons:
- Reduced tendon eccentricity requires more prestressing steel.
- Integrity of anchorages becomes critical.
- At the ultimate limit state tendons rarely yield. This results in much reduced moment capacity.
- Fretting at internal deviators can be a problem.
- Exposed tendons are more prone to accidental damage.
- Large prestressing units require substantial deviators which cancel out any saving in concrete weight by using thin webs.

There are examples of bridges where a combination of internal and external tendons has been used in order to combine the advantages of both systems.

16.17 REFERENCES TO EUROCODE 2 CLAUSES

In this chapter, reference has been made to the following clauses in Eurocode 2:
Effective width of flanges: 5.3.2.1
Anchorage zones of post-tensioned members: 8.10.3(4)
Design with strut - tie models: 6.5

CHAPTER 17

LOWER BOUND APPROACHES TO DESIGN AT ULTIMATE LIMIT STATE

17.1 INTRODUCTION

Nowadays in many cases, stress analysis is carried out by finite element method. Although nonlinear analysis is sometimes attempted, the vast majority of analyses are elastic analysis for applied loads at the ultimate limit state. The stress output from the analysis is used in designing the reinforcement using the lower bound limit state theorem.

17.2 THEORY OF PLASTICITY

According to the theory of plasticity, for a correct solution at collapse the following three conditions should be satisfied.

1. Equilibrium condition: The state of stress and the applied load must be in equilibrium. If a structure is subjected to the loads at the ultimate limit state but the stress analysis is carried out at the elastic state, the state of stress will be in equilibrium with the applied loads at the ultimate limit state.

2. Yield condition: The state of stress must not violate the yield criterion for the material.

3. Collapse condition: At the ultimate limit state sufficient regions must yield to convert the structure into a mechanism indicating that the structure can carry no additional loads.

However if only conditions 1 and 2 are satisfied, then there may not be enough plastic regions to cause collapse. In such a case the ultimate load capacity calculated will be equal to or less than the true collapse load. It is therefore called a lower bound to the true collapse load. In the case of steel structures, the yield stress and cross-sectional properties will define the plastic resistance capacity. However, in the case of reinforced and prestressed concrete structures, if the yield criterion is known then for a given state of stress, the necessary reinforcement can be calculated in order to cause yielding. If the calculated reinforcement properly matches the stress state, then all parts of the structure yield and the structure will satisfy the collapse condition. However, practical restrictions prevent the exact matching between the reinforcement provided and the corresponding state of stress, resulting in a solution which will give a collapse load larger than the applied load. In other words, the resulting solution will be a lower bound solution. Designs based on lower bound approach have proved to be a very powerful approach and have become very popular.

17.3 IN-PLANE STRESSES

Fig. 17.1 shows a membrane element subjected to in-plane forces per unit length. The normal forces per unit length in the x- and y-directions are respectively N_x and N_y and the shear force is N_{xy}. The sign convention adopted is normal forces are positive tensile and shear forces are positive as shown in Fig. 17.1. Note that if t is the thickness of the element,

$$N_x = \sigma_x\, t,\ N_y = \sigma_y\, t,\ Nxy = \tau_{xy}\, t$$

where σ_x, σ_y and τ_{xy} are the normal and shear stresses in the element.

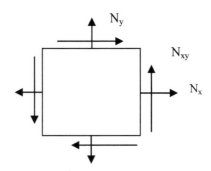

Fig. 17.1 In-plane stresses

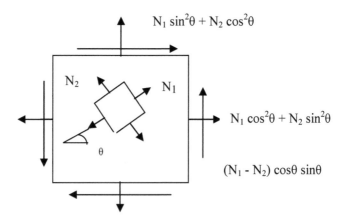

Fig. 17.2 Stresses in concrete

The applied stresses are resisted by a combination of stresses in concrete and steel reinforcement. Let the principal forces per unit length in concrete be N_1 and N_2 with N_1 inclined to the horizontal at an angle θ as shown in Fig. 17.2. The principal stresses result in normal and shear stresses as shown in Fig. 17.2. If the major and minor principal stresses in concrete are σ_1 and σ_2 respectively, then

$$N_1 = \sigma_1\, t,\ \ N_2 = \sigma_2\, t$$

If as shown in Fig. 17.3, reinforcements in the x- and y-directions of A_x and A_y respectively per unit length are provided, then the reinforcement will provide a resistance in the x- and y-directions equal to $A_x f_x$ and $A_y f_y$ respectively, where f_x and f_y are the axial stresses in steel.

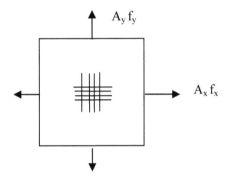

Fig. 17.3 Stresses in steel

Equilibrium requires that the resistance provided by steel and concrete should match the applied forces:

$N_x = (N_1 \cos^2 \theta + N_2 \sin^2 \theta) + A_x f_x$

$N_y = (N_2 \cos^2 \theta + N_1 \sin^2 \theta) + A_y f_y$

$N_{xy} = (N_1 - N_2) \cos\theta \, \sin \theta$ (17.1)

Four possible design situations need to be considered.

Case1: Stresses in concrete σ_1 and σ_2 are both compressive. In this case no reinforcement is required. The principal stresses are given by

$\sigma_1 = (\sigma_x + \sigma_y)/2 + \sqrt{[\{\sigma_x - \sigma_y \}/2\}^2 + \tau_{xy}{}^2]}$

$\sigma_2 = (\sigma_x + \sigma_y)/2 - \sqrt{[\{\sigma_x - \sigma_y \}/2\}^2 + \tau_{xy}{}^2]}$ (17.2)

The maximum compressive stress σ_2 should be limited to

$$\sigma_{cd\,max} = 0.85 f_{cd} \frac{(1 + 3.8\alpha)}{(1+\alpha)^2}$$ (17.3)

$\alpha = \sigma_1 / \sigma_2 \le 1$.

If $\sigma_2 > \sigma_{cd\,max}$, then the section thickness should be increased.

Case 2: If the major principal stress σ_1 is tensile, then ignoring the tensile strength of concrete, set σ_1 to zero. The equilibrium equations are then

$N_x = N_2 \sin^2 \theta + A_x f_x$,

$N_y = N_2 \cos^2 \theta + A_y f_y$

$N_{xy} = - N_2 \cos\theta \, \sin \theta$ (17.4)

Rearranging equation (17.4),

$(N_x - A_y\, f_x) = N_2 \sin^2 \theta$

$(N_y - A_y\, f_y) = N_2 \cos^2 \theta$

$N_{xy} = -\, N_2 \cos\theta\, \sin\theta$ (17.5)

Adding the first two equations from (17.4),

$N_x + N_y = N_2 + A_x\, f_x + A_y\, f_y$ (17.6)

Equating the product of the first two equations of (17.4) to the square of the third equation,

$(N_x - A_x\, f_x)\,(N_y - A_y\, f_y) = N_{xy}^2$ (17.7)

Equation (17.7) is known as the yield criterion for the in-plane case.

From (17.7), two cases can be considered.

Case 2a: No reinforcement in the x-direction is required.

Setting $A_x = 0$ and $f_y = f_{yd}$,

$$A_y\, f_{yd} = [\, N_y - \frac{N_{xy}^2}{N_x}]$$ (17.8)

Note $A_y = 0$, if $N_x\, N_y = N_{xy}^2$

Substituting $A_x = 0$ in (17.6)

$N_2 = N_x + N_y - A_y\, f_{yd}$

Substituting for $A_y\, f_{yd}$ from (17.8),

$$N_2 = N_x + N_y - [\, N_y - \frac{N_{xy}^2}{N_x}] = N_x + \frac{N_{xy}^2}{N_x}$$ (17.9)

$N_2 \le\ f_{cd}\, t$ and $N_{xy} < 0.5\, N_2$ (17.10)

Case 2b: No reinforcement in the y-direction is required.

Proceeding as in case 2a, setting $A_y = 0$ and $f_x = f_{yd}$,

$$A_x\, f_{yd} = [\, N_x - \frac{N_{xy}^2}{N_y}]$$ (17.11)

Note $A_x = 0$, if $N_x\, N_y = N_{xy}^2$

Substituting $A_y = 0$ in (17.6)

$N_2 = N_x + N_y - A_x\, f_{yd}$

Substituting for $A_x\, f_{yd}$ from (17.11),

$$N_2 = N_x + N_y - [N_x - \frac{N_{xy}^2}{N_y}] = N_y + \frac{N_{xy}^2}{N_y} \qquad (17.12)$$

$$N_2 \leq f_{cd} \, t \text{ and } N_{xy} \leq 0.5 \, N_2 \qquad (17.13)$$

Case 3: Both A_x and A_y are greater than zero.

Setting $f_x = f_y = f_{yd}$, minimize the total consumption of steel. Total steel S is

$$S = A_x \, f_{xd} + A_y \, f_{yd} \qquad (17.14)$$

Substituting for $A_y \, f_{yd}$ in (17.14) from equation (17.7)

$$S = A_x \, f_{yd} + N_y - \frac{N_{xy}^2}{(A_x \, f_{yd} - N_x)} \qquad (17.15)$$

Differentiating (17.15), with respect to $A_x \, f_{yd}$ and setting the differential coefficient to zero,

$$1 - \frac{N_{xy}^2}{(A_x \, f_{yd} - N_x)^2} = 0$$

$$A_x \, f_{yd} = N_x \pm N_{xy} \qquad (17.16)$$

Substituting (17.16) in (17.7),

$$A_y \, f_{yd} = N_y \pm N_{xy} \qquad (17.17)$$

Since $(A_x \, f_{yd})$ and $(A_y \, f_{yd})$ must both be positive,

$$A_x \, f_{yd} = N_x + |N_{xy}|, \quad A_y \, f_{yd} = N_y + |N_{xy}| \qquad (17.18)$$

Substituting equation (17.18) in equation (17.6) and simplifying,

$$N_2 = -2 \, |N_{xy}| \qquad (17.19)$$

Note from (17.18),

$$A_x = 0, \text{ when } \frac{N_x}{|N_{xy}|} = -1 \text{ and } A_y = 0, \text{ when } \frac{N_y}{|N_{xy}|} = -1 \qquad (17.20)$$

Fig. 17.4 summarises in a graphical form the design equations.

17.3.1 Examples of Reinforcement Calculations

In the following examples

$t = 200$ mm, $f_{yd} = 400$ MPa, $f_{ck} = 30$ MPa, $\gamma_m = 1.5$, $f_{cd} = f_{ck}/\gamma_m = 20.0$ MPa,

Maximum shear stress permitted = $f_{cd}/2 = 10.0$ MPa.

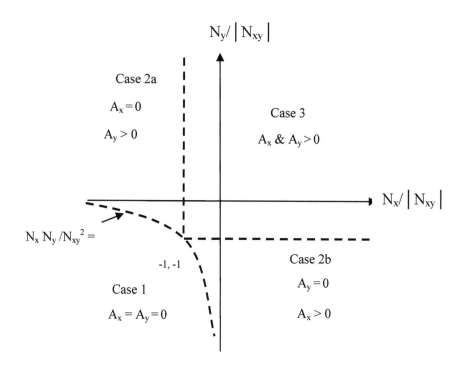

Fig. 17.4 Four cases of orthogonal reinforcement

Example 1a: Calculate the required reinforcement when $N_x = 4000$ N/mm, $N_y = -2000$ N/mm and $N_{xy} = -1200$ N/mm.

$N_x / |N_{xy}| = 3.3$, $N_y / |N_{xy}| = -1.7$, $N_x N_y / N_{xy}^2 = -5.6$

This combination falls into case 2b. Therefore

$$A_x \times 400 = [N_x - \frac{N_{xy}^2}{N_y}]t = [4000 - \frac{1200^2}{(-2000)}], A_x = 11.8 \, mm^2 / mm$$

Providing 25 mm bars at 40 mm c/c provides $A_x = 12.3$ mm²/mm.

$A_y = 0$

$N_2 = [N_y + N_{xy}^2/ N_y] = -2720$ N/mm, $\sigma_2 = N_2/ t$, $\sigma_2 = -13.6$ MPa

Maximum compressive stress σ_2 is less than f_{cd}.

Example 1b: Calculate the required reinforcement when $N_x = -2400$ N/mm, $N_y = 3000$ N/mm, $N_{xy} = 1600$ N/mm

$N_x / |N_{xy}| = -1.5$, $N_y / |N_{xy}| = 1.9$, $N_x N_y / N_{xy}^2 = -2.8$

This combination falls into case 2a. Therefore $A_x = 0$

$$A_y \times 400 = [N_y - \frac{N_{xy}^2}{N_x}] = [3000 - \frac{1600^2}{(-2400)}], \quad A_y = 10.2 \, \text{mm}^2 / \text{mm}$$

Providing 25 mm bars at 45 mm c/c provides $A_x = 10.9 \, \text{mm}^2/\text{mm}$.

$N_2 = [N_x + N_{xy}^2/ N_x] = -3467 \, \text{N/mm}$, $\sigma_2 = N_2/ t = -17.3 \, \text{MPa}$

Maximum compressive stress σ_2 is less than f_{cd}.

Example 1c: Calculate the required reinforcement when $N_x = 2000 \, \text{N/mm}$, $N_y = 3000 \, \text{N/mm}$, $\tau_{xy} = 1800 \, \text{N/mm}$

$N_x / |N_{xy}| = 1.1$, $N_y / |N_{xy}| = 1.7$, $N_x N_y / N_{xy}^2 = 1.85$

This combination falls into case 3. Therefore

$A_x \times 400 = [N_x + |N_{xy}|]$, $A_x = 9.5 \, \text{mm}^2/ \text{mm}$

$A_y \times 400 = [N_y + N_{xy}|]$, $A_x = 12.0 \, \text{mm}^2/ \text{mm}$

Providing 25 mm bars at 50 mm c/c provides $A_x = 9.8 \, \text{mm}^2/\text{mm}$ and at 40 mm c/c provides $A_y = 12.3 \, \text{mm}^2/\text{mm}$.

$N_2 = -2 |N_{xy}| = -3600 \, \text{N/mm}$, $\sigma_2 = N_2/ t = -18 \, \text{MPa}$

Maximum compressive stress σ_2 is less than f_{cd}.

Example 1d: Calculate the required reinforcement when $N_x = -2400 \, \text{N/mm}$, $N_y = -3000 \, \text{N/mm}$, $N_{xy} = 2000 \, \text{N/mm}$

$N_x / |N_{xy}| = -1.2$, $N_y / |N_{xy}| = -1.5$, $N_x N_y / N_{xy}^2 = 1.8$

This combination falls into case 1. Therefore $A_x = A_y = 0$

$$N_1 = (N_x + N_y)/2 + \sqrt{[\{(N_x - N_y)/2\}^2 + N_{xy}^2]} = -678 \, \text{N/mm}$$
$$N_2 = (N_x + N_y)/2 - \sqrt{[\{(N_x - N_y)/2\}^2 + N_{xy}^2]} = = -4722 \, \text{N/mm}$$

$\sigma_1 = N_1/ t = -3.4 \, \text{MPa}$, $\sigma_2 = N_2/ t = -23.6 \, \text{MPa}$

Example 2: Fig. 17.5 shows a box girder 1.2 m wide $\times$ 1.7 m deep with 200 mm thick walls. At a cross section, it is subjected to the following load combination: torsional moment T = 1000 kNm, bending moment M = 1500 kNm and

Shear force V = 900 kN.

Design the required reinforcement given that: $f_{yd} = 400 \, \text{MPa}$, $f_{ck} = 30 \, \text{MPa}$, $f_{cd} = 20.0 \, \text{MPa}$.

Carry out an elastic stress analysis

(a). Torsion:

Centre-line dimensions of the box:

Width = 1.2 – 0.2 = 1.0 m,

Depth = 1.7 – 0.2 = 1.5 m

Area A_k enclosed by the centre lines = 1.0 × 1.5 = 1.5 m²

From equation (9.27), the shear flow in the sides of the box, as shown in Fig. 17.5 is

$N_{xy} = q = T/ (2 \times A_k) = 1000 / (2 \times 1.5) = 333$ kN/m = 333 N/mm

(b). Shear force:

Second moment of area, $I = [1.2 \times 1.7^3 - (1.2 - 0.4) \times (1.7 - 0.4)^3]/ 12 = 0.345$ m⁴

From equation (9.2), determine the shear stress distribution at three levels.

In the top flange at the flange-web junction,

First moment of area S = 1.2 × 0.2 × (1.7/2 – 0.2/2) = 0.18 m³

$\tau = 900 \times 0.18/ (0.345 \times 1.2) = 391$ kN/m² = 0.4 MPa

In the web at the flange-web junction,

$\tau = 900 \times 0.18/ (0.345 \times 2 \times 0.2) = 1174$ kN/m² = 1.2 MPa

At the centroidal axis,

$S = 1.2 \times (1.7/2)^2 \times 0.5 - (1.2 - 0.4) \times \{ (1.7 - 0.4)/2\}^2 \times 0.5 = 0.2645$ m³

$\tau = 900 \times 0.2645/ (0.345 \times 2 \times 0.2) = 1725$ kN/m² = 1.7 MPa

The shear stress distribution in the cross section is shown in Fig. 17.6. As the shear stress distribution in the web is parabolic, the average stress in the web is approximately 1.2 + (1.7 – 1.2) × (2/3) = 1.5 MPa. The multiplication factor (2/3) comes from the fact that the area of the parabola is (2/3) the area of the enclosing rectangle.

$N_{xy} = 1.5 \times 200 = 300$ N/mm

(c). Bending stress:

The bending stress distribution in the cross section is shown in Fig. 17.6.

Stress at bottom and top = ±1500 × (1.7/2)/ 0.345 = 3696 kN/m² = ± 3.7 MPa

(d). Stress distribution in the web:

The shear stresses due to torsion and shear are of opposite signs in the web on the left and are of the same sign in the web on the right as shown in Fig. 16.93. Design will be based on the larger shear stress in order to keep the reinforcement symmetrical. As the bending stress varies from zero at the neutral axis to a maximum value at the top and bottom fibres, only an average value is used in design.

$N_{xy} = 333$ (torsion) + 300 (shear) = 633 N/mm

In the top half of the web, the average stress $\sigma_x \approx -3.7/2 = 1.85$ MPa

$N_x = -1.85 \times 200 = -370$ N/mm

Similarly in the bottom half of the web,

$N_x = 1.85 \times 200 = 370$ N/mm, $N_y \approx 0$

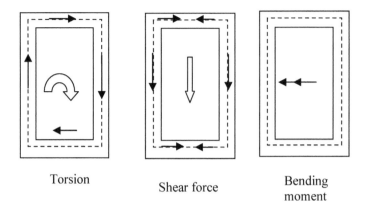

Torsion

Shear force

Bending moment

Fig. 17.5 Forces at the cross section

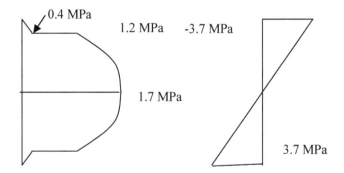

0.4 MPa

1.2 MPa   -3.7 MPa

1.7 MPa

3.7 MPa

Fig. 17.6 Shear and bending stress distribution in the cross section

(e). Reinforcement in the top half of the web:

$N_x / |N_{xy}| = -370/633 = -0.59$, $N_y / |N_{xy}| = 0$, $N_x N_y / N_{xy}^2 = 0$.

This combination falls into case 3.

$A_x \times 400 = N_x + |N_{xy}|$, $A_x = 0.66$ mm²/mm

Total reinforcement over a depth $1.7/2 = 0.85$m is $0.66 \times 0.85 \times 10^3 = 559$ mm².
Provide three number 16 mm bars giving a total area of 603 mm².

$A_y \times 400 = N_y + N_{xy}|$, $A_y = 1.6$ mm²/mm.

Provide 16 mm diameter links at 125 mm c/c giving A_y = 1.61 mm^2/mm.

$N_2 = -2 |N_{xy}| = -1266$ N/mm, $\sigma_2 = N_2/t = -6.3$ MPa

The maximum compressive stress is less than f_{cd}.

(f). Reinforcement in the bottom half of the web:

$N_x / |N_{xy}| = 370/633 = 0.59$, $N_y / |N_{xy}| = 0$, $N_x N_y /N_{xy}^2 = 0$.

This combination falls into case 3.

$A_x \times 400 = N_x + |N_{xy}|$, $A_x = 2.51$ mm^2/mm

Total reinforcement over a depth $1.7/2 = 0.85$m is $2.51 \times 0.85 \times 10^3 = 2131$ mm^2.
Provide five number 25 mm bars giving a total area of 2454 mm^2.

$A_y \times 400 = [N_y +N_{xy}|]$, $A_y = 1.6$ mm^2/mm.

Provide 16 mm diameter links at 125 mm c/c giving A_y = 1.61 mm^2/mm.

$N_2 = -2 |N_{xy}| = -1266$ N/mm, $\sigma_2 = N_2/t = -6.3$ MPa

(g). Stress distribution in the flanges:

The shear stresses due to torsion and shear force are of opposite signs on the right half of the flange and are of the same sign on the left half of the flange. Design will be based on the larger shear stress in order to keep the reinforcement symmetrical. As the shear stress due to shear force varies from zero at the centre to a maximum value of 0.4 MPa at the edge, only an average value is used in design.

(h). Reinforcement in the top flange:

$N_{xy} = 333$ (torsion) $+ (0.4/2) \times 200$ (shear) $= 373$ N/mm

$N_x \approx -3.7 \times 200 = -740$ N/mm $\sigma_y \approx 0$

$N_x / |N_{xy}| = -2.0$, $N_y / |N_{xy}| = 0$, $N_x N_y /N_{xy}^2 = 0$

This falls into case 2a. There isd no reinforcement in the axial direction.

$A_y \times 400 = N_y +N_{xy}|$, $A_y = 0.93$ mm^2/mm.

Provide 16 mm diameter links at 200 mm c/c giving A_y = 1.01 mm^2/mm.
However the link diameter and spacing determined from the reinforcement calculation in the web will govern the final value.

$N_2 = -2 |N_{xy}| = -746$ N/mm, $\sigma_2 = N_2/t = -3.7$ MPa

(i). Reinforcement in the bottom flange:

$N_{xy} = 333$ (torsion) $+ (0.4/2) \times 200$ (shear) $= 373$ N/mm

$N_x \approx 3.7 \times 200 = 740$ N/mm, $N_y \approx 0$

$N_x / |N_{xy}| = 2.0$, $N_y / |N_{xy}| = 0$, $N_x N_y /N_{xy}^2 = 0$

This falls into case 3.

$A_x \times 400 = N_x + |N_{xy}|$, $A_x = 2.8$ mm^2/mm

Over a width of 1.2 m, this requires $A_x = 2.8 \times (1200 - 400) = 2240$ mm^2. Provide five 25 mm bars giving a total area of 2450 mm^2.

$A_y \times 400 = N_y + |N_{xy}|$, $A_y = 0.93$ mm^2/mm.

Provide 16 mm diameter links at 200 mm c/c giving $A_y = 1.01$ mm^2/mm. However the link diameter and spacing determined from the reinforcement calculation in the web will govern the final value.

$N_2 = -2|N_{xy}| = -746$ N/mm, $\sigma_2 = N_2/t = -3.7$ MPa

Fig. 17.7 shows a possible arrangement of reinforcement.

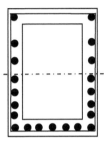

Fig. 17.7 Schematic representation of reinforcement

17.3.2 Presence of Prestressing Cables

If part of the in-plane stresses is provided by prestressing cables, then equations (17.1) are modified by replacing $A_x f_{yd}$ and $A_y f_{yd}$ by $\{A_x f_{yd} + A_{xp} \times (f_{pd} - f_{pex})\}$ and $\{A_y f_{yd} + A_{yp} \times (f_{pd} - f_{pey})\}$ respectively. In these equations, A_x and A_y refer to areas of unstressed steel in the x- and y-directions respectively and A_{xp} and A_{yp} refer to areas of stressed steel. f_{pex} and f_{pey} refer to prestress in the cables in x and y directions respectively and f_{yd} and f_{pd} refer to the design stresses in unstressed and stressed steel respectively.

Example 1: $t = 200$ mm, $\sigma_x = 10$ MPa, $\sigma_y = 15$ MPa, $\tau_{xy} = 9$ MPa

Prestress in the x-direction only is -2.0 MPa provided by $A_{xp} = 0.37$ mm^2/mm with the cables stressed to $f_{pe} = 1068$ MPa.

$f_{pd} = 1617$ MPa, $f_{yd} = 400$ MPa

$N_x = (10 - 2.0) \times 200 = 1600$ N/mm, $N_y = 15 \times 200 = 3000$ N/mm

$N_{xy} = 9 \times 200 = 1800$ N/mm

$N_x/|N_{xy}| = 0.89$, $N_y/|N_{xy}| = 1.7$, $N_x N_y/N_{xy}^2 = 1.48$

This combination falls into case 3.

$A_{xp} \times (f_{pd} - f_{pe}) = 0.37 \times (1617 - 1068) = 203$ mm^2/ mm

$A_x f_{yd} + A_{xp} \times (f_{pd} - f_{pe}) = N_x + |N_{xy}| = 3400$, $A_x = 8.0$ mm^2/mm

25 mm bars at 60 mm c/c provide $A_x = 8.18$ mm^2/mm

Note that in Example 3 in section 17.3.1, the required unstressed area was 25 mm bars at 40 mm c/c. Prestressing has reduced the required unstressed steel area.

$A_y f_y = N_y + |N_{xy}|$, $A_y = 12.0$ mm^2/mm

25 mm bars at 40 mm c/c provide $A_y = 12.3$ mm^2/mm.

17.4 DESIGNS FOR A COMBINATION OF IN-PLANE AND FLEXURAL FORCES

In general elements are subjected to a combination of

- In-plane forces N_x, N_y and N_{xy} (Fig. 17.1)

- Flexural forces M_x, M_y and M_{xy} (Fig. 17.8)

- Out-of-plane shear forces V_x and V_y (Fig. 17.9)

For design purposes, the slab is divided into three layers as shown in Fig. 17.10. The top and bottom layers are designed to resist the forces caused by in-plane and bending forces while the middle layer is designed to resist the out-of-plane shear forces.

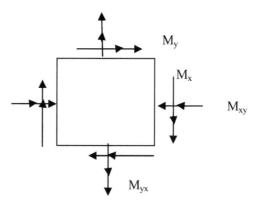

Fig. 17.8 Flexural forces

As shown in Fig. 17.11, the in-plane force N_x can be distributed to the top and bottom layers but keeping the resultant at the centroid of the whole slab. If N_{xs} and N_{xi} are the forces distributed into the top and bottom layers, and y_{xs} and y_{xi} are the distances to the centre of the steel layers in the y-direction in the top and bottom of the slab, then

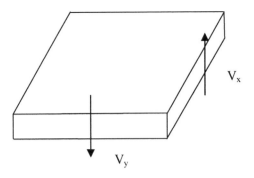

Fig. 17.9 Out-of-plane shear forces

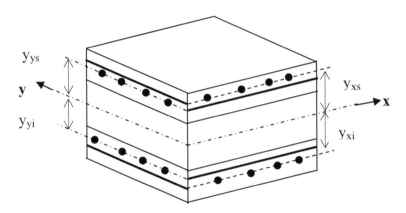

Fig. 17.10 Division of slab into layers

$$y_{xs} + y_{xi} = z_x, \ N_{xs} + N_{xi} = N_x \tag{17.21}$$

For the resultant of N_{xs} and N_{xi} to coincide with the centroidal axis of the slab,

$$N_{xs} \times y_{xs} = N_{xi} \times y_{xi}$$

Solving for N_{xs} and N_{xi}

$$N_{xs} = N_x \times y_{xi}/z_x = N_x \{z_x - y_{xs}\}/z_x$$

$$N_{xi} = N_x \times y_{xs}/z_x = N_x \{z_x - y_{xi}\}/z_x \tag{17.22}$$

Similarly, the moment M_x can be replaced by a compressive force $-M_x/z_x$ and a tensile force M_x/z_x in the top and bottom layers acting at a lever arm z_x.

Thus the combined effect of in-plane force N_x and bending moment M_x result in in-plane forces in the top and bottom layers given by

$$N_{Edxs} = N_x \{z_x - y_{xs}\}/z_x - M_x/z_x$$

$$N_{Edxi} = N_x \{z_x - y_{xi}\}/z_x + M_x/z_x \tag{17.23}$$

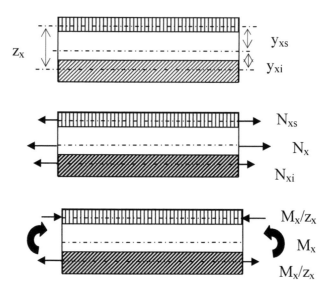

Fig. 17.11 Distribution of in-plane and moment forces to layers

Similarly, the force N_y and moment M_y can be replaced by

$$N_{Edys} = N_y \{z_y - y_{ys}\}/z_y - M_y/z_y$$
$$N_{Edyi} = N_y \{z_y - y_{yi}\}/z_y + M_y/z_y \qquad (17.24)$$

where y_{ys} and y_{yi} are respectively the distances to the centre of the steel layers in the y-direction in the top and bottom of the slab.

In a similar manner force combinations (N_{xy}, M_{xy}) and (N_{yx}, M_{yx}) can also be replaced by in-plane forces in the top and bottom layers in the x- and y-directions:

$$N_{Edxys} = N_{xy} \{z_{xy} - y_{xys}\}/z_{xy} - M_{xy}/z_{xy}$$
$$N_{Edxyi} = N_{xy} \{z_{xy} - y_{xyi}\}/z_{xy} + M_y/z_{xy}$$
$$N_{Edyxs} = N_{xy} \{z_{yx} - y_{yxs}\}/z_{yx} - M_{xy}/z_{yx}$$
$$N_{Edyxi} = N_{xy} \{z_{yx} - y_{yxi}\}/z_{yx} + M_y/z_{yx} \qquad (17.25)$$

Although (z_x, y_{xs}, y_{xi}) will not be equal to (z_y, y_{ys}, y_{yi}) because of the fact that the steel layers in the x- and y-directions cannot coincide, for practical calculations it is permissible to assume that the thicknesses of the outer layers are approximately equal to twice the concrete cover and adopt a common value for the lever arm values in the two directions. Thus the following simplifications are permitted:

$$y_s = y_{xs} = y_{ys} = y_{xys} = y_{yxs}, \, y_i = y_{xi} = y_{yi} = y_{xyi} = y_{yxi}, \, z = z_x = z_y = z_{xy} = z_{yx}$$

17.4.1 Example of Design for a Combination of In-plane and Flexural Forces

Example 1: Total thickness of slab = 300 mm, cover to reinforcement = 30 mm, reinforcing bars = 20 mm diameter, the design stress in steel f_{yd},= 400 MPa,

f_{ck} = 30 MPa, f_{cd} = 20.0 MPa

N_x = 1000 kN/m, N_y = 1400 kN/m, N_{xy} = 900 kN/m

M_x = 60 kNm/m, M_y = 100 kN/m, M_{xy} = 60 kN/m

Assume top and bottom layers = 80 mm thick

z = 300 – (80 + 80)/2 = 220 mm

y_s = y_i = (300 -80)/2 = 110 mm

Note: 1 kN/m = 1 N/mm, 1 kNm/m = 1 × 10^3 Nmm/mm

N_{xs} = 1000 (220 – 110)/220 - 60×10^3 /220 = 227 N/mm

N_{xi} = 1000 (220 – 110)/220 + 60×10^3 /220 = 773 N/mm

N_{ys} = 1400 (220 – 110)/220 - 100×10^3 /220 = 246 N/mm

N_{yi} = 1400 (220 – 110)/220 + 100×10^3 /220 = 1155 N/mm

N_{xys} = 900 (220 – 110)/220 - 60×10^3 /220 = 177 N/mm,

N_{xyi} = 900 (220 – 110)/220 + 60×10^3 /220 = 723 N/mm

Reinforcement design:

Top layer:

N_{xs} = 227 N/mm, N_{ys} = 246 N/mm, N_{xys} = 177 N/mm

$N_x/ \left| N_{xy} \right|$ = 1.3, $N_y/ \left| N_{xy} \right|$ = 1.4, $N_x N_y / N_{xy}^2$ = 1.8

This corresponds to case 3.

$A_x \times 400 = N_x + \left| N_{xy} \right|$, A_x = 1.0 mm^2/mm,

20 mm bars at 300 mm c/c provide 1.05 mm^2/mm of steel area.

$A_y \times 400 = N_y + \left| N_{xy} \right|$, A_y = 1.06 mm^2/mm,

20 mm bars at 290 mm c/c provide 1.08 mm^2/mm of steel area.

$N_2 = - 2 \left| N_{xy} \right|$ = 354, σ_2 = N_2/ t = −4.4 MPa

Bottom layer:

N_{xi} = 773 N/mm, N_{yi} = 1155 N/mm, N_{xyi} = 723 N/mm

$N_x/ \left| N_{xy} \right|$ = 1.08, $N_y/ \left| N_{xy} \right|$ = 1.6, $N_x N_y /N_{xy}^2$ = 1.72

This corresponds to case 3.

$A_x \times 400 = N_x + \left| N_{xy} \right|$, A_x = 3.74 mm^2/mm,

20 mm bars at 80 mm c/c provide 3.93 mm^2/mm of steel area.

$A_y \times 400 = N_y + |N_{xy}|$, $A_y = 4.68$ mm²/mm,

20 mm bars at 65 mm c/c provide 4.83 mm²/mm of steel area.

$N_2 = -2|N_{xy}| = -1446$, $\sigma_2 = N_2/t = -18.1$ MPa

17.5 CRITERION FOR CRACKING

For a given set of stresses in two-dimensional problems, normally it is sufficient to calculate the principal stresses and if the major principal stress is tensile, then the section can be considered as cracked. However, to judge cracking, Eurocode 2 uses a more elaborate yield criterion Φ expressed in terms of stress invariants (I_1, J_2 and J_3) as follows:

$$\Phi = \{\alpha\frac{J_2}{f_{cm}^2} + \lambda\frac{\sqrt{J_2}}{f_{cm}} + \beta\frac{I_1}{f_{cm}} - 1\} \leq 0 \tag{17.26}$$

Stress invariants:

$$I_1 = [\sigma_1 + \sigma_2 + \sigma_3], \text{ Average stress, } \sigma_m = I_1/3$$

$$J_2 = \frac{1}{6}[(\sigma_1 - \sigma_2)^2 + (\sigma_2 - \sigma_3)^2(\sigma_3 - \sigma_1)^2]$$

$$J_3 = [(\sigma_1 - \sigma_m)(\sigma_2 - \sigma_m)(\sigma_3 - \sigma_m)]$$

$$\cos 3\theta = [\frac{3\sqrt{3}}{2}]\frac{J_3}{J_2^{1.5}} \tag{17.27}$$

Material parameters:

$$k = \frac{f_{ctm}}{f_{cm}}, f_{cm} = f_{ck} + 8, f_{ctm} = 0.3 f_{ck}^{\frac{2}{3}}$$

$$\alpha = \frac{1}{9k^{1.4}}, \beta = \frac{1}{3.7k^{1.1}}, c_1 = \frac{1}{0.7k^{0.9}}, c_2 = 1 - 6.8(k - 0.07)^2$$

$$\lambda = c_1 \cos[\frac{1}{3}\cos^{-1}(c_2 \cos 3\theta)], \cos 3\theta \geq 0$$

$$\lambda = c_1 \cos[\frac{\pi}{3} - \frac{1}{3}\cos^{-1}(-c_2 \cos 3\theta)], \cos 3\theta < 0 \tag{17.28}$$

The following two examples make a comparison between using the yield criterion and principal tension criterion for judging cracked state.

Example 1: $\sigma_x = -12$, $\sigma_y = -8$, $\tau_{xy} = 12$ MPa

Concrete strength: $f_{ck} = 40$ MPa, $f_{cm} = f_{ck} + 8 = 48$, $f_{ctm} = 0.3 \times 40^{2/3} = 3.5$

The principal stresses are:

$\sigma_1 = (\sigma_x + \sigma_y)/2 + \sqrt{[\{(\sigma_x - \sigma_y)/2\}^2 + \tau_{xy}^2]} = -10.0 + 12.2 = 2.2$ MPa

$\sigma_2 = (\sigma_x + \sigma_y)/2 - \sqrt{[\{(\sigma_x - \sigma_y)/2\}^2 + \tau_{xy}^2]} = -10.0 - 12.2 = -22.2$ MPa

Comparing $\sigma_1 = 2.2$ to f_{ctm} the slab is uncracked.

Applying the criterion given by Eurocode 2,

$\sigma_1 = 2.2$, $\sigma_2 = -22.2$, $\sigma_3 = 0$ as it is a two-dimensional problem.

$I_1 = -20.0$, $\sigma_m = -6.67$, $J_2 = 181.33$, $J_3 = -912.59$,

$f_{cm} = 48$ MPa, $f_{ctm} = 0.3 \times 40^{2/3} = 3.5$, $k = 3.5/48 = 0.0732$,

$\alpha = 4.32$, $\beta = 4.796$, $c_1 = 15.078$, $c_2 \approx 1.0$

$\cos 3\theta = -0.9709$, $\cos^{-1}(-c_2 \cos 3\theta) = 13.86° = 0.077\,\pi$ radians

$[\frac{\pi}{3} - \frac{1}{3}\cos^{-1}(-c_2 \cos 3\theta)] = [\frac{\pi}{3} - \frac{1}{3} \times 0.077\,\pi] = 0.308\pi$ radians $= 55.6°$

$\cos(55.4°) = 0.568$, $\lambda = 15.078 \times 0.568 = 8.57$

$\Phi = -0.27 < 0$: The slab is uncracked.

Example 2: $\sigma_x = -8$, $\sigma_y = -8$, $\tau_{xy} = 12$ MPa

The principal stresses are:

$\sigma_1 = (\sigma_x + \sigma_y)/2 + \sqrt{[\{\sigma_x - \sigma_y)/2\}^2 + \tau_{xy}^2]} = -8.0 + 12.0 = 4.0$ MPa

$\sigma_2 = (\sigma_x + \sigma_y)/2 - \sqrt{[\{(\sigma_x - \sigma_y)/2\}^2 + \tau_{xy}^2]} = -8.0 - 12.0 = -20.0$ MPa

Comparing $\sigma_1 = 4.0$ to f_{ctm} the slab is cracked.

Applying the criterion given by Eurocode 2,

$\sigma_1 = 4.0$, $\sigma_2 = -20.0$, $\sigma_3 = 0$ as it is a two dimensional problem.

$I_1 = -16.0$, $\sigma_m = -5.33$, $J_2 = 165.33$, $J_3 = -730.07$

The rest of the parameters are as for Example 1.

$\cos 3\theta = -0.892$, $\cos^{-1}(-c_2 \cos 3\theta) = 26.9° = 0.149\,\pi$ radians

$[\frac{\pi}{3} - \frac{1}{3}\cos^{-1}(-c_2 \cos 3\theta)] = [\frac{\pi}{3} - \frac{1}{3} \times 0.149\,\pi] = 0.284\pi$ radians $= 51.04°$

$\cos(51.04°) = 0.629$, $\lambda = 15.078 \times 0.629 = 9.45$

$\Phi = 0.24 > 0$: The slab is cracked.

From these two examples it can be concluded that both the principal tensile stress criterion and the Eurocode 2 criterion predict correctly if the slab is cracked or not under the given state of two-dimensional stress system.

17.6 OUT-OF-PLANE SHEAR

Having calculated the reinforcement in the outer layers to resist in-plane forces and flexural and torsional moments, it is necessary to check the shear resistance of the section. It is assumed that only the core resists the out-of-plane shear forces. This is reasonable as the maximum shear stress occurs at the centroidal axis of the slab and almost towards the outer layers.

Example 1: Check the shear resistance of the slab given that

$V_{edx} = 120$ N/mm, $V_{edy} = 180$ N/mm

Calculate the resultant shear force V_{Ed0}:

$V_{Ed0} = \sqrt{(V_{edx}^2 + V_{edy}^2)} = 216$ N/mm

$\tan \varphi_0 = V_{edy}/ V_{edx} = 1.5$, $\varphi_0 = 56°$

From the example in section 17.4.1, the tension steel in the bottom layers comprises 20 mm bars at 80 mm c/c in the x-direction and 20 mm bars at 65 mm c/c in the y-direction. This gives $A_{sx} = 3.93$ mm^2/mm and $A_{sy} = 4.83$ mm^2/mm. Taking the effective depth as an average value of $300 - 40$ mm cover $= 260$ mm, the steel ratios in the x- and y-directions are

$\rho_x = 3.93 / 360 = 1.5\%$ and $\rho_y = 4.83 / 360 = 1.9\%$

$\cos \varphi_0 = 0.56$, $\sin \varphi_0 = 0.83$

$\rho_t = (\rho_x \cos^2\varphi_0 + \rho_y \sin^2\varphi_0) = 1.78\% < 2\%$

In the case of sections where the flexural tensile stress is greater than f_{ctd} and no shear reinforcement is provided, the shear capacity of the section is given by (9.6), Chapter 9:

$$V_{Rd,c} = [C_{Rd,c} k (100 \rho_t f_{ck})^{\frac{1}{3}} + k_1 \sigma_{cp}] b_w d \geq [v_{min} + k_1 \sigma_{cp}] b_w d$$

where $C_{Rd,c} = 0.12$
d = effective depth to steel in tensile zone = 260 mm (average)
b_w = width of web, $k_1 = 0.15$.
$\sigma_{cp} = 0$ as there is no axial compressive stress in the middle layer as this is distributed to top and bottom layers.

$$k = 1 + \sqrt{\frac{200}{(d=260)}} = 1.77 \leq 2.0$$

$$v_{min} = 0.035 k^{\frac{3}{2}} \sqrt{f_{ck}} = 0.035 \times 0.77^{1.5} \times \sqrt{40} = 0.15$$

$$V_{Rd,c} = [0.12 \times 1.77 \times (1.78 \times 40)^{\frac{1}{3}}] 260 \geq [0.15]260$$

$$= [0.88] \times 260 = 228 N/mm > V_{Ed0} = 216$$

Therefore no shear reinforcement is required. The unreinforced core can resist the applied shear force.

17.7 STRUT-AND-TIE METHOD OF DESIGN

One very popular method of design at the ultimate limit state is the strut-and-tie method. It is based on the lower bound principle explained in section 17.2. In any structure subjected to loads, the stress system consists of a complex system of tensile and compressive stresses in the principal stress directions. The complex stress system can be replaced by a system of concrete struts and steel ties. From the point of view of the lower bound principle, all that is required is any set of stresses which can maintain equilibrium with the external loads. The state of stress needs to satisfy neither the material laws nor continuity of strains, etc. Having replaced the complex stress system by a set of struts and ties and further if one can specify the dimensions of the struts, then all one has to do is to ensure that the compressive stress in the struts does not exceed the permissible value and sufficient steel is provided to ensure the steel ties can resist the required force. In principle the method is very simple and has the great advantage that the designer can visualize the flow of forces in the structure.

One important point to observe is that the method concentrates on ensuring safety but does not allow any control over the serviceability aspects of design.

17.7.1 B and D Regions

In regions away from concentrated loads and discontinuities, the stress state due to axial force, bending moment and shear force can be determined on the basis of the 'plane sections remain plane' assumption, also known as Bernoulli assumption. Such regions are known as Bernoulli or B-regions. However, in areas near concentrated loads or geometric discontinuities, the state of stress cannot be calculated on the basis of Bernoulli assumption. Such regions in the structure are called discontinuity or D-regions. Fig. 17.12 shows some of the common D-regions.

Fig. 17.13 shows a D-region near a concentrated load. In the D-region, the principal compressive stress follows a curved path. The vertical compressive stresses are larger towards the centre and smaller towards the edges. In addition the curved compressive stresses require tensile stresses normal to the compressive stresses to maintain equilibrium. However, away from the concentrated load, the vertical normal stress is uniform across the width. Near the concentrated load both compressive as well as tensile stresses exist.

Fig. 17.13b shows the state of stress at a joint in a frame structure. In the D-region, the resultant normal forces can be approximately represented by horizontal and vertical arrows as shown. This leads to the conclusion that there must exist a diagonal compressive force at the corner. If the directions of the moments are reversed, then the diagonal force becomes tensile and has to be resisted by steel reinforcement. It is in the design of D-regions the strut-and-tie method is most commonly used.

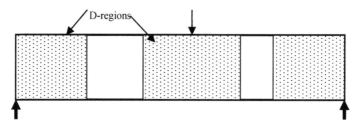

Fig. 17.12a D-regions near concentrated loads

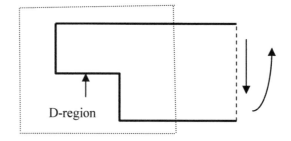

Fig. 17.12b D-region at a 'half' joint

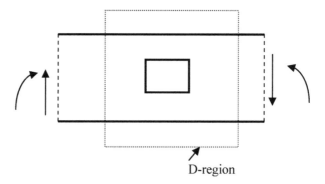

Fig. 17.12c D-region at a hole

7.7.2 Saint Venant's Principle

Saint Venant's principle gives a method of estimating the size of a D-region. For example as shown in Fig. 17.13, away from the concentrated load, the effect of the localized disturbance dies away and the state of stress becomes uniform. Similar effects have been noticed for many types of local disturbance, and Saint Venant's principle states that the localized effect of any type of disturbance dies away at a distance of about one member depth. In other words, the D-regions extend to about one member depth from the point of the disturbance. In all the cases illustrated in Figs. 17.12a to e, the structure is larger than the D-regions indicated. However, there are many cases where the D-region can occupy the entire structure such as a deep beam (a beam where span/depth ≈ 1.0) as shown in Fig. 17.14.

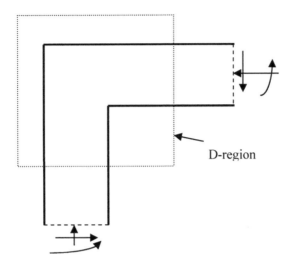

Fig. 17.12d D-region at a frame joint

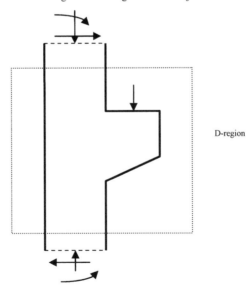

Fig. 17.12e D-region at a corbel

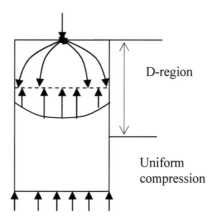

Fig. 17.13 Stresses near a concentrated load

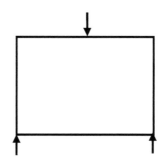

Fig. 17.14 A deep beam

17.7.3 An Example of Strut-Tie Modelling

Fig. 17.15 shows a single span deep under a mid-span concentrated load. A possible strut-tie model consists of two struts with a force F and a tie with a force of T. If the struts are inclined at an angle θ to the horizontal, then

$$F = \frac{W}{2\sin\theta}, \quad T = \frac{W}{2\tan\theta}$$

If b is the width of the bearing plate at the top, from geometry the width of the struts is $b/(2\sin\theta)$. The compressive stress σ in a strut is

$\sigma = W/(b\,t)$, t = thickness of the beam

The required area of tension steel in the tie is

$$A_s = \frac{W}{2\tan\theta} \times \frac{1}{f_{yd}}, \quad f_{yd} = \text{permissible tensile stress in steel.}$$

Apart from ensuring the safe design of struts and ties, it is also important to ensure that the stress inside the three nodal zones where forces meet is also safe.

The above is a brief description of the essentials of the strut-tie method of design. However, there are many aspects to consider which are detailed in the following sections.

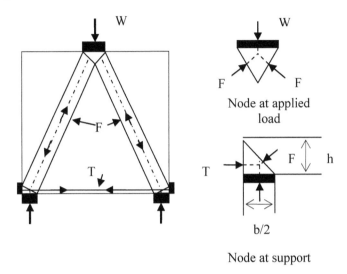

Fig. 17.15 Strut-tie model for a deep beam

17.7.4 Design of Struts

The struts in Fig. 17.15 are shown as prismatic struts. In fact this is only an idealization. The actual strut has variable width along its length as shown in Fig. 17.16. When the compressive stress path is curved, there are always tensile stresses normal to the compression stress path. If adequate transverse reinforcement is not provided, this seriously decreases the compression capacity of the strut. The required tension reinforcement can be determined by replacing the prismatic strut by a more elaborate modelling of the strut as shown in Fig. 17.17.

Eurocode 2 limits the compressive stress in a strut with transverse tension to

$$\sigma_{RD, max} = 0.6 \times (1 - f_{ck}/250) \, f_{cd}$$

On the other hand, in the case of struts subjected to pure compression the compression stress is limited to $\sigma_{RD, max} = f_{cd}$. For example if $f_{ck} = 40$ MPa, $\gamma_c = 1.5$, in the case of struts with and without transverse tension, the permissible compression stresses are 26.7 MPa and 13.4 MPa respectively.

For struts requiring transverse reinforcement, two cases are considered. If the bottle-shaped regions do not merge as shown in Fig. 17.18a, then if the largest width b_{ef} of the strut is equal to less than half the length H of the strut, the total tension T to be resisted by smeared reinforcement is given by

$T = 0.25 (1 - a/b_{ef}) F$

where F = applied compressive force, a = width of loaded area

The reinforcement is distributed over the length where the compression trajectories are curved.

On the other hand, if the bottle-shaped regions merge as shown in Fig. 17.18b, then if the largest width b_{ef} of the strut is equal to less than half the length H of the strut, the total tension T to be resisted by smeared reinforcement distributed over the length where the compression trajectories are curved is given by

$T = 0.25 (1-1.4 \ a/H) F$

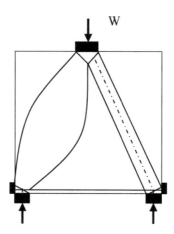

Fig. 17.16 Strut-Tie model with a bottle shaped strut

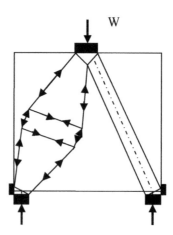

Fig. 17.17 Strut-Tie model with a modified bottle shaped strut

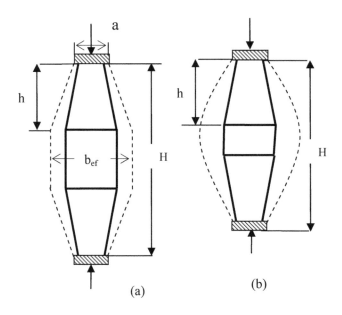

Fig. 17.18 Two different strut configurations

17.7.5 Types of Nodal Zones

Nodal zones are where forces meet. They are classified according to the types of forces that meet at the node. There are many different types of nodes. The main ones are:

(a). CCC node: If three compressive forces meet then this is called a CCC node. Fig. 17.19 shows such a node. The top node in Fig. 17.15 is an example of such a node. Note that the stress in the struts is W/ (bt) and this is also true of the bearing pressure from the vertical load W. In other words, the nodal zone is subjected to a hydrostatic pressure equal to W/ (bt). If W/ (bt) is less than the permissible compressive stress in concrete $\sigma_{RD, max} = (1 - f_{ck}/250) f_{cd}$, then the nodal zone is safe.

(b). CCT node: If two compressive forces and a tensile force meet at a node, then this is called a CCT node. Fig. 17.20 and Fig. 17.21 show such a node. In fact if the steel tie is anchored external to the nodal zone by a bearing plate as shown in Fig. 17.20 or by a bond in the zone outside the nodal zone as shown in Fig. 17.21, then the nodal zone is in compression and should in reality be treated as a CCC zone. However, if the tension force is transmitted via bond forces as shown in Fig. 17.22, then it becomes a true CCT zone. The maximum stress in the struts in a true CCT node is limited to $\sigma_{RD, max} = 0.85 (1-f_{ck}/250) f_{cd}$. The two side nodes in Fig. 17.15 are examples of nodes that are not true CCT node.

(c). CTT node: If two tensile forces and a compressive force meet at a node as shown in Fig. 17.23, then this is called a CTT node. Such a node often occurs in

frame corners as shown in Fig. 17.24.   The compressive stress in the strut is
limited to $\sigma_{RD, max} = 0.75 \ (1-f_{ck}/250) \ f_{cd}$

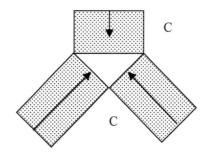

Fig. 17.19 CCC node

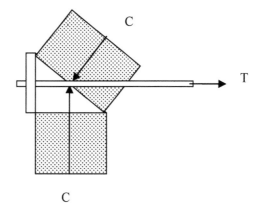

Fig. 17.20 CCT node

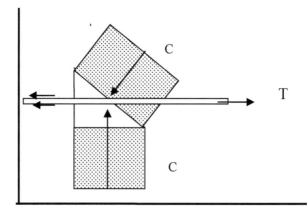

Fig. 17.21 CCT node with reinforcement anchored outside the node

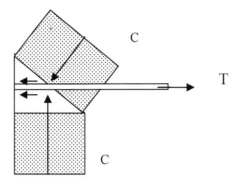

Fig. 17.22 True CCT node

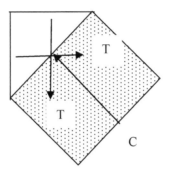

Fig. 17.23 CTT node

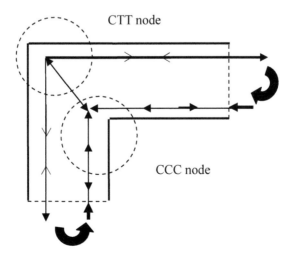

Fig. 17.24 CTT and CCC nodes in a frame corner

Increase in the permissible stress in nodal zones: Eurocode 2 permits 10% increase in the permitted design stress given in (a) to (c), provided the following conditions are met:

- Triaxial compression is assured

- All angles between struts and ties are $\geq 55°$

- The stresses applied at supports and point loads are uniform and the node is confined by stirrups as shown in Fig. 17.25.

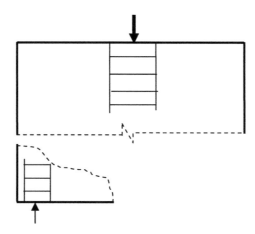

Fig. 17.25 Nodes confined by stirrups.

17.7.6 Correct Layout of Struts and Ties

The basic ideas behind the theory of plasticity assume that the material has unlimited ductility. When applying these ideas to a material like reinforced concrete with limited ductility, it is necessary to ensure that the stress distribution assumed in the strut-tie modelling does not depart greatly from the elastic stress distribution. If this is done, it reduces the ductility demand on the material. It is a good idea that before attempting to postulate a strut-tie model for a specific problem, an elastic finite element analysis is carried out and a vector or a contour plot of the stresses is obtained. By studying the plots one can come up with good strut-tie models.

17.7.6.1 Correct layout of struts and ties: deep beam

In order to demonstrate this fact, an elastic finite element analysis of a simply supported deep beam 3 m span × 4 m deep with a thickness of 300 mm and loaded at the top was done. Fig. 17.26 shows a vector plot of the principal stress σ_1. The plot clearly shows the need for a tie near the soffit. The plot also shows the direction of the struts which lie normal to the direction of the tensile stress. This is also reflected in the vector plot for σ_2 shown in Fig. 17.27. In order not to increase the demand on ductility, it is recommended that the direction of the struts should not depart by more than about $\pm 15°$ from the direction of the compression vectors.

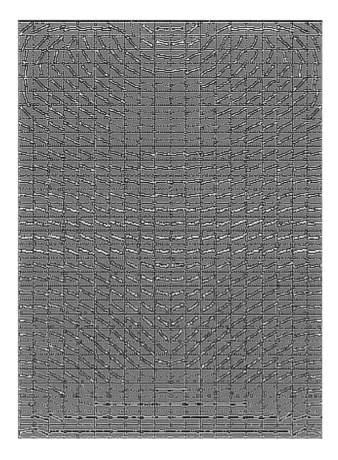

Fig. 17.26 Vector plot of stress σ_1 in a deep beam

17.7.6.2 Correct layout of struts and ties: corbel

A corbel is a short cantilever carrying a concentrated load which is attached to a column. The results of elastic stress analysis are shown in Fig. 17.28 and 17.29. Comparing the strut-tie model with the direction of the principal stresses σ_1 shown in Fig. 17.28 and σ_2 shown in Fig. 17.29, one can be confident that the strut-tie model in Fig. 17.30 is a suitable one. Note that at the applied load, the nodal zone is a CCT node. The node on the top left is a CTT node while the third node near the middle is a CCC node. Fig. 17.31 shows a possible schematic arrangement of reinforcement for such a structure.

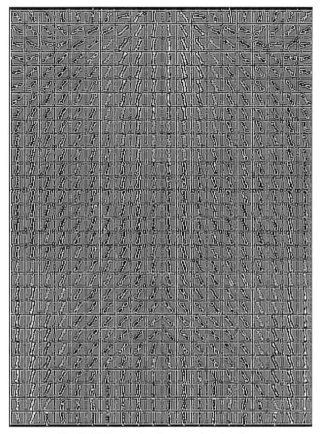

Fig. 17.27 Vector plot of stress σ_2 in a deep beam

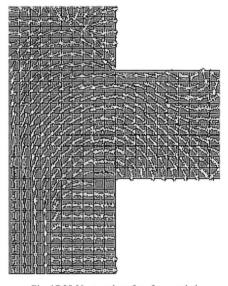

Fig. 17.28 Vector plot of σ_1 for a corbel

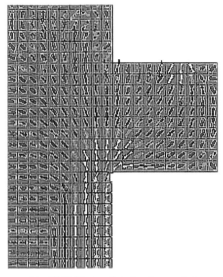

Fig. 17.29 Vector plot of σ_2 for a corbel

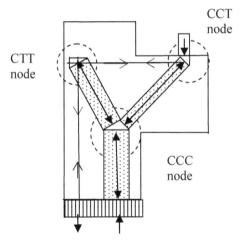

Fig. 17.30 Strut-tie model for a corbel

17.7.6.2.1 Code recommendation for design of corbel

Eurocode 2 considers two cases as shown in Fig. 17.32 and Fig. 17.33.

Case 1: Deep corbel where the distance a_c from the face of the column is less than half the depth h_c of the corbel as shown in Fig. 17.32. In this case shear force dominates. The main reinforcement $A_{s\,Main}$ is calculated from the strut-tie model. The additional horizontal reinforcement $\Sigma A_{s\,Links} \geq 0.25\,A_{s\,Main}$

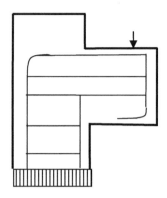

Fig. 17.31 Schematic arrangement of reinforcement

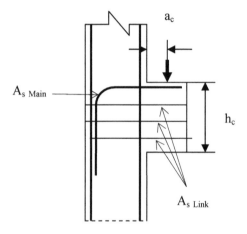

Fig. 17.32 Deep corbel ($a_c \leq 0.5\ h_c$)

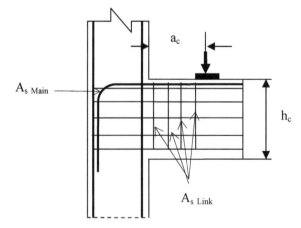

Fig. 17.33 Moderately deep corbel ($a_c > 0.5\ h_c$)

Case 2: Moderately deep corbel where the distance ac from the face of the column is greater than half the depth hc of the corbel as shown in Fig. 17.31and the applied load FED is greater than the shear capacity $V_{RD, c}$ of the section due to concrete alone, the main reinforcement $A_{s\ main}$ is calculated from the strut-tie model. The closed vertical link reinforcement $A_{s\ Links} \geq 0.5\ FED/\ f_{yd}$

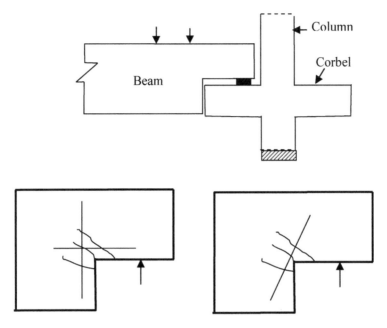

Fig. 17.34 Half-joint/Dapped end

17.7.6.3 Correct layout of struts and ties: half-joint

Fig. 17.34 shows a half-joint also called a dapped end. This is a common way of supporting a beam on a corbel. Fig. 17.35 and Fig. 17.36 show the vector plots for the principal stresses σ_1 and σ_2 respectively. As is to be expected at a re-entrant corner, Fig. 17.35 shows clearly tensile stresses which need to be resisted by tensile reinforcement. Two options are possible. As shown in Fig. 17.34, either one can provide both horizontal and vertical reinforcement at the re-entrant corner or an inclined bar, which is likely to be more effective in resisting tensile stresses. Fig. 17.37 and Fig. 17.38 shows possible strut-tie models for the two options. Full lines indicate ties and chain dotted lines indicate struts.

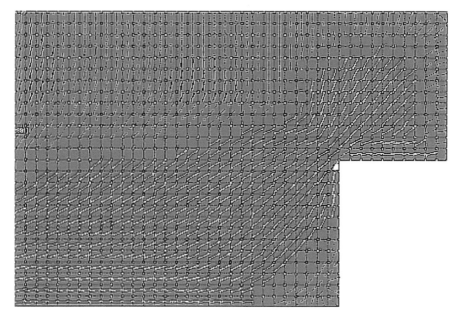

Fig. 17.35 Vector plot of principal stress σ_1 for a half-joint

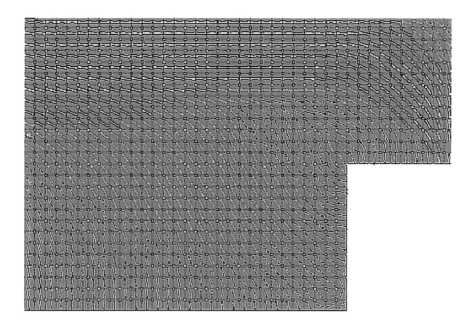

Fig. 17.36 Vector plot of principal stress σ_2 for a half-joint

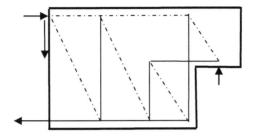

Fig. 17.37 Strut-tie model-1 for half-joint with orthogonal reinforcement

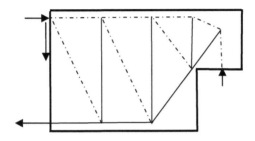

Fig. 17.38 Strut-tie model 2 for half joint with inclined tension reinforcement

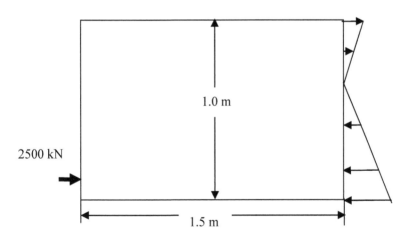

Fig. 17.39 An end block

17.7.6.4 Correct layout of struts and ties: end block

Fig. 17.39 shows a portion of a post-tensioned beam with an anchor. The anchor applies a force of 2500 kN. The thickness of the end block is 450 mm. Fig. 17.40 and Fig. 17.41 show vector plots of elastic principal stresses σ_1 and σ_2 respectively.

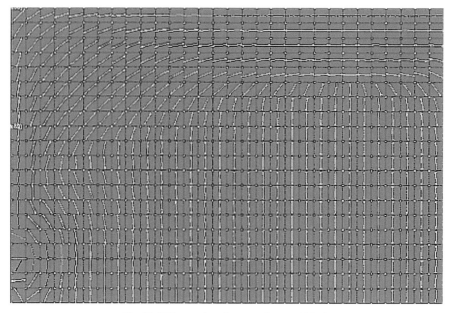

Fig. 17.40 Vector plot of stress σ_1 in an end block

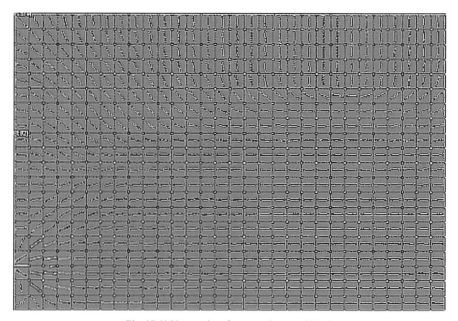

Fig. 17.41 Vector plot of stress σ_2 in an end block

Fig. 17.42 shows a very basic strut-tie model. The American Concrete Institute code suggests the strut-tie models for single anchors as shown in Fig. 17.43 for two cases viz. the case where the beam has a rectangular cross section for a considerable length and the case where the beam becomes a flanged beam after a distance of half the depth of the beam.

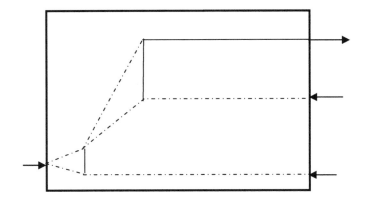

Fig. 17.42 Strut-tie model for an end block

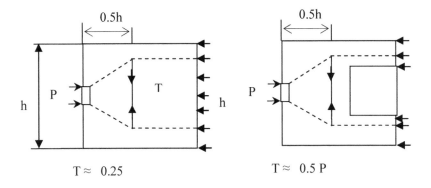

Fig. 17.43 Strut-tie model in rectangular and flanged beams

17.7.6.5 Reinforcement at frame corners

Eurocode 2 gives some guidance on suitable strut-tie models. The strut-tie models and also the corresponding reinforcement depend on whether the frame corner is subjected to closing or opening moments.

(i). Frame corners with closing moments: In this case no check of link reinforcement or anchorage length within the beam column joint is required. However, reinforcement should be provided for transverse tensile forces perpendicular to an in-plane node. Depending on the ratio of the depth of the column h_2 to depth of the beam h_1, two cases are considered as follows:

Case 1a: Almost equal depths of beam and column: Fig. 17.44 shows the strut-tie model to be used if $2/3 < h_2/h_1 < 3/2$.

Case 1b: Very different depths of beam and column: Fig. 17.45 shows the strut-tie model to be used if $h_2/h_1 < 2/3$. The angle θ of the inclination of the strut to the horizontal should be in the range $0.4 \leq \tan \theta \leq 1.0$.

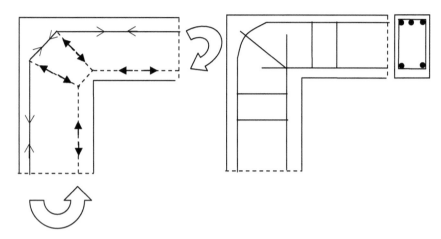

Fig. 17.44 Approximately equal depths: closing moments

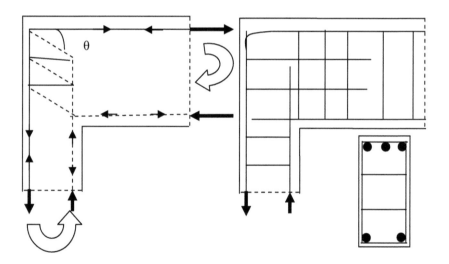

Fig. 17.45 Very different depths: closing moments

(ii). Frame corners with opening moments: Fig. 17.46 shows the strut-tie model and the reinforcement arrangement. Two overlapping U-bars satisfy the steel requirements. If the moment is large, an additional diagonal bar can be introduced to resist tensile stresses at the re-entrant corner.

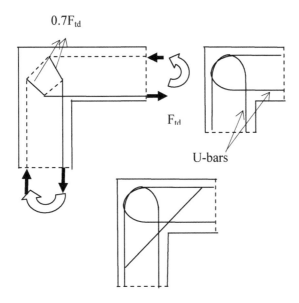

$0.7F_{td}$

F_{td}

U-bars

Fig. 17.46 Frame corner with opening moment

17.8 REFERENCE TO EUROCODE 2 CLAUSES

In this chapter the following clauses in Eurocode 2 have been referred to:

Analysis with strut and tie models: 5.6.4

Design with strut and tie models: 6.5

Struts, permissible stress: 6.5.2

Ties, permissible stress: 6.5.3

Nodes: 6.5.4

Half joints: 10.9.4.6

Tension reinforcement expressions for in-plane stress conditions: Annex F

Frame corners: Annex J.2

Frame corners with closing moments: J2.2

Frame corners with opening moments: J2.3

Corbels: Annex J.3

CHAPTER 18

DESIGN FOR EARTHQUAKE RESISTANCE

18.1 INTRODUCTION

Earthquakes cause serious loss of life and damage to property. The intensity of earthquakes at any particular area is highly unpredictable and a reasonably conservative approach to design is recommended. It is uneconomical to design structures to remain structurally undamaged during large earthquakes. However, in order to prevent loss of life, it is equally important that under large earthquakes, structures although suffering damage beyond repair, still remain intact without serious loss of resistance to lateral forces. Structures should therefore be designed to have this high energy absorbing capacity and the most important property that ensures is that the structures behave in a ductile manner.

In Eurocode 8, Part 1, for the design of buildings, there is no specific mention of prestressed concrete and its behaviour. However, in Part 2, there is limited advice on the design of bridges. This chapter gives a very brief introduction to the design of earthquake resistant building structures according to Eurocode 8. The reader should refer to the code and the excellent guide to the code for complete details.

The code provides for a two-level design against earthquake hazards:

- Ultimate limit states: This is mainly directed towards saving life. It is accepted that the structure might be severely damaged and repair might prove uneconomic. However, the structure needs to preserve its integrity and the aim is to design to prevent collapse of the parts or of the whole structure.

- Damage limitation state: This is mainly directed towards reducing property loss by limiting structural and non-structural damage during frequent earthquakes.

The intensity of earthquake naturally depends on many factors, among which the main ones are:

- Geographical location of the area, which is characterized by a reference ground acceleration a_{gR}. Earthquake intensity maps of the area will provide information on this.

- Type of ground as shown in Table 18.1.

Table 18.1 Ground types

Ground type	Description of stratigraphical profile	Parameters		
		$V_{s, 30}$ * m/sec	$N_{SPT}$$	C_u#, kPa
A	Rock or other rock-like geological formation, including at most 5 m of weaker material at the surface.	> 800		
B	Deposits of very dense sand, gravel, or very stiff clay, at least several tens of metres in thickness, characterized by a gradual increase in mechanical properties with depth.	360 to 800	> 50	> 250
C	As B except thickness from several tens to many hundreds of metres.	180 to 360	15 to 50	70-250
D	Deposits of loose-to-medium cohesion less soil (with or without some soft cohesive layers), or of predominantly soft-to-firm cohesive soil.	< 180	< 15	< 70
E	A soil profile consisting of a surface alluvium layer with average shear wave velocity values of type C or D and thickness varying between about 5 m and 20 m, underlain by stiffer material with v_s > 800 m/s			
S_1	Deposits consisting of or containing a layer at least 10 m thick, of soft clays/silts with a plasticity index > 40 and high water content.	<100 (indicative)		10-20
S_2	Deposits of liquefiable soils, of sensitive clays, or any other soil profile not included in types A to E or S_1.			

$V_{s, 30}$ = shear wave velocity, $ standard penetration test: No. of blows per 30 cm,

C_u = undrained shear strength

The acceptable damage depends on the importance of the structure which is taken into account by the importance factor γ_1 as given in Table 18.2. The design ground acceleration a_g is given by

$$a_g = \gamma_1 \times a_{gR} \qquad (18.1)$$

Table 18.2 Importance classes for buildings

Importance class	Buildings	γ_1
I	Buildings of minor importance for public safety, e.g. Agricultural buildings, etc.	0.8
II	Ordinary buildings not belonging to other categories.	1.0
III	Buildings whose seismic resistance is of importance in view of the consequences associated with a collapse, e.g. schools, assembly halls, cultural institutions, etc.	1.2
IV	Buildings whose integrity during earthquakes is of vital importance for civil protection, e.g. hospitals, fire stations, power plants, etc.	1.4

18.2 DUCTILITY

Fig. 18.1 shows the load versus displacement relationship. The area under the curve is an indication of the total energy absorbed.

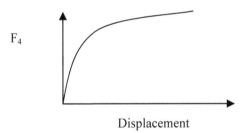

Fig. 18.1 Load-displacement relationship

Structures designed to resist earthquake forces do not remain in the elastic state. They are allowed to deform inelastically through cracking of concrete and yielding of steel, thereby dissipating the energy input by the earthquake. It is important to remember that in order to ensure overall ductility, attention should be paid to all areas where plastification can occur and to ensure that these regions are in predictable positions. This in summary is the philosophy behind design of earthquake resistant structures. For example in the design of frames, the design concept adopted is 'strong column - weak beam'. The idea behind this concept is to confine inelastic deformation resulting in plastic hinges to the ends of the beams. The reason for this is that columns are invariably less ductile compared to beams and in addition the consequence of a column failing is far more serious compared to the local failure of a beam.

18.3 TYPES OF STRUCTURAL SYSTEMS

The three common types of structural systems considered are:

(a). Frame systems: Fig. 18.2 shows a typical frame. The global overturning is resisted mainly by the axial forces in the columns and to a very small extent by the moments in the columns at the base. The storey shears are resisted by columns bending in double curvature. The mode of collapse is a sway mode with plastic hinges at the ends of beams and at the base of the columns. In this system at least 65% of base shear is resisted by the frames.

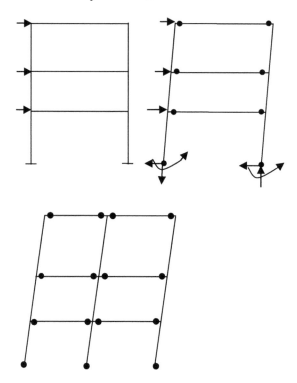

Fig. 18.2 Frame system and collapse mechanism

(b). Wall systems: Wall systems as shown in Fig. 18.3 resist overturning moment by bending action as vertical cantilevers rather than through the axial force as in the case of frames. In this system the base shear resisted by the walls should be at least 65% of the total base shear.

(c). Coupled wall system: Coupled walls as shown in Fig. 18.4 are a set of walls connected by beams. Depending on the stiffness of the connecting beams, their behaviour is intermediate between that of a frame and a wall. However, they generally have a greater energy dissipating capacity than a wall but if the coupling beams are 'stocky' they are unlikely to form plastic hinges at their ends. For a wall system to be considered as a coupled wall system, the coupling beams should be able to reduce the base moment in the walls by at least 25% of what it would be if the walls were acting independently. One important difference between a plain wall and a coupled wall is that under the action of lateral loads due to earthquake, in the former the cross section is subjected to moment and shear force. On the other hand, in a coupled wall, individual parts of the wall are subjected to bending moment, axial force and shear force.

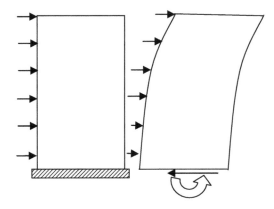

Fig. 18.3 Wall system

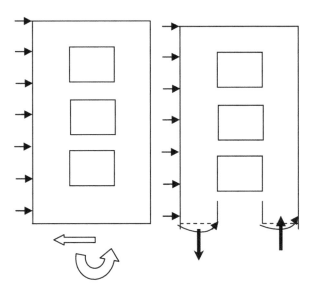

Fig. 18.4 Coupled wall

(d). Dual systems consisting of frames, and walls or coupled walls. In the frame - equivalent dual system the frame system should resist at least 50% of total base shear and in the case of wall-equivalent dual system; the walls should resist at least 50% of total shear force. Fig. 18.5a shows the deflection pattern of a frame which is in 'shearing mode' and the deflection pattern of a wall which is in 'bending mode'. Fig. 18.5b shows the typical collapse mechanism of a wall-frame dual system.

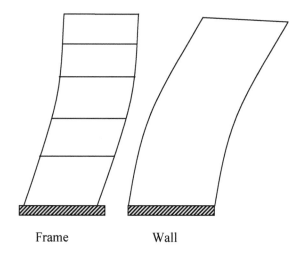

Fig. 18.5 a Deflection patterns of a frame and a wall.

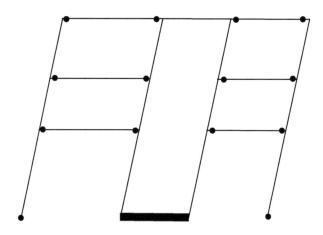

Fig. 18.5 b Wall-frame system collapse mode

18.4 BEHAVIOUR FACTOR, q

As already mentioned, during an earthquake, structures do not behave elastically. Because nonlinear behaviour is allowed for, one needs to design for forces smaller than that obtained from elastic analysis. Fig. 18.6 shows the results of a pushover analysis taking into account non-linear behaviour.

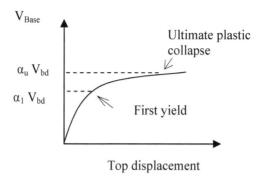

Fig. 18.6 Base shear vs. Top displacement

The base shear V_{Base} is plotted on the vertical axis and top displacement on the horizontal axis. The forces considered during the pushover analysis include both seismic forces as well as gravity loads considered to act simultaneously with the seismic action. If elastic analysis is carried out at design seismic loads resulting in a base shear value equal to V_{Design}, the base shear value when plastification takes place anywhere in the structure is a multiple of V_{Design} equal to $\alpha_1 \times V_{Design}$. Similarly the base shear value when plastification is sufficient to cause the collapse of the structure in a ductile manner is set equal to $\alpha_u \times V_{Design}$ The code introduces a factor called the behaviour factor q which is approximately proportional to the ratio α_u / α_1. Table 18.3 gives the value of q_0 for various types of structures and for two classes of ductility.

Table 18.3 Basic values of behaviour factor q_0 for systems regular in elevation

Structural type	DCM (medium ductility)	DCH (high ductility)
Frame system, dual system, coupled wall system	$3.0\ \alpha_u/\alpha_1$	$4.5\ \alpha_u/\alpha_1$
Uncoupled wall system	3.0	$4.0\ \alpha_u/\alpha_1$
Torsionally flexible system	2.0	3.0
Inverted pendulum system	1.5	2.0

α_u = multiplier on horizontal seismic action in order to cause collapse

α_1 = multiplier on horizontal seismic action in order to the first yield

These factors can only be obtained from non-linear analysis. In the absence of analysis data, the code recommends the following values to be used:

(a). Frames or frame-equivalent dual systems:

$\alpha_u/\alpha_1 = 1.2$ for multistorey single-bay frames

$\alpha_u/\alpha_1 = 1.3$ for multistorey multi-bay frames or frame-equivalent dual structures.

(b). Wall or wall-equivalent dual systems:

$\alpha_u/\alpha_1 = 1.0$ for wall systems with only two uncoupled walls per horizontal direction

$\alpha_u/\alpha_1 = 1.1$ for other uncoupled wall systems

$\alpha_u/\alpha_1 = 1.2$ wall-equivalent dual or coupled wall systems

For design,

$$q = q_0\, k_w \geq 1.5 \qquad\qquad\qquad\qquad (18.2)$$

The factor k_w takes into account the prevailing mode of failure in structural systems with wall.

$k_w = 1$ for frames and frame-equivalent dual systems. For wall, wall-equivalent and torsionally flexible systems, refer to the code for full details.

18.5 DUCTILITY CLASSES

Structures, in order to resist seismic action causing repeated reversed loading during an earthquake, need both strength and ductility to dissipate energy through hysteretic damping. It is possible to design a structure for a higher design load and a lower level of ductility and vice versa. Euroode 8 allows two classes of ductility viz. DCM (**D**issipation **C**apacity, **M**edium) and DCH (**D**issipation **C**apacity, **H**igh). As can be seen from Table 18.3, structures designed to DCH classification have a higher q_0 value, indicating that they are designed to a smaller load compared to buildings designed to DCM classification, which are required to carry almost 50% more load. Euroode 8 provides very detailed rules for the design of structures according to the two levels of ductility class. The detailing provisions for DCH are very stringent. It has been suggested that DCM designs perform better in moderate earthquakes while DCH designs provide higher safety margins against local or overall collapse under earthquake action much stronger than the design seismic action.

18.6 A BRIEF INTRODUCTION TO STRUCTURAL DYNAMICS

In order to understand some of the fundamental concepts involved in earthquake analysis, in the following sections, a brief introduction to fundamental concepts of structural dynamics is given.

18.6.1 Single-Degree-of-Freedom System

Consider the simple 'structure' shown in Fig. 18.7. The mass M is attached to the support through a 'weightless' spring of stiffness K and a dashpot simulating damping normally present in all structures. It is assumed for simplicity that the mass can move only horizontally. In other words the structure has only one degree of freedom of movement.

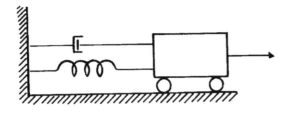

Fig. 18.7 Mass-spring-damper System

The equation of equilibrium for the system is

$$M x'' + D x' + K x = F \qquad (18.4)$$

where

$x'' =$ acceleration of the mass M

$x' =$ velocity of the mass M

$D =$ damping constant

In (18.4), Mx'' is the inertial force, Dx' is the damping force, Kx is the elastic force and F is an externally applied force which is a function of time. It is assumed that the damping present is viscous damping whose value depends on the velocity x'.

In the absence of any external force, if the system is disturbed from its rest position, it continues to vibrate at a frequency of ω cycles/second. It can be shown that

$$\omega = \sqrt{(K/M)}, \ T = 2\pi/\omega \qquad (18.5)$$

where T is the time for one complete cycle.

As the damping is increased, the vibratory motion keeps decreasing and at a value of damping called the critical damping, all vibratory motion is damped out. It can be shown that the value of critical damping D_{cr} is given by

$$D_{cr} = 2\sqrt{(K \times M)} \qquad (18.6)$$

In earthquake analysis, there is no external force F applied but the forces are created due to the acceleration of the base, x''_{Base}. In this case the equilibrium equation becomes

$$M x'' + D x' + K x = -M x''_{Base} \qquad (18.7)$$

where the displacement of the base is x_{Base} and the relative displacement of the mass with respect to the foundation is x

As can be seen, the modified equation is identical to the original equation of equilibrium (18.4), except that the force F is replaced by $-$ M x''$_{Base}$.

If D = ξ D$_{cr}$, where ξ is a constant, using (18.6),

$$\frac{D}{M} = \xi\frac{D_{cr}}{M} = \xi\frac{2\sqrt{K\ M}}{M} = 2\xi\sqrt{\frac{K}{M}} = 2\xi\omega \tag{18.8}$$

Dividing (18.7), by M and substituting for K/M = ω^2 and D/M = 2$\eta\omega$, the differential equation becomes

$$x'' + 2\xi\ \omega\ x' + \omega^2 x = -\ x''_{Base} \tag{18.9}$$

18.6.2 Multiple-Degree-of-Freedom System

A multistorey two-dimensional structure has many degrees of freedom as at each joint two translations and one rotation are possible. As a simple example consider a multi-storey shear frame with rigid beams and deformable columns shown in Fig. 18.8. Because the beams are rigid, there are no rotations at the joint and each storey deforms in a shearing mode. The independent displacements are the horizontal displacements x_1 to x_4 at storey levels. k_1, k_2, k_3 and k_4 are the shear stiffness (12 EI/h^3) of columns of storeys 1 to 4 respectively and m_1, m_2, m_3 and m_4 are the masses at levels 1 to 4 respectively.

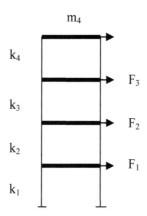

Fig. 18.8 Four-storey shear frame

Consider the equilibrium of the mass m_2 at second storey level. As shown in Fig 18.9 the forces acting on the mass are
- inertial force = M_2 x''$_2$,
- shear forces from the column below = 2 k_2 ($x_2 - x_1$)
- shear forces from the column above = $-2k_3$ ($x_3 - x_2$)
- external force f_2
- damping force

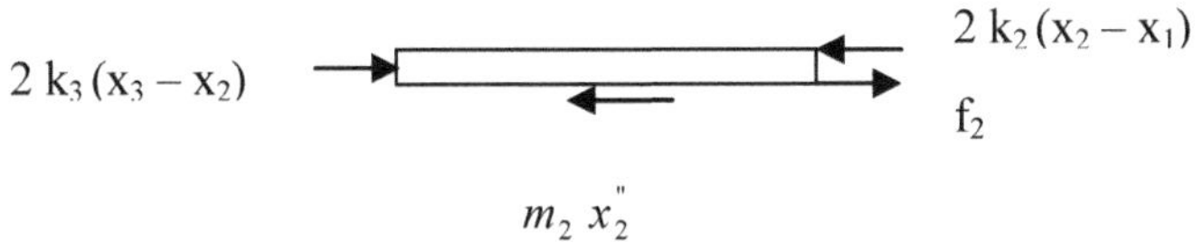

Fig. 18.9 Forces acting on mass m_2.

Ignoring damping, the equilibrium equation for mass m_2 is given by
$m_2 x''_2 + 2 k_2 (x_2 - x_1) - 2 k_3 (x_3 - x_2) = f_2$
Simplifying
$m_2 x''_2 - 2 k_2 x_1 + 2 (k_2 + k_3) x_2 - 2 k_3 x_3 = f_2$

Similarly, the equations of equilibrium of the individual masses are
$m_1 x''_1 + 2 (k_1 + k_2) x_1 - 2 k_2 x_2 = f_1$
$m_2 x''_2 + 2 (k_2 + k_3) x_2 - 2 k_2 x_1 - 2 k_3 x_3 = f_2$
$m_3 x''_3 + 2 (k_3 + k_4) x_3 - 2 k_3 x_2 - 2 k_4 x_4 = f_3$
$m_4 x''_4 + 2 k_4 x_4 - 2 k_4 x_3 = f_4$ (18.10)

Equations (18.10) can be expressed in matrix form:

$$\begin{bmatrix} m_1 & 0 & 0 & 0 \\ 0 & m_2 & 0 & 0 \\ 0 & 0 & m_3 & 0 \\ 0 & 0 & 0 & m_4 \end{bmatrix} \begin{bmatrix} x''_1 \\ x''_2 \\ x''_3 \\ x''_4 \end{bmatrix} + 2 \begin{bmatrix} k_1 + k_2 & -k_2 & 0 & 0 \\ -k_2 & k_2 + k_3 & -k_3 & 0 \\ 0 & -k_3 & k_3 + k_4 & -k_4 \\ 0 & 0 & -k_4 & k_4 \end{bmatrix} \begin{bmatrix} x_1 \\ x_2 \\ x_3 \\ x_4 \end{bmatrix} = \begin{bmatrix} f_1 \\ f_2 \\ f_3 \\ f_4 \end{bmatrix}$$
(18.11)

Let

$$K = 2 \begin{bmatrix} k_1 + k_2 & -k_2 & 0 & 0 \\ -k_2 & k_2 + k_3 & -k_3 & 0 \\ 0 & -k_3 & k_3 + k_4 & -k_4 \\ 0 & 0 & -k_4 & k_4 \end{bmatrix}, M = \begin{bmatrix} m_1 & 0 & 0 & 0 \\ 0 & m_2 & 0 & 0 \\ 0 & 0 & m_3 & 0 \\ 0 & 0 & 0 & m_4 \end{bmatrix}$$

Similarly the acceleration, velocity and displacement and force vectors X'', X', X and F respectively are given by

$$X'' = \begin{bmatrix} x''_1 \\ x''_2 \\ x''_3 \\ x''_4 \end{bmatrix}, X' = \begin{bmatrix} x'_1 \\ x'_2 \\ x'_3 \\ x'_4 \end{bmatrix}, X = \begin{bmatrix} x_1 \\ x_2 \\ x_3 \\ x_4 \end{bmatrix}, F = \begin{bmatrix} f_1 \\ f_2 \\ f_3 \\ f_4 \end{bmatrix}$$

Equation (18.11) can be expressed in matrix form as
$M X'' + K X = F$ (18.12)

Equation (18.12) is similar to the equilibrium equation for a single-degree-of-freedom system given by (18.4) with zero damping. Equation (18.12) can be modified to include a damping matrix D and rewritten as

$$M X'' + DX' + K X = F \tag{18.13}$$

In practice there is not enough information to specify the coefficients of the damping matrix. It is usually assumed that

$$D = \alpha M + \beta K \tag{18.14}$$

where the coefficients α and β are constants.

This is known as Rayleigh damping. The manner in which the constants α and β affect the damping ratio can be understood by considering a single-degree-of-freedom system. From (18.6), the critical damping $D_{cr} = 2\sqrt{(KM)}$ and from (18.5), the frequency of vibration $\omega = \sqrt{(K/M)}$. Therefore

$$\xi = \frac{D}{D_{cr}} = \alpha \frac{M}{2\sqrt{(K M)}} + \beta \frac{K}{2\sqrt{(K M)}} = \frac{1}{2}[\frac{\alpha}{\omega} + \beta \omega] \tag{18.15}$$

Equation (18.15) shows that the first part due to the parameter α has the effect of having lower damping for higher frequency while the second parameter β has exactly the opposite effect. In other words the damping matrix D can be manipulated by adjusting just two constants to fit the desired damping for any two frequencies. Unfortunately, in a multi-degree-of-freedom system, the structure has many more than just two frequencies. In practice the constants α and β are calculated such that the required values of damping are obtained for the first two or for any two desired frequencies and this is assumed to hold good for other frequencies also.

18.6.3 Response to an Acceleration of the Base

If instead of force input, the frame is subjected to an acceleration of the base as would happen in an earthquake, then the problem can be analysed in a similar way to force input. As in the case of a single-degree-of-freedom system, acceleration of the masses is governed by absolute displacement but the elastic forces are governed by displacements measured relative to the base. Ignoring damping and external force, the equilibrium of the masses at the four storeys are given by (18.10) modified as follows:

$$m_1 (x''_1 + x''_{base}) + 2 (k_1 + k_2) x_1 - 2 k_2 x_2 = f_1$$
$$m_2 (x''_2 + x''_{base}) + 2 (k_2 + k_3) x_2 - 2 k_2 x_1 - 2 k_3 x_3 = f_2$$
$$m_3 (x''_3 + x''_{base}) + 2 (k_3 + k_4) x_3 - 2 k_3 x_2 - 2 k_4 x_4 = f_3$$
$$m_4 (x''_4 + x''_{base}) + 2 k_4 x_4 - 2 k_4 x_3 = f_4 \tag{18.16}$$

Expressing equation (18.16) in matrix form,

$$M X'' + K X = F \tag{18.17}$$

This equation is identical to equation (18.12), except that the force vector F is given by.

$$F = -x''_{Base} \begin{bmatrix} m_1 & 0 & 0 & 0 \\ 0 & m_2 & 0 & 0 \\ 0 & 0 & m_3 & 0 \\ 0 & 0 & 0 & m_4 \end{bmatrix} \begin{bmatrix} 1 \\ 1 \\ 1 \\ 1 \end{bmatrix} = -x''_{Base} \, M \, I \qquad (18.18)$$

where I is a unit vector.

18.6.4 Vibration of an Undamped Free Multiple-Degree-of-Freedom System

In the case of undamped free vibration of a multiple-degree-of-freedom system, there are no external forces applied at the nodes. In other words, the force vector F is a null vector. The displacement of the structure results from giving an initial displacement and velocity at a particular storey level and letting the structure vibrate. The displacements at any level will be dependent on the value of the initial displacement, which is perfectly arbitrary. The displacement vector X and acceleration vector X'' can be expressed as

$$X = \begin{bmatrix} x_1 \\ x_2 \\ x_3 \\ x_4 \end{bmatrix} \cos(\omega t + \varsigma), \quad X'' = -\omega^2 \begin{bmatrix} x_1 \\ x_2 \\ x_3 \\ x_4 \end{bmatrix} \cos(\omega t + \varsigma)$$

where x_1, x_2, x_3, x_4 are the amplitudes of the storey displacements and ς is the phase angle. Substituting for X and X'', the differential equation (18.17) with the force vector F being null simplifies to
$$[K - \omega^2 M] X = 0 \qquad (18.19)$$
This equation has a non-zero solution for the vector X only if the determinant of [K $- \omega^2 M$] is zero. The value of ω^2 can be determined by setting the determinant of [K $- \omega^2 M$] to zero.

18.6.5 Calculation of Eigenvalues

Example: Calculate the eigenvalues, i.e. ω^2 for the frame structure shown in Fig. 18.8. Assume
$k_1 = 4, k_2 = 3, k_3 = k_4 = 2$
$m_1 = m_2 = 4, m_3 = m_4 = 2,$
Substituting in equation (18.11), the stiffness and mass and damping matrices are

$$K = \begin{bmatrix} 14 & -6 & 0 & 0 \\ -6 & 10 & -4 & 0 \\ 0 & -4 & 8 & -4 \\ 0 & 0 & -4 & 4 \end{bmatrix}, \quad M = \begin{bmatrix} 4 & 0 & 0 & 0 \\ 0 & 4 & 0 & 0 \\ 0 & 0 & 2 & 0 \\ 0 & 0 & 0 & 2 \end{bmatrix}$$

$$K - \omega^2 M = \begin{bmatrix} (14-\omega^2 4) & -6 & 0 & 0 \\ -6 & (10-\omega^2 4) & -4 & 0 \\ 0 & -4 & (8-\omega^2 2) & -4 \\ 0 & 0 & -4 & (4-\omega^2 2) \end{bmatrix}$$

The determinant is given by

$$\left|K - \omega^2 M\right| = (-1)^{(1+1)}(14-\omega^2 4) \times \mathrm{Det} \begin{bmatrix} (10-\omega^2 4) & -4 & 0 \\ -4 & (8-\omega^2 2) & -4 \\ 0 & -4 & (4-\omega^2 2) \end{bmatrix}$$

$$+ (-1)^{(1+2)}(-6) \times \mathrm{Det} \begin{bmatrix} -6 & -4 & 0 \\ 0 & (8-\omega^2 2) & -4 \\ 0 & -4 & (4-\omega^2 2) \end{bmatrix}$$

$$\left|K - \omega^2 M\right| = (14 - 4\omega^2)[(10 - 4\omega^2)\{(8 - 2\omega^2)(4 - 2\omega^2) - 16\}$$

$$+ (-1)(-4)\{-4(4 - 2\omega^2)\} + (-1)(-6)[(-6)\{(8 - 2\omega^2)(4 - 2\omega^2) - 16\}]$$

$$\left|K - \omega^2 M\right| = 64\omega^8 - 768\omega^6 + 2848\omega^4 - 3328\omega^2 + 768$$

The determinant is a fourth order polynomial in ω^2.   Fig. 18.10 shows a plot of the determinant versus ω^2.

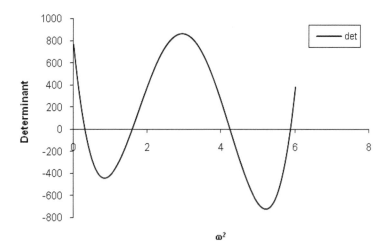

Fig. 18.10 Variation of the determinant for the four-storey shear frame

The four values of ω^2 for which the determinant is zero are
$\omega^2 = 0.3035$, 1.5917, 4.2410 and 5.8642. The structure thus has four eigenvalues which are the natural frequencies of the system. The corresponding periods are given by $T = 2\pi/\omega$:
$T = (11.4, 5.0, 3.1, 2.6)$ seconds
Note: In large-scale structures, the algebraic equation for the determinant is not explicitly formed. But other more efficient procedures are used for calculating the eigenvalues.

18.6.6 Eigenvectors of $[K - \omega^2 M]$

Eigenvectors are the deformation shape of the structure corresponding to each eigenvalue. The displacements at any level will be dependent on the value of the initial displacement, which is perfectly arbitrary. Therefore all one can do is to determine the relative rather than the absolute value of displacements. It is conventional to M-normalize the eigenvectors, i.e. $X^T MX = 1$.

i. Eigenvector for $\omega^2 = 0.3032$

$$K - \omega^2 M = \begin{bmatrix} 12.7874 & -6 & 0 & 0 \\ -6 & 8.7874 & -4 & 0 \\ 0 & -4 & 7.3937 & -4 \\ 0 & 0 & -4 & 3.3937 \end{bmatrix}$$

Let $x_1 = 1$.
Row 1: $12.7874\,x_1 - 6\,x_2 = 0$, $x_2 = 2.1312$
Row 2: $-6\,x_1 + 8.7874\,x_2 - 4\,x_3 = 0$, $x_3 = 3.1820$
Row 3: $-4\,x_2 + 7.3937\,x_3 - 4\,x_4 = 0$, $x_4 = 3.7505$

$$X = \begin{bmatrix} 1.0 \\ 2.1312 \\ 3.1820 \\ 3.7505 \end{bmatrix}$$

$$X^T MX = \begin{bmatrix} 1.0 & 2.1312 & 3.1820 & 3.7505 \end{bmatrix} \begin{bmatrix} 4 & 0 & 0 & 0 \\ 0 & 4 & 0 & 0 \\ 0 & 0 & 2 & 0 \\ 0 & 0 & 0 & 2 \end{bmatrix} \begin{bmatrix} 1.0 \\ 2.1312 \\ 3.1820 \\ 3.7505 \end{bmatrix} = 70.5514$$

M-normalizing X, the eigenvector for $\omega^2 = 0.3032$ is given by

$$\phi_1 = \frac{1}{\sqrt{70.5514}} \begin{bmatrix} 1.0 \\ 2.1312 \\ 3.1820 \\ 3.7505 \end{bmatrix} = \begin{bmatrix} 0.1191 \\ 0.2537 \\ 0.3788 \\ 0.4465 \end{bmatrix}$$

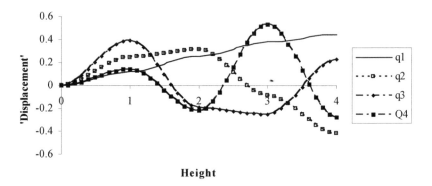

Fig. 18.11 Mode shapes of the four-storey frame

In a similar way, the rest of the eigenvectors are given as follows:
(ii). Eigenvector for $\omega^2 = 1.5917$:

$$X = \begin{bmatrix} 1.0 \\ 1.2722 \\ -0.3444 \\ -1.6870 \end{bmatrix}, \quad X^T MX = 16.4030, \quad \phi_2 = \frac{1}{\sqrt{16.4030}} \begin{bmatrix} 1.0 \\ 1.2722 \\ -0.3444 \\ -1.6870 \end{bmatrix} = \begin{bmatrix} 0.2469 \\ 0.3141 \\ -0.0850 \\ -0.4165 \end{bmatrix}$$

iii. Eigenvector for $\omega^2 = 4.2410$:

$$X = \begin{bmatrix} 1.0 \\ -0.4940 \\ -0.6399 \\ 0.5711 \end{bmatrix}, \quad X^T MX = 6.4476, \quad \phi_3 = \frac{1}{\sqrt{6.4476}} \begin{bmatrix} 1.0 \\ -0.4940 \\ -0.6399 \\ 0.5711 \end{bmatrix} = \begin{bmatrix} 0.3938 \\ -0.1946 \\ -0.2520 \\ 0.2249 \end{bmatrix}$$

iv. Eigenvector for $\omega^2 = 5.8642$:

$$X = \begin{bmatrix} 1.0 \\ -1.5761 \\ 3.8023 \\ -1.9680 \end{bmatrix}, \quad X^T M X = 50.5981, \quad \phi_4 = \frac{1}{\sqrt{50.5941}} \begin{bmatrix} 1.0 \\ -1.5761 \\ 3.8023 \\ -1.9680 \end{bmatrix} = \begin{bmatrix} 0.1406 \\ -0.2216 \\ 0.5345 \\ -0.2767 \end{bmatrix}$$

Fig. 18.11 shows the four mode shapes. Each storey deforms in a shearing mode.

18.6.7 Properties of Eigenvectors

Two of the most important properties of eigenvectors are:
a. Orthogonal property of eigenvectors: If the eigenvectors are M-normalized,

$$\phi_j^T M \phi_i = 0, \text{ if } i \neq j, \phi_i^T M \phi_i = 1 \tag{18.20}$$

b. Any arbitrary vector Y can be expressed in terms of eigenvectors.

$$Y = \sum_1^N C_i \phi_i \tag{18.21}$$

Multiplying (18.21) by $\phi_i^T M$ and making use of the orthogonality property in (18.20),

$$C_i = \phi_i^T M Y \tag{18.22}$$

As an example, for the four-storey shear frame shown in Fig. 18.16, if Y is a unit vector and the mass matrix and eigenvectors are

$$M = \begin{bmatrix} 4 & 0 & 0 & 0 \\ 0 & 4 & 0 & 0 \\ 0 & 0 & 2 & 0 \\ 0 & 0 & 0 & 2 \end{bmatrix}, Y = \begin{bmatrix} 1 \\ 1 \\ 1 \\ 1 \end{bmatrix}$$

$$\phi_1 = \begin{bmatrix} 0.1191 \\ 0.2537 \\ 0.3788 \\ 0.4465 \end{bmatrix}, \phi_2 = \begin{bmatrix} 0.2469 \\ 0.3141 \\ -0.0850 \\ -0.4165 \end{bmatrix}, \phi_3 = \begin{bmatrix} 0.3938 \\ -0.1946 \\ -0.2520 \\ 0.2249 \end{bmatrix}, \phi_4 = \begin{bmatrix} 0.1406 \\ -0.2216 \\ 0.5345 \\ -0.2767 \end{bmatrix},$$

then $C_1 = 3.1418, C_2 = 1.2410, C_3 = 0.7426, C_4 = 0.1916$

18.6.8 Mode Superposition: Undamped Forced Response

Equation (18.12) gives the equilibrium equation for the forced vibration of an undamped multi-degree-of-freedom system. Expressing the displacement vector X which is a function of time, in terms of the eigenvectors ϕ_i of the system,

$$X(t) = \sum_{i=1}^N C_i(t) \phi_i \tag{18.23}$$

where the coefficients $C_i(t)$ are functions of time and ϕ_i is the ith eigenvector:

$$C_i(t) = \phi_i^T M X(t) \tag{18.24}$$

Substituting for X and its derivatives in terms of ϕ and C from (18.23), the equilibrium equation (18.12) becomes

$$\sum_1^N C(t)_i'' M \phi_i + \sum_1^N C(t)_i K \phi_i = F \tag{18.25}$$

Multiplying (18.25) by Φ_j^T,

$$\sum_1^N C_i'' \phi_j^T M \phi_i + \sum_1^N C_i \phi_j^T K \phi_i = \phi_j^T F \tag{18.26}$$

Using (18.20) and K $\phi_i = \omega_i^2 M\phi_i$, (18.26) simplifies to

$$C_i'' + \omega_i^2 C_i = \phi_i^T F, \quad i = 1 - N \tag{18.27}$$

In other words, (18.27) reduces the solution of the N-degree-of-freedom equation (18.12) to solving N single-degree-of-freedom equations similar to (18.4). Depending on the forcing function F(t), this differential equation can be solved by a numerical method such as the Duhamel integral approach.

18.6.9 Mode Superposition: Damped Forced Response

Proceeding as in section 18.6.8, the differential equation for the case of damped forced motion transforms to

$$C_i^{''} + 2\,\beta_i\,C_i^{'} + \omega_i^2\,C_i = \phi_i^T\,F, I = 1 - N$$

where $2\beta_i = (\alpha + \beta\omega_i^2)$ \hfill (18.28)

Thus as in the case of an undamped system, Rayleigh damping allows the mode superposition method to be applied to damped forced system analysis as well, resulting in the N simultaneous differential equations being reduced to N differential equations in single unknowns.

18.6.10 Mass Participation Factors and Effective Mass

Equation (18.18) gives the force vector $F = -x''_{base}\,M\,I$
when the structure is subjected to an acceleration of the base during an earthquake. The force F_j for the jth mode is given by

$$F_j = \phi_j^T\,F = -x''_{base}\,\phi_j^T\,M\,I$$ \hfill (18.29)

As an example, for the four-storey frame considered in section 18.6.3, using the eigenvectors given in section 18.6.6,

$F_1 = \{m_1 \times 0.1191 + m_2 \times 0.2547 + m_3 \times 0.3788 + m_4 \times 0.4465\}$
$F_2 = \{m_1 \times 0.2469 + m_2 \times 0.3141 + m_3 \times (-0.0850) + m_4 \times (-0.4165)\}$
$F_3 = \{m_1 \times 0.3938 + m_2 \times (-0.1946) + m_3 \times (-0.2520) + m_4 \times 0.2249\}$
$F_4 = \{m_1 \times 0.1406 + m_2 \times (-0.2216) + m_3 \times 0.5345 + m_4 \times (-0.2767)\}$

If the displacement of all the masses is unity, the displacement vector Y is a unit vector. In this case the entire mass moves as a rigid body with an acceleration equal to the acceleration of the base. The rigid body mass M_R is defined as

$$M_R = Y^T\,M\,Y$$ \hfill (18.30)

From (18.21),

$Y = \Sigma\,C_i\,\phi_i = C_1\,\phi_1 + C_2\,\phi_2 + C_3\,\phi_3 + C_4\,\phi_4$
$Y^T = \Sigma\,C_i\,\phi_i^{\,T} = C_1\,\phi_1^{\,T} + C_2\,\phi_2^{\,T} + C_3\,\phi_3^{\,T} + C_4\,\phi_4^{\,T}$
$M_R = Y^T\,M\,Y = [\Sigma C_i \phi_i^T]M[\Sigma C_i \phi_i] = \Sigma C_i^2 \{\phi_i^T\,M\,\phi_i\}$

As the eigenvectors are M-normalized, $\phi_i^T\,M\,\phi_i = 1$. Therefore

$$M_R = \Sigma\,C_i^2$$ \hfill (18.31)

C_j^2 can be interpreted as the mass participating in the jth mode of vibration. In earthquake analysis the term earthquake participation factor is used. For the jth mode, the earthquake participation factor C_j is defined as

$$C_j = \phi_j^T M I \tag{18.32}$$

The term Effective Mass M_{Eff}^j for the jth mode is defined as

$$M_{Eff}^j = C_j^2, \quad \Sigma M_{Eff}^j = \Sigma C_j^2 = M_R \tag{18.33}$$

18.6.10.1 Mass Participation Factors: Example

For the four storey frame, earthquake participation factors and the effective masses for all four modes are

$M_{total} = 4 + 4 + 2 + 2 = 12$

$C_i = \phi_i^T M I$

$C_1 = 3.1418, C_2 = 1.2410, C_3 = 0.7426, C_4 = 0.1916$

$\Sigma C_i^2 = M_{total} = 12$

$$M_{Eff}^1 = C_1^2 = 9.871, \frac{M_{Eff}^1}{M_{Total}} = 82.2\%$$

$$M_{Eff}^2 = C_2^2 = 1.563, \frac{M_{Eff}^2}{M_{Total}} = 13.0\%$$

$$M_{Eff}^3 = C_3^2 = 0.552, \frac{M_{Eff}^3}{M_{Total}} = 4.6\%$$

$$M_{Eff}^4 = C_4^2 = 0.037, \frac{M_{Eff}^4}{M_{Total}} = 0.31\%$$

It is clear from the values of the participation factors and effective mass, that their values decrease as the mode number increases. The practical significance of this is that in general it is not necessary to include all the modes in a calculation. Only a few significant modes need to be included in order to obtain accurate enough results for practical problems. It is therefore not necessary to determine all the eigenvalues and the corresponding eigenvectors. This results in significant savings in practical computations.

18.7 RESPONSE ACCELERATION SPECTRUM

The differential equation for a single-degree-of-freedom system subjected to base acceleration is given by equation (18.9). For a given earthquake acceleration record x"$_{Base}$ and for fixed values of frequency ω and the ratio of damping to critical damping ξ, the differential equation can be solved by numerical methods and the displacement and acceleration of the mass can be determined. From the results of the analysis, maximum values of acceleration and displacement can be recorded. The analysis can be repeated for a range of ω and ξ. Fig. 18.12 shows a typical plot. As the analysis is carried out assuming elastic behaviour, it is known as the elastic acceleration spectrum.

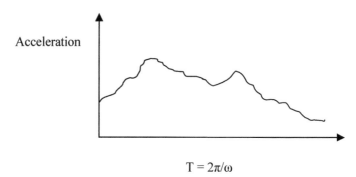

$$T = 2\pi/\omega$$

Fig. 18.12 An acceleration response spectrum

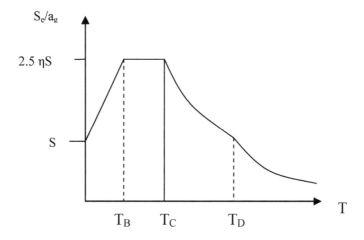

Fig. 18.13 Code elastic design acceleration spectrum

18.7.1 Design Elastic Response Acceleration Spectrum

Fig. 18.13 shows the acceleration spectrum according to Eurocode 8. The y-axis is the ratio of spectrum acceleration S_e/ design acceleration a_g and the x-axis is the period T of the single-degree-of-freedom structure:

$$\eta = \sqrt{\frac{10}{(5+\xi)}} \geq 0.55, \ \xi=\% \text{ critical damping present.}$$

Note that $\eta = 1$ for damping 5% of critical damping.

Table 18.4 gives numerical values for two types of earthquakes, called type 1 and type 2. The type 1 spectrum corresponds to a high-seismicity context and type 2 to moderate seismicity context. The type 2 corresponds to the site where the earthquake which contributes possesses a surface wave magnitude less than 5.5.

Table 18.4 Parameters of the acceleration spectrum

Ground type	S	T_B Seconds	T_C Seconds	T_D Seconds
A	1.0	0.15	0.40	2.0
	(1.0)	**(0.05)**	**(0.25)**	**(1.2)**
B	1.20	0.15	0.50	2.0
	(1.35)	**(0.05)**	**(0.25)**	**(1.2)**
C	1.15	0.20	0.60	2.0
	(1.50)	**(0.10)**	**(0.25)**	**(1.2)**
D	1.35	0.20	0.80	2.0
	(1.80)	**(0.10)**	**(0.30)**	**(1.2)**
E	1.40	0.15	0.50	2.0
	(1.60)	**(0.05)**	**(0.25)**	**(1.2)**

Bold vales in parenthesis refer to a type 2 spectrum.

18.7.2 Elastic Design Spectrum: Eurocode 8

For design purposes, the elastic response spectrum values are reduced by using the behaviour factor q which is an indication of the ductility of the structure. The more ductile the structure greater is the reduction. The equations for the design spectrum are

$$0 \le T \le T_B: \quad S_d(T) = a_g S[\frac{2}{3} + \frac{T}{T_B}(\frac{2.5}{q} - \frac{2}{3})]$$

$$T_B \le T \le T_C: \quad S_d(T) = a_g S[\frac{2.5}{q}]$$

$$T_C \le T \le T_D: \quad S_d(T) = a_g S[\frac{2.5}{q} \frac{T_C}{T}] \ge 0.2 a_g$$

$$T_D \le T: \quad S_d(T) = a_g S[\frac{2.5}{q} \frac{T_C T_D}{T^2}] \ge 0.2 a_g \quad (18.34)$$

18.8 METHODS OF ANALYSIS

In order to determine the forces generated due to seismic and other related forces, the structures are analysed as linearly elastic, but to allow for cracking only 50% of the gross stiffness is used. Very often additional simplifications are introduced. One common simplification is to analyse the structure as two independent planar structures in two orthogonal directions and finally, using combination rules (see section 18.8.2.2), results from the two analyses are combined to obtain the resultant forces on the entire structure. Two methods of analysis commonly employed are the lateral force method and modal response spectrum analysis. These methods are described in detail in the next sections.

18.8.1 Lateral Force Method of Analysis

This is a simple method commonly used. It is based on two fundamental assumptions. They are
- Only the first mode of vibration needs to be taken into account.
- A simple approximation to the mode shape is possible.

The code specifies that the fundamental period of vibration T_1 should satisfy
$T \le \{4\ T_C$ (see table 18.4), 2.0 seconds$\}$
The total base shear F_b is given by
$F_b = S_d\ (T_1)\ m\ \lambda$
m= total mass of the building above the foundation
$\lambda = 0.85$ if $T_1 \le 2\ T_C$ and the building has more than two storeys, otherwise $\lambda = 1$
For moment-resistant reinforced concrete space frames up to height $H = 40$ m,

$$T_1 = 0.075 \times H^{\,0.75} \hspace{4cm} (18.35)$$

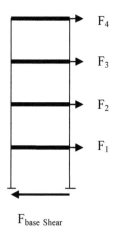

Fig. 18.14 Storey level forces.

18.8.1.1 Lateral force method: example

Determine the design earthquake forces for a four-storey single-bay concrete frame building, with a storey height of 3 m as shown in Fig. 18.14. The following data may be assumed:
(i). $a_{gR} = 1.2\ g = 1.2 \times 9.81 = 11.77$ m/sec^2
(ii). Importance class: III
(iii). Type 2 earthquake classification (moderate seismicity context)
(iv). Ground type: B
(v). Design for DCM ductility (lower ductility, higher design load)
(vi). Mass at storey levels: Storey 1 = 300 $\times 10^3$ kg, Storey 2 = 250 $\times 10^3$ kg,
                                   Storey 3 = 200 $\times 10^3$ kg, Storey 4 = 150 $\times 10^3$ kg

Solution:

$a_{gR} = 11.77$ m/sec^2

Importance class is III, $\gamma_1 = 1.2$ (from Table 18.2)

From (18.1), $a_g = \gamma_1 \times a_{gR} = 1.2 \times 11.77 = 14.13$ m/sec^2

Total height of building, $H = 4 \times 3 = 12$ m

From (18.35), the approximate period T_1 for the first mode for moment-resistant reinforced concrete space frames is

$T_1 = 0.075 \times H^{0.75} = T_1 = 0.075 \times 12^{0.75} = 0.48$ seconds

Behaviour factor q_0 for DCM level of design: from Table 18.3 $q_0 = 3 \ \alpha_u/\alpha_1$

For single-bay multistorey frames, $\alpha_u/\alpha_1 = 1.2$

$q_0 = 3 \times 1.2 = 3.6$

$k_w = 1.0$ for frames

From (18.2), $q = q_0 \, k_w = 3.6 \times 1.0 = 3.6 > 1.5$ (o.k.)

Design acceleration spectra S_d (t) for ground type B and a type 2 earthquake:

From Table 18.2,

$S = 1.35$, $T_B = 0.05$ seconds, $T_C = 0.25$ seconds, $T_D = 1.2$ seconds

$(T_C = 0.25) < (T_1 = 0.48) < (T_D = 1.2)$.

From (18.34),

$$T_C \leq T \leq T_D: \quad S_d(T) = a_g \, S[\frac{2.5}{q} \frac{T_C}{T}] \geq 0.2 \, a_g$$

S_d (T) $= a_g \times S \times (2.5/q) \times (T_C/T) = 14.13 \times 1.35 \times (2.5/3.6) \times (0.25/0.48)$
$= 6.90$ m/sec$^2 \geq (0.2 \, a_g = 2.83)$

S_d (T) $= 6.90$ m/sec^2

Total mass m $= (300 + 250 + 200 + 150) \times 10^3 = 9.0 \times 10^5$ kg

$\lambda = 0.85$ as the building is more than two storeys high

$(T_1 = 0.48) < (2T_C = 0.50)$

Total base shear force, $F_{base} = S_d \times m \times \lambda$

$F_{base} = 6.90 \times 9.0 \times 10^5 \times 0.85 = 5279 \times 10^3$ kg m/sec$^2 = 5279$ kN

$$F_i = F_{Base \ shear} \frac{m_i h_i}{\sum\limits_{j=1}^{N} m_j h_j} \tag{18.36}$$

Distribution of total base shear force to the forces at storeys: Assume that for the first mode of vibration, the displacement of the ith storey is proportional to height h_i above the base. Therefore, force at the ith level F_i is given by

$\sum m_i \, h_i = (300 \times 3 + 250 \times 6 + 200 \times 9 + 150 \times 12) \times 10^3 = 60 \times 10^5$

$F_1 = 0.15 \, F_{Base} = 792$ kN

$F_2 = 0.25 \, F_{Base} = 1320$ kN

$F_3 = 0.30 \, F_{Base} = 1584$ kN

$F_4 = 0.30 \, F_{Base} = 1584$ kN

Overturning moment M $= \sum F_i \, h_i = 43560$ kNm

Fig. 18.14 shows the storey level forces.

18.8.2 Modal Response Spectrum Method of Analysis

Modal spectrum analysis is a general method applicable to all types of structures. In this section, the method is illustrated by a simple example of a four-storey shear building shown in Fig. 18.8. It requires the values of the eigenvalues and corresponding eigenvectors for the structure in question. For the frame shown in Fig.18.8 the following data can be assumed.
   (i). Stiffness of columns: Note that to allow for cracking, it is suggested only 50% of the gross stiffness should be used in analysis.
Storey 1, 2, 3, 4 = (132, 100, 74, 54} $\times 10^6$ N/m.
(ii). Mass on beams:
Storey 1, 2, 3, 4 = (300, 250, 200, 150) $\times 10^3$ kg
Storey height = 3 m

$$\text{Stiffness matrix: K} = 10^8 \begin{bmatrix} 4.64 & -2.0 & 0 & 0 \\ -2.0 & 3.48 & -1.48 & 0 \\ 0 & -1.48 & 2.56 & -1.08 \\ 0 & 0 & -1.08 & 1.08 \end{bmatrix}$$

$$\text{Mass matrix: M} = 10^5 \begin{bmatrix} 3.0 & 0 & 0 & 0 \\ 0 & 2.5 & 0 & 0 \\ 0 & 0 & 2.0 & 0 \\ 0 & 0 & 0 & 1.5 \end{bmatrix}$$

The determinant of $(K - \omega^2 M) = 0$, when
$\omega^2 = (130.4, 746.4, 1542.1, 2335.8)$
Period of oscillation, $T = 2\pi/\omega = (0.55, 0.23, 0.16, 0.13)$ seconds
   Table 18.8 shows the eigenvectors for the four modes. The modes are similar to that shown in Fig. 18.11.

Table 18.8 Eigenvectors

$\varphi_1 \times \sqrt{10^5}$	$\varphi_1 \times \sqrt{10^5}$	$\varphi_3 \times \sqrt{10^5}$	$\varphi_1 \times \sqrt{10^5}$
0.1322	0.3031	0.3538	0.3144
0.2812	0.3716	−0.0456	−0.4252
0.4214	0.0064	−0.4595	0.3336
0.5131	−0.5027	0.3624	−0.1390

Modal participation factors C_i can be determined using the equation (18.32)
$C_i = \phi_i^T M I$, $I^T = [1\ 1\ 1\ 1]$
The modal participation factors are:
   $(C_1, C_2, C_3, C_4) = (2.7121, 1.0971, 0.5719, 0.3390) \times \sqrt{10^5}$
Modal masses are calculated from (18.33): they are
$(C_1^2, C_2^2, C_3^2, C_4^2) = (7.3537, 1.2035, 0.3271, 0.1149) \times 10^5$ kg
$\sum C_i^2$ = Total mass = 9.0×10^5 kg
Modal masses as a ratio of total mass are:
   $C_i^2 / \sum C_i^2 = (81.7, 13.4, 3.6, 1.3)$ %

Eurocode 8 suggests that

- All modes with effective modal masses greater than 5% of the total mass should be taken into account.
- The sum of the effective modal masses for the modes taken into account amounts to at least 90% of the total mass of the structure.

In this case therefore at least the first two modes should be taken into account.

18.8.2.1 Displacement spectrum

From (18.17), if the displacement x (t) is

$x(t) = x_0 \cos(\omega t + \varsigma)$

where x_0 = amplitude, t = time and ς = phase angle, the acceleration x" is

$x'' = -\omega^2 x_0 \cos(\omega t + \varsigma) = -\omega^2 x$

$$x = -\frac{1}{\omega^2} x'' = -(\frac{T}{2\pi})^2 x''$$

The design displacement spectrum $S_{d\,Dis}(T)$ is therefore obtained from the design acceleration spectrum $S_d(T)$ as

$$S_{d\,Dis}(T) = S_d(T) \times (T/2\pi)^2$$

From (18.34), the elastic design acceleration spectrum values are

$$T_B \leq T \leq T_C: \quad S_d(T) = a_g S [\frac{2.5}{q}]$$

$$T_C \leq T \leq T_D: \quad S_d(T) = a_g S[\frac{2.5}{q} \frac{T_C}{T}] \geq 0.2 a_g$$

In order to compare the results of the lateral forces determined for the example in section 18.8.1, choose $a_g = 14.13$ m/sec^2, q = 3.6.

From the eigenvalue analysis, T = (0.55, 0.23, 0.16, 0.13) seconds

The design elastic acceleration spectrum values are

S = 1.35, (T_B, T_C, T_D) = (0.05, 0.25, 1.2) seconds

From the above data, the value of displacement spectrum values $S_{d\,Dis}$ can be calculated as shown in Table 18.6. The storey level displacements are shown in Table 18.7.

Table 18.6 Values of the displacement spectrum

T	$S_d(T)$	$S_{d\,Dis}(T) \times 10^2$	$C_i / \sqrt{10^5}$	$S_{d\,Dis}(T) \times C_i/\sqrt{10^5}$
0.55	6.021	4.614	2.7118	0.1251
0.23	13.247	1.775	1.0971	0.0195
0.16	13.247	0.859	0.5719	0.0049
0.13	13.247	0.567	0.3390	0.0019

Table 18.7 Values of storey level displacements

$S_{d\,Dis}(T) \times C_i \times \varphi_i$			
i = 1	i = 2	i = 3	i=4
0.016538	0.005902	0.001738	0.000604
0.035176	0.007236	−0.00022	−0.00082
0.052718	0.000125	−0.00226	0.000641
0.064194	−0.00979	0.001781	−0.00027

18.8.2.2 Combining modal values: SRSS AND CQC rules

Although the displacements have been computed for the four modes, they cannot be algebraically added because the maximum values do not occur at the same time for all the modes. Two different ways of getting a more realistic value of the maximum displacements are used in practice. They are known as SRSS (**S**quare **r**oot of the **S**um of the **S**quares) and CQC (**C**omplete **Q**uadratic **C**ombination) rules. The corresponding equations are

$$E_{SRSS} = \sqrt{\sum_{i=1}^{N} E_i^2} \tag{18.37}$$

$$E_{CQC} = \sqrt{\sum_{i=1}^{N}\sum_{j=1}^{N} r_{ij} E_i E_j} \tag{18.38}$$

where E_i = result from mode i
Note that the results of CQC will be the same as results for SRSS if
$r_{ij} = 0$ if $i \neq j$ and $r_{ij} = 1$ if $i = j$
r_{ij} is known as correlation coefficient of the two modes i and j

$$r_{ij} = \frac{8\sqrt{\xi_i\xi_j}(\xi_i + \rho\xi_j)\rho^{1.5}}{[(1-\rho^2)^2 + 4\xi_i\xi_j\rho(1+\rho^2) + 4(\xi_i^2 + \xi_j^2)\rho^2]} \tag{18.39}$$

where ρ = ratio of periods T_i/T_j, ξ_i = damping ratio for mode i
The code suggests that the response in two modes i and j may be taken as independent of each other i.e. $r_{ij} = 0$, if their periods T_i and T_j satisfy the condition $T_j \leq 0.9\,T_i$, $T_j \leq T_i$.
If all the modes can be treated as independent of each other, i.e. $r_{ij} = 0$, $i \neq j$, then the SRSS combination can be used.

18.8.2.3 Rayleigh damping

In order to carry out a CQC calculation, it is necessary to calculate the damping coefficients for different modes. Generally, because of less cracking, damping in prestressed concrete structures is much smaller than in reinforced concrete structures. It is generally assumed to be of the order of 2% to 5%.

If $\xi = 0.05$ (5% of critical damping) for the first two frequencies $\omega = 11.42$ and 27.32, then, assuming Rayleigh damping given by (18.15),

$$\xi_1 = 0.05 = 0.5(\alpha/11.42 + \beta \times 11.42)$$
$$\xi_2 = 0.05 = 0.5(\alpha/27.32 + \beta \times 27.32)$$

Solving for α and β: $\alpha = 0.8054$, $\beta = 2.5813 \times 10^{-3}$

ξ for the remaining frequencies are given by:

$$\xi_3 = 0.5(\alpha/39.27 + \beta \times 39.27) = 0.06$$
$$\xi_4 = 0.5(\alpha/48.33 + \beta \times 48.33) = 0.07$$

Table 18.8 shows the values of ξ and ρ for all four modes and Table 18.9 shows the values of the correlation coefficients.

Table 18.8 Values of η and ρ

i	ξ	T_i	ρ			
			T_1/T_i	T_2/T_i	T_3/T_i	T_4/T_i
1	0.05	0.55	1	0.420	0.281	0.229
2	0.05	0.23	2.381	1	0.668	0.546
3	0.06	0.16	3.562	1.496	1	0.817
4	0.07	0.13	4.362	1.832	1.225	1

Table 18.9 Values of correlation coefficient r_{ij}

i	r_{1i}	r_{2i}	r_{3i}	r_{4i}
1	1	0.0113	0.0057	0.0047
2	0.0113	1	0.0681	0.0363
3	0.0057	0.0681	1	0.2909
4	0.0047	0.0363	0.2909	1

Table 18.10 Values of storey level displacements

Storey level	Storey level displacement, m					
	Mode 1	Mode 2	Mode 3	Mode 4	SRSS	CQC
1	0.0165	0.0059	0.0017	0.0006	0.0177	0.0176
2	0.0352	0.0072	−0.0002	−0.0008	0.0359	0.0360
3	0.0527	0.0001	−0.0023	0.0006	0.0528	0.0527
4	0.0642	−0.0098	0.0018	−0.0003	0.0650	0.0648

18.8.2.4 'Resultant' storey level displacements

Table 18.10 shows the storey level displacements for the four modes (from Table 18.7) as well as the 'resultant' displacement according to SRSS and CQC rules.

18.8.2.5 'Resultant' storey level forces

Knowing the storey level displacements, the storey level forces can be obtained as the product of the stiffness matrix and the displacement vector of the storey level displacement. Table 18.11 shows the storey level forces according to SRSS and

CQC rules. The forces obtained from the lateral force method from section 18.8.1 are also given for comparison. The base shear is simply the sum of the storey level forces. In this case there is not much difference between SRSS and CQC values, mainly because the correlation coefficients r_{ij} are very small if $i \neq j$. Compared with the CQC prediction, prediction by the lateral force method overestimates the base shear force by 14%.

Using the CQC storey displacements, shear and moments in the columns can be determined. The value of drift is the relative displacement between top and bottom of columns. They are shown in Table 18.12.

Table 18.11 Values of storey level forces

Storey level	Storey level force, kN		
	SRSS	CQC	Lateral force
1	1008	967	792
2	1160	1205	1320
3	1177	1166	1584
4	1317	1308	1584
Base shear	4662	4646	5280

Table 18.12 Values of Moments and shear forces in columns

Storey level	Column forces		
	Drift, mm	Shear, kN	Moments, kNm
1	17.6	2322	3483
2	18.4	1839	2759
3	16.7	1237	1856
4	12.1	654	981

18.9 COMBINATION OF SEISMIC ACTION WITH OTHER ACTIONS

The seismic forces will have to be combined with the gravity loads of all masses appearing in the following equation:

$$\sum G_{k,j} \text{ "+" } \sum \psi_{E,i} \times Q_{k,i} \tag{18.40}$$

where $\psi_{E,i}$ is the combination coefficient for variable action i.

$\psi_{E,i} = \varphi \times \psi_{2i}$

ψ_{2i} values are given in Table 14.1 of Chapter 14. φ values are given in Table 18.13.

Table 18.13 Values of φ

Type of variable action	Storey	φ
Categories A to C*	Roof	1.0
	Storeys with correlated occupancies	0.8
	Independently occupied storeys	0.5
Categories D to F** and archives		1.0

*Category: A = areas of domestic and residential activities, B = office areas, C = areas where people may congregate
**D= shopping areas, E_1 = areas susceptible to accumulation of goods, including access areas, E_2= industrial use, F = actions induced by forklifts.

18.10 BASIC PRINCIPLES OF CONCEPTUAL DESIGN

Some of the basic principles to be observed in the design of earthquake resistant buildings are as follows:
- Structural simplicity characterized by clearly identifiable and direct paths of transmission of forces.
- Distribution of the structural elements along the height of the building and a symmetric layout of structural elements in plan. This also minimizes torsional forces generated by earthquake forces.
- Since horizontal seismic action is bi-directional, it is necessary that the building should be able to resist the forces in any direction.
- The floor slabs should be stiff enough to transmit the horizontal forces to the vertical elements.
- Force-resisting elements should be placed close to the periphery in plan in order to provide high torsional resistance.

In addition to the above points, it is important to reduce the torsional moment acting on the structure. This is done by keeping the centre of mass and centre of rigidity close to each other.

Centre of mass refers to the resultant position of the earthquake forces, which are proportional to the mass at any storey level. Similarly, centre of rigidity refers to the resultant position of the stiffness of the elements of the structure.

Fig. 18.15 shows some typical plans of structures with the following comments on their structural performance:
- Plan (a) is symmetrical but its torsional resistance will be low because the stair/lift well is located at the centre and only columns at the periphery provide torsional resistance.
- Plan (b) is also symmetrical but its torsional resistance will be good because the stair/lift well is located towards the periphery.
- Plan (c) is also symmetrical but its torsional resistance will be high because the stair/lift well is located towards the periphery and in addition on the longer side some of the columns have been replaced by a shear wall.

- Plan (d) is unsymmetrical and seismic forces will induce large twisting moments.

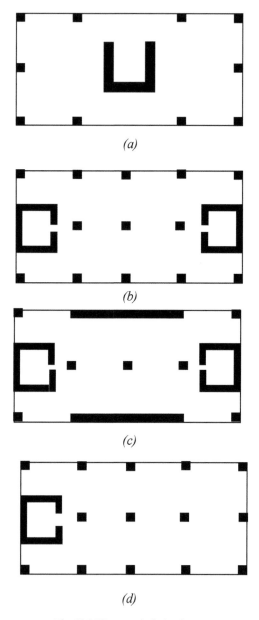

(a)

(b)

(c)

(d)

Fig. 18.15 Some typical plan forms

Fig. 18.16 shows some possible elevations of a coupled wall, with the following comments on their structural performance:

- Elevation (a) is uniform along the height, a very desirable feature.

- Elevation (b) has an abrupt weakness at the lower level where the whole of the base shear acts. This is a highly undesirable feature. The code explicitly forbids 'soft storey' design as it will lead to sway failure at the lower storey level.
- Elevation (c) has an abrupt discontinuity but it is not as undesirable as (b).
- Elevation (d), although it shows a smooth transition in variation in elevation with height, has the very undesirable feature of having a larger mass towards the top, where the earthquake forces will be larger than that at the lower level.

Table 18.14 shows the consequences of structural regularity on analysis and design. In this connection a planar model refers to the structure being analysed in two orthogonal directions as a set of two-dimensional structures. Note that an increase in the value of q_0 results in a greater force, for which the structure needs to be designed.

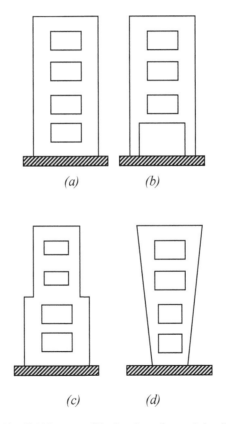

Fig. 18.16 Some possible elevations of a coupled wall

Table 18.14 Consequences of structural regularity

Regularity		Allowed simplification		Behaviour factor, q_0 $
Plan	Elevation	Model	Linear elastic analysis	(For linear analysis)
Yes	Yes	Planar	Lateral force	Reference value
Yes	No	Planar	Modal	0.8 ×Reference value
No	Yes	Spatial	Lateral force	Reference value
No	No	Spatial	Modal	0.8 ×Reference value

$See Table 18.3 for q_0 factors.

18.11 DETAILING FOR LOCAL DUCTILITY: BEAMS

In the design of frames the 'Weak beam-strong column' concept is a fundamental design criterion. This requirement is satisfied by the condition

$$\Sigma M_{Rc} \geq 1.3 \Sigma M_{Rb} \tag{18.41}$$

where
ΣM_{Rc} = sum of design values of moment of resistance of columns framing at the joint
ΣM_{Rb} = sum of design values of moment of resistance of beams framing at the joint

The regions in the beam where plastic hinges form need to be reinforced to provide a ductile region. The ductility of the region depends on
 • ductility of steel reinforcement,
 • ductility of concrete

One of the most effective ways of increasing the ductility of concrete is to provide closely spaced stirrups which aid in generating a three-dimensional compression state. It also has additional advantages in the sense that it improves shear strength and limits shear deformation and prevents the buckling of the main reinforcement during reversal of moment direction.

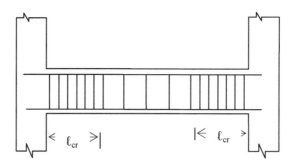

Fig. 18.17 Critical regions for local ductility

As shown in Fig. 18.17, the critical length ℓ_{cr} equal to the depth of the beam is where plastic hinges are expected to form. The reinforcements in this region should satisfy the following conditions:

- At the compression zone, the longitudinal reinforcement is not less than half of that provided in the tension maximum zone. This is in addition to any compression reinforcement needed to resist ULS forces.
- The maximum ratio of the tension zone steel ratio ρ does not exceed ρ_{max} given by

$$\rho_{max} = \rho' + \frac{0.0018}{\mu_\varphi \, \varepsilon_{sy,d}} \frac{f_{cd}}{f_{yd}} \tag{18.42}$$

where tension steel ratio $\rho = A_{st}/(bd)$, compression steel ratio $\rho' = A_{sc}/(bd)$
b = width of compression flange (***not of the web***), d = effective depth
$\varepsilon_{sy,d}$ = design value of strain at yield in steel
f_{yd} = design value of yield stress in steel
f_{cd} = design value of concrete compressive strength
μ_φ = curvature ductility factor

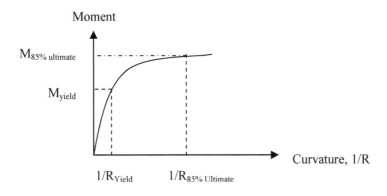

Fig. 18.18 Moment-curvature relationship for primary beams

As shown in Fig. 18.18, μ_φ is the ratio of curvature at 85% of ultimate moment of resistance/curvature at yield moment. The minimum value of μ_φ is given by

$$\begin{aligned} \mu_\varphi &= 2q_0 - 1, & T_1 \geq T_c \\ \mu_\varphi &= 1 + 2(q_0 - 1) \, T_c/T_1, & T_1 < T_c \end{aligned} \tag{18.43}$$

See Table 18.3 for q_0 values and Table 18.4 and Fig. 18.13 for values of T_c.
T_1 = fundamental period of vibration.

- Along the entire length of the (primary) beam, the tension steel ratio ρ shall not be less than ρ_{min}:

$$\rho_{min} = 0.5 \frac{f_{ctm}}{f_{yk}}$$

where f_{yk} = characteristic tensile strength of reinforcement
f_{ctm} = mean value of axial tensile strength of concrete

- Within the critical regions of primary beams, hoops satisfying the following conditions shall be provided:
    1. The minimum diameter d_{bw} of the hoops should be 6 mm.
    2. The maximum spacing s in mm of the hoops should not exceed
  s = Min {depth of beam/4; 24 × hoop dia.; 8 × min. longitudinal bar dia.}
    3. The first hoop shall be placed at distance ≤ 50 mm from the face of the column

18.12 DETAILING FOR LOCAL DUCTILITY: COLUMNS

In columns, the value $v_d \leq 0.65$, where

$$v_d = \frac{\text{Axial force}, N_{ED}}{\text{Area of cross section}, A_c \times f_{cd}} \qquad (18.44)$$

The longitudinal reinforcement ratio ρ should satisfy the condition
$$0.01 \leq \rho \leq 0.04$$
In addition, at least one intermediate bar shall be provided between corner bars along each side.

   In the case of columns, if plastic hinges are likely to form, the regions up to a distance of ℓ_{cr} from both ends of the column are considered critical regions as shown in Fig. 18.19.

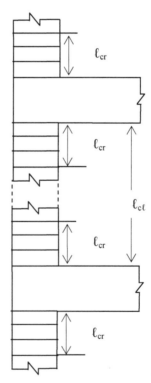

Fig. 18.19 Critical zones in columns

ℓ_{cr} = Max [h_c, clear height of column/6, 0.45] in meters
h_c = Largest cross sectional dimension of column in meters.
The hoops provided should satisfy the conditions $\omega_{wd} \geq 0.08$

$$\alpha\, \omega_{wd} \geq 30\, \mu_\phi\, v_d\, \varepsilon_{sy,d}\, \frac{b_c}{b_0} - 0.035$$

$$\omega_{wd} = [\frac{\text{Volume of confining hoops}}{\text{Volume of concrete core}}] \times [\frac{f_{yd}}{f_{cd}}]$$

$$v_d = \frac{\text{Axial force, } N_{ED}}{\text{Area of cross section, } A_c \times f_{cd}}$$

$$\alpha_n = 1 - \frac{1}{6 b_0 h_0} \sum_1^n b_i^2, \quad \alpha_s = (1 - \frac{s}{2 b_0})(1 - \frac{s}{2 h_0})$$

$$\alpha = \alpha_n\, \alpha_s$$

where
h_0, b_0 = depth and width of confined core (to the centreline of the hoops)
h_c, b_c = gross cross sectional depth and width
b_i = distance between consequtive engaged bars (see Fig. 18.20)
n = No. of longitudinal bars laterally engaged by hoops or cross ties
s = spacing of loops in the along the column
The maximum spacing should not exceed, s_{max}
where s_{max} = Min ($b_0/2$, 175, 8 × minimum diameter of longitudinal bar)

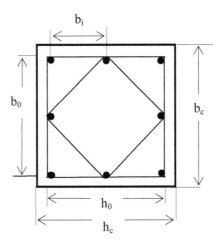

Fig. 18.20 Confinement of concrete

The hoop reinforcement needs to be provided where plastic hinges are likely to form. In design, because of the strong column-weak beam concept, this will happen only at the base of columns.

Example: Fig. 18.21 shows a column 500 mm square with twelve 25 mm bars. It carries an axial load of 3300 kN along with bi-axial moments of 500 kNm. Check the adequacy of the spacing for the hoops at 100 mm c/c.

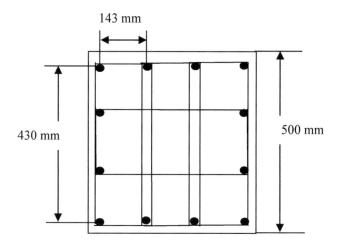

Fig. 18.21 Column cross section

Cover to steel: 30 mm, hoop bar diameter = 10 mm, f_{yk} = 500 MPa,
f_{ck} = 40 MPa, E_{steel} = 195 GPa, $f_{yd} = f_{yk}/ (\gamma = 1.15) = 435$ MPa,
$f_{cd} = f_{ck}/ (\gamma = 1.5) = 26.7$ MPa, $\varepsilon_{sy,\,d} = f_{yd}/ E_{steel} = 2.23 \times 10^{-3}$
N_{ED} =3300 kN,
$v_d = 3300 \times 10^3/ (500 \times 500 \times 26.7) = 0.49 < 0.65$
$b_c = h_c =$ (Gross cross section width and depth) = 500 mm
$b_0 = h_0 =$ (Core section width and depth) = $(500 - 30 - 30 - 10) = 430$ mm
$q_0 = 3.6$ for a frame system,
$T_1 = 0.48$ seconds, $T_c = 0.25$ seconds
then $\mu_\varphi = 2\times 3.6 - 1 = 6.2$

$$30\,\mu_\phi\, v_d\, \varepsilon_{sy,d}\, \frac{b_c}{b_0} - 0.035 = 30\times 6.2 \times 0.49 \times 2.23 \times 10^{-3} \times \frac{500}{430} - 0.035 = 0.20$$

$b_i = 143$ mm, N = 12, s = 100 mm

$$\alpha_n = 1 - \frac{1}{6 b_0\, h_0} \sum_1^n b_i^2 = 1 - \frac{8\times 143^2}{6\times 430\times 430} = 0.85$$

$$\alpha_s = (1 - \frac{s}{2\, b_0})(1 - \frac{s}{2\, h_0}) = (1 - \frac{100}{2\times 430}) \times (1 - \frac{100}{2\times 430}) = 0.78$$

$$\alpha = \alpha_n\, \alpha_s = 0.85 \times 0.78 = 0.66$$

$$(\alpha = 0.66) \times \omega_{wd} \ge [30\,\mu_\phi\, v_d\, \varepsilon_{sy,d}\, \frac{b_c}{b_0} - 0.035 = 0.20]$$

$$\omega_{wd} \ge 0.30$$

Volume of core = 430 × 430 × (spacing = 100 mm) = 18.5×10^6 mm^3
Area of cross section of 10 mm bar = 78.5 mm^2
There are four rectangular hoops of dimension $b_o \times b_i$
Volume of hoops = 4 × Area of cross section × 2 × ($b_0 + b_i$) = 0.36×10^6 mm^3
ω_{wd} = (0.36/18.5) × (f_{yd} = 435)/ (f_{cd} = 26.7) = 0.32 > 0.30 required.
Check: Minimum ω_{wd}: $\omega_{wd} \geq 0.08$, o.k.
Check maximum spacing:
b_0 = 430 mm, minimum diameter of longitudinal bar = 25 mm,
s_{max} = min (430/2, 175, 8 × 25) = 175 mm
Spacing provided 100 mm < s_{max}
Four 10 mm hoops at a vertical spacing of 100 mm are adequate.

18.13 DESIGN SHEAR FORCE IN BEAMS AND COLUMNS

Fig. 18.21 shows the equations for calculating design shear force in beams due to earthquake forces. These have to be combined with shear forces due to other forces acting on the beams along with the seismic forces.

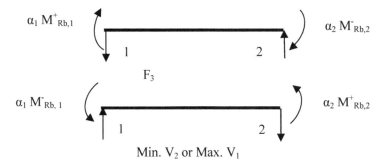

Fig. 18.21 Shear force calculation in beams

$$\alpha_1 = \gamma_{Rd} \times Min[1, \frac{\sum M_{Rc}}{\sum M_{Rb}}]_1, \quad \alpha_2 = \gamma_{Rd} \times Min[1, \frac{\sum M_{Rc}}{\sum M_{Rb}}]_2$$

where
$M^+_{Rb, i}$ and $M^-_{Rb, i}$ are the design resistance of beam at end i in the sagging and hogging sense respectively.
$[\sum M_{Rc}/ \sum M_{Rb}]1$ = (Sum of design resistance of columns at joint 1)/(Sum of design resistance of beams at joint 1)
$[\sum M_{Rc}/ \sum M_{Rb}]2$ = (Sum of design resistance of columns at joint 2)/(Sum of design resistance of beams at joint 2)

γ_{Rd} = factor allowing for over-strength due to strain hardening of steel and confinement of concrete, taken as equal to 1.0 for DCM designs

Fig. 18.22 shows the equations for calculating design shear force in columns due to earthquake forces.   These have to be combined with shear forces due to other forces acting on columns along with the seismic forces.

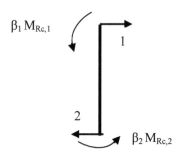

Fig. 18.22 Shear force calculation in columns

$$\beta_1 = \gamma_{Rd} \times Min[1, \frac{\Sigma M_{Rb}}{\Sigma M_{Rc}}]_1, \quad \beta_2 = \gamma_{Rd} \times Min[1, \frac{\Sigma M_{Rb}}{\Sigma M_{Rc}}]_2$$

$M_{Rc, 1}$ and $M_{Rc, 2}$ are the design resistance of column at end 1 and end 2 respectively.

$[\Sigma M_{Rb}/ \Sigma M_{Rc}]1$ = (Sum of design resistance of beams at joint 1)/(Sum of design resistance of columns at joint 1)

$[\Sigma M_{Rb}/ \Sigma M_{Rc}]2$ = (Sum of design resistance of beams at joint 2)/(Sum of design resistance of columns at joint 2)

γ_{Rd} = A factor allowing for over strength due to strain hardening of steel and confinement of concrete taken as equal to 1.1

18.14 DESIGN PROVISIONS FOR DUCTILE WALLS

Eurocode 8 provides detailed rules for redistribution of forces in walls, including coupled walls. In designing the walls, from the analysis of the structure the moment envelope is calculated. It is permissible to assume that the moment envelope is linear if the structure does not exhibit significant discontinuities of mass and stiffness over its height. The design envelope is obtained by vertically shifting the envelope by an amount equal to the tension shift as shown in Fig. 18.23.

The idea of 'tension shift' arises from purely static considerations. Consider the equilibrium of forces across a diagonal section of a wall as shown in Fig. 18.24. Taking moments at the centroid of compression at section 1-2, the moment M_1 at section 1-1 is given by

$$M_1 = T_2 \times Z,$$

where Z = lever arm and T_2 = the resultant tensile force at section 2-2.

Assuming that Z is approximately equal to the width ℓ_w of the wall and the inclination to the horizontal of the diagonal crack is θ, the moment M_2 at section 2-2 is related approximately to M_1 through the shear force V at section 1-1 as follows:

$$M_2 = M_1 - V \times \ell \tan \theta$$

Replacing M_1 by $T_2 \times Z$,

$$M_2 = T_2 \times Z - V \times \ell \tan \theta$$
$$T_2 \times Z = M_2 + V \times \ell \tan \theta \approx M_1$$

The tension force T_2 at section 2-2, which is the top of the diagonal crack, is closely related to the moment at the moment at 1-1, which is the bottom of the crack. The distance $\ell_w \tan \theta$ is known as the tension shift.

The design shear force is taken as 50% higher than that obtained from analysis.

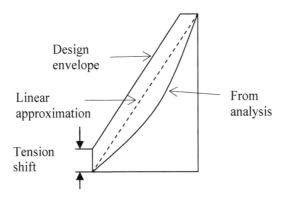

Fig. 18.23 Design moment envelope

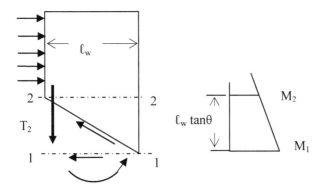

Fig. 18.24 Tension shift

18.15 REFERENCE TO EUROCODE 8 CLAUSES

In this chapter, the following clauses in Eurocode 8 have been referred to:

CHAPTER 19

MISCELLANEOUS TOPICS

19.1 INTRODUCTION

In this chapter some miscellaneous design topics will be discussed briefly.

19.2 UNBONDED DESIGN

In Chapters 4 to 6, design using bonded cables was discussed. In bonded cables, there is a full bond between steel and concrete and therefore the strains in concrete and steel are equal. However in the case of unbonded design, the contact between steel and concrete is only at the anchors. The strain in steel is therefore the average of the strain in concrete between the anchors. Because of this averaging of strain, the tensile force developed will be very much smaller than in the case of bonded tendons and consequently leads to a much smaller value of the ultimate moment. In addition, without the presence of unstressed steel, only a few large cracks occur.

19.3 DESIGN OF A POST-TENSIONED BOX GIRDER BRIDGE

Fig. 16.63 shows the cross section of the box girder. In order to keep the calculations simple, it is assumed that there is no variation in the cross section along the span, although in practice often the cross sections at mid-span and at support tend to be different. It is used as a simply supported beam of 30 m.

The cross-sectional properties of the bridge are

Area of cross section, $A = 9.12 \times 10^6$ mm^2

Self weight $= 9.12 \times 10^6 \times 10^{-6} \times 24 = 218.88$ kN/m

$y_b = 1315$ mm, $y_t = 1185$ mm
Second moment of area, $I = 9.168 \times 10^{12}$ mm^4

Section moduli: $Z_t = 7737 \times 10^6$ mm^3, $Z_b = 6972 \times 10^6$ mm^3

19.3.1 Calculation of Live Loadings at SLS

(i). Live loading: The live loadings to be used are as detailed in section 16.5.8. All γ factors are taken as unity. The ψ_0 values for UDL due to vehicular traffic and pedestrian load is 0.40. For a tandem load $\psi_0 = 0.75$.

For tandem load, $\alpha_Q = 0.9$ for lane 1 and $\alpha_Q = 1.0$ for lanes 2 and 3
(a). Super dead load on foot path due to 120 mm black top weighing 24 kN/m^3
    SDL = $120 \times 24 \times 10^{-3} = 2.88$ kN/m^2
(b). Super dead load on lanes due to 50 mm black top weighing 24 kN/m^3,
    SDL = $50 \times 24 \times 10^{-3} = 1.2$ kN/m^2
(c). Footpath load = 3 kN/m^2 × ($\psi_0 = 0.40$) = 1.2 kN/m^2
(d). Lane A: UDL = = 9 kN/m^2 × ($\psi_0 = 0.40$) = 3.6 kN/m^2,
Tandem load: four wheel loads each W = $(300/2) \times (\alpha_Q = 0.9) \times (\psi_0 = 0.75)$
                             = 101.25 kN.
(e). Lane B: UDL = = 2.5 kN/m^2 × ($\psi_0 = 0.40$) = 1.0 kN/m^2,
Tandem load: four wheel loads each W = $(200/2) \times (\alpha_Q = 1.0) \times (\psi_0 = 0.75)$
                             = 75.0 kN.
(f). Lane C: UDL = = 2.5 kN/m^2 × ($\psi_0 = 0.40$) = 1.0 kN/m^2,
Tandem load: four wheel loads each W = $(100/2) \times (\alpha_Q = 1.0) \times (\psi_0 = 0.75)$
                             = 37.5 kN.

19.3.2 Calculation of Total Loads at SLS

(a). Self weight = 218.88 kN/m
(b). Uniformly distributed live loads over the whole span:
(i). 120 mm black top over the two cantilevers 2 m wide
    = $2 \times 2.88 \times 2.0 = 11.52$ kN/m
(ii). 50 mm black top over the 9 m wide box
    = $1.20 \times 9.0 = 10.80$ kN/m
(iii). Foot path: 1.2 kN/m^2 acts over two cantilevers each 2 m wide
    = $2 \times 1.2 \times 2.0 = 4.80$ kN/m
(iv). UDL over lanes 1, 2 and 3: $(3.6 + 1.0 + 1.0) \times 3 = 16.8$ kN/m
Total = $11.52 + 10.80 + 4.80 + 16.81 = 43.92$ kN/m
(c). Two wheel loads separated by 1.2 m:
Axle loads over lanes 1, 2 and 3:
Each wheel load = $2 \times (101.25 + 75 + 37.5) = 427.5$ kN
Total uniformly distributed load = 218.88 (self weight) + 43.92 (Live load)
                             = 262.8 kN/m

19.3.3 Calculation of Bending Moments at SLS

Fig. 19.1 shows the influence line for the bending moment. A uniformly distributed load covers the entire span and wheel loads are positioned for maximum effect.
(i). Bending moment at mid-span: A uniformly distributed load covers the whole span and the wheels either straddle the mid-span or there is one wheel at mid-span and the other to the left or right at 1.2 m from mid-span. At any section at a distance x from the left hand support,
M = $0.5 \times 262.8 \times x \times (L - x) + \{427.4 / L\} \times x \times (L - x) \times [1 + 1 - 1.2 / (L - x)]$

At mid-span,
$$M = 0.5 \times 262.8 \times 15 \times (30 - 15)$$
$$+ \{427.4 / 30\} \times 15 \times (30-15) \times [2 - 1.2/ (30-15)]$$
$$= 29565 + 6154.56 = 35719.6 \text{ kNm.}$$

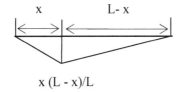

$$x \quad \quad L\text{-}x$$

$$x (L - x)/L$$

Fig. 19.1 Influence line for bending moment

19.3.4 Bending Stresses at SLS

Using the bending moments calculated, the stresses at mid-span are
$M_{sw} = 218.88 \times 30^2/ 8 = 24624$ kNm
$\sigma_{top} = M_{sw}/Z_t = 3.18$ MPa, $\sigma_{bottom} = M_{sw}/Z_b = 3.53$ MPa
$M_{service} = 35719.6$ kNm
$\sigma_{top} = M_{service}/Z_t = 4.62$ MPa, $\sigma_{bottom} = M_{service}/Z_b = 5.12$ MPa

19.3.5 Thermal Stresses: Heating and Cooling

From sections 16.14.2 and 16.14.3, the stresses due to heating and cooling are
Heating: $\sigma_{Top} = -3.07$ MPa, $\sigma_{Bottom} = -0.95$ MPa,
Cooling: $\sigma_{Top} = 1.66$ MPa, $\sigma_{Bottom} = 2.44$ MPa,

19.4 DETERMINATION OF PRESTRESS

The prestress required at mid-span is calculated by drawing the Magnel diagram in the usual manner.
(i). From the moment calculations in section 19.3.4, stresses due to self weight and SLS moments are:
$M_{sw}/ Z_t = 3.18$ MPa, $M_{sw}/ Z_b = 3.53$ MPa
$M_{service}/ Z_t = 4.62$ MPa, $M_{service}/ Z_b = 5.12$MPa
(ii). From the thermal stresses due to cooling calculated in section 19.3.5, tensile thermal stresses are:
 1.66 MPa at top and 2.44 MPa at bottom
(iii). Taking concrete strength as $f_{ck} = 40$ MPa and $f_{cki} = 30$ MPa, the permissible stresses are: $f_{tt} = 2.9$ MPa, $f_{tc} = -18.0$ MPa, $f_{st} = 3.5$ MPa, $f_{sc} = -24.0$ MPa
(iv). Load factors on prestressing force are

$\gamma_{superior} = 1.1$ and $\gamma_{Inferior} = 0.9$

(v). Cross-sectional properties are:

$A = 9.12 \times 10^6 \text{ mm}^2$, $Z_t = 7737 \times 10^6 \text{ mm}^3$, $Z_b = 6972 \times 10^6 \text{ mm}^3$

Substituting in equations 4.5 and 4.6 at transfer and 4.9 and 4.10 at service, the Magnel equations are:

Transfer:

$-10.965 + 0.013 \text{ e} \leq 4.61 \text{ y}$

$-10.965 - 0.014 \text{ e} \geq -16.31 \text{ y}$

Service:

$-10.965 + 0.013 \text{ e} \geq -23.38 \text{ y}$

$-10.965 - 0.014 \text{ e} \leq -4.50 \text{ y}$

Where $y = 10^8 / P_s$

Fig. 19.2 shows the Magnel diagram

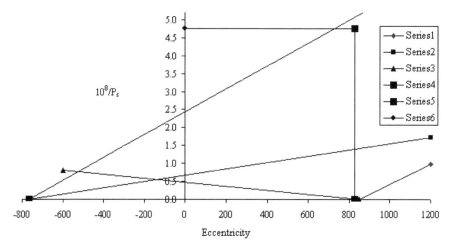

Fig. 19.2 Magnel diagram

Using on each side two cables containing thirty two 7-wire strands each contained in a 130 mm duct with a shift of 22 mm, the maximum eccentricity available is

$e = y_b - 400 \text{ (bottom flange)} - 130/2 \text{ (duct)} - 22 \text{ (shift)} = 828 \text{ mm}$

From the Magnel diagram, for e = 828 mm, $10^8 / P_s = 4.75$

$P_s = 21053 \text{ kN}$, $P_{Jack} = P_s / 0.75 = 28070 \text{ kN}$

For a strand with a cross-sectional area of 150 mm², the maximum force at jacking is 215 kN.

The number of strands N required is

$N = P_{Jack} / 215 = 28070 / 215 = 132$

Keeping this force constant and substituting in equations 5.1 to 5.4, the permissible cable zone can be calculated. Fig. 19.3 shows the feasible cable zone

for a symmetrical half of the span.   Fig. 19.4 shows the final cable arrangement.
Two deviators at third points along the span are used.   The cable is anchored at the
support at the centroidal axis.

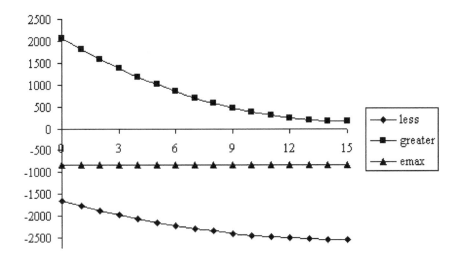

Fig. 19.3 Feasible cable zone

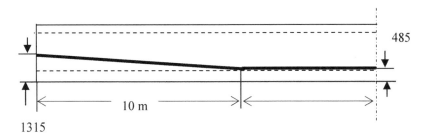

Fig. 19.4 Cable arrangement

Fig. 19.5 Forces acting on the cable

The inclination of the cable to the horizontal is
$\tan \theta = (1315 - 485) \times 10^{-3} / 10.0 = 0.083$, $\sin \theta = 0.083$, $\cos \theta = 0.997$

$P_s \sin \theta = 28070 \times 0.083 = 2330$ kN

Fig. 19.5 shows the forces acting on the cable. Fig. 19.6 shows the bending moment distribution in the beam due to the reversed loads acting on the cable. This represents the bending moment due to prestressing forces.

23298

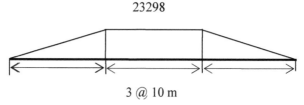

3 @ 10 m

Fig. 19.6 Bending moment distribution due to prestress

Fig. 19.7 shows the bending moment distributions due to
- self weight and external loads
- prestress force
- net bending moment

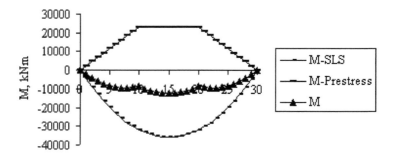

Fig. 19.7 Distribution of bending moments

19.5 CRACKING MOMENT

After having decided on the value of the prestressing force, the moment to cause cracking can be determined as follows:

$f_{ck} = 40$ MPa, From (3.2), $f_{ctm} = 0.30 \times f_{ck}^{0.667} = 3.5$ MPa

From (3.3), $f_{ctk, \, 0.05} = 0.7 \times f_{ctm} = 2.5$ MPa

$f_{ctd} = f_{ctk, \, 0.05} / (\gamma = 1.5) = 1.6$ MPa.

$$0.90 \times [-\frac{28070 \times 10^3}{9.12 \times 10^6} - \frac{23298 \times 10^6}{6972 \times 10^6}] + \frac{M_{Crack} \times 10^6}{6972 \times 10^6} + 2.44 = 1.6$$

$$-2.77 - 3.01 + \frac{M_{Crack} \times 10^6}{6972 \times 10^6} + 2.44 = 1.6$$

$M_{crack} = 34442$ kNm

19.6 ULTIMATE MOMENT CAPACITY

The moment distribution at ULS is calculated in a manner similar to that at SLS. The main difference is the inclusion of load factors as detailed in section 16.5.9. In the following, SLS loads from section 19.3.2 are multiplied by appropriate γ factors.

(i). Self-weight loading:
Self weight = (γ_G = 1.35) × 218.88 = 295.5 kN/m
(ii). Live loading: Using the loads calculated in section 19.3.2 and applying load factor of γ_Q = 1.35,

Uniformly distributed load = 43.92 × 1.35 = 59.29 kN/m

Total uniformly distributed load = 295.5 (self weight) + 59.29 (live load)
= 354.29 kN/m
Tandem loads: Using a load factor of 1.35, two concentrated wheel loads W are
Lane A: W = 2 × (γ_Q = 1.35) × 101.25 =273.38 kN
Lane B: W =2 × (γ_Q = 1.35) × 75.0 = 202.5 kN
Lane C: W = 2 × (γ_Q = 1.35) × 37.5 = 101.26 kN
Total = 273.38 + 202.5 + 101.26 = 576.14 kN

Bending moment at mid-span: Arranging loads as in section 19.3.3, and noting that the influence line ordinate at mid-span = 15 × 15/30 = 7.5
M = 354.29 × 30^2/ 8 + 576.14 × 7.5 × {1 + (15.0 – 1.2)/ 15}
= 39858 + 8296 = 48154 kNm.
In the case of unbonded tendons, at the ultimate limit, the steel strains are insufficient to cause yielding. Eurocode 2 suggests that an increase of only 100 MPa above the stress at service can be assumed. The stress at service is
σ_{pe} = P_s/ A_{ps}= (21053×10^3)/ (132 × 150) = 1063 MPa
At the ultimate limit, σ_{ult}= 1063 + 100 = 1163 MPa
P_{ult} = 1163 × (132 × 150) × 10^{-3} = 23033 kN
Using the rectangular stress block as shown in Fig. 7.4, from equations (7.1) and (7.3),
η = 1, λ = 0.8, f_{cd} = f_{ck}/1.5 = 40/1.5 = 26.7 MPa
Assuming that the stress block is inside the top flange,
Total compression C in concrete = 26.7 × 13000 × (0.8 × x) × 10^{-3} = P_{ult}
x = 83 mm

Lever arm = e + y_t – 0.4x = 1980 mm
M_{ult} = P_{ult} × 1980 ×10⁻³ = 45601 kNm
Provided ultimate moment of 45601 kNm is less than the required value of
48154 kNm. The difference of 2553 kNm can be made up by using unstressed
steel. The unstressed steel will help to spread the cracking over the span.
Using f_{yk} = 500 MPa, f_{yd} = 500/ 1.15 = 435 MPa
Using 25 mm bars at a cover of 30 mm,
Effective depth d = 2500 -30 – 25/2 – 10 (links) = 2447 mm
Lever arm ≈ 2447 – 0.4x = 2414
   The required area of steel is
435 × A_s × 2414 = 2553 × 10⁶, A_s = 2431 mm².
Provide six 25 mm bars. A_s =2945 mm².
With unstressed steel, ultimate moment capacity is
M_{ult} = 45601 + 435 × 2945 × 2414 × 10⁻⁶ = 48694 kNm
(M_{ult} = 48694) > [1.15 × (M_{Crack} = 34442) = 39608]
Design is satisfactory.

19.7 ULTIMATE SHEAR CAPACITY

The influence line for shear force at a section in a simply supported beam is shown
in Fig. 19.8.

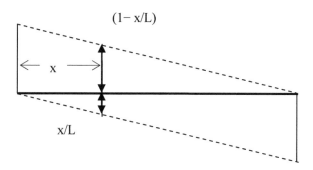

Fig. 19.8 Influence line for shear force

The shear forces at a section x from the different loadings are
V_{Dead} = q_{Dead} × (0.5 L – x)
V_{Live} = 0.5 × q_{live} × L × (1- x/L)²
V_W = W × (1 – x/L) + W × {1 – (x – 1.2)/L} = 2 × W × (1 – x/L) – W × 1.2/L
V = V_{Dead} + V_{Live} + V_W
where q_{dead} = 295.5 kN/m, q_{live} = 59.29 kN/m, W = 576.16 kN, L = 30 m
Fig. 19.9 shows the shear force envelope for one half of the span.

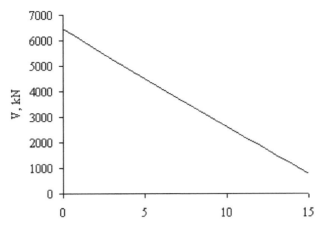

Fig. 19.9 Shear force envelope

(i). Shear capacity of a section without shear reinforcement and uncracked in flexure

$\sigma_{cp} = P_s/A = 21053 \times 10^3 / (9.12 \times 10^6) = 2.3$ MPa

$f_{ck} = 40$ MPa. From (3.2), $f_{ctm} = 0.30 \times f_{ck}^{0.667} = 3.5$ MPa

From (3.3), $f_{ctk, 0.05} = 0.7 \times f_{ctm} = 2.5$ MPa

$f_{ctd} = f_{ctk, 0.05} / (\gamma = 1.5) = 1.6$ MPa

$\tau = \sqrt{(f_{ctd}^2 + \sigma_{cp} f_{ctd})} = 2.5$ MPa

First moment of area, S about neutral axis:

$S = 13000 \times 300 \times (y_t - 300/2) + 2 \times 450 \times 0.5 \times (y_t - 300)^2$, $y_t = 1185$ mm

$= (585 + 352.45) \times 10^6 = 937.45 \times 10^6$ mm^3

$b_w = 2 \times 450 = 900$ mm

$I = 9.168 \times 10^{12}$ mm^4

From (9.2), $\tau = V \times \dfrac{S}{I \times b_w}$

$V = [2.5 \times 9.168 \times 10^{12} \times 900/ (937.45 \times 10^6)] \times 10^{-3} = 22004$ kN

(ii). Check start of cracked section

Assuming that the section might crack beyond 10 m from the supports,

$\gamma_{Inferiorr} [-\dfrac{P_s}{A} - \dfrac{M_{prestress}}{Z_b}] + \dfrac{M}{Z_b} = (f_{ctd} = 1.6)$

$$0.90 \times [-\frac{28070 \times 10^3}{9.12 \times 10^6} - \frac{23298 \times 10^6}{6972 \times 10^6}] + \frac{M \times 10^6}{6972 \times 10^6} = 1.6$$

$$-2.77 - 3.01 + \frac{M \times 10^6}{6972 \times 10^6} = 1.6$$

$$M = 51523 \text{ kNm}$$

At mid-span, at ULS, M =48154 kNm
The beam will not crack at the ultimate limit state. The beam needs only nominal shear steel.
Using equation, 9.19, the minimum steel is given by

$$\frac{A_{sw}}{s \times b_w} \geq 0.08 \frac{\sqrt{f_{ck}}}{f_{yk}}$$

Assuming 8 mm diameter bars, for a two-leg link, A_{sw} = 100.5 mm^2
b_w = 450 mm per web, f_{ck} = 40 MPa, f_{yk} = 500 MPa.
Substituting the values in the formula,
s = 221 mm
Provide in each web two-legged links at a spacing of 220 mm c/c.

19.8 CALCULATION OF DEFLECTION

In practice, the limit on the value of deflection at SLS is satisfied by maintaining the minimum value of the span/ effective depth ratio. It is very rare to actually calculate the value of deflection. The maximum deflection under quasi-permanent loads is normally limited to span/250. Eurocode 2 gives the following two formulae to calculate the span/effective depth ratio in the case of reinforced concrete members without axial compression. The implication quite clearly is that in the case of fully prestressed concrete members, because they are likely to crack, higher values than that permitted for reinforced concrete members can be used. However, Eurocode 2 gives no guidance on this matter.

$$\frac{L}{d} = K[11 + 1.5\sqrt{f_{ck}} \frac{\rho_0}{\rho} + 3.2\sqrt{f_{ck}} (\frac{\rho_0}{\rho} - 1)^{1.5}], \ \rho \leq \rho_0 \tag{19.1}$$

$$\frac{L}{d} = K[11 + 1.5\sqrt{f_{ck}} \frac{\rho_0}{(\rho - \rho')} + \frac{1}{12}\sqrt{f_{ck}} \sqrt{\frac{\rho'}{\rho}}], \ \rho > \rho_0 \tag{19.2}$$

$\rho_0 = 10^{-3} \times \sqrt{f_{ck}}$
ρ = required tension reinforcement ratio at mid-span to resist design loads
ρ' = required compression reinforcement ratio at mid-span to resist design loads
Note: In the case of cantilevers, the above ratios are calculated at support section.
K = factor shown in Table 19.1.
Values of K have been calculated on the assumption that at SLS, the maximum stress in steel is 300 MPa for f_{yk} = 500 MPa.

Table 19.1 L/d ratios for reinforced slabs without axial compression

Structural system	K
Simply supported beam, one- or two-way spanning simply supported slab	1.0
End span of continuous beam or one-way continuous slab or two-way spanning slab continuous over one long side	1.3
Interior span of continuous beam or one-way or two-way spanning slab	1.5
Flat slabs based on longer span	1.2
Cantilever	0.4

Note: For two-way spanning slabs use the shorter span.

Example: Fig. 19.10 shows an L-beam used as a simply supported beam over a span of 7 m. It carries a live load of 5 kN/m² and a screed 30 mm thick. Calculate the permissible span/effective depth ratio.

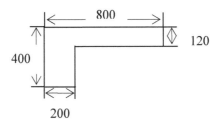

Fig. 19.10 An L-beam

(i). Self weight = $[400 \times 200 + 120 \times 600] \times 10^{-6} \times 25 = 3.8$ kN/m
(ii). Screed = $30 \times 800 \times 10^{-6} \times 25 = 0.6$ kN/m
(iii). Live load = $5 \times 800 \times 10^{-3} = 4.0$ kN/m
$g_k = 3.8 + 0.6 = 4.4$ kN/m
$q_k = 4.0$ kN/m

(a). Design for ULS
At ULS using load factors of 1.5 for g_k and 1.35 for q_k, the total load is
$q = 1.5 \times 4.4 + 1.35 \times 4.0 = 12.0$ kN/m
Effective depth d: Using 20 mm bars,
$d = 400 - 30$ (cover) $- 8$ (link) $- 20/2 = 352$ mm
At mid-span, $M = 12.0 \times 7^2/ 8 = 73.5$ kNm
Using $f_{ck} = 30$ MPa, $f_{cd} = 30/ 1.5 = 20$ MPa,
$73.5 \times 10^6 = 20 \times 800 \times s \times (d - 0.5s)$
where $s = 0.8x$ = stress block depth
Solving the quadratic equation, $s = 13.3$ mm, $x = 16.6$ mm
Lever arm = $d - 0.5s = 345$ mm

Using f_{yk} = 500 MPa, f_{yd} = f_{yk}/ 1.15 = 435 MPa
$73.5 \times 10^6 = 435 \times A_s \times 345$, $A_{s\,Required}$ = 490 mm^2
Provide two 20 mm bars.
$A_{s\,Provided}$ = 628 mm^2
$\rho = A_{s\,provided}$ /(b × d) = 628 /(200 × 352) = 0.89%
$\rho_0 = 10^{-3} \times \sqrt{30}$ = 0.55%
$\rho' = 0$
$\rho_0 / \rho = 0.62$
K = 1.0
Substituting in (19.1),

$$\frac{L}{d} = 1.0 \times [11 + 1.5\sqrt{f_{ck}}\frac{\rho_0}{(\rho - \rho')} + \frac{1}{12}\sqrt{f_{ck}}\sqrt{\frac{\rho'}{\rho}}\ ]$$

$$= [11.0 + 1.5 \times 5.48 \times 0.62] = 16.1$$

$A_{s\,provided}$ /$A_{s\,required}$ = 628/ 490 = 1.28
L/d required = 16.1 × 1.28 = 20.6
L/d provide = 7 × 10^3/ 352 = 19.9
The design satisfies the deflection check.

(b). Checks at SLS

(i). Young's modulus, E_{cm} = 22 (0.1 × f_{cm})$^{0.3}$
f_{ck} = 30 MPa, f_{cm} = f_{ck} + 8 = 38 MPa
E_{cm} = 32.8 GPa

(ii). Creep coefficient $\varphi(\infty, t_0)$
Area of cross section, A_c = [400 × 200 + 120 × 600] = 1.52 × 10^5 mm^2
Assume all sides except the flange depth is exposed to atmosphere.
Perimter of exposure, u = 200 + 400 + 800 + (400 − 120) + (800 −200) = 2280 mm
h_0 = 2 × A_c/ u = 133 mm
$\varphi(\infty, t_0) \approx 1.8$
Long-term E_c = 1.05 × E_{cm}/ $\varphi(\infty, t_0)$ = 19.1 GPa
Young's moduls for steel, E_s = 195 GPa
Modular ratio α_e = E_s/ E_c = 10.2

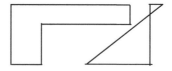

Fig. 19.11 Strain distribution at SLS

(iii). Stress analysis at SLS
Assuming that the neutral axis is inside the flange as shown in Fig. 19.11, taking moments about the neutral axis,

$0.5 \times 800 \times x^2 = \alpha_e \times A_{s\ Provided} \times (d - x)$
$x = 68$ mm
Lever arm $= d - x/3 = 352 - 68/3 = 329$ mm
M at SLS $= (4.4 + 4.0) \times 7^2 / 8 = 51.45$ kNm
Stress in steel $=$ M at SLS$/ (A_{s\ provided} \times$ Lever arm$)$
$\qquad\qquad = 51.45 \times 10^6 / (628 \times 329) = 249$ MPa
Equation (19.2) is derived on the basis that the stress in steel is 300 MPa for the required steel. For the provided steel, stress in steel is
$300 \times A_{s\ required}/ A_{s\ Provided} = 300 \times (490/ 628) = 234$ MPa
The calculated value of 249 MPa is close to the value assumed in the Eurocode 2 formula.

19.9 REFERENCES TO CLAUSES IN EUROCODE 2

In this chapter, the following clauses in Eurocode 2 have been referred to:
Unbonded tendons, increase in stress at ULS: 5.10.8 (2)
General considerations: 7.4.1
Cases where calculations may be omitted: 7.4.2
$M_{ult} > 1.15\ M_{crack}$: 9.2.1.1(4)

REFERENCES

(A). General References

1. Bazant, Z.P. (Editor), 1988, *Mathematical modelling of creep and shrinkage*, (Wiley).
2. Bhatt, Prab., MacGinley, Thomas 1998, J. and Choo, Ban Seng., 2006, *Reinforced concrete: Design, Theory and Examples*, (Taylor & Francis)
3. Collins, Michael P. and Mitchell, Denis., 1991, *Prestressed concrete structures*, (Prentice Hall).
4. Dilger, W. and Neville, Adam., 1970, *Creep of concrete: Plain, reinforced and prestressed,* (Elsevier).
5. Gerwick, B.C., 1993, *Construction of prestressed concrete*, (Wiley).
6. Hurst, Melvin Keith., 1998, *Prestressed concrete design (2nd Edition)*, (E. & F.N. Spon)
7. Kotsovos, Michael D. and Pavlovic, M.N., 1995, *Structural Concrete: Finite element analysis for limit state design,* (Thomas Telford).
8. Naaman, Antoine E., 2004, *Prestressed concrete analysis and design fundamentals,* (Techno Press).
9. Nawy, Edward G., 2000*, Prestressed concrete: A fundamental approach*, (Prentice Hall)
10. O'Brien, Eugene J. and Dixon, Andrew S., 1995, *Reinforced and prestressed concrete design: The complete process*, (Longman Scientific and Technical).
11. Post-tensioning Institute., 2000, *Post-tensioning Manual, (5th Edition).*
12. Prestressed Concrete Institute, 1999, *PCI Design Handbook.*
13. Rombach, Gunter Axel, 2004, *Finite element design of concrete structures: Practical problems and their solutions,*(Thomas Telford).

(B). Eurocodes and Guides

1. BSI, 2004, *Eurocode 2: Design of concrete structures, Part 1-1: General rules and rules for buildings*, (British Standards Institution)
2. Beeby, A.W. and Narayanan, R.S. 2005, *Designers' guide to EN 1992-1-1 and EN 1992-1-2, Eurocode 2: Design of concrete structures*, (Thomas Telford).
3. British standards Institution, 2005, Eurocode 2-Design of concrete structures, Part 2-Concrete bridges-Design and detailing rules.
4. Hendy, C.R. and Smith, D.A. 2007, *Designers' guide to EN 1992-2, Eurocode 2: Design of concrete structures, part 2: Concrete Bridges*, (Thomas Telford).
5. British Standards Institution., 2010, *Eurocode 8: Design of structures for earthquake resistance.*
6. Fardis, Michael., Carvalho, E., Elnashai, A., Faccioli, E., Pino, P. and Plumier, A., 2005, *Designers' guide to EN 1998-1 and EN 1998-5 Eurocode 8: Design of Structures for earthquake resistance.*,(Thomas Telford).
7. BSI, 2003, *Eurocode 1: Action on Structures, Part 2: Traffic loads on bridges*, (British Standards Institution)

8. Calagro, J.A., Tschumi, M., Shetty, N. and Gulvanessian, H., 2007, *Designers' guide to EN 1992-2, 1991-1.3 and 1991-1.5 to 1.7, Eurocode 1: Actions on Structures-Traffic loads and other action on bridges,* (Thomas Telford).
9. BSI, 2003, *Eurocode 1: Action on Structures, parts 1-5: General actions-Thermal actions,* (British Standards Institution)
10. BSI, 2002, *Eurocode 1: Action on Structures, part 1-1: General actions-General actions-Densities, self-weight, imposed loads for buildings.*, (British Standards Institution)

(C). References to Chapter 12

1. Aalami, Bijan O., 1989, *Design of post-tensioned floor slabs*, Concrete International, June 1989, p. 59-67.
2. Aalami, Bijan O. and Kelley, Gail S., 2001, Design of concrete floors with particular reference to post-tensioning, Post-tensioning Institute Technical notes, Issue 11, January, 2001.
3. Andrew, Arthur E., 1987, *Unbonded tendons in post-tensioned construction*, (Thomas Telford).
4. Concrete Society. *Post-tensioned concrete floors: Design Handbook,* (Concrete Society, U.K.)
5. Khan, Sami. And Williams, Martin., 1995, *Post-tensioned concrete floors,* (Butterworh Heinemann)
6. Moss, R. and Brooker, O., 2006, *Flat slabs, (*British Cement Association).
7. Stevenson, A.M., 1994, *Post-tensioned concrete floors in multi-storey buildings*, (British Cement Association).

(D). References to Chapter 16

1. Benaim, Robert., 2008, *The Design of Prestressed Concrete Bridges*, (Taylor & Francis).
2. Daly, A.F. and Jackson, P., 1999, *Design of bridge with external prestressing,* (TRL Report 391).
3. Hambly, E.C.,1999, *Bridge Deck Behaviour*, (2nd Edition, E&FN Spon).
4. Hewson, Nigel., 2003, *Prestressed Concrete Bridges: Design and Construction*, (Thomas Telford).
5. Mathivat, Jacques. , 1979, *The Cantilever Construction of Prestressed Concrete Bridges*, (Wiley).
6. Menn, Christian. , 1990, *Prestressed Concrete Bridges*, (Birkhauser Verlag).
7. Nicholson, B.A., 1997, *Simple Bridge Design Using Prestressed Beams*, (Prestressed Concrete Association).
8. O'Brien, Eugene J. and Keogh, Damien L., 1999, *Bridge Deck Analysis*, (E&FN Spon).
9. Podolny, Walter and Muller, Jean M., 1982, *Construction and design of segmental bridges*, (Wiley).
10. *Post-tensioned concrete bridges.*, 1999, (Thomas Telford)
11. Rasignoli, Marco., 2002, *Bridge Launching*, (Thomas Telford).

12. Ryall, M. J., Parke, G.A.R. and Harding, J.E. (2008), *I.C.E. Manual of bridge engineering,* (Thomas Telford).
13. Taly, Narendra., 2008, *Design of modern highway bridges*, (McGraw-Hill).
14. Wang, Chi-The., 1953, *Applied Elasticity*, (McGraw-Hill).

(E). References to Chapter 17

1. MacGregor, James, G. and Wight, James K., (2005), *Reinforced Concrete: Mechanics and Design*, (Pearson Prentice Hall).
2. Muttoni, A., Schwartz, J. and Thurlimann, B., 1997, Design of concrete structures with stress fields, (Birkhauser Verlag)

(F). References to Chapter 18

1. Ambrose, James and Vergun, Dimitry., 1990, *Simplified building design for wind and earthquake forces*, (2nd Edition, Wiley).
2. Bhatt, P.,2002, *Programming the dynamic analysis of structures,* (Taylor & Francis)
3. Booth, Edmund and Key, David., 2006, *Earthquake design practice for buildings*,(2nd Edition, Thomas Telford).
4. Dowrick, David J.,2003, *Earthquake risk reduction*, (Wiley).
5. Green, Norman B., 1987, *Earthquake resistant building design and construction*, (Elsevier).
6. Penelis, George G and Kappos, Andreas. J., 1997, *Earthquake resistant concrete structures*, (E & F.N. Spon).
7. Paulay, Thomas. and Priestley, M.J.N., 1992, *Seismic design of reinforced concrete and masonry*, (Wiley).
8. Priestley, M.J.N., Seible, F and Calvi, G.M.,1996, *Seismic design and retrofit of bridges*, (Wiley).

INDEX